Pharmaceutical Microbiology

EDITED BY

W. B. HUGO
BPharm PhD FRPharmS
Formerly Reader in Pharmaceutical Microbiology
University of Nottingham

AND

A. D. RUSSELL
BPharm DSc PhD FRPharmS FRCPath
Professor of Pharmaceutical Microbiology
University of Wales College of Cardiff
Cardiff

FIFTH EDITION

OXFORD

Blackwell Scientific Publications
LONDON EDINBURGH BOSTON

MELBOURNE PARIS BERLIN VIENNA

First published 1977
Second edition 1980
Third edition 1983
Reprinted 1986
Fourth edition 1987
Reprinted 1989, 1991
Italian Edition 1991
Fifth edition 1992
Reprinted 1993, 1994

Set by Excel Typesetters Company Ltd,
Hong Kong
Printed and bound in Great Britain at
The Alden Press, Oxford

DISTRIBUTORS

Marston Book Services Ltd
PO Box 87
Oxford OX2 0DT
(*Orders*: Tel: 0865 791155
        Fax: 0865 791927
        Telex: 837515)

USA
  Blackwell Scientific Publications, Inc.
  238 Main Street
  Cambridge, MA 02142
  (*Orders*: Tel: 800 759-6102
          617 876-7000)

Canada
  Times Mirror Professional Publishing, Ltd
  130 Flaska Drive
  Markham, Ontario L6G 1B8
  (*Orders*: Tel: 800 268-4178
          416 470-6739)

Australia
  Blackwell Scientific Publications Pty Ltd
  54 University Street
  Carlton, Victoria 3053
  (*Orders*: Tel: 03 347-5552)

A catalogue record for this book
is available from the British Library.

ISBN 0-632-03428-9

Contents

Contributors, vii

Preface to Fifth Edition, ix

Preface to First Edition, x

Part 1   Biology of microorganisms, 1

1   Bacteria
    W. B. Hugo, 3

2   Mycology
    R. C. Rees & A. M. Dickinson, 43

3   Viruses
    D. J. Stickler, 60

4   Principles of microbial pathogenicity and epidemiology
    P. Gilbert, 82

Part 2   Antimicrobial agents, 97

5   Types of antibiotics and synthetic antimicrobial agents
    A. D. Russell, 99

6   Clinical uses of antimicrobial drugs
    R. G. Finch, 134

7   Manufacture of antibiotics
    G. Fare, 154

8   Principles of methods of assaying antibiotics
    D. S. Reeves & L. O. White, 166

9   Mechanisms of action of antibiotics
    P. A. Lambert, 189

10   Bacterial resistance to antibiotics
     T. J. Franklin, 208

11   Chemical disinfectants, antiseptics and preservatives
     E. M. Scott & S. P. Gorman, 231

12   Evaluation of non-antibiotic antimicrobial agents
     W. B. Hugo & A. D. Russell, 258

13   Mode of action of non-antibiotic antibacterial agents
     W. B. Hugo, 288

v

14 Resistance to non-antibiotic antimicrobial agents
   P. A. Lambert, 295

15 Fundamentals of immunology
   J. R. Furr, 305

16 The manufacture and quality control of immunological products
   F. W. Sheffield, 332

Part 3   Microbiological aspects of pharmaceutical processing, 351

17 Ecology of microorganisms as it affects the pharmaceutical industry
   E. Underwood, 353

18 Microbial spoilage and preservation of pharmaceutical products
   E. G. Beveridge, 369

19 Contamination of non-sterile pharmaceuticals in hospital and
   community environments
   R. M. Baird, 391

20 Principles and practice of sterilization
   S. P. Denyer, 403

21 Sterile pharmaceutical products
   M. C. Allwood, 428

22 Factory and hospital hygiene and good manufacturing practice
   S. P. Denyer, 445

23 Sterilization control and sterility testing
   S. P. Denyer, 458

24 Production of therapeutically useful substances by recombinant DNA
   technology
   S. B. Primrose, 471

25 Additional applications of microorganisms in the pharmaceutical
   sciences
   P. Williams, 487

Index, 503

Contributors

M. C. ALLWOOD BPharm, PhD, MRPharmS *Director, Medicines Research Unit, Derbyshire College, Mickleover, Derby*

R. M. BAIRD BPharm, PhD, MRPharmS *Consultant in Pharmaceutical Microbiology, Summerlands House, Summerlands, Yeovil*

E. G. BEVERIDGE BPharm, PhD, MRPharmS, MIBiol, CBiol, MIQA *Principal Lecturer in Pharmaceutical Microbiology, School of Health Sciences, University of Sunderland, Sunderland*

S. P. DENYER BPharm, PhD, MRPharmS *Professor of Pharmacy and Head, Department of Pharmacy, Brighton Polytechnic, Moulsecoomb, Brighton*

A. M. DICKINSON BSc, PhD *Senior Scientific Officer, Department of Haematology, The Royal Victoria Infirmary, Queen Victoria Road, Newcastle-upon-Tyne*

G. FARE BSc, PhD *Technical Information Manager, Glaxochem Ltd, Ulverston, Cumbria*

R. G. FINCH MB, ChB, FRCP, FRCPath *Consultant Physician in Microbial Diseases, City Hospital, Nottingham, and Professor of Infectious Diseases, Department of Microbiology, Faculty of Medicine, Queen's Medical Centre, The University, Nottingham*

T. J. FRANKLIN BSc, PhD *Senior Research Associate, ICI Pharmaceuticals, Alderley Park, Macclesfield, Cheshire*

J. R. FURR MPharm, PhD, MRPharmS *Lecturer in Pharmaceutical Microbiology, Welsh School of Pharmacy, University of Wales College of Cardiff, Cardiff*

P. GILBERT BSc, PhD *Senior Lecturer in Pharmacy, Department of Pharmacy, University of Manchester, Manchester*

S. P. GORMAN BSc, PhD, MPSNI *Senior Lecturer in Pharmaceutics, School of Pharmacy, Medical Biology Centre, Queen's University of Belfast, Belfast*

W. B. HUGO BPharm, PhD, FRPharmS *Formerly Reader in Pharmaceutical Microbiology, University of Nottingham. Present address: 618 Wollaton Road, Nottingham*

P. A. LAMBERT BSc, PhD, DSc *Senior Lecturer in Pharmaceutical Microbiology, Department of Pharmaceutical Sciences, Aston University, Aston Triangle, Birmingham*

S. B. PRIMROSE BSc, PhD, *Director of New Business Development, Amersham International plc. Amersham Laboratories, Amersham*

R. C. REES BSc, PhD, MRCPath, FIBiol, CBiol *Reader in Tumour Biology, Institute for Cancer Studies, University of Sheffield Medical School, Sheffield*

D. S. REEVES MD, FRCPath *Consultant Medical Microbiologist, Department of Medical Microbiology, Southmead Health Services NHS Trust, Westbury-on-Trym, Bristol*

A. D. RUSSELL BPharm, DSc, PhD, FRPharmS, FRCPath *Professor of Pharmaceutical Microbiology, Welsh School of Pharmacy, University of Wales College of Cardiff, Cardiff*

E. M. SCOTT BSc, PhD, MPSNI *Senior Lecturer in Pharmaceutics, School of Pharmacy, Medical Biology Centre, Queen's University of Belfast, Belfast*

F. W. SHEFFIELD MB, ChB *The Limes, Wilcot, Pewsey, Wiltshire. Formerly Head of the Division of Bacterial Products, National Institute for Biological Standards & Control, South Mimms, Hertfordshire*

D. J. STICKLER BSc, MA, DPhil *Senior Lecturer in Microbiology, School of Pure and Applied Biology, University of Wales College of Cardiff, Cardiff*

E. UNDERWOOD BSc, PhD *Nutritional Product & Process Development Manager, Wyeth Laboratories, New Lane, Havant, Hants*

L. O. WHITE BSc, PhD *Clinical Scientist Grade C, Department of Medical Microbiology, Southmead Health Services NHS Trust, Westbury-on-Trym, Bristol*

P. WILLIAMS BPharm, PhD, MRPharmS *Lecturer in Biochemistry, School of Pharmacy, University of Nottingham, Nottingham*

Preface to Fifth Edition

We were gratified that our publishers have called for a fifth edition of this book which was first published in 1977. We were equally pleased that our contributors agreed to update their chapters.

Data from the publishers indicate that *Pharmaceutical Microbiology* is reaching a much wider audience than the student of pharmacy for which it was first intended and we have tried to keep this in mind when revising the texts.

We again thank our contributors for their cooperation and our publishers for their support and expertise.

W. B. Hugo
A. D. Russell

Preface to First Edition

When we were first approached by the publishers to write a textbook on pharmaceutical microbiology to appear in the spring of 1977, it was felt that such a task could not be accomplished satisfactorily in the time available.

However, by a process of combined editorship and by invitation to experts to contribute to the various chapters this task has been accomplished thanks to the cooperation of our collaborators.

Pharmaceutical microbiology may be defined as that part of microbiology which has a special bearing on pharmacy in all its aspects. This will range from the manufacture and quality control of pharmaceutical products to an understanding of the mode of action of antibiotics. The full extent of microbiology on the pharmaceutical area may be judged from the chapter contents.

As this book is aimed at undergraduate pharmacy students (as well as microbiologists entering the pharmaceutical industry) we were under constraint to limit the length of the book to retain it in a defined price range. The result is to be found in the following pages. The editors must bear responsibility for any omissions, a point which has most concerned us. Length and depth of treatment were determined by the dictate of our publishers. It is hoped that the book will provide a concise reading for pharmacy students (who, at the moment, lack a textbook in this subject) and help to highlight those parts of a general microbiological training which impinge on the pharmaceutical industry.

In conclusion, the editors thank most sincerely the contributors to this book, both for complying with our strictures as to the length of their contribution and for providing their material on time, and our publishers for their friendly courtesy and efficiency during the production of this book. We also wish to thank Dr H. J. Smith for his advice on various chemical aspects, Dr M. I. Barnett for useful comments on reverse osmosis, and Mr A. Keall who helped with the table on sterilization methods.

W. B. Hugo
A. D. Russell

Part 1   Biology of microorganisms

Pharmaceutical microbiology is one of the many facets of applied micro-biology, but very little understanding of its posed and potential problems will be achieved unless the basic properties of microorganisms are understood.

This section considers, in three separate chapters, the anatomy and physiology of bacteria, fungi and yeasts, and viruses, together with a survey of the characters of individual members of these groups likely to be of importance to the applied field covered by this book. Chapter 4 considers the principles of microbial pathogenicity and epidemiology.

The treatment is perforce brief, but it is hoped that the material will give an understanding of the essentials of each group which may be amplified as required from the bibliographic material listed at the end of each section.

1 Bacteria

1	**Introduction**	5.5	Growth curves
2	**Structure and form of the bacterial cell**	**6**	**Properties of selected bacterial species**
2.1	Size and shape	6.1	Gram-positive cocci
2.2	Structure	6.1.1	*Staphylococcus*
2.2.1	Cell wall	6.1.2	*Streptococcus*
2.2.2	Cytoplasmic membrane	6.1.3	*Diplococcus (now Streptococcus)*
2.2.3	Cytoplasm	6.2	Gram-negative cocci
2.2.4	Appendages to the bacterial cell	6.2.1	*Neisseria* and *Branhamella*
2.2.5	Capsules and slime	6.3	Gram-positive rods
2.2.6	Pigments	6.3.1	*Bacillus*
2.3	The bacterial spore	6.3.2	*Clostridium*
2.3.1	The process of spore formation	6.3.3	*Corynebacterium*
2.3.2	Spore germination and out-growth	6.3.4	*Listeria*
2.3.3	Parameters of heat resistance	6.4	Gram-negative rods
		6.4.1	*Pseudomonas*
3	**Toxins**	6.4.2	*Vibrio*
		6.4.3	*Yersinia* and *Francisella*
4	**Reproduction**	6.4.4	*Bordetella*
4.1	Binary fission	6.4.5	*Brucella*
4.2	Reproduction involving genetic exchange	6.4.6	*Haemophilus*
		6.4.7	*Escherichia*
4.2.1	Transformation	6.4.8	*Salmonella*
4.2.2	Conjugation	6.4.9	*Shigella*
4.2.3	Transduction	6.4.10	*Proteus*
		6.4.11	*Serratia marcescens*
5	**Bacterial growth**	6.4.12	*Klebsiella*
5.1	The growth requirements of bacteria	6.4.13	*Flavobacterium*
		6.4.14	*Acinetobacter*
5.1.1	Consumable determinants	6.4.15	*Bacteroides*
5.1.2	Environmental determinants	6.4.16	*Campylobacter*
5.1.3	Culture media	6.4.17	*Helicobacter*
5.2	Energy provision	6.4.18	*Chlamydia*
5.2.1	Carbohydrate metabolism	6.4.19	*Rickettsia*
5.2.2	Protein and amino acid metabolism	6.4.20	*Legionella*
		6.5	Acid-fast organisms
5.2.3	Comment	6.5.1	*Mycobacterium*
5.3	Identification of bacteria	6.6	Spirochaetes
5.3.1	Selective and diagnostic media	6.6.1	*Borrelia*
5.3.2	Examples of additional biochemical tests	6.6.2	*Treponema*
		6.6.3	*Leptospira*
5.4	Measurement of bacterial growth		
5.4.1	Mean generation time	**7**	**Further reading**

1 Introduction

Bacteria share with the blue–green algae a unique place in the world of living organisms. Formerly classified with the fungi, bacteria were considered as primitive members of the plant kingdom, but they are now called *prokaryotes*, a name which means primitive nucleus. All other living organisms are called *eukaryotes*, a name implying a true or proper nucleus. This important division does not invalidate classification schemes within the world of bacterial, animal and plant life.

This subdivision is not based on the more usual macroscopic criteria; it was made possible when techniques of subcellular biology became sufficiently refined for many more fundamental differences to become apparent. Some of the criteria differentiating eukaryotes and prokaryotes are given in Table 1.1.

Recently, a third class of prokaryotes must be added to the bacteria and blue–green algae. These have been named the Archaebacteria; they differ from bacteria and blue–green algae in their wall and membrane structure and pattern of metabolism. They are thought to be the first of the prokaryotes and indeed the first living organisms to appear on earth.

2   Structure and form of the bacterial cell

2.1   Size and shape

The majority of bacteria fall within the general dimensions of 0.75 to 4 μm. They are unicellular structures which may occur as cylindrical (rod-shaped) or spherical (coccoid) forms. In one or two genera, the cylindrical form may be modified in that a single twist (vibrios) or many twists like a corkscrew (spirochaetes) may occur.

Another feature of bacterial form is the tendency of coccoid cells to grow in aggregates. Thus, there exist assemblies (i) of pairs (called diplococci); (ii) of groups of four arranged in a cube (sarcinae); (iii) in a generally unorganized array like a bunch of grapes (staphylococci); and (iv) in chains like a string of beads (streptococci). The aggregates are often so characteristic as to give rise to the generic name of a group, e.g. *Diplococcus* (now called *Streptococcus*) *pneumoniae*, a cause of pneumonia; *Staphylococcus aureus*, a cause of boils and food poisoning; and *Streptococcus pyogenes*, a cause of sore throat.

Rod-shaped organisms occasionally occur in chains either joined end to end or branched.

2.2   Structure

Three fundamental divisions of the bacterial cell occur in all species: cell wall, cell or cytoplasmic membrane, and cytoplasm.

2.2.1   Cell wall

Extensive chemical studies have revealed a basic structure of alternating *N*-acetylglucosamine and *N*-acetyl-3-*O*-1-carboxyethyl-glucosamine molecules, giving a polysaccharide backbone. This is then cross-linked by peptide chains, the nature of which varies from species to species. This

Table 1.1 The main features distinguishing prokaryotic and eukaryotic cells

Feature	Prokaryotes	Eukaryotes
Nucleus	No enclosing membrane	Enclosed by a membrane
Cell wall	Peptidoglycan	Cellulose
Mitochondria	Absent	Present
Mesosomes	Present	Absent
Chloroplasts	Absent	Present

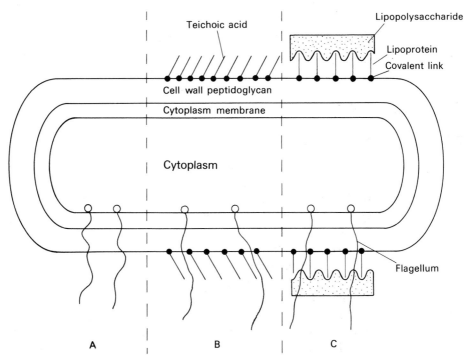

Fig. 1.1 Diagram of the bacterial cell. A, the generalized structure of the bacterial cell; B, Gram-positive structure; C, Gram-negative structure.

structure (Fig. 1.1) possesses great mechanical strength and is the target for a group of antibiotics which, in different ways, inhibit the biosynthesis occurring during cell growth and division (Chapter 9).

This basic peptidoglycan (sometimes called murein or mucopeptide) also contains other chemical structures which differ in two types of bacteria, Gram-negative and Gram-positive. In 1884, Christian Gram discovered a staining method for bacteria which bears his name. It consists of treating a film of bacteria, dried on a microscope slide, with a solution of a basic dye, such as gentian violet, followed by application of a solution of iodine. The dye complex may be easily washed from some types of cells which, as a result, are called Gram-negative whereas others, termed Gram-positive, retain the dye despite alcohol washing. These

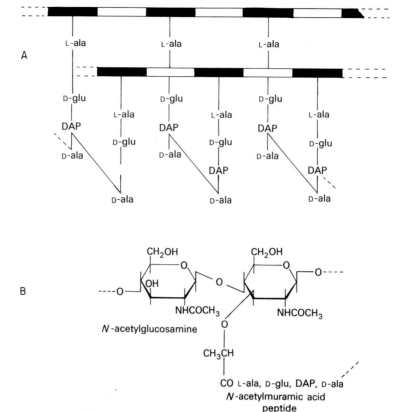

Fig. 1.2 A, peptidoglycan of *Escherichia coli*. ■, *N*-acetylmuramic acid; □, *N*-acetylglucosamine. B, repeating unit of peptidoglycan of *E. coli*. L-ala, L-alanine; D-glu, D-glutamine; DAP, diaminopimelic acid; D-ala, D-alanine.

marked differences in behaviour, discovered by chance, are now known to be a reflection of different wall structures in the two types of cell. These differences reside in the differing chemistry of material attached to the outside of the peptidoglycan (Fig. 1.2).

In the walls of Gram-positive bacteria, molecules of a polyribitol or polyglycerolphosphate are attached by covalent links to the oligo-saccharide backbone; these entities are teichoic acids. The glycerol teichoic acid may contain an alanine residue; the ribitol teichoic acid contains a glucose residue (Fig. 1.3). Teichoic acids do not confer additional rigidity on the cell wall, but as they are acidic in nature they may function by sequestering essential metal cations from the media on which the cells are growing. This could be of value in situations where cation concentration in the environment is low.

The Gram-negative cell envelope (Fig. 1.4) is even more complicated; essentially, it contains lipoprotein molecules attached covalently to the oligosaccharide backbone and in addition, on its outer side, a layer of lipopolysaccharide (LPS) and protein attached by hydrophobic inter-

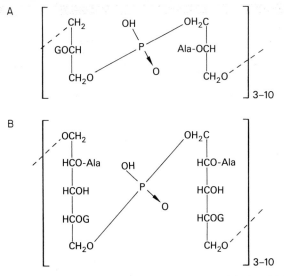

A

B

Fig. 1.3 A, glycerol teichoic acid; B, ribitol teichoic acid. G, glycosyl; Ala, D-alanyl.

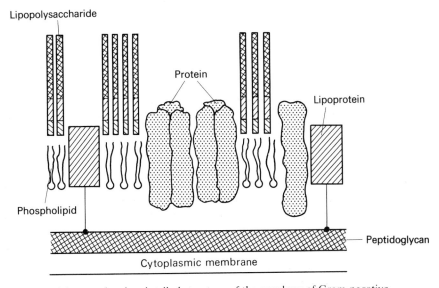

Fig. 1.4 Diagram showing detailed structure of the envelope of Gram-negative bacteria.

actions and divalent metal cations, Ca^{2+} and Mg^{2+}. On the inner side is a layer of phospholipid (PL).

The LPS molecule consists of three regions, called lipid A, core polysaccharide and O-specific side chain (Fig. 1.5). The O-specific side chain comprises an array of sugars that are responsible for specific serological reactions of organisms, which are used in identification. The

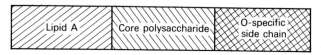

Fig. 1.5 Lipopolysaccharide structure in Gram-negative bacteria.

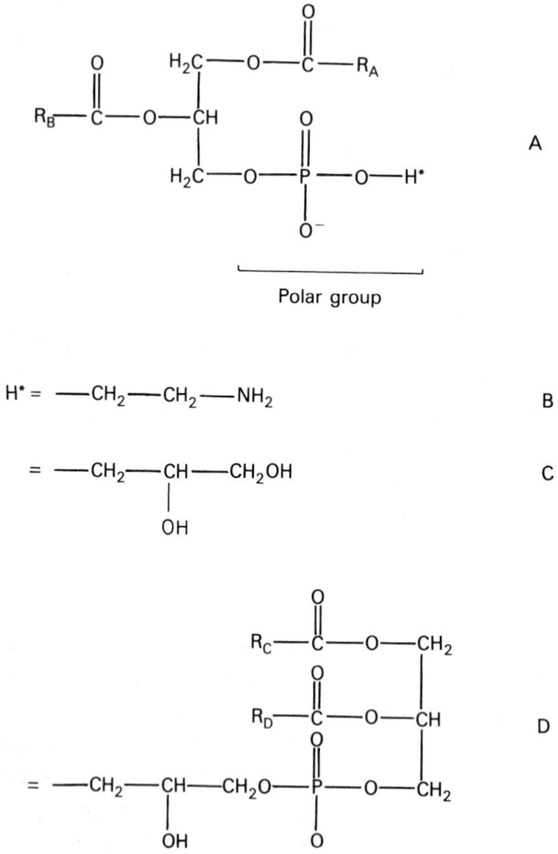

A

Polar group

H* = —CH₂—CH₂—NH₂   B

= —CH₂—CH—CH₂OH   C

OH

D

= —CH₂—CH—CH₂O—

OH

Fig. 1.6 The structure of some phospholipids found in *E. coli*. A, the structure of phosphatidic acid. H* of this structure is replaced by grouping B–D to give the following phospholipids: B, phosphatidyl ethanolamine; C, phosphatidyl glycerol; D, diphosphatidyl glycerol (cardiolipin). R_A . COO and R_B . COO are fatty acid residues.

lipid A region is responsible for the toxic and pyrogenic (fever-producing) properties of this group (see Chapter 18).

The complex outer layers beyond the peptidoglycan in the Gram-negative species, the outer membrane, protect the organism to a certain extent from the action of toxic chemicals (see Chapter 14). Thus, disinfectants are often effective only at concentrations higher than those affecting Gram-positive cells and these layers provide unique protection to the cells from the action of benzylpenicillin and lysozyme.

Part of the LPS may be removed by treating the cells with

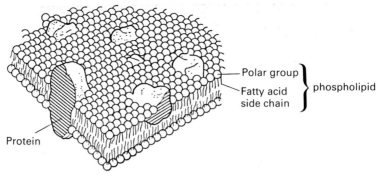

Fig. 1.7 Membrane structure.

ethylenediamine tetra-acetic acid (EDTA) or related chelating agents (Chapter 14).

The proteins of the outer membrane, many of which traverse the whole structure, are currently the subject of active study. Some of the proteins consist of three subunits, and these units with a central space or pore running through them are known as porins. They are thought to act as a mechanism of selectivity for the ingress or exclusion of metabolites and antibacterial agents (see Chapter 10).

2.2.2   *Cytoplasmic membrane*

The chemistry and structure of this organelle have been the subject of more than a century of research, but it is only during the last 20 years that some degree of finality has been realized.

Chemically, the membrane is known to consist of phospholipids and proteins, many of which have enzymic properties. The phospholipid molecules are arranged in a bimolecular layer with the polar groups directed outwards on both sides. The structures of some phospholipids found in bacteria are shown in Fig. 1.6. Earlier views held that the protein part of the membrane was spread as a continuous sheet on either side of the phospholipid bilayer. The current view is that protein is distributed in local patches in the bilayer, the mosaic structure (Fig. 1.7).

Unlike the wall, which has great mechanical strength, determines the characteristic shape of the cell and is metabolically inert, the membrane is structurally a very delicate organelle and is highly active metabolically.

It acts as a selective permeability barrier between the cytoplasm and the cell environment; the wall acts only as a sieve to exclude molecules larger than about 1 nm. Certain enzymes, and especially the electron transport chain, that are located in the membrane are responsible for an elaborate active transport system which utilizes the electrochemical potential of the proton to power it.

An interesting experiment serves to illustrate the differing mechanical strengths of the wall and membrane. The wall of some Gram-positive bacteria may be partially dissolved by treatment of cells with lysozyme or in the case of Gram-negative cells with EDTA plus lysozyme. Upon doing this, cells so treated burst due to the fact that the cytoplasm contains a large number of solutes giving it an effective osmotic pressure of 608–2533 kPa (6–25 atm). Water enters the cell, now no longer protected by the peptidoglycan, causing the naked protoplast to swell and burst. If this experiment is conducted in a medium containing 0.33 M sucrose, a non-penetrating solute, the osmotic pressure inside and outside the protoplast is equalized, thus no bursting occurs and forms free of cell wall (protoplasts) may be observed in the medium.

2.2.3   *Cytoplasm*

The cytoplasm is a viscous fluid and contains within it systems of paramount importance. These are the nucleus, responsible for the genetic make-up of the cell, and the ribosomes, which are the site of protein synthesis. In addition are found granules of reserve material such as polyhydroxybutyric acid, an energy reserve, and polyphosphate or volutin granules, the exact function of which has not yet been elucidated. The prokaryotic nucleus or bacterial chromosome exists in the cytoplasm in the form of a loop and is not surrounded by a nuclear membrane. Bacteria carry other chromosomal elements: episomes, which are portions of the main chromosome that have become isolated from it, and plasmids, which may be called miniature chromosomes. These are small annular pieces of DNA which carry a limited amount of genetic information, often associated with the expression of resistance to antimicrobial agents (Chapters 10 and 14).

Despite the differences in nuclear structures between prokaryotes and eukaryotes, the genetic code, i.e. the combination of bases which codes for a particular amino acid in the process of protein synthesis, is the same as it is in all living organisms.

2.2.4   *Appendages to the bacterial cell*

Three types of thread-like appendages may be found growing from bacterial cells: flagella, pili (fimbriae) and F-pili (sex strands).

Flagella are threads of protein often 12 μm long which start as a small basal organ just beneath the cytoplasmic membrane. They are responsible for the movement of motile bacteria. Their number and distribution varies. Some species bear a single flagellum, others are flagellate over their whole surface.

Pili are responsible for haemagglutination in bacteria and also for intercellular adhesiveness giving rise to clumping. At the moment a clear role for these structures has not been formulated.

F-pili or sex strands are part of a primitive genetic exchange system in some bacterial species. Part of the genetic material may be passed from one cell to another through the hollow pilus, thus giving rise to a simple form of sexual reproduction.

Capsules and slime

Some bacterial species accumulate material as a coating of varying degrees of looseness. If the material is reasonably discrete it is called a capsule, if loosely bound to the surface it is called slime.

Recently a phenomenon of resistance to biocide solutions has been recognized (see also Chapters 10 and 14) in which bacteria adhere to a container wall and cover themselves with a carbohydrate slime called a glycocalyx; thus, doubly protected (wall and glycocalyx), they have been found to resist biocide attack.

Bacillus anthracis, the causative organism of anthrax, possesses a capsule composed of polyglutamic acid; the slime layers produced by other organisms are of a carbohydrate nature.

An extreme example of slime production is found in *Leuconostoc dextranicum* and *L. mesenteroides* where so much carbohydrate, called dextran, may be produced that the whole medium in which these cells are growing becomes almost gel-like. This phenomenon has caused pipe blockage in sugar refineries and is deliberately encouraged for the production of dextran as a blood substitute (Chapter 25).

2.2.6   *Pigments*

Some bacterial species produce pigments during their growth which give the colonies a characteristic colour.

Thus, *Staph. aureus* produces a golden yellow pigment, *Serratia marcescens* a bright red pigment. There appears to be no valid function for these pigments but they may afford the cell some protection from the toxic effects of sunlight.

2.3   **The bacterial spore**

In a few bacterial genera, notably *Bacillus* and *Clostridium*, a unique process takes place in which the vegetative cell undergoes a profound biochemical change to give rise to a structure called a spore or endospore (Fig. 1.8). This process is not part of a reproductive cycle, but the bacterial endospore is highly resistant to adverse environments such as lack of moisture or essential nutrients, toxic chemicals and radiations and high temperatures. Because of their heat resistance all sterilization processes have to be designed to destroy the bacterial spore.

2.3.1   *The process of spore formation*

In general, an adverse environment, and in particular the absence or limited presence of one component, induces spore formation. Examples of such components are alanine, Zn^{2+}, Fe^{2+}, PO_3^- and, in the case of the aerobic (oxygen-requiring) *Bacillus* species, oxygen. Equally, certain substances, for instance Ca^{2+} and Mn^{2+}, have to be present for the process of spore formation to proceed to completion.

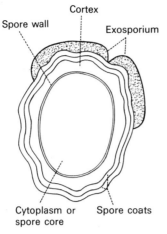

Fig. 1.8 Diagram of a transverse section of a bacterial spore.

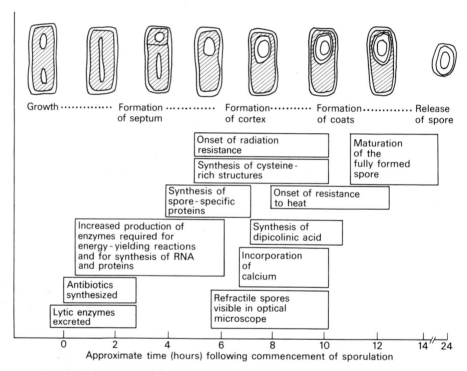

Fig. 1.9 Changes occurring during spore formation. The position and length of the boxes represent the approximate time and duration of the various activities.

If the conditions for spore formation are fulfilled the sequence of events shown in Fig. 1.9 occurs.

The essential genetic material of the original vegetative bacterium is retained in the core or protoplast; around this lies the thick cortex which contains the murein or peptidoglycan already encountered as a cell wall

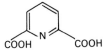

Fig. 1.10 Pyridine 2,6-dicarboxylic acid, dipicolinic acid (DPA).

component (see Fig. 1.2). The outer coats which are protein in composition are distinguished by their high cysteine content. In this respect they resemble keratin, the protein of hair and horn.

Another feature of the spore is the presence of pyridine 2,6-dicarboxylic acid (Fig. 1.10) occurring as a complex with calcium, which at one time was implicated in heat resistance. The isolation of heat-resistant spores containing no Ca-DPA has refuted this hypothesis.

The reason for heat resistance is thought to lie in the fact that the core or spore cytoplasm becomes dehydrated during sporulation. The mechanism for this dehydration is the mechanical expulsion of water by the expansion of the peptidoglycan network which comprises the cortex—the expanded cortex theory.

Dehydration of the core by means of concentrated sucrose solution also results in heat resistance.

The tough keratin-like spore coats probably help to protect the spore core or protoplast from the harmful effects of chemicals. Radiation resistance has not been fully explained.

The same generally impervious properties make spores difficult to stain by simple stains. However, if a slide preparation of spores is warmed with a stain the spores are dyed so effectively that dilute acid will not wash out the colour. This is the basis of the acid-fast stain for spores.

2.3.2 *Spore germination and outgrowth*

In nature, spores can revert to the vegetative form by a process called either 'germination' or 'germination and outgrowth'. The process of germination may be triggered by specific germination stimulants, such as L-alanine or glucose, by the physical processes of shaking with small glass beads, or by sublethal heating (e.g. at 60°C for 1 hour). Outgrowth and subsequent growth depends on the presence of the necessary nutrients for the particular organism concerned. The stages of germination and outgrowth, and also the action of inhibitors of the process are shown in Fig. 1.11.

2.3.3 *Parameters of heat resistance*

The existence and possible presence of bacterial spores determines the parameters, i.e. time and temperature relationships, of thermal sterilization processes which are used extensively by the food and pharmaceutical industry. These are defined below (see also Chapters 20 and 23).

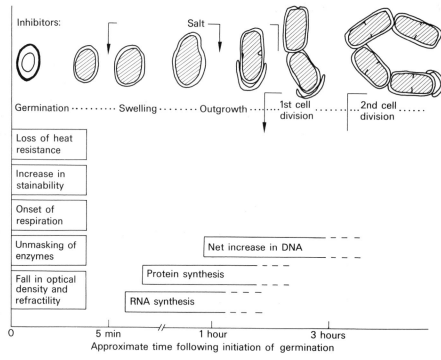

Fig. 1.11 Spore germination, outgrowth and division of the outgrown cells.

1 *D*-value (decimal reduction time, DRT) is the time in minutes required to destroy 90% of a population of cells. The *D*-value has little relevance to the sterilization of medicines for injection, surgical instruments or dressings, where a process designed to kill all living spores must be developed. The *D*-value is used extensively in the food industry.

2 The F_0-value is a process-describing unit expressed in terms of minutes at 121.1°C (originally 250°F) or a corresponding time–temperature relationship to produce the same complete spore-killing effect.

3 The *z*-value is the increase in temperature (°C) to reduce the *D*-value to one-tenth.

3 Toxins

Although bacteria are associated with the production of disease, only a few species are disease-producing or pathogenic.

The mechanism whereby they produce the disease with its attendant symptoms is often due to the cells' ability to produce specific poisons, toxins or aggressins (Chapter 15). Many of these are tissue-destroying enzymes which can damage the cellular structure of the body or destroy red blood cells. Others (neurotoxins) are highly specific poisons of the central nervous system, for example the toxin produced by *Clostridium botulinum* is, weight for weight, one of the most poisonous substances known.

4   Reproduction

4.1   Binary fission

The majority of bacteria reproduce by simple binary fission; the circular chromosome divides into two identical circles which segregate at opposite ends of the cell. At the same time, the cell wall is laid down in the middle of the cell, which finally grows to produce two new cells each with its own wall and nucleus. Each of the two new cells will be an exact copy of the original cell from which they arose and no new genetic material is received and none lost.

4.2   Reproduction involving genetic exchange

For many years, it was thought that binary fission was the only method of reproduction in bacteria, but it is now known that there are three methods of reproduction in which genetic exchange can occur between pairs of cells, and thus a form of sexual reproduction is exhibited. These processes are transformation, conjugation and transduction. Further details of these processes as they affect antibiotic resistance will be found in Chapter 10 (section 2.2).

4.2.1   *Transformation*

In 1928, long before the role of DNA, the genetic code and the mechanics of genetics and gene expression were known, Griffith found that a culture of *Streptococcus pneumoniae* deficient in capsular material could be made to produce normal capsulated cells by the addition of the cell-free filtrate from a culture in which a normal capsulated strain had been growing. The state of knowledge at that time was insufficient for the great significance of this experiment to be realized and developed. It was not until 16 years later that the material in the culture filtrate responsible for the reestablishment of capsulated cells was shown to be DNA.

4.2.2   *Conjugation*

Conjugation, discovered in 1946, is a natural process found in certain bacterial genera and involves the active passage of genetic material from one cell to another by means of the sex pili (p. 10). However, despite the resemblance of this process to the complete genetic exchange found in eukaryotes, it is not possible to designate the male and female bacteria. Bacteria which are able to effect transfer contain in their genetic make-up a fertility factor and are designated F^+ strains. These are able to transfer part, and in some cases all, of their genetic material to F^- strains.

It should be realized that this is an extremely brief and incomplete account of conjugation. The importance of bacterial conjugation in antibiotic resistance will be considered later (Chapter 10).

Viruses are discussed more fully elsewhere (Chapter 3). However, there are certain groups of viruses, called bacteriophages (phages), which can attack bacteria. This attack involves the injection of viral DNA into bacterial cells which then proceed to make new virus particles. Some viruses, known as temperate viruses, do not cause this catastrophic event when they infect their host, but can pass genetic material from one cell to another. By this process again, genetic material may be exchanged between bacterial cells.

In summary, then, conjugation is a natural process representing the early stages in a true sexually reproductive process. Transformation involving autolysis of the culture with loss of genetic material, and transduction arising out of an infective process, are secondary processes which are not known to occur in eukaryotes; nevertheless, they must have taken their part in microbial evolution.

5   Bacterial growth

The preceding account has been concerned with the single bacterial cell and the information has been obtained by various forms of microscopy and by chemical analysis. Further bacteriological information has been, and is being, obtained by observing bacteria in very large numbers either as a culture in liquid growth medium or as colonies on a solid growth medium. Under these circumstances bacteria can be seen but the behaviour of aggregates is really a statistical average behaviour of its individuals. A somewhat fanciful analogy is that whereas a molecule of a chemical substance is invisible, molecules in mass, i.e. a chemical specimen, are visible and macroscopic properties are determinable.

5.1   The growth requirements of bacteria

The determinants of microbial growth are described as consumable and environmental.

5.1.1   *Consumable determinants*

The consumables represent the essential food or nutritional requirements. Conventionally they include sugars, starches, protein, vitamins, trace elements, oxygen, carbon dioxide and nitrogen; but bacteria are probably the most omnivorous of all living organisms and to the above list may be added plastic, rubber, kerosene, naphthalene, phenol and cement. One is left feeling that there is no substance which is immune from microbial attack. It is easy, too, to overlook the importance of water; bacteria cannot grow without water and, besides a milieu in which to thrive, water also provides hydrogen as part of reaction sequences for the metabolism of the substrates.

Some bacteria have very simple growth requirements, and the following medium (expressed as gl^{-1}) will support the growth of a wide range of species: glucose, 20; $(NH_4)_2HPO_4$, 0.05. On the other hand, some species may need the addition of some 20 amino acids and perhaps 8–10 vitamins or growth factors (thiamine or vitamin B_1 is an example of the latter) before growth will occur, and if follows that these requirements have to be present in natural environments also. Between the extremes of the nutritionally non-exacting and the nutritionally highly exacting, a whole range of intermediate requirements are found.

The requirements of a microorganism for an amino acid or vitamin can be used to determine the amount of that substance in foods or pharmaceutical products by growing the organism in a medium containing all the essential requirements and measured doses of the substance to be determined.

Mention has been made of gases as part of the bacteria consumables list. Some bacteria cannot grow unless oxygen is present in their immediate atmosphere; in practical terms this means that they grow in air. Such organisms are called obligate aerobes. Another group is actually inhibited in the presence of oxygen, this gas behaving almost as an intoxicant, and such bacteria are known as obligate anaerobes. A large number of species can grow both in the presence and absence of oxygen and these are termed facultative bacteria. These organisms, however, make much better use of foodstuffs, i.e. their consumables, when growing in air. A fourth group is named microaerophilic; these grow best in the presence of oxygen at slightly lower concentrations than that found in air. Special techniques are needed to grow anaerobic bacteria which, briefly, consist of cultivation in oxygen-free atmospheres or growth in culture media containing a reducing agent; sometimes a combination of both methods is used.

5.1.2   *Environmental determinants*

The main environmental determinants of microbial growth are pH and temperature. The availability of water may be lowered when certain solutes are present in high concentration; thus concentrated salt and sugar solutions may either slow down or prevent growth.

Most bacteria grow best at pH values of 7.4–7.6, on the alkaline side of neutrality, but some bacterial species are able to grow at pH 1–2 or 9–9.5, although they are exceptional.

Bacteria also show a wide range of growth temperatures. Those organisms which cause disease in man and other mammals, and in consequence have been extensively studied, grow best at the temperature of the mammalian body, i.e. 37–39°C. However, viable bacteria have been recovered from hot springs and from the polar seas, and certainly there are bacteria which can grow in domestic refrigerators. Bacteria which grow best at 15–20°C are called psychrophiles, at 25–40°C mesophiles, and at 55–75°C thermophiles.

The growth of bacteria, as with other living organisms, can be inhibited or prevented. Antiseptics, disinfectants, antibiotics and chemotherapeutic agents are the names given to special chemicals developed to combat infection. They are discussed in later chapters.

5.1.3 *Culture media*

Mention has already been made of the wide variety of consumable nutrients which may be required by bacteria, and also how some bacteria can grow in simple aqueous solution containing an energy source, such as glucose, and a few inorganic ions.

For the routine cultivation of bacteria, a cheap source of all likely nutrients is desirable, and it should also be remembered that even bacteria whose minimum requirements are very simple grow far better on more highly nutritious media.

The media usually employed are prepared from protein by acid or enzymic digestion. Typical sources are muscle tissue (meat), casein (milk protein) and blood fibrin. Their digestion provides a supply of the natural amino acids and, because of their origin as living tissue, they will also contain vitamins or growth factors and mineral traces. Solutions of these digests, with the addition of sodium chloride to optimize the tonicity, comprise the common liquid culture media of the bacteriological laboratory. If it is required to study the characteristic colony appearance of cultures, the above media may be solidified by a natural carbohydrate gelling agent, agar, which is derived from seaweed.

In addition, a vast array of special culture media have been developed containing chemicals which by either their selective inhibitory properties or characteristic changes act as selective and diagnostic agents to pick out and identify bacterial species from specimens containing a mixture of microorganisms. The examination of faeces for pathogens is a good example.

5.2 Energy provision

The growth requirements outlined above express themselves in growth itself through the less tangible but fundamental necessity of energy which is provided by metabolism. Not all metabolic reactions, however, provide energy; esterase activity is an example of one that does not.

The energy provision by carbohydrate metabolism has been extensively studied from the beginning of this century, chiefly in an attempt to understand the basic biochemistry of alcohol production from carbohydrate. However, many laboratory culture media contain only nitrogenous compounds and their metabolism is of importance as it clearly provides energy for growth and maintenance.

In addition, living cells need a system of energy storage and this is provided by 'bond energy', strictly the free energy of hydrolysis of a diphosphate bond in the compound adenosine triphosphate (ATP).

Energy-yielding reactions and energy-storage systems form a common pattern found in all living systems and may be depicted thus:

$$\text{Reactant} \rightarrow \text{products} + \text{energy} \begin{cases} \nearrow \text{stored} \\ \searrow \text{utilized as produced.} \end{cases}$$

A fundamental characteristic of the overall reaction is that it proceeds by a series of steps, each catalysed by a separate enzyme. This ensures gentle and not explosive release of energy and also provides a useful set of intermediates for the biosynthetic reactions which are concomitant to growth.

It is the complexity of the array of enzymes, coenzymes and inter-mediates which at first sight provides a daunting barrier to those wishing to try to understand cellular energetics.

As stated in section 5.1.1, some bacteria derive energy from food sources without the use of oxygen, whereas others are able to use this gas. The pathway of oxygen utilization itself is also a stepwise series of reactions and thus the overall picture emerges of cellular metabolism characterized by multistep reactions.

Although bacteria (the prokaryotes) differ in many fundamental ways from all other living organisms (see Table 1.1), their metabolic pathways do not. The handling of carbohydrates by the Embden–Meyerhof pathway and the Krebs citric acid cycle and many of the reactions of the metabolism of nitrogen-containing compounds are common to both eukaryotes and prokaryotes. The enzymes and coenzymes for handling molecular oxygen are also strikingly similar in both classes. A full treatment of these pathways is given in the third edition of this book and in textbooks of biochemistry and microbial chemistry.

5.2.1   *Carbohydrate metabolism*

Polysaccharides such as starch and cellulose are broken down to their monomer, glucose. A variety of other sugars may also be used and the varying ability of organisms to break down individual carbohydrates is used as a diagnostic tool in bacteriology (section 5.3).

The prime reaction sequence for glucose catabolism or breakdown is known as the Embden–Meyerhof pathway after the biochemists who elucidated it in the early 1900s, and is shown in Fig. 1.12. This proceeds without oxygen, i.e. in an anaerobic reaction sequence, and glucose is converted to pyruvic acid.

There are several points to note in this sequence. First, reactions proceed via phosphate esters of the various components (in Fig. 1.12 —P represents —$PO(OH)_2$) and the origin of the —P is ATP. Second, there is one oxidation reaction, when glyceraldehyde is oxidized to glyceric acid, both as their phosphate esters.

Pyruvic acid is a junction point in carbohydrate metabolism. In the alcoholic fermentation by yeast studied by Embden and Meyerhof,

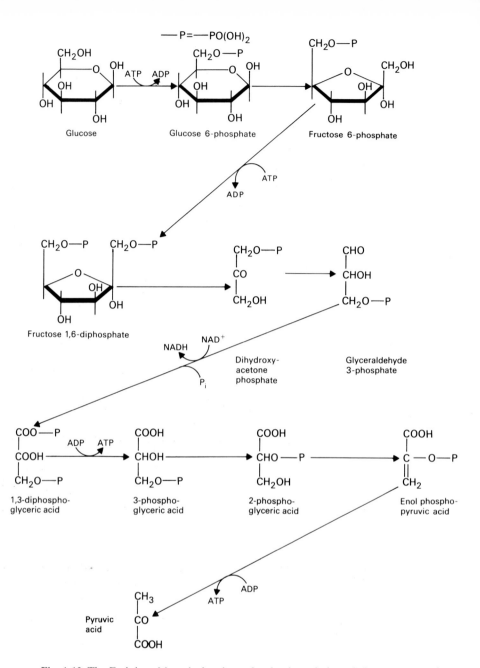

Fig. 1.12 The Embden–Meyerhof pathway for the degradation of glucose to pyruvic acid.

pyruvic acid is reduced to ethanol, this reduction being coupled to the oxidation of glyceraldehyde to glyceric acid discussed above.

Figure 1.13 shows the variety of pathways by which pyruvic acid is subsequently metabolized by different species of microorganism. The metabolism of pyruvic acid to water and carbon dioxide proceeds via the

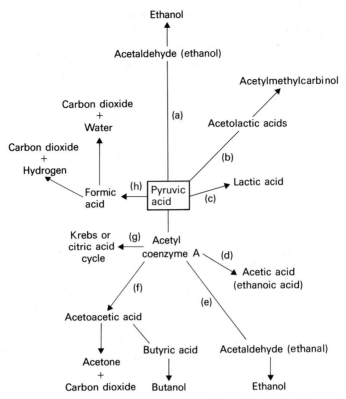

Fig. 1.13 Products of the decomposition of pyruvic acid by microorganisms.
(a) Production of ethanol by yeast. The completed Embden–Meyerhof pathway.
(b) Production of acetylmethylcarbinol (used to distinguish *Klebsiella pneumoniae*
subsp. *aerogenes* from *E. coli*—the former produces it, the latter does not). (c) Lactic
acid fermentation. (d) Production of acetic acid. (e) Production of ethanol by *E. coli*,
differs from (a) in that acetyl coenzyme A is an intermediate. (f) The acetone–butanol
fermentation. (g) The Krebs cycle (see Fig. 1.14). (h) Acid and gas production in
enterobacterias.

Krebs, tricarboxylic acid or citric acid cycle, and this is shown in Fig.
1.14. Again the feature of the Krebs sequence is its stepwise nature and
the provision of intermediates useful for other biosynthetic reactions.

This reaction is aerobic, and oxygen is metabolized through a reaction
sequence shown very simply in Fig. 1.15.

5.2.2   *Protein and amino acid metabolism*

Protein metabolism is diverse, varying between species of microbe.

The process of putrefaction is due to the breakdown of protein, the
principal nitrogenous components of plants and animals, with the production of, amongst other things, evil-smelling amines. Many bacteria
possess the ability to decompose proteins and the products derived from
protein breakdown.

Proteolytic activity is of practical significance in bacteriology in that it

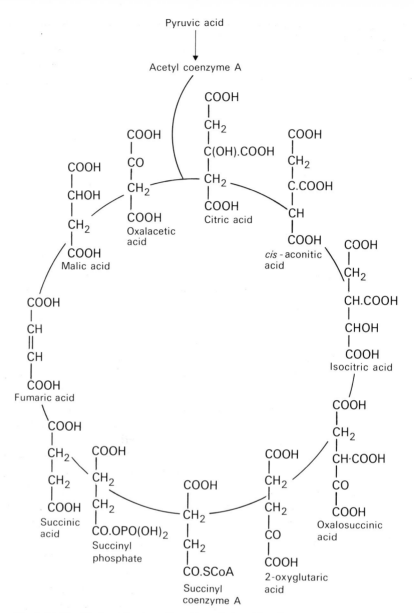

Fig. 1.14 The Krebs, tricarboxylic acid or citric acid cycle.

is used as a diagnostic tool. Many pathogenic bacteria produce toxins which are proteolytic enzymes, and cause tissue damage by the exertion of this activity. Proteolytic activity can cause spoilage to foodstuffs composed of protein materials and is of importance in the nitrogen cycle in nature.

Proteins consist of an array of α-amino acids linked through their amino and carboxylic acid groups by elimination of water to form the peptide bond.

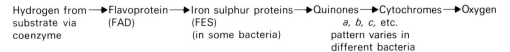

Hydrogen from ⟶ Flavoprotein ⟶ Iron sulphur proteins ⟶ Quinones ⟶ Cytochromes ⟶ Oxygen
substrate via   (FAD)   (FES)   *a, b, c,* etc.
coenzyme   (in some bacteria)   pattern varies in
                                                                                 different bacteria

Fig. 1.15 Simplified diagram of pathway of hydrogen from substrate (food) to oxygen in the air.

$$R.CH.CO \overbrace{[OH + H]}^{} HN.CH.R \rightarrow R.CH.CO.NH.CH.R + H_2O$$

Peptide link

$$
\begin{array}{llll}
R.CH.CO\,[OH + H]\,HN.CH.R & \rightarrow & R.CH.CO.NH.CH.R + H_2O \\
\quad | & & \quad | & \quad | \\
\quad NH_2 & COOH & NH_2 & COOH
\end{array}
$$

1st amino acid   2nd amino acid   Peptide

Thus protein decomposition must involve initially the hydrolysis of this peptide link. A more fundamental understanding of the nature of proteolytic enzymes was made possible when a technique for synthesizing peptides was developed.

Originally proteolytic enzymes were classified according to whether they decomposed proteins or protein subunits such as peptones and peptides. This idea has now been reformulated and proteolytic enzymes are classified as *exopeptidases* if they hydrolyse peptide bonds at or near the ends of polypeptides or proteins, and *endopeptidases* if they are able to hydrolyse peptide bonds near the middle. In addition, the activity of proteolytic enzymes was found to be affected by the nature of the amino acid adjacent to the peptide bond and by the presence of a free —COOH or —NH₂ group near the bond. Thus, among the exopeptidases are recognized the carboxypeptidases, which are only active against peptide links provided that there is a free carboxyl group in the vicinity, and the aminopeptidases, which function only if there is a free amino group near the peptide bond concerned. A third type of exopeptidase, called a dipeptidase, is active against dipeptides.

Proteolytic activity in routine bacteriology is tested for by measuring the ability of organisms to liquefy gelatin or to digest clotted milk. Ability to liquefy gelatin is tested by inoculating a nutrient medium solidified with gelatin and, after incubation, chilling the medium in ice water and determining by inversion of the tube whether the gelatin has been liquefied. Hydrolysis of the peptide links of the macromolecular protein causes a progressive reduction in viscosity.

It should be remembered, however, that gelatin is an artificial product and may have undergone chemical change during its manufacture. In fundamental studies of protein breakdowth by microorganisms, a wider variety of substrates should be used, including natural or native protein (collagen is the natural protein from which gelatin is derived). Furthermore, evidence of the attack on protein should be sought by a determination of the free amino or carboxyl groups produced as the protein is hydrolysed.

Many proteolytic enzymes are extracellular and can be recovered from the culture medium in which the cells are grown. The production of proteolytic enzymes under artificial conditions by bacteria is affected by the composition of the medium in which the cells are grown and by the growth temperature. The main factors appear to be the correct balance of inorganic salts, of which a source of calcium ions seems to be the most important, and in some instances the presence of a utilizable source of carbohydrates.

It has been seen that the result of the action of proteolytic enzymes on proteins is the production of amino acids. Amino acids undergo a diversity of catabolic processes, the more important of which will be considered below. In addition, amino acids produced by the breakdown of proteins and peptides may be used by the cell in synthetic or anabolic processes. Catabolic reactions of amino acids may involve either the —NH$_2$ group or the —COOH group.

Oxidative deamination. This type of reaction is catalysed by a group of enzymes called the amino acid oxidases. The result of the reaction is the conversion of an amino acid to an oxo acid and it has been proved in some cases that the reaction proceeds via an intermediary imino acid.

$$
\begin{array}{ccc}
\text{R} & \text{R} & \text{R} \\
| & | & | \\
\text{CH.NH}_2 \xrightarrow{\text{Amino acid oxidase}} & \text{C:NH} \xrightarrow[\text{+H}_2\text{O}]{\text{Hydrolysis}} & \text{C:O + NH}_2 \\
| & | & | \\
\text{COOH} & \text{COOH} & \text{COOH}
\end{array}
$$

Thus the amino acid, glutamic acid, yields the oxo acid 2-oxoglutaric acid and alanine yields pyruvic acid.

The formation of indole. Tests for the formation of indole have long been used in diagnostic bacteriology and various techniques have been devised to demonstrate its presence. Indole arises from the breakdown of the amino acid tryptophan, and its formation depends on the presence of this compound in the culture medium and the presence in the bacteria of the necessary enzymes. The reaction proceeds according to the simplified scheme at the top of p. 25.

The overall reaction is catalysed by an enzyme complex called tryptophanase, which has been found in various species of the bacterial genera *Escherichia*, *Proteus* and *Vibrio*. Tryptophanase has been the subject of much study and it has been found that a high level of the enzyme can be induced if cells capable of developing tryptophanase activity are grown in the presence of tryptophan and under aerobic conditions.

Indole is detected in bacterial cultures by adding *p*-dimethyl-amino-benzaldehyde solution, when a characteristic pink colour is produced.

Not all bacteria produce indole from tryptophan; members of the genera *Bacillus* and *Pseudomonas* may break down this compound via a

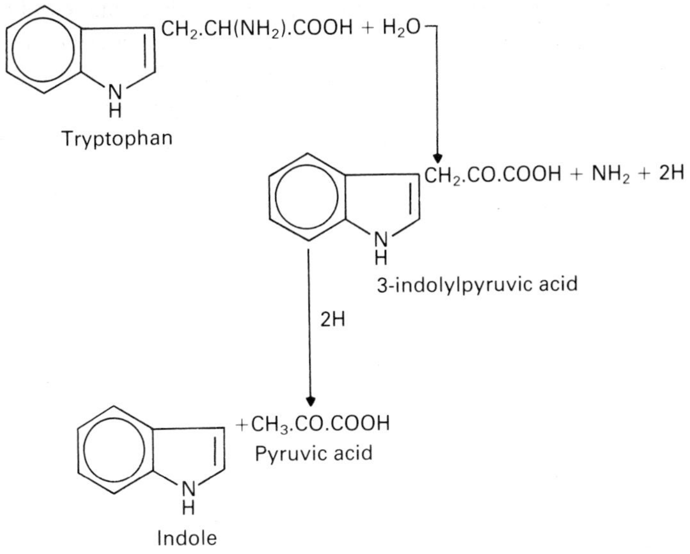

Tryptophan

$CH_2.CH(NH_2).COOH + H_2O$

$CH_2.CO.COOH + NH_2 + 2H$

3-indolylpyruvic acid

2H

$+CH_3.CO.COOH$
Pyruvic acid

Indole

complex series of steps to 3-oxoadipic acid, the nitrogen being lost as ammonia.

Transamination. This reaction involves the exchange of an amino group between an amino acid and an oxo acid and may best be illustrated by examples.

$$
\begin{array}{cccc}
\text{COOH} & & & \text{COOH} \\
| & & & | \\
\text{CH}.\text{NH}_2 & \text{CH}_3 & \text{CH}_3 & \text{CO} \\
| & | & | & | \\
\text{CH}_2 \quad + & \text{CO} \longrightarrow & \text{CH}.\text{NH}_2 + & \text{CH}_2 \\
| & | & | & | \\
\text{CH}_2 & \text{COOH} & \text{COOH} & \text{CH}_2 \\
| & & & | \\
\text{COOH} & & & \text{COOH}
\end{array}
$$

Glutamic   Pyruvic   Alanine   2-oxoglutaric
acid (amino acid)   acid (oxo acid)   acid

Similarly,

Aspartic acid + pyruvic acid → oxalacetic acid + alanine
Glutamic acid + oxalacetic acid → 2-oxoglutaric acid + aspartic acid.

    In an extensive study with *E. coli, Pseudomonas aeruginosa* and *Bacillus subtilis* it was shown that alanine, aspartic acid, leucine, methionine, norleucine, phenylalanine, tryptophan and tyrosine are able to transfer their amino group to 2-oxoglutaric acid with formation of glutamic acid. It should be appreciated that this reaction forms a route for the biosynthesis of amino acids as well as for their decomposition. All transaminases found so far in the bacteria are specific for the L-isomer of the amino acids involved.

The Stickland reaction. Stickland, in 1934, showed that amino acids could take part in anaerobic oxidation–reduction reactions, certain acids acting as hydrogen donors and others as hydrogen acceptors. The reaction was first studied in the anaerobic organism *Clostridium sporogenes*, for which amino acids could be divided into two categories. Hydrogen *donors* were histidine, alanine, serine, valine, tyrosine and leucine; and hydrogen *acceptors* were tryptophan, proline, ornithine, hydroxyproline, arginine and glycine.

The reaction between alanine and proline was found to be as follows:

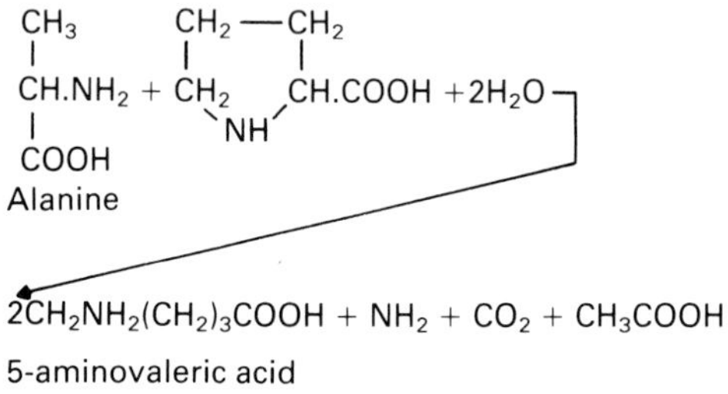

$$
\begin{array}{ll}
CH_3 & CH_2\!-\!CH_2 \\
| & | \quad\;\; | \\
CH.NH_2 + CH_2 \quad CH.COOH + 2H_2O \\
| \qquad\qquad\; \diagdown NH \diagup \\
COOH \\
\text{Alanine}
\end{array}
$$

$2CH_2NH_2(CH_2)_3COOH + NH_2 + CO_2 + CH_3COOH$

5-aminovaleric acid

The Stickland reaction yields energy to the cell and is an example of an anaerobic energy-yielding reaction.

Formation of hydrogen sulphide from sulphur-containing amino acids. Hydrogen sulphide production in culture media occurs in heterotrophic bacteria as a result of the decomposition of sulphur-containing amino acids, which themselves may be produced by prior decomposition of proteins or peptones. Hydrogen sulphide production is used in diagnostic bacteriology; a culture medium containing an iron salt is often used to detect the gas by the formation of black iron sulphides. Its production may also be demonstrated by adding a thick, washed suspension of the bacterium to a solution of the sulphur-containing amino acid cysteine in a tube and trapping a piece of lead acetate paper in the cotton-wool plug of

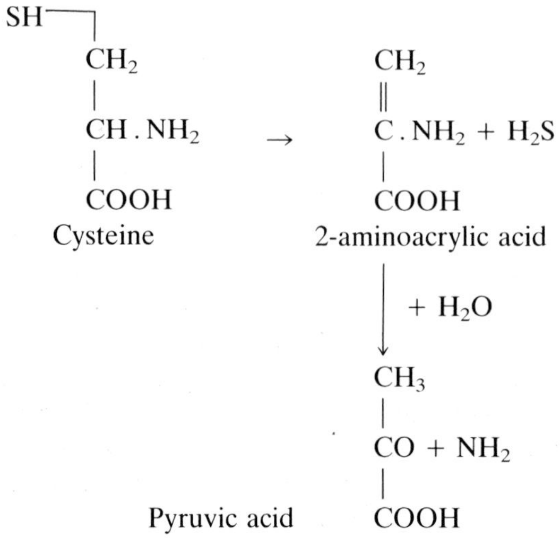

$$
\begin{array}{ccc}
SH & & \\
| & & \\
CH_2 & & CH_2 \\
| & & \| \\
CH.NH_2 & \rightarrow & C.NH_2 + H_2S \\
| & & | \\
COOH & & COOH \\
\text{Cysteine} & & \text{2-aminoacrylic acid}
\end{array}
$$

$$+ H_2O \downarrow$$

$$
\begin{array}{c}
CH_3 \\
| \\
CO + NH_2 \\
| \\
COOH
\end{array}
$$

Pyruvic acid

the tube. If hydrogen sulphide is evolved, the paper will be blackened within 3–4 hours after incubation was started. In one mechanism of hydrogen sulphide production an enzyme called cysteine desulphydrase decomposes cysteine to hydrogen sulphide and 2-aminoacrylic acid. The latter compound then hydrolyses spontaneously to form pyruvic acid and ammonia.

Reaction of amino acids involving the carboxylic acid group: the amino acid decarboxylases. This group of enzymes catalyses reactions of the following type:

$$
\begin{array}{ccc}
R & & R \\
| & & | \\
CH.NH_2 & \rightarrow & CH_2 + CO_2 \\
| & & | \\
COOH & & NH_2 \\
\text{Amino acid} & & \text{Amine}
\end{array}
$$

As can be seen, this reaction results in the production of an amine from a carboxylic acid; thus, on decarboxylation, histidine yields histamine, tyrosine yields tyramine and glutamic acid yields α-aminobutyric acid.

5.2.3   *Comment*

The foregoing is a very simple outline of the main metabolic activities of bacteria growing in laboratory culture media. No attempt has been made to depict the structures and roles of the coenzymes associated with the reaction sequences. Textbooks of microbial chemistry should be consulted where amplification is desired.

5.3   **Identification of bacteria**

The varying metabolic activities of bacteria and their response to immediate environmental factors have been exploited in the design of special diagnostic and selective media. Recipes for these run into many hundreds; such media are used in hospital and public health laboratories for identifying organisms found in samples believed to be contaminated by them, and as an aid to diagnosis and treatment. In addition they are used to detect contaminants in pharmaceutical products (*British Pharmacopoeia* 1988). A few examples will be given to illustrate the principle.

Mention has already been made of varying abilities to ferment carbohydrate; differences in metabolism of pyruvate are exploited to distinguish *K. pneumoniae* subsp. *aerogenes* and *E. coli*. The ability to produce indole is also used. The varying responses of bacteria to inhibitors are considered in Chapters 11 and 13. The exploitation of these differences is elaborated below.

27   *Bacteria*

MacConkey's medium was introduced in 1905 to isolate Enterobacteriaceae from water, urine, faeces, foods, etc. Essentially, it consists of a nutrient medium with bile salts, lactose and a suitable indicator. The bile salts function as a natural surface-active agent which, while not inhibiting the growth of the Enterobacteriaceae, inhibits the growth of Gram-positive bacteria which are likely to be present in the material to be examined.

E. coli and *K. pneumoniae* subsp. *aerogenes* produce acid from lactose on this medium, altering the colour of the indicator, and also adsorb some of the indicator which may be precipitated around the growing cells. The organisms causing typhoid and paratyphoid fever and bacillary dysentery do not ferment lactose, and colonies of these organisms appear transparent.

Many modifications of MacConkey's medium exist; one employs a synthetic surface-active agent in place of bile salts.

The selectivity of MacConkey's medium may be further increased by the addition of inhibitory dyes such as neutral red and crystal violet. These further suppress the growth of Gram-positive organisms such as staphylococci.

Bismuth sulphite agar.   This medium was developed in the 1920s for the identification of *Salmonella typhi* in water, faeces, urine, foods and pharmaceutical products. It consists of a buffered nutrient agar containing bismuth sulphite, ferrous sulphate and brilliant green.

E. coli (which is also likely to be present in material to be examined) is inhibited by the concentration (0.0025%) of brilliant green used, while *Sal. typhi* will grow luxuriantly. Bismuth sulphite also exerts some inhibitory effect on *E. coli*.

Sal. typhi, in the presence of glucose, reduces bismuth sulphite to bismuth sulphide, a black compound; the organism can produce hydrogen sulphide from sulphur-containing amino acids in the medium and this will react with ferrous sulphate to give a black deposit of ferrous sulphide (Table 1.2).

Selective media for staphylococci.   It is often necessary to examine pathological specimens, food and pharmaceutical products for the presence of staphylococci, organisms which can cause food poisoning as well as systemic infections.

In media selective for enterobacteria a surface-active agent is the main selector, whereas in staphylococcal medium, sodium and lithium chloride are the selectors; staphylococci are tolerant of 'salt' concentrations to around 7.5%. Mannitol salt, Baird-Parker (BP) and Vogel-Johnson (VJ) media are three examples. The other principles are the use of a selective carbon source, mannitol or sodium pyruvate together with a buffer plus acid–base indicator for visualizing metabolic activity and, by inference,

Table 1.2 Appearance of bacterial colonies on bismuth sulphite agar

Organism	Appearance of colonies on bismuth sulphite agar
Salmonella typhi *Salmonella enteritidis* *Salmonella schotmülleri*	Black with blackened extracolonial zone
Salmonella paratyphi *Salmonella typhimurium* *Salmonella choleraesuis*	Green
Shigella flexneri *Shigella sonnei*	Brown
Other shigellae *Escherichia coli*	No growth

growth. BP medium also contains egg yolk; the lecithin (phospholipid) in this is hydrolysed by staphylococcal (esterase) activity so that organisms are surrounded by a cleared zone in the otherwise opaque medium. The *United States Pharmacopeia* (1990) includes a test for staphylococci in pharmaceutical products, whereas the *British Pharmacopoeia* (1988) does not.

Selective media for pseudomonads. These media depend on the relative resistance of pseudomonads to the quaternary ammonium disinfectant cetrimide. In some recipes the antibiotic nalidixic acid (Chapter 5) is added, to which pseudomonads are also resistant.

Media for fungi. Most fungi encountered as contaminants in pharmaceutical products will grow on media similar to that used to grow bacteria. Growth is favoured, however, if the proportion of carbohydrate is increased in relation to that of nitrogenous constituents.

Thus, media for the cultivation of fungi often contain additional glucose, malt, sucrose or wort. The optimum pH for mould growth is usually on the acid side of neutrality and so the pH of culture media for moulds is usually 5–6. This, while entirely suitable for most common moulds, at the same time discourages bacterial growth and thus renders the medium selective. Examples of such media are Sabouraud maltose or dextrose agar, malt extract agar and soya tryptone agar.

The optimum temperature varies widely from species to species but in general the common moulds will grow better at 22–25°C than most human pathogenic and commensal bacteria. It is customary, therefore, to incubate mould cultures at lower temperatures then bacterial cultures.

5.3.2   *Examples of additional biochemical tests*

E. coli and *K. pneumoniae* subsp. *aerogenes* may be distinguished by their differing abilities to produce indole when grown in peptone water, by their action on glucose in peptone water buffered in pH 7.5 and containing methyl red, and by their ability to utilize citric acid as sole carbon source. As shown in Fig. 1.13h, *E. coli* produces acid substances from glucose in sufficient quantity that, despite the buffer capacity of the medium, the pH is reduced to below 4.2 and the methyl red indicator is turned from yellow to red—such a result is termed methyl red (MR) positive. *K. pneumoniae* subsp. *aerogenes* produces acetylmethylcarbinol (see Fig. 1.13b) from glucose; this product is not acid, the pH is not reduced to below 4.2, and the methyl red indicator remains yellow or orange. This organism is described as MR negative. The acetylmethylcarbinol can be detected by suitable reagents (the Voges–Proskauer (VP) reaction). *K. pneumoniae* subsp. *aerogenes* is VP positive.

Another physiological difference which may be exploited is growth temperature (section 5.1.2). *K. pneumoniae* subsp. *aerogenes* will grow at 44°C, *E. coli* will not. Incubation at 44°C will therefore distinguish these—the Eijkman (E) test. The test can be memorized by a mnemonic, IMViC or, if the Eijkman test is included, IMVEC, and many texts refer to the IMViC or IMVEC characteristics of these organisms.

In summary, therefore, the behaviour of the two organisms is as given in Table 1.3.

Table 1.3 Comparison of *E. coli* and *K. pneumoniae* subsp. *aerogenes*

	Indole	MR	VP	Citrate	44°C
E. coli	+	+	−	−	−
K. pneumoniae subsp. *aerogenes*	−	−	+	+	+

The differing ability to ferment sugars, glycosides and polyhydric alcohols is widely used to differentiate the Enterobacteriaceae and in diagnostic bacteriology generally. The test is usually carried out by adding the reagent aseptically to sterilized peptone water and a suitable indicator, contained in a 5-ml bottle closed with a rubber-lined screw cap and containing a small inverted tube filled with the medium. Acid production is indicated by a change in colour of the indicator, and gas production by gas collecting in the inverted tube.

It is possible to buy ingenious testing devices which consist of a plastic strip containing cavities in which dried reagents are placed. Such a strip may contain some 50 different tests and is used by depositing in the cavity a culture medium containing a suspension of bacteria from the colony to be investigated. The strip is then incubated. This is the API system. Another useful device consists of a plastic tube with a number of compartments of about $1.2 \, \text{cm}^3$, each containing agar medium. These are inoculated by means of a still wire run through their centre; this enables some 11 tests to be carried out. It is known as the Enterotube.

In addition to the widely used biochemical tests described above, there are a number of tests which involve a determination of DNA, RNA and protein sequences. They are not used for routine laboratory diagnoses but are invaluable for settling questions in taxonomy. Examples of such methods include DNA and RNA fingerprinting, Southern blotting (named after E. M. Southern), immunoblot fingerprinting and pulsed field gel electrophoresis.

Both plasmids and bacteriophages may also be used for typing.

Techniques involving comparison of DNA and RNA sequences are known in popular parlance as genetic fingerprinting.

Physical methods have also been used for identification. These include examination of bacterial lipids by gas/liquid chromatography and the use of flow cytometry.

5.4   Measurement of bacterial growth

The quantification of the growth response to the total environment may be determined by counting the bacterial population to see if it changes with the passage of time. The most direct method is literally to count the bacterial cells placed on a calibrated microscope slide. This slide has a grid of 0.05-mm squares ruled on it and is so arranged that when a microscope slide is placed in position on two ledges raised by 0.02 mm, a known volume ($0.00005 \, mm^3$) is spread over each square. From the counts per unit of known volume, the total count may be calculated. This method cannot distinguish between living and dead bacteria, however, and to determine the number of living bacteria in a culture it is necessary to perform what is known as a viable count. In this method, an aliquot of the culture, suitably diluted, is mixed with, or placed on the surface of, a suitable solid culture medium and the mixture incubated. Viable colonies appear in or on the medium and are counted. It will be realized here that a single bacterium in the original culture being plated is assumed to give rise to a single viable colony—this may not always be true, and aggregates of two or more cells may give rise to a single colony. Ideally, this situation should be avoided, but in order to present some notion of scientific correctness or semantic perfection, the viable count may be referred to as the number of colony-forming units (cfu) rather than as 'number of bacteria'.

A third method of determining the changes in a viable population is to take advantage of the fact that bacteria in suspension scatter or absorb light. By shining a light beam through a bacterial suspension and calculating changes in light intensity by allowing the emergent beam to fall on a photoelectric cell connected to a galvanometer, the bacterial population observed as light-scattering or light-absorbing units may be determined. This method is rapid but it counts both living and dead bacteria and, for that matter, non-bacterial particles. A calibration curve relating bacterial numbers to galvanometer reading must be produced for each experimental circumstance.

Great care, skill and understanding are required to determine the value of a bacterial population whether growing, stationary or dying. In addition to these time-honoured methods, newer techniques involving bioluminescence, monoclonal antibodies and fluorescent dyes (epifluorescence) are being developed. A feature being sought in these methods is rapidity.

5.4.1 *Mean generation time*

The time interval between one cell division and the next is called the generation time. When considering a growing culture containing many thousands of cells, a mean generation time is usually calculated.

If a single cell reproduces by binary fission, then the number of bacteria n in any generation will be as follows:

1st generation $n = 1 \times 2 = 2^1$
2nd generation $n = 1 \times 2 \times 2 = 2^2$
3rd generation $n = 1 \times 2 \times 2 \times 2 = 2^3$
yth generation $n = 1 \times 2^y = 2^y$

For an initial inoculum of n_0 cells, as distinct from one cell, at the yth generation the cell population will be:

$$n = n_0 \times 2^y$$

This equation may be rewritten thus:

$$\log n = \log n_0 + y \log 2$$

whence

$$y = \frac{\log n - \log n_0}{\log 2} = \frac{\log n - \log n_0}{0.3010}$$

where y is the number of generations that have elapsed in the time interval between determining the viable count n_0 and the population reaching n.

If this time interval is t, then the mean generation time G is given by the expression:

$$G = \frac{t}{y} = \frac{t}{\dfrac{\log n - \log n_0}{0.3010}} = \frac{t \times 0.3010}{\log n - \log n_0}$$

5.5 Growth curves

When a sample of living bacteria is inoculated into a medium adequate for growth, the change in viable population with time follows a characteristic pattern (Fig. 1.16).

The first phase, A, is called the lag phase. It will be short if the culture medium is adequate, i.e. not necessarily minimal, and is at the optimum temperature for growth. It may be longer if the medium is minimal or has to warm up to the optimum growth temperature, and prolonged if toxic substances are present; other things being equal, there is a relationship between the duration of the lag phase and the amount of the toxic inhibitor.

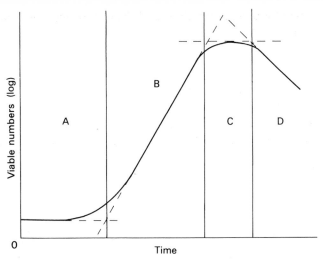

Fig. 1.16 Typical bacterial growth curve: A, lag phase; B, log phase; C, stationary phase; D, phase of decline.

In phase B it is assumed that the inoculum has adapted itself to the new environment and growth then proceeds, each cell dividing into two. Cell division by binary fission may take place every 15–20 minutes and the increase in numbers is exponential or logarithmic, hence the name log phase. Phase C, the stationary phase, is thought to occur as a result of the exhaustion of essential nutrients and possibly the accumulation of bacteriostatic concentrations of wastes. Growth will recommence if fresh medium is added to provide a new supply of nutrients and to dilute out toxic accumulations.

In phase D, the phase of decline, bacteria are actually dying due to the combined pressures of food exhaustion and toxic waste accumulation.

6 Properties of selected bacterial species

In this section, no attempt will be made to follow the modern classification system; the reader is referred to the works of Bergey (Buchanan & Gibbons, 1974), and Cowan and Steel (Cowan, 1974), for an overview of classification.

6.1 Gram-positive cocci

6.1.1 *Staphylococcus*

The spheres grow characteristically in aggregates which have been likened to a bunch of grapes. The organisms are non-motile and non-sporing; they can grow aerobically or anaerobically. *Staph. aureus* produces a golden yellow pigment. It is a cause of skin lesions such as boils, and can affect bone tissue in the case of staphylococcal osteomyelitis. It produces a toxin which, if ingested with food in which the organism has been growing, can give rise to food poisoning. A common manifestation of its infection is the production of pus, i.e. the organism is pyogenic. Other common conditions associated with staphylococcal infections are styes, impetigo and conjunctivitis.

6.1.2   *Streptococcus*

These also are non-sporing, spherical organisms which grow characteristically in chains like a string of beads, and can grow aerobically or anaerobically.

Streptococcus pyogenes can be an extremely dangerous pathogen; it produces a series of toxins, including an erythrogenic toxin which induces a characteristic red rash, and a family of toxins which destroy the formed elements of blood.

Typical diseases caused by *Strep. pyogenes* are scarlet fever and acute tonsillitis (sore throat), and the organism is a dangerous infective agent in wounds and in blood poisoning after childbirth (puerperal sepsis). Rheumatic fever and acute inflammation of the kidney are serious sequelae of streptococcal infection.

6.1.3   *Diplococcus (now Streptococcus)*

As their name implies, these organisms grow in pairs, otherwise they are similar to streptococci and are now referred to as streptococci. *Strep. pneumoniae* is the causal agent of acute lobar pneumonia and also of meningitis, peritonitis and conjunctivitis.

6.2   Gram-negative cocci

6.2.1   *Neisseria and Branhamella*

The Gram-negative pathogenic cocci belong to the genus *Neisseria*. The cells are slightly curved rather than true spheres and have been likened to a kidney bean in shape. They often occur in pairs and embedded in pus cells. *N. gonorrhoeae* is the causal organism of the venereal disease gonorrhoea. The organism can also affect the eyes, causing a purulent ophthalmia. *N. meningitidis* is a cause of cerebrospinal fever or meningococcal meningitis. *Branhamella catarrhalis* (formerly *N. catarrhalis*) is a harmless member of the genus and is often isolated from sputum.

6.3   Gram-positive rods

The genera of importance in this group are *Bacillus*, *Clostridium* and *Corynebacterium*.

6.3.1   *Bacillus*

Members of this genus are widespread in air, soil and water, and in animal products such as hair, wool and carcasses. It occurs characteristically as a large rod with square ends; it is aerobic and spore-forming. The most dangerous member of the group, *B. anthracis*, is the causal organism of anthrax. *B. cereus* has been implicated during recent years as a cause of food poisoning, *B. polymyxa* is the source of the antibiotic polymyxin, *B. brevis* of tyrothricin and *B. subtilis* and *B. licheniformis* of bacitracin.

6.3.2   *Clostridium*

Clostridia are anaerobic, spore-forming rods. The genus contains a number of dangerous pathogens.

Cl. septicum, *Cl. perfringens* (*welchii*) and *Cl. novyi* (*oedematiens*) cause serious damage to tissue if they are able to develop in wounds where the oxygen supply is limited. Tissue may be destroyed and carbon dioxide produced from muscle glycogen gives rise to the condition known as gas gangrene.

Cl. botulinum secretes an extremely toxic nerve poison and ingestion of food in which this organism has grown is fatal. Cooking rapidly destroys the poison but cold meats, sausages and pâtés, which contain the organism and which are eaten uncooked, are a possible source of botulism. *Cl. tetani* also produces a powerful central nervous system poison and give rise to the condition known as lockjaw or tetanus. *Cl. sporogenes* is a non-pathogenic member of the genus and is sometimes used as a control organism for anaerobic culture media in sterility testing (although the *European Pharmacopoeia* specifies *Cl. sphenoides*: Chapter 23).

Cl. difficile, described in older texts as of little significance as a pathogen if present in the gut, may, after therapy with antibiotics such as clindamycin or ampicillin, remain uninhibited, grow and produce toxins which give rise to a serious condition known as pseudomembranous colitis. The organism will usually succumb to vancomycin.

6.3.3   *Corynebacterium*

C. diphtheriae, which is non-sporing, is the causal organism of diphtheria, a disease which has largely been eradicated by immunization (Chapter 16).

Gardnerella vaginalis (previously named *C. vaginale* or *Haemophilus vaginalis*), although often part of the normal flora of the vagina, can be a cause of vaginitis. It has been suggested that the condition is expressed in association with anaerobes. It responds to treatment with metronidazole (Chapter 5).

Listeria

L. monocytogenes has been known as a pathogen since the 1920s. It has achieved prominence and some notoriety lately as a contaminant in dairy products. It occurs as a non-sporing Gram-positive coccobacillus or rod-shaped organism, and is able to survive and multiply at low temperatures. Thus it is essential that freezer cabinets in retail outlets should be maintained at temperatures low enough to prevent growth of the organism.

Ingestion of *L. monocytogenes* can cause abortion in humans and animals and in the case of listeriosis a prime characteristic is an increase in monocytes.

Listeriosis may be treated with a combination of ampicillin and gentamicin.

6.4   **Gram-negative rods**

6.4.1   *Pseudomonas*

Ps. aeruginosa (*pyocyanea*) has, in recent years, assumed the role of a dangerous pathogen. It has long been a troublesome secondary infector of wounds, especially burns, but was not necessarily pathogenic. With the advent of immunosuppressive therapy following organ transplant, systemic infections including pneumonia have resulted from infection by this organism. It has also been implicated in eye infections resulting in the loss of sight.

It is resistant to many antibacterial agents (Chapter 14) and is biochemically very versatile, being able to use many disinfectants as food sources.

6.4.2   *Vibrio*

V. cholerae (*comma*) is often seen in the form of a curved rod (comma), hence its alternative specific name. It is the causal organism of Asiatic cholera. This disease is still endemic in India and Burma, and was in the UK until the nineteenth century, the last epidemic occurring in 1866. It is a water-borne organism and infection may be prevented in epidemics by boiling all water and consuming only well-cooked foodstuffs. *V. parahaemolyticus* occurs in sea water and has been implicated in food poisoning following consumption of raw fish. It accounts for more than half the cases of food poisoning in Japan, where raw fish, suchi, is an important dietary item. Food poisoning from this organism also occurs in the UK.

6.4.3   *Yersinia and Francisella*

Y. pestis (formerly *Pasteurella pestis*) is the causal organism of plague or the Black Death which ravaged the UK at various times, the Great Plague occurring in 1348. It infects the lymphatic system to give bubonic plague, the more usual form, or the respiratory system, giving the rapidly fatal pneumonic plague.

F. tularensis (formerly *Pasteurella tularensis*) causes tularaemia in man, a disease endemic in the American Midwest and contracted from infected animals.

6.4.4 Bordetella

Bord. pertussis is the causal organism of whooping-cough, a disease which has been largely eradicated by a successful immunization programme (Chapter 16).

6.4.5 Brucella

This genus is found in many domesticated animals and in some wild species.

Br. abortus is a cause of spontaneous abortion in cattle. In humans it causes undulant fever, i.e. a fever in which temperature undulates with time. *Br. melitensis* infects goats; it causes an undulant fever called Malta fever, which is common in Mediterranean countries where large flocks of goats are kept.

Br. suis is found in pigs; it too manifests itself in humans as undulant fever and occurs frequently in North America.

6.4.6 Haemophilus

H. influenzae owes its specific name to the fact that it was thought to be the causal organism of influenza (now known to be a virus disease) as it was often isolated in cases of influenza. It can cause infantile meningitis and conjunctivitis and is one of the most important causes of chronic bronchitis.

6.4.7 Escherichia

E. coli and the organisms listed below (sections 6.4.8–6.4.12) are members of a group of microorganisms known as the enterobacteria, so called because they inhabit the intestines of humans and animals. Many selective and diagnostic media and differential biochemical reactions are available to isolate and distinguish members of this group, as they are of great significance in public health.

E. coli is a cause of enteritis in young infants and the young of farm animals, where it can cause diarrhoea and fatal dehydration. It is a common infectant of the urinary tract and bladder in humans, and is a cause of pyelitis, pyelonephritis and cystitis.

6.4.8 Salmonella

Sal. typhi is the causal organism of typhoid fever, *Sal. paratyphi* causes paratyphoid fever, whilst *Sal. typhimurium*, *Sal. enteritidis* and very many other closely related organisms are a cause of bacterial food poisoning.

6.4.9   *Shigella*

Sh. shiga, *Sh. flexneri*, *Sh. sonnei* and *Sh. boydii* are the causes of bacillary dysentery.

6.4.10   *Proteus*

Pr. vulgaris and *Pr. morganii* can infect the urinary tract of humans. They are avid decomposers of urea, producing ammonia and carbon dioxide. These organisms occasionally cause wound infection.

6.4.11   *Serratia marcescens*

This very small organism, $0.5–1.0\,\mu m$ long, is used to test bacterial filters. It is not to be regarded as non-pathogenic, although infections arising from it are rare.

6.4.12   *Klebsiella*

K. pneumoniae subsp. *aerogenes* is found in the gut and respiratory tract of man and animals, and in soil and water. It may be distinguished from *E. coli* by a pattern of biochemical tests (Table 1.3). It can give rise to acute bronchopneumonia in humans but is not a common pathogen.

6.4.13   *Flavobacterium*

Various species of this characteristically pigmented genus occur in water and soil and can contaminate pharmaceutical products.

6.4.14   *Acinetobacter*

This genus has the same distribution and the same opportunities for causing contamination as *Flavobacterium*. These organisms are not pigmented.

6.4.15   *Bacteroides*

The characteristic of this genus is that its members are anaerobes. They occur in the alimentary tract of humans and animals and have been associated with wound infections, especially after surgery. *Bacteroides fragilis* is a frequently encountered member of the genus.

6.4.16   *Campylobacter*

Campylobacters are thin, Gram-negative organisms which are in essence rod-shaped but often appear in culture with one or more spirals or as 'S' and 'W' (gull-winged) shaped cells. They are microaerophilic or anaerobic and move by means of a single polar flagellum. They are unable to grow below 30°C.

C. jejuni has emerged during the last few years as a major cause of enteritis in man and is mainly transmitted by contaminated food, in other words it is a food-poisoning microorganism.

| 6.4.17 | *Helicobacter* |

This genus, originally grouped with the campylobacters (section 6.4.16), is now considered a separate genus. *H. pylori* is of interest as a possible cause of peptic ulcer.

| 6.4.18 | *Chlamydia* |

The diseases associated with chlamydias (e.g. psittacosis) were at one time thought to be due to what were regarded as large viruses.

Chlamydias, however, are bacteria and have been shown to possess a cell wall containing muramic acid (section 2.2.1), to contain ribosomes of the bacterial (prokaryotic) type, to reproduce themselves by binary fission and to be inhibited by antibiotics active against bacteria.

They are coccoid-shaped organisms and the feature which at one time consigned them to the virus class was the fact that they would only reproduce in living tissue.

Ch. psittaci is the casual organism of psittacosis or ornithosis and occurs mainly in the parrot family (hence psittacosis), but it is now known to be found in other avian species (hence ornithosis). It is often found in persons who work in pet shops selling parrots and budgerigars, and can be fatal.

Ch. trachomatis can cause a variety of diseases in man, for example trachoma, conjunctivitis and non-gonococcal urethritis. It is sensitive to the rifampicins, the tetracyclines and erythromycin.

| 6.4.19 | *Rickettsia* |

This group of microorganisms shares with chlamydias the property of growing only in living tissue. Rickettsiae occur as small ($0.3 \times 0.25\,\mu m$) rod-shaped or coccoid cells. They can be stained by special procedures. Division is by binary fission. They may be cultivated in the blood of laboratory animals or in the yolk sac of the embryo of the domestic fowl, and it is by this method that the organism is grown to produce vaccines.

Infection with rickettsiae gives rise to a variety of typhus infections in humans, the intermediate carriers being lice, fleas, ticks or mites. Rickettsiae can occur without harm to these arthropod hosts.

Amongst the diseases caused by rickettsiae are epidemic typhus, trench fever and murine typhus, caused by *R. prowazeki*, *R. quintana* and *R. typhi* respectively. Q-fever is caused by *Coxiella burneti*.

Few people can have failed to have heard of Legionnaires' disease or legionellosis. The causal organism of this disease, which must have existed undetected from time immemorial, was isolated and verified in 1977 and called *L. pneumophila*.

It causes an influenza-like fever which is accompanied by pneumonia in 90% of cases and which was usually diagnosed as atypical or viral pneumonia.

L. pneumophila is a rod-shaped, Gram-negative organism which grows on a conventional laboratory medium provided the concentrations of cysteine and iron are optimal. The organism will grow on a medium of sterilized tap water. This is in keeping with its known habitat of water supplies, especially water maintained in storage tanks, and must rely on the correct nutrients being present in the water.

The organism is sensitive to the antibiotic erythromycin (Chapter 5).

In addition to *L. pneumophila*, 16 other species of *Legionella* of proven pathogenicity have been described.

6.5   Acid-fast organisms

These comprise a group of organisms which, like the Gram-positive and Gram-negative groups, have been named after a staining reaction.

Due to a waxy component in the cell wall these organisms are difficult to stain with ordinary stain solutions, the hydrophobic nature of the wall being stain repellent; however, if the bacterial smear on the slide is warmed with the stain, the cells are dyed so strongly that they are not decolorized by washing with dilute acid, hence the term acid-fast. Many bacterial spores exhibit the phenomenon of acid fastness.

6.5.1   *Mycobacterium*

M. tuberculosis is the causal organism of tuberculosis in humans. Allied strains cause infections in animals, e.g. bovine tuberculosis and tuberculosis in rodents. Due to the waxy nature of the cell wall this organism will resist desiccation and will survive in sputum. Tuberculosis has been largely eliminated by immunization and chemotherapy.

M. leprae is the cause of leprosy.

6.6   Spirochaetes

Spirochaetes have a unique shape, structure and mode of locomotion. They are not stained easily by normal staining methods and thus cannot be designated either Gram-negative or Gram-positive. They are best observed by dark-ground illumination. They are slender rods in the form of spirals, like a corkscrew, and may be as long as $500\,\mu$m. Examples of spirochaete genera follow.

Borrelia

Borr. recurrentis causes a relapsing fever in humans. *Borr. vincenti* is the cause of Vincent's angina in humans, an ulcerative condition of the mouth and gums. *Borr. burgdorferi* is the causal organism of the tick-borne Lyme disease.

6.6.2   *Treponema*

T. pallidum is the causal organism of syphilis. *T. pertenue* causes the tropical disease called yaws.

6.6.3   *Leptospira*

L. icterohaemorrhagiae is the cause of a type of jaundice in man called Weil's disease. The disease is carried by rats and is encountered in sewer workers. Other species of *Leptospira*, with hosts ranging from domestic animals such as the pig to wild animals such as opossums and jackals, give rise to a variety of fevers encountered locally or widely across the world.

7   Further reading

Buchanan R.E. & Gibbons N.E. (Eds) (1974) *Bergey's Manual of Determinative Bacteriology*, 8th edn. Baltimore: Williams & Wilkins.

Collee J.G., Duguid J.P., Fraser A.G. & Marmion B.P. (1989) *Mackie & McCartney Practical Microbiology*, 13th edn. Edinburgh: Churchill Livingstone.

Cowan S.T. (1974) *Cowan and Steel's Manual for the Identification of Medical Bacteria*, 2nd edn. Cambridge: Cambridge University Press.

Davis B.D., Dulbecco R., Eisen H. & Ginsberg H.S. (1990) *Microbiology*, 4th edn. Philadelphia: J.B. Lippincott.

Dawes I.W. & Sutherland I.W. (1991) *Microbial Physiology*, 2nd edn. Oxford: Blackwell Scientific Publications.

Gould G.W. (1983) Mechanisms of resistance and dormancy. In *The Bacterial Spore* (Eds A. Hurst & G.W. Gould), vol. 2, pp. 173–209. London: Academic Press.

Gould G.W. (1985) Modification of resistance and dormancy. In *Fundamental and Applied Aspects of Bacterial Spores* (Eds G.J. Dring, D.J. Ellar & G.W. Gould), pp. 371–382. London: Academic Press.

Hugo W.B. (1972) *An Introduction to Microbiology*, 2nd edn. London: Heinemann Medical Books.

Olds R.J. (1975) *A Colour Atlas of Microbiology*. London: Wolfe Publishing.

Parker M.T. & Collier L.H. (Eds) (1990) *Topley and Wilson's Principles of Bacteriology, Virology and Immunity*, 8th edn, vols 1–5. London: Edward Arnold.

Rose A.H. (1976) *Chemical Microbiology*, 3rd edn. London: Butterworths.

Russell A.D. (1982) *The Destruction of Bacterial Spores*. London: Academic Press.

Skerman V.B.D., McGowan V. & Sneath P.H.A. (1980) Approved list of bacterial names. *Int J Syst Bacteriol*. **30**, 225–240.

Spivey M.J. (1978) The acetone/butanol/ethanol fermentation. *Process Biochem*. **13**, 2–6.

Stokes E.J. & Ridgway G.L. (1987) *Clinical Microbiology*, 6th edn. London: Edward Arnold.

Stryer L. (1987) *Biochemistry*, 4th edn. San Francisco: W.H. Freeman & Co.

The following references are included because, although of an advanced nature, they concern the interaction of drugs and bacteria.

Chopra I. (1988) Efflux of antibacterial agents from bacteria. *FEMS Symposium No. 44: Homeostatic Mechanisms of Microorganisms*, pp. 146–158. Bath: Bath University Press.

Costerton J.W., Cheng K.-J., Geesey G.G., Ladd T.I., Nickel J.C., Dasgupta M. & Marrie T.J. (1987) Bacterial biofilms in nature and disease. *Annu Rev Microbiol.* **41**, 435–464.

Hammond S.M., Lambert P.A. & Rycroft A.N. (1984) *The Bacterial Cell Surface.* London: Croom Helm.

Hinkle P.C. & McCarty R.E. (1976) How cells make ATP. *Sci Am.* **238**, 104–123.

Nikaido H. & Vaara T. (1986) Molecular basis of bacterial outer membrane permeability. *Microbiol Rev.* **49**, 1–32.

Russell A.D. & Chopra I. (1990) *Understanding Antibacterial Action and Resistance.* Chichester: Ellis Horwood.

2 Mycology

1   **Introduction**

2   **Structure and habitat**
2.1   General structure of moulds and
       yeasts
2.2   Habitat
2.3   Subcellular composition
2.3.1 The cell wall
2.3.2 The cell membrane
2.3.3 The cytoplasm

3   **Reproduction**
3.1   Asexual reproduction
3.2   Sexual reproduction

4   **Industrial importance of fungi**

5   **Medically important fungi**
5.1   Superficial mycoses
       (dermatophytoses)

5.2   Deep mycoses

6   **Plant pathogens**

7   **Toxic fungi**
7.1   Mycetismus (mushroom
       poisonings)
7.2   Mycotoxicoses
7.2.1 Ergotism
7.2.2 Alimentary toxic aleukia
7.2.3 Human aflatoxins
7.3   Pharmaceutical uses of toxin
       derivatives

8   **Molecular biology of fungi**

9   **Further reading**

1   Introduction

Moulds and yeasts constitute an extremely large and diverse group, and their biological significance to man varies from being harmful to being indispensible. They can be defined as nucleated, spore-bearing micro-organisms, do not possess chlorophyll, generally reproduce both asexually and sexually, and have somatic structures which are surrounded by cell walls consisting of polysaccharides, cellulose and/or chitin, mannan or glucan.

Fungi are mostly saprophytic, utilizing inanimate organic matter as an energy source, and are important natural organic decomposers, as well as destroyers of foodstuffs. The majority are facultative parasites able to utilize live or dead organic matter, although some species can only survive on living cells and are obligate parasites, causing disease of plants, animals and man. Consequently these are of economic and medical importance. Their industrial importance and usefulness (or utility) is emphasized in the manufacture of bread, beer and wines (exemplified by fermentation processes), and in the production of cheese, vitamins, organic acids (Chapter 25) and antibiotics (Chapter 7). In addition, fungi have proved increasingly useful to geneticists and biochemists in biological research due to their unique reproductive cycles and relatively simple metabolism, and more recently in molecular biology as an expression system for genetic sequences. Table 2.1 lists some of the more significant uses of fungi.

43

Table 2.1 Biological significance of fungi

Natural decomposers of organic matter—saprophytic

Plant pathogens
    Obigate parasites, species dependent, e.g. rusts and mildews
    Facultative parasites, e.g. smut fungi

Industrial importance
    Fermentation processes, e.g. yeasts in beers and wines
    Foodstuffs, e.g. *Penicillium roqueforti* in cheese
    Antibiotic production, e.g. *Penicillium* sp.
    Gene cloning

Human and animal pathogens—superficial and systemic mycoses and allergies

Research—genetics, biochemistry, molecular biology

2   Structure and habitat

2.1   General structure of moulds and yeasts

Fungi are classified according to structure and life cycles, taking account of their mode of sexual and asexual reproduction. The fungal body may either consist of a single unicellular structure or more commonly be filamentous. Progression from the simple species through to the more advanced members allows us to observe distinct progressions in development. First, there is an obvious advance from an aquatic habitat, through semi-aquatic, to entirely terrestrial environments; this is exemplified by comparing species of the chytridiomycetes with those belonging to the most advanced fungi, the basidiomycetes. Second, changes in their mode of asexual and sexual reproductive cycles occur with the appearance of characteristic reproductive structures. Additionally, there is an obvious change in morphology of the fungal thallus.

The more primitive unicellular fungi (chytrids) show some affinity with protozoa, but possess a definite cell wall for the major part of their life cycle; they may be classified apart from other fungi by their production of motile, flagellated cells or zoospores, and their simple life style. The chytridiomycetes are found in aquatic habitats, some being parasitic on algae, and often causing infections of epidemic proportion. Most, however, are of little direct economic importance to humans and their life is simple, since they have no branching filamentous stage and during early development their cells may lack a rigid wall.

The majority of true fungi consist of branching filaments known as hyphae, with septa which divide the fungal structure into distinct compartments. These compartments form behind the growing tip of the hyphae in a manner characteristic of the species. Other fungi have aseptate (coenocytic) hyphae, where the whole thallus (vegetable body) is a single multinucleated cell, and cross-walls are formed only in connection with reproductive structures or to detach living portions of the thallus from dead or injured portions of the organism. Septa have pores which allow cytoplasmic flow from one portion of the thallus to another, but also give added rigidity to the hyphae, enabling the formation of complex fruiting bodies. The manner of growth of most fungi is by extension of the hyphal mass just behind the growing tip, and branches develop from points further back along the hyphae, which in turn grow and extend the apical portions.

2.2 Habitat

Fungi utilize most forms of organic matter, their basic requirements being an organic source of carbon, in the form of sugar or starch, and an organic or inorganic source of nitrogen. Inorganic ions, such as potassium and magnesium, together with phosphate, sulphate and trace elements of copper, zinc and manganese, may also be necessary. Additional factors may be needed by those species unable to synthesize their own vitamins. Primitive aquatic fungi have evolved in morphology and nutritional requirements to inhabit the soil, where diversity into yeast forms and more advanced species occurred. The soil fungi rely on a variety of organic matter for their energy source, including leaf litter and animal or vegetable organic debris. Some predacious (carnivorous) soil fungi bear specialized structures for trapping nematodes and rotifers, and many are parasitic on plants, animals and humans.

2.3 Subcellular composition

2.3.1 *The cell wall*

The cell wall gives the thallus rigidity and structure, enabling constant passage of nutrients to the cytoplasm and waste matter into the environment. The chemical composition of fungal cell walls varies between species, and with the age and morphology of the fungus. The wall is composed of 80–90% carbohydrate, the rest being protein and lipid, and in some cases sterol. The main structural polysaccharides include cellulose, chitin, mannan and glucan; these integral structures differ from the cell walls of bacteria, resembling those of higher plants.

Chitin is composed of N-acetylglucosamine linked by $(1 \rightarrow 4)$-β-glucosidic bonds, its structure resembling cellulose, which is the main constituent found in plants. Glucan, or polyglucose, is made up of D-glucose residues with $(1 \rightarrow 3)$-β- or $(1 \rightarrow 6)$-β-linkages. Mannan is a $(1 \rightarrow 6)$-α-linked polymer of D-mannose with $(1 \rightarrow 2)$-α- and $(1 \rightarrow 3)$-α-branches. These structures are illustrated in Fig. 2.1. The arrangement of these linkages in the side chains of fungal polysaccharides gives these microorganisms their antigenic specificity; the cross-reactivity which occurs among species is also determined by antigenic polysaccharides. Several yeasts have a pleomorphic structure and may exist either in a yeast or mycelial form. Such diversity and morphological change is an

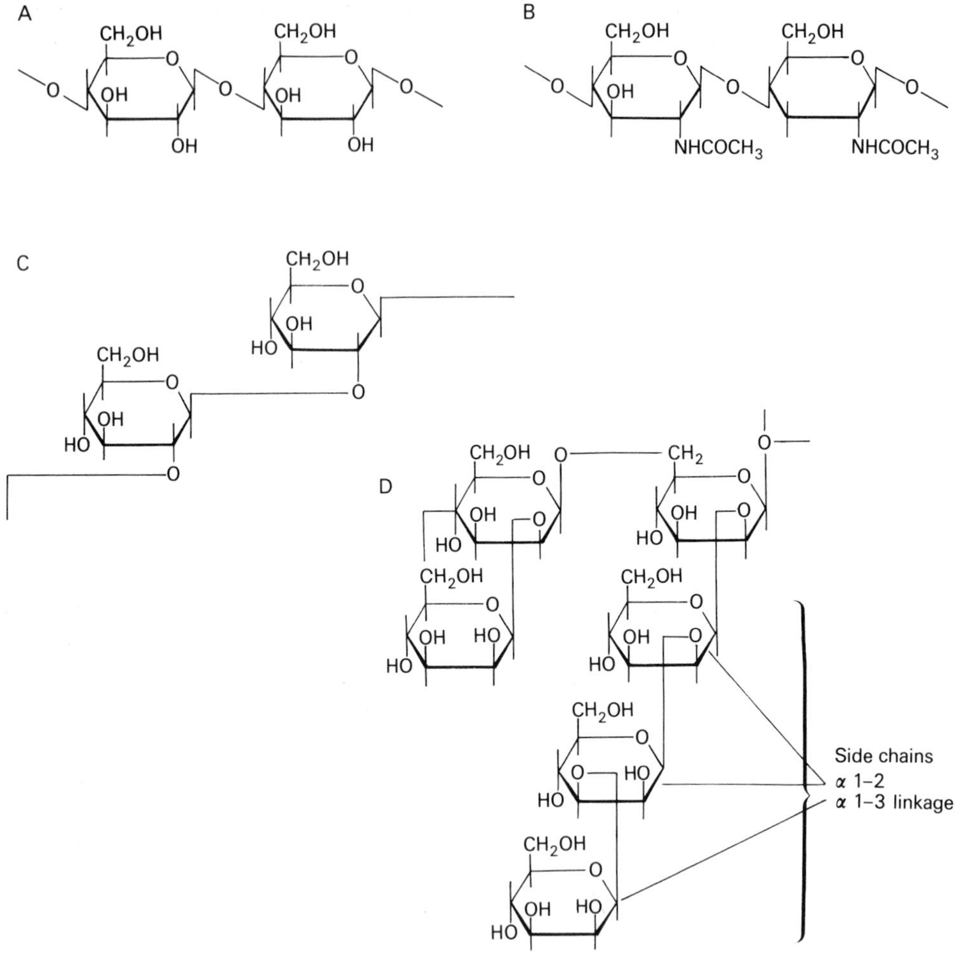

Fig. 2.1 Structure of: A, cellulose; B, chitin; C, glucan; D, mannan.

important property of certain pathogenic species, and is associated with changes in cell wall composition, which may result in increased pathogenicity, as with *Candida albicans*.

The composition of the fungal cell wall may also aid in classification. For example, some moulds have a chitin–glucan composition, whereas yeast forms have mannan as the most predominant component giving mannan–chitin or mannan–glucan structures. In addition, with many fungi, the polysaccharides are linked to protein enzymes which are important during cell division, and degrading substrates before passage across the cell membrane into the cytoplasm.

2.3.2   *The cell membrane*

Fungi possess a distinct cell membrane which resembles the plasmalemma of higher organisms and is composed of lipids and protein, 6% sterols, traces of nucleotides and the carbohydrates glucan and mannan. The major proportion of fungal lipids occur in membrane components, although some can be found as triglyceride storage droplets in the cytoplasm. The presence of membrane sterols in moulds and yeasts, but not in bacteria, accounts for the specific toxicity of polyene antibiotics against superficial infections in humans. Antibiotics such as amphotericin B (Chapter 5) bind to the sterol in the membrane (Chapter 9) and induce permeability changes, thereby enabling a rapid release of small molecules and ions from the cytoplasm.

The main functions of the membrane are those of regulating the osmotic pressure of the cell and the diffusion of substances. This is especially important during growth, when there is high internal osmotic pressure due to salt accumulation enabling expansion of the protoplasm when the cell wall is weakened by enzyme action. Enzyme localization in the membrane also enables resynthesis of cell wall components.

2.3.3   *The cytoplasm*

Fungi resemble higher plants and animals in their structural organization and this can also be seen in their eukaryotic cytoplasmic composition. A well-defined nuclear membrane surrounds the nucleus, which is present together with membranous mitochondria, endoplasmic reticulum, ribosomes and glycogen storage granules. Several cytoplasmic vacuoles exist, especially at the hyphal tip during elongation, and may form part of the secretory system (exocytosis) transporting materials and enzymes required for growth. This process may result in the fusion of these vacuoles with the plasma membrane.

3   Reproduction

Most fungi reproduce by asexual as well as sexual processes, resulting in the production of unique reproductive structures and cycles which are used in their classification.

3.1   Asexual reproduction

The most common method of asexual reproduction is by means of spores, most of which are colourless (hyaline), although some are pigmented, being green, yellow, orange, red, brown or black. Their size may vary from minute to large, and their shape from globose through oval, oblong and needle-shaped to helical. This infinite variation in spore appearance and arrangement is used in identification. They may be borne in a sac-like structure called a sporangium, the spores being referred to as sporangiospores; if not, they are termed conidia (Fig. 2.2).

The simplest form of fungal spore is called a zoospore, and is associated with the primitive aquatic or semi-aquatic species, having no rigid cell wall, and is motile by means of flagella. The flagellum is more

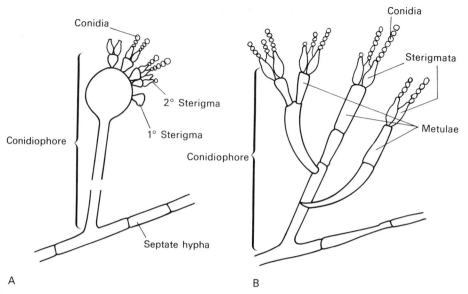

Fig. 2.2 Asexual reproductive spores of A, *Aspergillus* and B, *Penicillium*.

complex than that found in bacteria, being composed of eleven parallel fibrils, nine forming a cylinder and two placed centrally. The base of the flagellum enters the cells, and is attached to the nucleus by a structure called a rhizoplast. The flagellum structure (9 + 2 fibrils) is consistent with that demonstrated for other flagellated organisms, and zoospores may possess either one or two of these structures. The sporangium is the asexual reproductive structure of these aquatic fungi, and in its early stage consists entirely of nuclei and protoplasm. Eventually, cleavage occurs and the numerous sections develop into uninucleated spores. Following a motile phase, the zoospore encysts, withdrawing or losing its flagellum, and rests prior to germination.

Terrestrial species possess non-motile spores which are usually distributed by air movement. Spores of higher fungi are borne on conidiophores and readily disperse from the parent hyphae in a similar manner. Spore germination takes place under suitable conditions and an initial outgrowth of hyphae can be seen within a few hours of germination.

Hyphal fragmentation is a common means of propagation. The resulting cells are known as oidia or arthrospores which behave as conventional spores. In addition, hyphal cells may become enveloped in a thick wall, being referred to as chlamydospores. With yeasts, two distinct types of asexual reproduction occur. Budding is the more common method; a small outgrowth is produced from the parent cell, which undergoes mitotic division, and one daughter nucleus migrates into the bud, which continues to grow in size and eventually becomes free of the parent cell. This process occurs either as a phase of life cycle, or under conditions of restricted growth. Binary fission, the simple division of one cell into two

Table 2.2 Sexual reproduction by fungi

Planogametic copulation	Occurs in lower fungi, the gametes being motile and usually of similar size and morphology
Gametangial contact	Male and female organs make contact, and intact nuclei are transferred into the oogonium, which continues to develop whilst the antheridium disintegrates
Gametangial copulation	Fusion of the gametangia occurs, resulting in the formation of a zoospore
Spermatization	Male spermatia are transformed to receptive hyphae which assumes the role of the female gametangia; nuclei from the spermatia enter via a pore and fusion occurs
Somatogamy	Somatic hyphal cells assume the role of sexual organs, and nuclei are transferred from one portion of the hyphae to another, this process occurs in higher fungi such as the basidiomycetes

daughter cells by the formation of a dividing cell wall, is an additional method of reproduction for some fungi.

3.2 Sexual reproduction

Sexual reproduction is characterized by the union of two compatible nuclei, and for convenience the process can be divided into three distinct phases. First, the union of the gametangia (sex organs) occurs bringing the nuclei into close proximity within the same protoplast. This stage is referred to as *plasmogamy*. Second, *karyogamy* takes place, involving the fusion of the two nuclei. In the lower fungi these two processes may occur in immediate sequence, whereas in the higher fungi they are distinct, occurring at different times during the life cycle. Nuclear fusion is followed by the final phase, *meiosis*, where chromosomes are reduced to their original haploid number.

Some species are hermaphrodite and produce distinguishable male and female organs on the same thallus. Other species are dioecious, having either male or female gametangia, and are therefore not self-fertilizing. With some fungi the male and female organs are morphologically identical whilst, in others, the gametangia and gametes (sex cells) have distinguishing features, the male being referred to as the antheridium and the female as the oogonium. The most common methods by which plasmogamy and karyogamy takes place are listed in Table 2.2.

4 Industrial importance of fungi

Many fungi are important to industry and are used in the production of beer and wine. Natural yeasts may be used to ferment fruit juices or cereal products such as malt; however, industrial manufacture usually involves pasteurization of the liquid, followed by selective inoculation with a special strain of yeast *Saccharomyces cerevisiae*. In wine and beer manufacture, the lower temperature favours yeast fermentation, and bacteria are discouraged either by the acidity of the medium or by the addition of hops, which are mildly inhibitory to bacterial growth. Fermentation occurs under anaerobic conditions, and this results in the production of alcohol. Alcohol, produced by fruit or cereal fermentation, can be distilled to produce spirits; brandy, for example, is obtained from wine, whisky from cereals, and rum from fermented molasses.

Baker's strains of *S. cerevisiae* are selected for their high production of carbon dioxide under aerobic conditions. Baker's yeast is manufactured specifically for bread-making, the carbon dioxide forming a raising agent, and is supplied as pressed or dried yeast, and may also be used as a vitamin supplement for food.

Certain fungi are particularly important in the manufacture of cheeses; for example, the mould *P. roqueforti* is used in the production of blue-veined cheeses. Spores of the fungus are used to inoculate the cheese, which is then ripened at low temperatures of +9°C to discourage the growth of organisms other than the *Penicillium*. Since these moulds are aerobic, perforations are made in the cheese to allow air to enter. Decomposition of fat takes place to give the cheeses their characteristic flavour. In contrast to *P. roqueforti*, which grows within the cheese, *P. camemberti*, which is used in the manufacture of Camembert cheese, grows on the surface and develops inwards producing the characteristic liquefaction and softening of the surface.

Besides the processes outlined briefly in this section, one of the main industrial uses of fungi in recent years has been in antibiotic production, which is discussed in detail in Chapter 7, and in many other pharmaceutical processes such as steroid transformation (see Chapter 25).

Recently fungi have been used in the treatment of effluent. Several species of fungi, including *Fusarium aquaeductum*, *Geotrichium candidium* and *Sabbaromyces splendes*, have been shown to grow actively in wastewater treatment systems. These filamentous fungi, as well as certain species of filamentous bacteria, are extremely tolerant to the fluctuations in conditions and pH in effluent plants. Also of current interest is the use of fungi as insecticides. There are over 400 species of fungi that can attack insects and mites, and this method of insect control may in the future gain widespread use and prove to be more effective than chemical pesticides, where insects may develop drug resistance.

5   Medically important fungi

Approximately 55 species of fungi are capable of infecting humans and causing disease, most belonging to the class Deuteromycetes, more commonly known as Fungi Imperfecti (fungi where the sexual reproductive cycle is unknown). Infections range from subclinical or mild to serious and fatal, and may conveniently be divided into superficial infections and deep mycoses. In addition, fungi may cause hypersensitivity reactions (e.g. farmer's lung disease), where fungal antigens (allergens) induce hypersensitivity once inhaled. This is usually an immediate hypersensitivity reaction, and as such is similar to the allergies produced by house dust and pollen antigens. Secondary eruptions on the skin are often due to allergic reactions, and may follow spontaneous inflammation of a primary lesion.

5.1   **Superficial mycoses (dermatophytoses)**

The dermatophytes form a closely related group of fungi which infect the keratinized tissues (hair, skin and nails) of the body; they seldom invade deeper tissues. They are classified into three genera: *Epidermophyton*, *Microsporum* and *Trichophyton*. Some species are associated with infections in particular areas of the world, e.g. *T. schoenleinii* is found in Mediterranean areas, and *T. rubrum* in tropical climates. Domestic animals may have dermatophyte infections which can be transmitted to humans. Laboratory identification of these organisms relies on colony appearance upon isolation on medium such as Sabouraud's glucose agar, and on the morphology of the spores. Asexual reproduction in tissue occurs mostly by fragmentation of the hyphae, producing arthrospores. Some species, such as *T. mentagrophytes*, which causes athlete's foot, occur worldwide, whilst others such as *T. violaceum* and *T. ferrugineum*, causing ringworm, are limited to particular areas.

Infected areas of the body are treated by removing as much of the dead epithelial layers as possible, followed by topical application of antifungal drugs. Spread of dermatophyte infections results from contact with infected individuals, and prevention of spreading can only be accomplished by successfully treating infected individuals together with attempts to minimize the number of contacts. Athlete's foot may be contracted through the use of communal showers and dressing rooms, where desquamated skin serves as a source of infection. In many instances, athlete's foot may be asymptomatic and infection becomes apparent following irritation, for example by footwear or excessive heat. In student populations, and among inmates of institutions and the armed forces, the incidence of infection may be as high as 65%.

5.2   **Deep mycoses**

A number of fungi are capable of infecting humans and producing systemic infections. Some, like *C. albicans*, are opportunistic pathogens and reside as part of the normal flora of the mucous membranes of the respiratory, gastrointestinal and female genital tracts. Serious disease conditions may result where the patient is in some way debilitated; for example, dissemination of the organism through the bloodstream can occur in cases of lymphoma or during immunosuppression, and in patients undergoing antibiotic therapy for an unrelated infection.

A great many of the fungi causing deep mycoses in humans have a biphasic growth cycle, being yeast-like in the host but reverting to mycelial forms when cultured *in vitro*. *Blastomyces dermatitidis* is such an organism, occurring mainly in North America and producing chronic granulomatous disease (blastomycosis). The route of infection is via the respiratory tract, and the disease is usually limited to the skin and lungs, but dissemination throughout the body can occur. The biphasic nature of this organism can be shown by culturing isolates on different media and at different incubation temperatures. On blood agar incubated at 37°C, soft, waxy colonies are formed, and the cells are morphologically similar to the microbes found in tissues. However, when grown at room temperature (20°C) on Sabouraud's glucose agar, brownish-white filamentous colonies develop, consisting of branching septate hyphae. *B. brasiliensis*, which causes blastomycosis in South and Central America, and *Histoplasma capsulatum*, which causes histoplasmosis, also show biphasic modes of growth.

One yeast-like organism capable of inducing a fatal meningitis in humans is *Cryptococcus neoformans*. The organism is surrounded by a wide mucoid capsule, which can be seen in the spinal fluid of infected individuals. The capsules become visible when the cells are mounted in Indian ink. Like so many fungi which cause systemic infections of humans, *C. neoformans* originates from the soil, flourishing particularly well on bird droppings, and so making control of the disease extremely difficult.

Other fungi frequently associated with infections in different parts of the world are to be found. *Coccidioides immitis* (endemic in southwest America) causes coccidioidomycosis, a disease of the lungs; *Geotrichum candidum* is a yeast-like fungus infecting the bronchi, lungs and mucous membranes; and *Sporothrix schenckii* causes sporotrichosis, a chronic granulomatous infection of the skin, lymphatics and other tissues. Madura foot (or maduromycosis) is a disease of the tropics and may be caused by several filamentous fungi; deep-seated abscesses or nodules form following infection with this agent.

6   Plant pathogens

Although not dealt with here in detail, fungi which parasitize cereal and root plants, causing destruction of crops, are widespread and of great economic importance worldwide. They may be obligate parasites, unable to exist outside the host plant, such as rusts and powdery mildew fungi, or they may be facultative parasites like the smut fungi. Organisms such as *Botrytis cinerea* (belonging to the Fungi Imperfecti) and *Pythium* sp. (class Oomyces) are normally saprophytic and are found in most soils, but can infect plants and cause disease when suitable conditions prevail, being non-specific in their host range and entirely dependent on their host for their food supply.

Such fungi have lost the ability to utilize simple nutrients and lack some of the essential enzymes necessary for metabolism; they rely on host enzymes to supplement these deficiencies, and are therefore difficult to cultivate in the laboratory. The hyphae of the majority of specialized fungi are confined to the intercellular spaces of the host, nourishment being obtained through specialized structures known as haustoria, which invade the host's cell wall without rupturing their plasma membranes. Certain parasites are known to interfere with the growth-regulatory mechanisms of the plant, causing multiplication of host cells (hypertrophy) and the formation of definite galls. This is characteristic of black warts of potatoes, club root of crucifers and many rust infections.

7 Toxic fungi

There are several species of fungi which produce toxins causing diseases of plants, animals and humans. This can be effected either by consumption of the poisonous fungi, as in mushroom and toadstool poisoning, or by eating foodstuffs contaminated with the toxin-producing fungus. Actual contact or inhalation of relatively small amounts of certain toxins may also give rise to toxicosis. Fungal toxins (mycotoxins) are secondary metabolites produced by fungi, and cause diseases of animals and humans when ingested on foodstuffs. Many names are given to the fungal toxins causing plant diseases, for example phytopathogenic toxins, vivotoxins or pathotoxins. Toxins may also include those substances produced by the plant itself in response to the fungus (phytoalexins), which in turn may cause diseases of livestock and humans.

The chemical composition and quantity of toxins produced varies greatly between different species of fungi, environmental conditions of growth and the biological activity of the fungus contributing to the amount produced. One species or fungus may produce a number of different toxins which act synergistically in producing their effects, and it may therefore be difficult to assess the significance of a toxin or fungus to a disease. Fungi produce a variety of toxic effects in animals and humans, including nephrotoxicity and damage to the skin and gastrointestinal and respiratory tracts. Several toxins are potent carcinogens, causing cancer when tested in laboratory animals, and correlations exist between concentrations of certain toxins in foodstuffs and the incidence of some human cancers.

In contrast to their destructive properties, many compounds derived from fungal toxins have therapeutic use against a wide variety of physiological disorders, and some show activity against bacteria, viruses and fungi, and produce cytostasis of tumour cell growth.

7.1 Mycetismus (mushroom poisonings)

Severe gastrointestinal symptoms or neurological effects, or both, can occur following ingestion of the toxin. *Amanita phalloides*, *A. muscaria* and related species are the most renowned of the poisonous mushrooms, and differ morphologically from the edible forms. Amanita toxins found in *A. phalloides* include five cyclopeptides, namely phalloidin, phalloin, and α-, β-, and γ-amantium, which give rise to the chronic intestinal symptoms. The parasympathetic nervous system as well as the gastro-intestinal tract is affected in *A. muscaria* poisoning due to the toxin muscarine, being fatal in some cases.

Mycotoxicoses

Mycotoxins are produced by many species of fungi and are now recognized as important causes of human and animal disease. Attention has been focused on animal diseases, mainly because acute animal intoxications occur more frequently and are more often diagnosed. Moulds such as *Fusarium*, *Papulosporia*, *Chaetomium* and *Sordaria* sp. invade crops and, during long-term storage (months to years), produce toxins which on ingestion cause severe intoxicating disease. *Fusarium* species are particularly important; *F. graminearum*, for example, produces a toxin (zearalenone) that induces oestrogenic syndrome in swine, resulting in abortion.

It is now well recognized that contamination of foodstuffs with fungal toxins is important in relation to public health. Serious outbreaks of disease have occurred in man and animals, and include the following.

1 Mycotoxicoses which occurred in horses in the Ukraine in 1931 due to the fungal growth of *Stachybotrys alternans* on hay and straw.

2 Re-occurrence of alimentary toxic aleukia in Russia in 1943–47 due to toxin production in overwintered grain such as millet, barley and wheat. The disease affects humans, and is mainly due to *F. sporotrichoides* and some *Cladosporium* species. Toxic trichothecenes are produced causing severe haemorrhagic disease, with fever, rash and exhaustion of the bone marrow.

3 In 1959, facial eczema of sheep occurred in New Zealand due to sporesdesmin from the spores of *Pithomyces chastarum* growing on grass refuse.

4 In 1960, an outbreak of turkey X disease occurred in Britain resulting from aflatoxin production by *Aspergillus flavus* and *A. parasiticus* growing on the groundnut meal used as food.

There are three main mycotoxicoses of importance in man, and these are associated with ergotism, alimentary toxic aleukia and liver cancer; these will now be considered in further detail.

Ergotism

Ergotism was one of the first mycotoxicoses to be studied. It is caused by the contamination of rye with the moulds *Claviceps purpurea* and *C. paspali* and the subsequent production of the ergot alkaloids. Ergot itself is the resistant stage of the fungus (the sclerotium) which resides in overwintered grain. The pharmacological effects of low doses of these

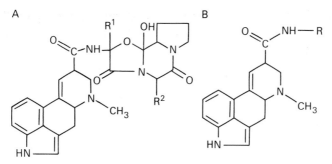

Fig. 2.3 A, chemical structure of ergot alkaloids and B, peptide-type ergot alkaloids derived from lysergic acid. R^1, R^2 and R represent a range of carbohydrate substitution groups.

alkaloids have for centuries been exploited in therapy, and derivatives have been used to control certain vascular disorders and to induce uterine contractions. Disease can be associated with consumption of bread made from the flour of rye or other grains overgrown with toxigenic strains of moulds. The alkaloids are derivatives of ergoline and are classified as lysergic acid derivatives or clavine alkaloids. The general structures of ergot alkaloids and peptide-type ergot alkaloids derived from lysergic acid are given in Fig. 2.3.

Although several alkaloids can be used to treat human diseases, the pharmacological effects of these compounds in the body vary. Ergot alkaloids produce an α-adrenergic blockade, inhibiting responses of adrenergic nerve activity to adrenaline and 5-hydroxytryptamine, which can result in vasoconstriction and the development of gangrene. Ergot alkaloids can stimulate the hypothalamus and other sympathetic portions of the midbrain, are capable of stimulating smooth muscle cell contraction, and have been used to induce uterine contractions.

7.2.2 Alimentary toxic aleukia

The disease known as alimentary toxic aleukia (ATA) is caused by the ingestion of wintered grain contaminated with toxigenic species of the mould *Fusarium*. Outbreaks have been reported, chiefly in Russia, and mortality rates have reached levels as high as 60%. Susceptibility to these fungal toxins may, in part, be due to the undernourished condition of the communities affected. *F. sporotrichoides* and *F. poae* produce sporofusarin and poaefusarin respectively, which when given to animals mimic the disease syndrome seen in humans. The fungi produce toxic glucosides which affect the haematopoietic (blood cell formation) system of the body. The disease is characterized by fever, haemorrhagic rash, bleeding from the nose, throat and gums, necrotic angina, leucopenia, agranulocytosis, sepsis and exhaustion of the bone marrow. Direct toxin injury occurs to the haematopoietic, autonomic nervous and endocrine systems. The clinical condition may last several weeks and can result in death or incomplete recovery.

In rural Japan in the latter half of the nineteenth century, a disease known as 'acute cardiac beriberi' was described, and was thought to represent a human mycotoxicosis. The disease was causally associated with consumption of polished rice, and the decrease in incidence which occurred was thought to be due to the exclusion of mouldy grain as a result of more stringent rice inspection. Clinical symptoms characterizing the disease appear within days, varying from chest pains, palpitation and abnormal breathing, to anguish, pain, restlessness and attacks of violence. The heart rhythm is affected due to dilation of the heart, and eventually the patient loses consciousness. *P. citreo-viride* is a likely microorganism causing this disease, through the production of neuro-mycotoxins; this, however, has not been firmly established, since no animal model exists for this hypothesis to be tested. Vitamin B_1 (aneurine or thiamine) deficiency is thought to be a factor contributing to the syndrome.

7.2.3   *Human aflatoxins*

Aflatoxins are heat-stable metabolites of certain strains of *A. flavus*, although other species of *Aspergillus* and *Penicillium* contaminate stored food, and are of potential significance as causative agents of human mycotoxicoses. The products of fungal metabolism have been designated aflatoxin B_1, B_2, G_1 and G_2 in accordance with their blue or green fluorescence in ultraviolet light. Of these, aflatoxin B_1 is the most potent, and a correlation between the ingestion of this compound with hepatic tumours has been suggested in humans. Most of the evidence associating aflatoxin B_1 with hepatocellular carcinoma stems from observations on the epidemiology of the disease relative to food contamination, and experimental evidence showing that administration of the toxin produces liver carcinoma in laboratory animals. Continuous feeding of rats with aflatoxin B at a concentration of $1 \mu g \, kg^{-1}$ bodyweight per day is sufficient to produce gross changes in hepatic cells, and the eventual development of liver tumours.

Contamination of foodstuffs occurs on prolonged storage; this coincides with areas where hepatic cancer is prevalent, such as certain parts of Africa and South East Asia. Food contamination with aflatoxin B has been shown to be as high as $100 \mu g \, kg^{-1}$. In Thailand, contaminated foodstuffs can include peanuts, dried corn, millet, garlic, dried chilli peppers and dried fish. Cooked rice left over from a previous meal and subsequently infected with a toxigenic strain of fungus may also be a source of aflatoxin.

It is interesting to speculate that aflatoxins, besides acting directly on hepatic cells and inducing malignant transformation, may lower the body's general defence barriers, by causing depression of both cellular and humoral immune responses. Fungal compound capable of subverting the natural immune defence/surveillance mechanism may allow cell transformation and progressive tumour growth to occur unchecked. It has recently been suggested that, rather than acting as a primary carcinogen, aflatoxin suppresses the host's cell-mediated immune mechanisms, and in turn allows hepatitis B virus, which is highly endemic in certain populations, to proliferate and so cause chronic infection, cirrhosis and in many cases hepatocellular carcinoma.

Besides the association of aflatoxins with liver cancer, chronic toxicity by mycotoxins is thought to lead to childhood liver cirrhosis in parts of India. The disease is characterized by fatty infiltration of liver cells with liver degeneration and fibrosis. Eventually, jaundice and hepatic coma occur. Recently, compounds with similar chemical properties to aflatoxin B_1 have been found in the milk of some mothers whose children were diagnosed as having Indian childhood cirrhosis. These compounds are absent from the milk of mothers with normal, healthy children.

Acute toxicity due to ingestion of high single doses of aflatoxin B_1 may also occur. Such intoxication causes mild to severe disease, with abdominal pain, vomiting and a palpable liver; in many cases death ensues. It is strongly suspected that in northeast Thailand, Reye's syndrome occurring in children up to adolescence is due to toxicity by aflatoxin B_1 and B_2, although a viral aetiology for the disease has also been proposed. The disease is a distinct clinical entity, characterized by acute brain swelling and fatty changes in the liver, being preceded by a mild infection of the upper respiratory tract. In many cases autopsy specimens from patients with Reye's syndrome have been found to be heavily contaminated with aflatoxin B_1 and B_2. Involvement of mycotoxin in the disease may be inferred, although current evidence indicates the aetiology to be multifactorial.

7.3 Pharmaceutical uses of toxin derivatives

A large number of compounds derived from fungi are useful therapeutic agents. Many derivatives have antimicrobial activities; these include rugulosin, palatin, penicillin, cephalosporin and penicillinic acid. Others are antifungal agents, for example griseofulvin (Chapters 5 and 6) from *P. griseofulvum*, and trichodermin, used against *C. albicans* infections.

The coumarin compounds, chemically similar to the aflatoxins, are renowned for their sedative and hypnotic properties and also for their anticoagulant activity. The latter has been tested for use in the treatment of certain thrombotic diseases. Other examples include the ergot alkaloids, which have a number of effects on the central nervous system, including contraction of the smooth muscle of the uterus, and derivatives such as ergometrine which are used on obstetrics.

8 Molecular biology of fungi

Eukaryotic genes can now be isolated by recombinant DNA technology and expressed in microorganisms. One of the first proteins produced by eukaryotic gene expression in *E. coli* was somatostatin (see Chapter 24).

E. coli, however, is not always the most suitable host for expression of eukaryotic genes since the concomitant production of toxic and pyrogenic cell wall components along with food or pharmaceutical products may not be desirable. More recently, the yeast *S. cerevisiae* has been used for the carriage and expression of eukaryotic genes. The yeast has several advantages over *E. coli*.

1 *S. cerevisiae* has a secretion system similar to that of higher eukaryotic systems, and is easily maintained in large-scale production and fermentation processes.

2 This yeast is non-pathogenic and its products are free from toxic contaminants.

Certain aspects of gene cloning in yeasts, using *S. cerevisiae* as an example, will therefore be discussed, although the techniques have also now been successfully used in non-*Saccharomyces* yeasts and some filamentous fungi.

Transfection of *S. cerevisiae* initially involves the enzymatic removal of the cell wall to produce spheroplasts which take up DNA on treatment with polyethylene glycol and calcium ions. The uptake of exogenous DNA is detected by the use of a genetic marker. The usual procedure is to use an auxotrophic host strain for incorporation of the exogenous DNA containing a wild type gene. The transformants which have taken up the wild type gene are then selected against a background of non-transformed auxotrophs.

A variety of gene vectors (plasmids) exist which carry the genes for eukaryotic proteins and their expression in *S. cerevisiae*. These vectors include shuttle vectors which have the ability to be selected in *E. coli* as well as *S. cerevisiae*. In general, these shuttle vectors contain a yeast genetic marker for selectivity transformants, a yeast DNA sequence for high frequency transformation (autonomous replicating segment, ARS), and a bacterial marker and DNA sequence for selection and propagation in *E. coli*. This ability to 'shuttle' between hosts allows the gene of interest to be replicated in *E. coli*, and the plasmid to be conserved by allelic rescue. The shuttle vector can also be initially prepared and altered using *E. coli* with only the final plasmid contraction being introduced into *S. cerevisiae*. As in the bacterial system, yeast vectors either replicate independently of the yeast chromosome, by way of the ARS, or lack these replicating sequences and are maintained only by integration into the *S. cerevisiae* chromosome. These later plasmids are known as integration vectors.

In order for efficient expression of a heterologous gene in *S. cerevisiae* it is necessary and important to have a *S. cerevisiae* promoter region in order to initiate and regulate transcription of the gene, and a stop region for efficient termination of transcription. Furthermore, for many proteins, especially pharmaceuticals, it is important that a completely authentic protein is produced in order to retain both biological activity and antigenicity. Many eukaryotic proteins are normally secreted and produced as precursor proteins, with an amino acid terminal sequence which is proteolytically cleaved to produce the mature protein. Using *S. cerevisiae*, several studies have shown that heterologous signal sequences are necessary in order to allow and direct the secretion of the protein product. Using human DNA secretor sequences, together with *S. cerevisiae* promotors and DNA sequences coding for the precursor proteins, interferons (IFN-α and IFN-γ) have been secreted and produced. In addition, a completely synthetic IFN-α has also been produced in *S. cerevisiae*, which has been shown to have a higher antiviral activity than that of the naturally occurring interferon.

In recent years a range of biologically active new and naturally occurring proteins have been produced. These have included insulin, α-antitrypsin and interleukin-2; in 1985 *S. cerevisiae* strains containing plasmids expressing both the heavy and light chains of a mouse immunoglobulin were shown to secrete active antibody. One example of a novel protein is that of a modified α-antitrypsin. The DNA encoding this protein was modified by *in vitro* mutagenesis in order to produce a more stable protein than normal when secreted from *S. cerevisiae*. This feature may enable it to be used in the clinical treatment of emphysema.

Further research into the production of novel proteins from fungal strains by means of genetic engineering may bring a new and exciting era in the development of pharmaceutical products.

9   Further reading

Alexopoulos C.J. (1962) *Introductory Mycology*, 2nd edn. New York: John Wiley & Sons.

Austwick P.K.C. (1975) Mycotoxins. *Br Med Bull.* **31**, 222–229.

Howard D.E. (Ed.) (1983) *Fungi Pathogenic for Humans and Animals*, 3 vols. New York & Basel: Marcel Dekker.

Jawetz E., Melnick J.L. & Adelberg E.A. (1974) *Review of Medical Microbiology*, 12th edn. Los Altos, California: Lange Medical Publications.

Kadis S., Ciegler A. & Aji S.J. (1971) *Microbial Toxins*, vols VI–VIII. New York: Academic Press.

Lutwick L.I. (1979) Relation between aflatoxin, hepatitis B virus, and hepatocellular carcinoma. *Lancet*, **i**, 755–757.

Riviere J. (1977) *Industrial Applications of Microbiology*. London: Surrey University Press.

Rose A.H. (Ed.) (1977) *Economic Microbiology Series. Vol 1: Alcoholic Beverages*. London & New York: Academic Press. (Other volumes in this series should be consulted where relevant.)

Rose A.H. & Harrison J.S. (Eds) (1987) *The Yeasts*, 2nd edn. London: Academic Press.

Ryan N.J., Hogan G.R., Hayes A.W., Unger P.D. & Siraj M.Y. (1979) Aflatoxin B1: its role in the etiology of Reye's syndrome. *Pediatrics*, **64**, 71–75.

Stevenson G. (1970) *The Biology of Fungi, Bacteria and Viruses*, 2nd edn. London: Edward Arnold.

Wylie T.D. & Morehouse L.G. (Eds) (1978) *Mycotoxic Fungi, Mycotoxins, Mycotoxicoses. An Encyclopedic Handbook*, vols 1–3. New York & Basel: Marcel Dekker.

3 Viruses

1 Introduction

2 General properties of viruses
2.1 Size
2.2 Nucleic acid content
2.3 Metabolic capabilities

3 Structure of viruses
3.1 Helical symmetry
3.2 Icosahedral symmetry

4 The effect of chemical and physical agents on viruses

5 Virus–host cell interactions

6 Bacteriophages
6.1 The lytic growth cycle
6.2 Lysogeny
6.3 Epidemiological uses

7 Human viruses
7.1 Cultivation of human viruses

7.1.1 Tissue cultures
7.1.2 The chick embryo
7.1.3 Animal inoculation

8 Multiplication of human viruses
8.1 Adsorption
8.2 Penetration and uncoating
8.3 Vegetative growth
8.4 Maturation and release

9 The problems of viral chemotherapy
9.1 Interferon

10 Tumour viruses

11 Acquired immune deficiency syndrome (AIDS)

12 The agents of spongiform encephalopathy

13 Further reading

1 Introduction

Following the demonstration by Koch and his colleagues that anthrax, tuberculosis and diphtheria were caused by bacteria, it was thought that similar organisms would, in time, be shown to be responsible for all infectious diseases. It gradually became obvious, however, that for a number of important diseases no such bacterial cause could be established. Infectious material from a case of rabies, for example, could be passed through special filters which held back all particles of bacterial size, and the resulting bacteria-free filtrate still proved to be capable of inducing rabies when inoculated into a susceptible animal. The term virus had, up until this time, been used quite indiscriminately to describe any agent capable of producing disease, so these filter-passing agents were originally called filterable viruses. With the passage of time the description 'filterable' has been dropped and the name virus has come to refer specifically to what are now known to be a distinctive group of micro-organisms different in structure and method of replication from all others.

2 General properties of viruses

All forms of life, animal, plant and even bacterial, are susceptible to infection by viruses. Three main properties distinguish viruses from their various host cells: size, nucleic acid content and metabolic capabilities.

2.1 Size

Whereas a bacterial cell like a staphylococcus might be 1000 nm in diameter, the largest of the human pathogenic viruses, the poxviruses, measure only 250 nm along their longest axis, and the smallest, the poliovirus, is only 28 nm in diameter. They are mostly, therefore, beyond the limit of resolution of the light microscope and have to be visualized in the electron microscope.

2.2 Nucleic acid content

Viruses contain only a single type of nucleic acid, either DNA or RNA.

2.3 Metabolic capabilities

Virus particles have no metabolic machinery of their own. They cannot synthesize their own protein and nucleic acid from inanimate laboratory media and thus fail to grow on even the most nutritious media. They are obligatory intracellular parasites, only growing within other living cells whose energy and protein-producing systems they redirect for the purpose of manufacturing new viral components. The production of new virus particles generally results in death of the host cell and as the particles spread from cell to cell, for example within a tissue, disease can become apparent in the host.

3 Structure of viruses

In essence, virus particles are composed of a core of genetic material, either DNA or RNA, surrounded by a coat of protein. The function of the coat is to protect the viral genes from inactivation by adverse environmental factors, such as tissue nuclease enzymes which would otherwise digest a naked viral chromosome during its passage from cell to cell within a host. In a number of viruses the coat also plays an important part in the attachment of the virus to receptors on susceptible cells, and in many bacterial viruses the coat is further modified to facilitate the insertion of the viral genome through the tough structural barrier of the bacterial cell wall. The morphology of a variety of viruses is illustrated in Fig. 3.1.

The viral protein coat, or *capsid*, is composed of a large number of subunits, the *capsomeres*. This subunit structure is a fundamental property and is important from a number of aspects.

1 It leads to considerable economy of genetic information. This can be illustrated by considering some of the smaller viruses, which might, for example, have as a genome a single strand of RNA composed of about 3000 nucleotides and a protein coat with an overall composition of some 20 000 amino acid units. Assuming that one amino acid is coded for by a triplet of nucleotides, such a coat in the form of a single large protein would require a gene some 60 000 nucleotides in length. If, however, the

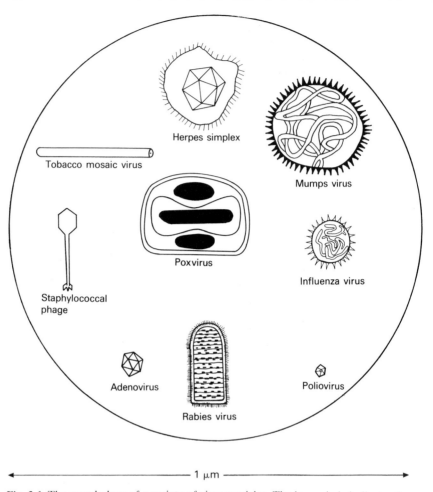

Fig. 3.1 The morphology of a variety of virus particles. The large circle indicates the relative size of a staphylococcus cell.

viral coat was made up from repeating units each composed of about 100 amino acids, only a section of about 300 nucleotides long would be required to specify the capsid protein, leaving genetic capacity for other essential functions.

2   Such a subunit structure permits the construction of the virus particles by a process in which the subunits self-assemble into structures held together by non-covalent intermolecular forces as occurs in the process of crystallization. This eliminates the need for a sequence of enzyme-catalysed reactions for coat synthesis. It also provides an automatic quality-control system, as subunits which may have major structural defects fail to become incorporated into complete particles.

3   The subunit composition also means that the intracellular release of the viral genome from its coat involves only the dissociation of non-covalently bonded subunits, rather than the degradation of an integral protein sheath.

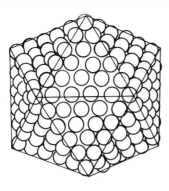

An icosahedral virus particle composed of 252 capsomeres 240 being hexons and 12 being pentons

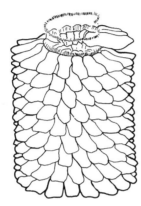

A helical virus partially disrupted to show the helical coil of viral nucleic acid embedded in the capsomeres

Fig. 3.2 Icosahedral and helical symmetry in viruses.

In addition to the protein coat many animal virus particles are surrounded by a lipoprotein envelope which has generally been derived from the cytoplasmic membrane of their last host cell.

The geometry of the capsomeres results in their assembly into particles exhibiting one of two different architectural styles—helical or icosahedral symmetry (Fig. 3.2).

There is a third structural group comprising the poxviruses and many bacterial viruses, in which a number of major structural components can be identified and the overall geometry of the particles is complex.

3.1 Helical symmetry

Some virus particles have their protein subunits symmetrically packed in a helical array, forming hollow cylinders. The virus of tobacco mosaic disease (TMV) is the classic example. X-ray diffraction data and electron micrographs have revealed that 16 subunits per turn of the helix project from a central axial hole that runs the length of the particle. The nucleic acid does not lie in this hole, but is embedded into ridges on the inside of each subunit and describes its own helix from one end of the particle to the other.

Helical symmetry was thought at one time to exist only in plant viruses. It is now known, however, to occur in a number of animal virus particles. The influenza and mumps viruses, for example, which were first seen in early electron micrographs as roughly spherical particles, have now been observed as enveloped particles; within the envelope, the capsids themselves are helically symmetrical and appear similar to the rods of TMV, except that they are more flexible and are wound like coils of rope in the centre of the particle.

3.2 Icosahedral symmetry

The viruses in this architectural group have their capsomeres arranged in the form of regular icosahedra, i.e. polygons having 12 vertices, 20 faces and 30 sides. At each of the 12 vertices or corners of these icosahedral particles is a capsomere, called a *penton*, which is surrounded by five neighbouring units. Each of the 20 triangular faces contains an identical number of capsomeres which are surrounded by six neighbours and called *hexons*. In plant and bacterial viruses exhibiting this type of symmetry, the hexons and pentons are composed of the same polypeptide chains; in animal viruses, however, they may be distinct proteins. The number of hexons per capsid varies considerably in different viruses. Adenovirus, for example, is constructed from 240 hexons and 12 pentons, while the much smaller poliovirus is composed of 20 hexons and 12 pentons.

4 The effect of chemical and physical agents on viruses

Heat is the most reliable method of virus disinfection. Most human pathogenic viruses are inactivated following exposure at 60°C for 30 minutes. The virus of serum hepatitis can, however, survive this temperature for up to 4 hours. Viruses are stable at low temperatures and are routinely stored at −40 to −70°C. Some viruses are rapidly inactivated by drying, others survive well in a desiccated state. Ultraviolet light inactivates viruses by damaging their nucleic acid and has been used to prepare viral vaccines. These facts must be taken into account in the storage and preparation of viral vaccines (Chapter 16).

Viruses that contain lipid are inactivated by organic solvents such as chloroform and ether. Those without lipid are resistant to these agents. This distinction has been used to classify viruses. Many of the chemical disinfectants used against bacteria, e.g. phenols, alcohols and quaternary ammonium compounds (Chapter 11), have minimal virucidal activity. The most generally active agents are chlorine, the hypochlorites, iodine, aldehydes and ethylene oxide.

5 Virus−host cell interactions

The precise sequence of events resulting from the infection of a cell by a virus will vary with different virus−host systems, but they will be variations of four basic themes.

1 Multiplication of the virus and destruction of the host cell.

2 Elimination of the virus from the cell and the infection aborted without a recognizable effect on the cells occurring.

3 Survival of the infected cell unchanged, except that it now carries the virus in a latent state.

4 Survival of the infected cell in a dramatically altered or transformed state, e.g. transformation of a normal cell to one having the properties of a cancerous cell.

6   Bacteriophages

Bacteriophages, or as they are more simply termed, phages, are viruses that have bacteria as their host cells. The name was first given by D'Herelle to an agent which he found could produce lysis of the dysentery bacillus *Shigella shiga*. D'Herelle was convinced that he had stumbled across an agent with tremendous medical potential. His phage could destroy *Sh. shiga* in broth culture so why not in the dysenteric gut of humans? Similar agents were found before long which were active against the bacteria of many other diseases, including anthrax, scarlet fever, cholera and diphtheria, and attempts were made to use them to treat these diseases. It was a great disappointment, however, that phages so virulent in their antibacterial activity *in vitro* proved impotent *in vivo*. A possible exception was cholera, where some success seems to have been achieved, and cholera phages were apparently used by the medical corps of the German and Japanese armies during the Second World War to treat this disease. Since the development of antibiotics, however, phage therapy has been totally abandoned.

Interest in bacterial viruses did not cease with the demise of phage therapy. They proved to be very much easier to handle in the laboratory than other viruses and had conveniently rapid multiplication cycles. They have, therefore, been used extensively as the experimental models for elucidating the biochemical mechanisms of viral replication. A vast amount of information has been collected about them and many of the important advances in molecular biology, such as the discovery of messenger RNA (mRNA), the understanding of the genetic code and the way in which genes are controlled, have come from work on phage–bacterium systems.

It is probable that all species of bacteria are susceptible to phages. Any particular phage will exhibit a marked specificity in selecting host cells, attacking only organisms belonging to a single species. A *Staphylococcus aureus* phage, for example, will not infect *Staph. epidermidis* cells. In most cases, phages are in fact strain specific, only being active on certain characteristic strains of a given species.

Most phages are tadpole-shaped structures with heads which function as containers for the nucleic acid and tails which are used to attach the virus to its host cell. There are, however, some simple icosahedral phages and others that are helically symmetrical cylinders. The dimensions of the

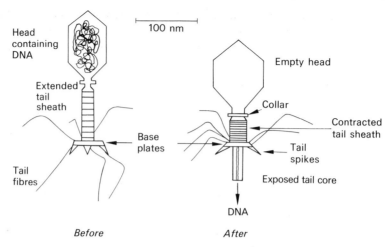

Fig. 3.3 T-even phage structure before and after tail contraction.

phage heads vary from the large T-even group (Fig. 3.3) of *Escherichia coli* phages (60 × 90 nm) to the much smaller ones (30 × 30 nm) of certain *Bacillus* phages. The tails vary in length from 15 to 200 nm and can be quite complex structures (Fig. 3.3). While the majority of phages have double-stranded DNA as their genetic material, some of the very small icosahedral and the helical phages have single-stranded DNA or RNA.

On the basis of the response they produce in their host cells, phages can be classified as *virulent* or *temperate*. Infection of a sensitive bacterium with a virulent phage results in the replication of the virus, lysis of the cell and release of new infectious progeny phage particles. Temperate phages can produce this lytic response, but they are also capable of a symbiotic response in which the invading viral genome does not take over the direction of cellular activity, the cell survives the infection and the viral nucleic acid becomes incorporated into the bacterial chromosome, where it is termed *prophage*. Cells carrying viral genes in this way are referred to as *lysogenic*.

6.1   **The lytic growth cycle**

The replication of virulent phage was initially studied using the T-even-numbered (T_2, T_4 and T_6) phages of *E. coli*. These phages adsorb, by their long tail fibres, on to specific receptors on the surface of the bacterial cell wall. The base plate of the tail sheath and its pins then lock the phage into position on the outside of the cell. At this stage, the tail sheath contracts towards the head, while the base plate remains in contact with the cell wall and, as a result, the hollow tail core is exposed and driven through to the cytoplasmic membrane (Fig. 3.3). Simultaneously, the DNA passes from the head, through the hollow tail core and is deposited on the outer surface of the cytoplasmic membrane, from where it finds its own way into the cytoplasm. The phage protein coat remains on the outside of the cell and plays no further part in the replication cycle.

Within the first few minutes after infection, transcription of part of the viral genome produces 'early' mRNA molecules, which are translated into a set of 'early' proteins. These serve to switch off host cell macromolecular synthesis, degrade the host DNA and start to make components for viral DNA. Many of the early proteins duplicate enzymes already present in the host, concerned in the manufacture of nucleotides for cell DNA. However, the requirement for the production of 5-hydroxymethylcytosine-containing nucleotides, which replace the normal cytosine derivatives in T-even phage DNA, means that some of the early enzymes are entirely new to the cell. With the build-up of its components, the viral DNA replicates and also starts to produce a batch of 'late' mRNA molecules, transcribed from genes which specify the proteins of the phage coat. These late messages are translated into the subunits of the capsid structures, which condense to form phage heads, tails and tail fibres, and then together with viral DNA are assembled into complete infectious particles. The enzyme digesting the cell wall, lysozyme, is also produced in the cell at this stage and it eventually brings about the lysis of the cell and liberation of about 100 progeny viruses, some 25 minutes after infection.

As other phage systems have been studied, it has become clear that the T-even model of virulent phage replication is atypical in a number of respects. The large T-even genomes, with their coding capacity for about 200 proteins, give these phages a relatively high degree of independence from their hosts. Although relying on the host energy and protein-synthesizing systems they are capable of specifying a battery of their own enzymes. Most other phages have considerably smaller genomes. They tend to disturb the host-cell metabolism to a much lesser extent than the T-even viruses, and also rely to a greater degree on pre-existing cell enzymes to produce components for their nucleic acid.

The lytic activity of the virulent phages can be demonstrated by mixing phage with about 10^7 sensitive indicator bacteria in 5 ml of molten nutrient agar. The mixture is then poured over the surface of a solid nutrient agar plate. On incubation, the phage particles will infect bacteria in their immediate neighbourhood, lysing them and producing a burst of progeny viruses. These particles then infect bacteria in the vicinity, producing a second generation of progeny and this sequence is repeated many times. In the meantime the uninfected bacteria produce a thick carpet or lawn of growth over the agar. As the lawn develops, clear holes or 'plaques' become obvious in it at each site of virus multiplication (Fig. 3.4). As each of these plaques is initiated by a single phage particle, they provide a means for titrating phage preparations.

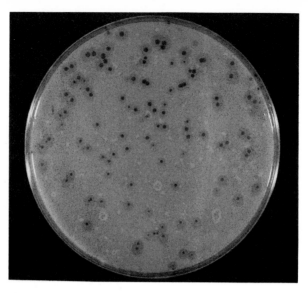

Fig. 3.4 Plaques formed by a phage on a plate seeded with *Bacillus subtilis*.

Lysogeny

When a temperate phage is mixed with sensitive indicator bacteria and plated as described above, the reaction at each focus of infection is generally a combination of lytic and lysogenic responses. Some bacteria will be lysed and produce phage, others will survive as lysogenic cells, and the plaque becomes visible as a partial area of clearing in the bacterial lawn. It is possible to pick off cells from the central areas of these plaques and demonstrate that they carry prophage.

The phage lambda (λ) of *E. coli* is the temperate phage that has been most extensively studied. When any particular strain of *E. coli*, say K12, is infected with λ, the cells surviving the infection are designated *E. coli* K12(λ) to indicate that they are carrying the λ-prophage.

The essential features of lysogenic cells and the phenomenon of lysogeny are listed below and summarized in Fig. 3.5.

1 Integration of the prophage into the bacterial chromosome ensures that on cell division each daughter cell will acquire the set of viral genes.

2 In a normally growing culture of lysogenic bacteria, the majority of bacteria manage to keep their prophages in a dormant state. In a very small minority of cells, however, the prophage genes express themselves. This results in the multiplication of the virus, lysis of the cells and liberation of infectious particles into the medium.

3 Exposure of lysogenic cultures to certain chemical and physical agents, e.g. hydrogen peroxide, mitomycin C and ultraviolet light, results in mass lysis and the production of high titres of phage. This process is called *induction*.

4 When a lysogenic cell is infected by the same type of phage as it

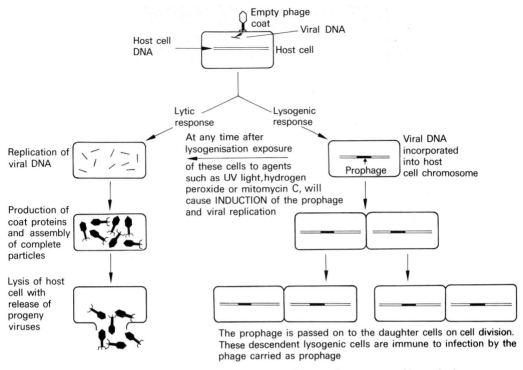

Fig. 3.5 Scheme to illustrate the lytic and lysogenic responses of bacteriophages.

Labels within figure:

Empty phage coat
Viral DNA
Host cell DNA
Host cell

Lytic response

Lysogenic response

Replication of viral DNA

Production of coat proteins and assembly of complete particles

Lysis of host cell with release of progeny viruses

At any time after lysogenisation exposure of these cells to agents such as UV light, hydrogen peroxide or mitomycin C, will cause INDUCTION of the prophage and viral replication

Prophage

Viral DNA incorporated into host cell chromosome

The prophage is passed on to the daughter cells on cell division. These descendent lysogenic cells are immune to infection by the phage carried as prophage

carries as prophage, the infection is aborted, the activity of the invading viral genes being repressed by the same mechanism that normally keeps the prophage in a dormant state.

5 Lysogeny is generally a very stable state, but occasionally a cell will lose its prophage and these 'cured' cells are once more susceptible to infection by that particular phage type.

Lysogeny is an extremely common phenomenon and it seems that most natural isolates of bacteria carry one or more prophages; some strains of *Staph. aureus* have been shown to carry four or five different prophages.

The induction of a lysogenic culture to produce infectious phages, followed by lysogenization of a second strain of the bacterial species by these phages, results in the transmission of a prophage from the chromosome of one type of cell to that of another. On this migration, temperate bacteriophages can occasionally act as vectors for the transfer of bacterial genes between cells. This process is called *transduction* and it can be responsible for the transfer of such genetic factors as those that determine resistance to antibiotics (Chapter 10). In addition, certain phages have the innate ability to change the properties of their host cell. The classic example is the case of the *β*-phage of *Corynebacterium diphtheriae*. The acquisition of the *β*-prophage by non-toxin-producing strains of this species results in their conversion to diphtheria-toxin producers.

6.3 Epidemiological uses

Different strains of a number of bacterial species can be distinguished by their sensitivity to a collection of phages. Bacteria which can be typed in this way include *Staph. aureus* and *Salmonella typhi*. The particular strain of, say, *Staph. aureus* responsible for an outbreak of infection is characterized by the pattern of its sensitivity to a standard set of phages and then possible sources of infection are examined for the presence of that same phage type of *Staph. aureus*.

More recently, the fact that many of the chemical agents which cause the induction of prophage are carcinogenic has led to the use of lysogenic bacteria in screening tests for detecting potential carcinogens.

7 Human viruses

Viruses are, of course, important and common causes of disease in humans, particularly in children. Fortunately, most infections are not serious and, like the rhinovirus infections responsible for the common cold syndrome, are followed by the complete recovery of the patient. Many viral infections are in fact so mild that they are termed 'silent', to indicate that the virus replicates in the body without producing symptoms of disease. Occasionally, however, some of the viruses that are normally responsible for mild infections can produce serious disease. This pattern of pathogenicity is exemplified by the enterovirus group. Most enterovirus infections merely result in the symptomless replication of the virus in the cells lining the alimentary tract. Only in a small percentage of infections does the virus spread from this site via the bloodstream and the lymphatic system to other organs, producing a fever and possibly a skin rash in the host. On rare occasions enteroviruses like poliovirus can progress to the central nervous system where they may produce an aseptic meningitis or paralysis. There are a few virus diseases, such as rabies, which are invariably severe and have very high mortality rates.

Human viruses will cause disease in other animals. Some are capable of infecting only a few closely related primate species, others will infect a wide range of mammals. Under the conditions of natural infection viruses generally exhibit a considerable degree of tissue specificity. The influenza virus, for example, replicates only in the cells lining the upper respiratory tract.

Table 3.1 presents a summary of the properties of some of the more important human viruses.

7.1 Cultivation of human viruses

The cultivation of viruses from material taken from lesions is an important step in the diagnosis of many viral diseases. Studies of the basic

Table 3.1 Important human viruses and their properties

Virus group	Virus	Characteristics	Clinical importance
DNA viruses			
Poxviruses	Variola Vaccinia	Large particles 200 × 250 nm: complex symmetry	Variola is the smallpox virus. It produces a systemic infection with a characteristic vesicular rash affecting the face, arms and legs, and has a high mortality rate. Vaccinia has been derived from the cowpox virus and is used to immunize against smallpox
Adenoviruses	Adenovirus	Icosahedral particles 80 nm in diameter	Commonly cause upper respiratory tract infections; tend to produce latent infections in tonsils and adenoids; will produce tumours on injection into hamsters, rats or mice
Herpesviruses	Herpes simplex	Enveloped, icosahedral particles 150 nm in diameter	The virus infects oral membranes in children. About 10% of those infected develop the primary symptoms, but the virus persists in a latent form in nerve cells and causes recurrent 'cold sores' in response to stimuli such as sunlight and hormonal changes. Recurrent herpes infections of genital membranes is an increasingly important sexually transmitted disease
	Varicella-zoster	Enveloped, icosahedral particles 180 nm in diameter	Causes chickenpox in children. After these primary infections the virus can apparently remain latent in ganglionic nerve cells. In adults it can be reactivated by trauma, drugs, tuberculosis and cancer, when it passes along the nerve fibres to the skin and cause shingles
	Cytomegalovirus	Enveloped, icosahedral particles 150 nm in diameter	About half of the human population carries the virus in a dormant state. In adults, it causes disease only when the normal resistance of the host is lowered, e.g. immunosuppression of renal transplant patients can activate the virus and produce fever, hepatitis and pneumonitis. Congenital infections occur and can result in microcephaly, motor disorders and mental retardation of the infant
	Epstein–Barr (EB) virus	Enveloped, icosahedral particles 150 nm in diameter	The virus is found associated with Burkitt's lymphoma, a malignant tumour common in African children. It also causes glandular fever in young adults in Europe and the USA
Hepatitis viruses	Hepatitis B	Small, spherical or tubular particles 30–40 nm in diameter	Responsible for serum hepatitis, where infected blood and its products are a major cause of infection. Drug addicts and homosexuals are at particular risk. There is also danger in handling infected blood

Contd over page

Table 3.1 *Continued*

Virus group	Virus	Characteristics	Clinical importance
Papovaviruses	Papilloma virus	Icosahedral symmetry 50 nm in diameter	Causes warts; there is evidence that the type causing genital warts has an aetiological role in genital cancer
RNA viruses Myxoviruses	Influenza virus	Enveloped particles, 100 nm in diameter with a helically symmetrical capsid; haemagglutinin and neuraminidase spikes project from the envelope	These viruses are capable of extensive antigenic variation, producing new types against which the human population does not have effective immunity. These new antigenic types can cause pandemics of influenza. In natural infections the virus only multiplies in the cells lining the upper respiratory tract. The constitutional symptoms of influenza are probably brought about by absorption of toxic breakdown products from the dying cells on the respiratory epithelium
Paramyxoviruses	Mumps virus	Enveloped particles, variable in size, 110–170 nm in diameter, with helical capsids	Infection in children produces characteristic swelling of parotid and submaxillary salivary glands. The disease can have neurological complications, e.g. meningitis, especially in adults
	Measles virus	Enveloped particles, variable in size, 120–250 nm in diameter, helical capsids	Very common childhood fever, immunity is life-long and second attacks are very rare
	Rubella	Enveloped particles 50–80 nm in diameter, capsid symmetry unknown	Causes German measles in children. An infection contracted in early pregnancy may cause congenital abnormalities in the fetus
Rhabdoviruses	Rabies virus	Bullet-shaped particles, 75–180 nm, enveloped, helical capsids	The virus has a very wide host range, infecting all mammals so far tested; dogs, cats and cattle are particularly susceptible. The incubation period of rabies is extremely varied, ranging from 6 days up to 1 year. The virus remains localized at the wound site of entry for a while before passing along nerve fibres to CNS, where it invariably produces a fatal encephalitis
Picornaviruses	Poliovirus	Small icosahedral particles 28 nm in diameter	One of a group of enteroviruses including Coxsackie and echoviruses, common in the gut of man. Primary site of multiplication is lymphoid tissue of the alimentary tract; only rarely do they cause serious neurological conditions like meningitis or poliomyelitis
	Rotavirus	Spherical particles 65 nm in diameter	A common cause of infantile gastroenteritis, it is particularly associated with low standards of hygiene and poor quality water supplies. In underdeveloped countries it is an important cause of infant mortality

Table 3.1 *Continued*

Virus group	Virus	Characteristics	Clinical importance
Picornaviruses (*contd*)	Rhinovirus	Small icosahedral particles	The common cold viruses; there are about 90 antigenically distinct types, hence the difficulty in preparing effective vaccines. Humans are the natural hosts. The gibbon ape is the only known laboratory animal susceptible to infection
Togaviruses	Yellow fever virus	Enveloped, icosahedral particles 30–60 nm in diameter	One of a very large group of insect-borne viruses. It can multiply in both human and invertebrate hosts. Yellow fever is a tropical disease passed on by both human and invertebrate hosts, passage to human beings by the mosquito
Hepatitis viruses	Hepatitis A	Small, spherical particles 30 nm in diameter	Responsible for infectious hepatitis spread by the oral–faecal route, especially in schoolchildren. Also associated with sewage contamination of food or water
Retroviruses	HIV	Enveloped, approx. 100 nm in diameter. Particles contain the enzyme reverse transcriptase	The virus responsible for AIDS (acquired immune deficiency syndrome), it specifically attacks T_4 lymphocytes and this has a profound effect on the immune system, producing a fatal susceptibility to a variety of opportunistic infections

biology and multiplication processes of human viruses also require that they are grown in the laboratory under experimental conditions. Human pathogenic viruses can be propagated in three types of cell systems.

7.1.1 Tissue cultures

Cells from human or other primate sources are obtained from an intact tissue, e.g. human embryo kidney or monkey kidney. The cells are dispersed by digestion with trypsin and the resulting suspension of single cells is generally allowed to settle in a vessel containing a nutrient medium. The cells will metabolize and grow and after a few days of incubation at 37°C will form a continuous film or monolayer one cell thick. These cells are then capable of supporting viral replication. Cell cultures may be divided into three types according to their history.

1 Primary cell cultures, which are prepared directly from tissues.

2 Secondary cell cultures, which can be prepared by taking cells from some types of primary culture, usually those derived from embryonic tissue, dispersing them by treatment with trypsin and inoculating some into a fresh batch of medium. A limited number of subcultures can be performed with these sorts of cells, up to a maximum of about 50 before the cells degenerate.

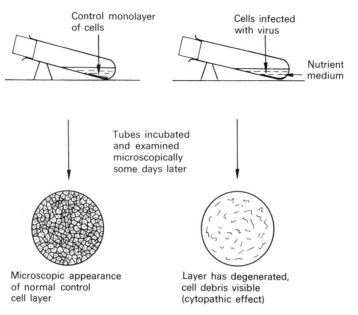

Control monolayer of cells

Cells infected with virus

Nutrient medium

Tubes incubated and examined microscopically some days later

Microscopic appearance of normal control cell layer

Layer has degenerated, cell debris visible (cytopathic effect)

Fig. 3.6 The cytopathic effect of a virus on a tissue culture cell monolayer.

3 There are now available a number of lines of cells, mainly originating from malignant tissue, which can be serially subcultured apparently indefinitely. These established cell lines are particularly convenient as they eliminate the requirement for fresh animal tissue for such sets or series of cultures. An example of these continuous cell lines are the famous HeLa cells, which were originally isolated from a cervical carcinoma of a lady called Helen Lane, long since dead but whose cells have been used in laboratories all over the world to grow viruses.

Inoculation of tissue cultures with virus-containing material produces characteristic changes in the cells. The replication of many types of viruses produces the cytopathic effect (CPE) in which cells degenerate and die and microscopic examination of the monolayer reveals an unstructured mass of cell debris (Fig. 3.6). The virus can be identified by inoculating a series of tissue cultures with mixtures of virus plus different known viral antisera. If the virus is the same as one of the types used to prepare any of the antisera, then its activity will be neutralized and CPE will not become apparent in that tube.

7.1.2

The chick embryo

Fertile chicken eggs, 10–12 days old, have been used as a convenient cell system in which to grow a number of human pathogenic viruses. Figure 3.7 shows that viruses generally have preferences for particular tissues within the embryo. Influenza viruses, for example, can be grown in the cells of the membrane bounding the amniotic cavity, while smallpox virus

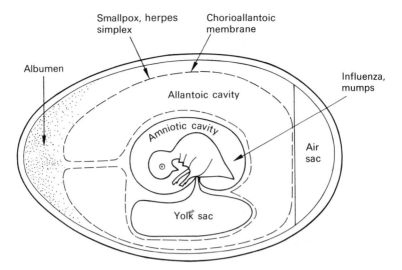

Fig. 3.7 A chick embryo showing the inoculation routes for virus cultivation.

will grow in the chorioallantoic membrane. The growth of smallpox virus in the embryo is recognized by the formation of characteristic pock marks on the membrane. Influenza virus replication is detected by exploiting the ability of these particles to cause erythrocytes to clump together. Fluid from the amniotic cavity of the infected embryo is titrated for its haemagglutinating activity.

7.1.3 Animal inoculation

Experimental animals such as mice and ferrets have to be used for the cultivation of some viruses. Growth of the virus is indicated by signs of disease or death of the inoculated animal.

8 Multiplication of human viruses

The long incubation times of many human virus diseases indicate that they replicate slowly in host cells. In tissue culture systems it has been shown that most human viruses take from 4 to 24 hours to complete a single replication cycle, contrasting with the 30 or so minutes for many bacterial viruses.

In general terms, four main stages can be recognized in the replication of human viruses: (i) adsorption; (ii) penetration and uncoating; (iii) vegetative growth; and (iv) maturation and release.

8.1 Adsorption

Most viruses seem to adsorb onto specific receptors on the host cell surface. The initial attachment to the host cells is probably an electrostatic attraction between a pattern of charged groups on the viral surface and a complementary pattern on the cell surface.

8.2 Penetration and uncoating

Non-enveloped particles appear to be engulfed by invaginations of the cell membrane. With enveloped viruses the virus envelope fuses with the cell membrane and the nucleocapsid passes into the cytoplasm. The protein capsids are then dismantled and the naked viral genome released.

8.3 Vegetative growth

During this phase most human viruses seem to bring host cell macro-molecular synthesis to a stop: the cell DNA, however, is generally not degraded. With the DNA-containing viruses, like adenovirus, the nucleic acid passes to the nucleus, where a host-cell RNA polymerase enzyme is used to transcribe part of the viral genome. These first messages are analogous to the 'early' messages of the T-even phages and are concerned in the production of enzymes for viral DNA synthesis. Viral DNA replication is then followed by formation of 'late' mRNA specifying capsid protein. The mRNA molecules are of course translated on the cytoplasmic ribosomes. The proteins produced are rapidly transported back to the nucleus, where capsid assembly takes place. An exception to this pattern of DNA virus replication is provided by the poxvirus, vaccinia. Within their complex structure these particles contain a DNA-dependent RNA polymerase enzyme, which is released during uncoating and proceeds to make mRNA molecules from the viral DNA. The whole of the replication of vaccinia takes place in the cell cytoplasm.

With some of the RNA viruses, for example poliovirus, the RNA strand from the virus particle can act directly as mRNA and is translated into viral proteins on the host cell ribosomes. With influenza and other myxoviruses, however, the RNA strand present in the particle is called *antimessenger*, to indicate that it will not act directly in viral protein synthesis. It has first to be copied by an RNA-dependent RNA polymerase enzyme, which is brought into the cell by the virus, and it is the complementary copy strands that the enzyme makes on the antimessage that function as messages.

8.4 Maturation and release

The assembly of the capsids and incorporation of the nucleic acid is followed by the release of the mature progeny particles, either by gradual extrusion through the membrane or as a sudden burst resulting from the lysis of the infected cell. Adenovirus particles mature in the nucleus and they accumulate there until cell lysis occurs. Poliovirus replication occurs entirely in the cytoplasm and while over 500 progeny particles can be produced per cell, they do not seem to accumulate in large numbers in

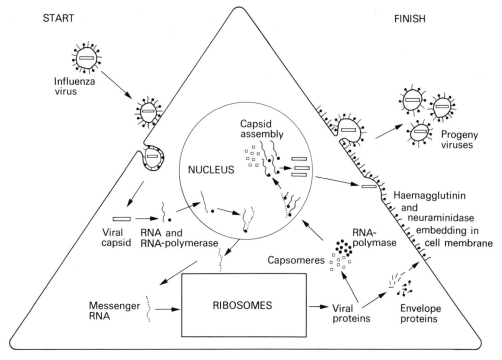

Fig. 3.8 Diagrammatic representation of the production and release of influenza virus particles from an infected cell.

infected cells. It appears that soon after maturation they are released by a mechanism which preserves the integrity of the cytoplasmic membrane, until the cells become rounded and die some 8 hours after infection.

The maturation and release of enveloped particles, such as influenza virus, is illustrated in Fig. 3.8. The capsid protein subunits are transported from the ribosomes to the nucleus, where they combine with new viral RNA molecules and are assembled into the helical capsids. The haemagglutinin and neuraminidase proteins that project from the envelope of the normal particles migrate to the cytoplasmic membrane where they displace the normal cell membrane proteins. The assembled nucleocapsids finally pass out from the nucleus, and as they impinge on the altered cytoplasmic membrane they cause it to bulge and bud off completed, enveloped particles from the cell. Virus particles are released in this way over a period of hours before the cell eventually dies.

9   The problems of viral chemotherapy

Bacteria are vulnerable to the selective attack of chemotherapeutic agents because of the many metabolic and molecular differences between them and animals cells. The biology of virus replication, with its considerable dependence on host-cell energy-producing, protein-synthesizing and biosynthetic enzyme systems, severely limits the opportunities for selective attack. Another problem is that many virus diseases only become apparent after extensive viral multiplication and tissue damage has been done.

Recently, however, there have been a number of encouraging developments in the field of antiviral therapy. For example, acycloguanosine (acyclovir: see Chapter 5) has been shown to be non-toxic to host cells while specifically inhibiting the replication of herpesviruses. Successful clinical trials have led to the introduction of this drug for the treatment of a variety of herpetic conditions.

The control over human viral diseases is exercised by active immunization (Chapter 16) of the population, together with general hygiene and physical and chemical disinfection procedures.

9.1   Interferon

Although it is difficult to obtain drugs capable of interrupting viral replication, it had been known for many years that infection of a host with one virus could sometimes prevent infections with a second, quite unrelated virus. This phenomenon was called *interference* and in many cases it proved to be due to the production of a substance called *interferon*.

Interferons are low molecular weight proteins produced by virus-infected cells. They have no direct antiviral activity. They bind to the cell membranes and induce the synthesis of secondary proteins. If interferon-treated cells are then infected with a virus, although adsorption, penetration and uncoating can take place, the interferon-induced proteins inhibit viral nucleic acid and protein synthesis and the infection is aborted. Interferons have major roles to play in protecting the host against natural virus infections. They are produced more rapidly than antibodies and the outcome of many natural viral infections is probably determined by the relative early titres of interferon and virus, protection being most effective when the infecting dose of virus is low.

Potentially, interferon is an ideal antiviral agent in that it acts on many different viruses and is not toxic to host cells. However, the exploitation of this agent in the treatment of viral infections has been delayed by a number of factors. For example, it has proved to be species specific and interferons raised in animal sources offered little protection to human cells. Human interferon is thus needed for the treatment of human infections and the production and purification of human interferon on a large scale has proved difficult. The insertion of human genes for interferon into *E. coli* has resolved the production problems (Chapter 24). Clinical trials have demonstrated that interferon prevents rhinovirus infection and has a beneficial effect in herpes, cytomegalovirus and hepatitis B virus infections.

Interferon does not only inhibit virus replication, it also has multiple effects on cell metabolism and slows down the growth and multiplication of treated cells. This is probably responsible for its widely reported antitumour effect. Encouraging results have been reported from clinical trials of interferon against several human tumours such as osteogenic sarcoma, myeloma, lymphoma and breast cancer.

10   Tumour viruses

Many viruses, both DNA and RNA containing, will cause cancer in animals. This so-called *oncogenic* activity of a virus can be demonstrated by the observation of tumour formation in inoculated experimental animals and by the ability of the virus to transform normal tissue culture cells into cells with malignant characteristics. These transformed cells are easily recognizable as they exhibit such properties as rapid growth and frequent mitosis, or loss of normal cell contact inhibition, so that they pile up on top of each other instead of remaining in a well-organized layer.

Studies on the transformation of tissue cultures with DNA-containing viruses have shown that, although complete virus particles cannot be found in the infected, transformed cells, viral DNA is present and is bound to the transformed cell DNA as *provirus*, analogous to the pro-phage of lysogenic bacteria.

RNA oncogenic viruses have an unusual enzyme, reverse transcriptase, which is capable of making DNA copies from an RNA template. Cells transformed by these retroviruses have been shown to possess DNA transcripts of the viral RNA. It appears that the transformation from normal to malignant is associated with the acquisition by the cell of viral DNA.

While human viruses like the adenoviruses can induce cancer in hamsters, rats and mice, the search for viruses causing human cancer is of course difficult because of the unacceptability of testing for oncogenic activity by infecting humans. In the last 10 years, however, it has been realized that viruses are a major cause of the disease in humans, being involved in the genesis of some 20% of human cancers worldwide. The characteristic features of the association between viruses and human cancers are that the incubation time between virus infection and development of the disease can be considerable, that less than 1% of infected individuals will develop the disease and that genetic and environmental cofactors are crucial for the progression to cancer. The Epstein–Barr (EB) virus, for example, is involved in the aetiology of Burkitt's lymphoma a malignant tumour of the jaw, found in African children. In fact this virus has a widespread distribution in the human population, being responsible for the condition of glandular fever which is common in young adults in Europe and America. The characteristic occurrence of Burkitt's lymphoma in hot humid areas of Africa where mosquitoes flourish has led to the hypothesis that infection with the EB virus has to be followed by malaria, which then induces immunosuppression and acts as the cofactor necessary for tumour formation.

The list of viruses involved in other human cancers includes hepatitis B, which is associated with hepatocellular carcinoma; human papilloma viruses with cervical, penile and some anal carcinomas; human T-cell lymphotropic virus type I associated with adult T-cell leukaemia/ lymphoma syndrome; and human immunodeficiency virus (HIV) with Kaposi's sarcoma.

11   Acquired immune deficiency syndrome (AIDS)

The human immunodeficiency virus (HIV) is an enveloped RNA-containing particle. In infected human cells, DNA copies of the viral RNA are made by a reverse transcriptase enzyme and these are then incorporated into host-cell chromosomes. The fate of these HIV-infected cells varies. The viral DNA can remain dormant, the cell surviving but carrying the latent HIV provirus. Alternatively, the viral DNA can be expressed and produce viral proteins and new infectious progeny particles. This process can either occur slowly allowing the survival of the host cell or rapidly resulting in the death of the infected cell. On first infecting a host, HIV often replicates prolifically and free virus particles can be found in the cerebrospinal fluid and the blood. Flu-like symptoms, neurological disorders and rashes can sometimes be produced during this initial period. Within a few weeks the virus titres in the body fluids drop markedly and the symptoms disappear. However, the virus persists and can be found in the provirus state in cells of the nervous system and intestine and especially in a subset of white blood cells, the T_4 lymphocytes. The latent state can be maintained for up to 10 years during which time the infected individual is asymptomatic. However, activation of the latent provirus induces viral replication and cell death. The loss of the T_4 cells in this way leads to the clinical symptoms of AIDS. The T_4 cells have a key role in the immunological defences of the body and their depletion causes the infected individual to become susceptible to a variety of infections by common microorganisms that ordinarily are unable to produce disease, but which in the immunodeficient patient are lethal. Many AIDS patients die of pneumocystosis caused by the yeast *Pneumocystis carinii*.

12   The agents of spongiform encephalopathy

Bovine spongiform encephalopathy (BSE) is one of a number of similar diseases of the central nervous systems of humans and animals. The causative agents of these diseases used to be referred to as slow viruses, but it now seems that they are a distinct class of infectious agents with unique and disturbing properties. These agents must exist because the diseases concerned, such as BSE, scrapie in sheep and Creutzfeldt–Jakob disease in humans, can be transmitted from one individual to another. The exact nature of these infectious agents is unknown: it has not been possible to see them in the electron microscope; they cannot be cultured in media or host cells; they do not stimulate the production of specific antibodies in infected animals. Whatever they are they seem to be incredibly tough agents as they are able to survive autoclaving. A common property is that the disease takes a long time to develop after infection, up to 20 years in humans. The disease is progressive, induces a gradual degeneration of the central nervous system and is virtually always fatal. The diagnosis of the diseases depends upon the clinical symptoms and the characteristic histopathology of the brain. Microscopic holes appear in the grey matter and the diseased brain has the appearance of a sponge with many holes of variable size scattered through the tissue. The causative agents have been called *prions* and recent evidence suggests that they are infectious proteinaceous particles that do not contain nucleic acid.

13   Further reading

Collier L.H. & Timbury M.C. (Eds) (1990) *Virology. Topley and Wilson's Principles of Bacteriology, Virology and Immunology*, 8th edn, vol. 4. London: Edward Arnold.

Crumpacker C.S. (1989) Molecular targets of antiviral therapy. *N Engl J Med.* **321**, 163–172.

Dalgleish A.G. (1991) Viruses and cancer. *Br Med Bull.* **47**, 21–46.

Dalgleish A.G. & Weiss R. (1990) *AIDS and the New Viruses*. London: Academic Press.

Hayden F.G., Albrecht J.K., Kaiser, D.L. & Gwaltney J.M. (1986) Prevention of natural colds by contact prophylaxis with intranasal α2-interferon. *N Engl J Med.* **314**, 71–75.

Weissmann C. (1991) The prion's progress. *Nature*, **349**, 569–571.

Zuckerman A.J., Banatvala J.E. & Pattison J.R. (Eds) (1990) *Principles and Practice of Clinical Virology*, 2nd edn. Chichester: J Wiley & Sons.

4 Principles of microbial pathogenicity and epidemiology

1   **Introduction**

2   **Portals of entry**
2.1   Respiratory tract
2.2   Intestinal tract
2.3   Urinogenital tract
2.4   Conjunctiva

3   **Consolidation**
3.1   Resistance to host's defences
3.1.1 Modulation of the inflammatory response
3.1.2 Avoidance of phagocytosis
3.1.3 Survival following phagocytosis
3.1.4 Killing of phagocytes

4   **Manifestation of disease**
4.1   Non-invasive pathogens

4.2   Partially invasive pathogens
4.3   Invasive pathogens
4.3.1 Active spread
4.3.2 Direct passive spread
4.3.3 Indirect passive spread

5   **Damage to tissues**
5.1   Direct damage
5.1.1 Specific effects
5.1.2 Non-specific effects
5.2   Indirect damage

6   **Recovery from infection: exit of microorganisms**

7   **Epidemiology of infectious disease**

8   **Further reading**

1   Introduction

The majority of microorganisms are free living and derive their nutrition from inert organic and inorganic materials. Their association with humans is generally harmonious, as the majority of microorganisms encountered are benign and often vital to a balanced ecosystem. In spite of the ubiquity of microorganisms the tissues of healthy animals and plants are essentially microbe-free. This is achieved through provision of a number of non-specific defences to those tissues, and specific defences such as antibodies (see Chapter 15) acquired after exposure to particular agents. Breach of these defences by microorganisms, either following accidental trauma, catheterization and implantation of medical devices or specific adaptation of the organisms to a pathogenic mode of life, may lead to the establishment of microbial infections.

The ability of organisms to establish infections varies considerably. Some are rarely, if ever, isolated from infected tissues, other opportunist pathogens (e.g. *Pseudomonas aeruginosa*) can establish themselves only in compromised tissues, yet a few are truly obligate pathogens, for which animals and humans are the only reservoirs for their existence (e.g. *Neisseria gonorrhoeae*). Even amongst the obligate pathogens the degree of virulence varies, in that some are able to coexist with certain individuals without causing the disease state (e.g. staphylococci), whilst others will always produce the disease condition. Organisms such as these invariably produce their effect by actively growing on or in the host tissues.

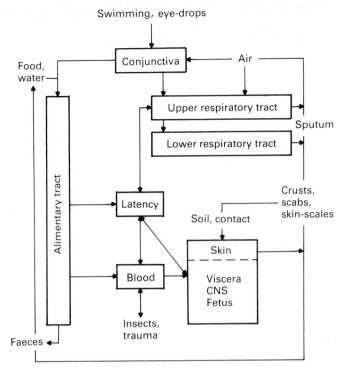

Fig. 4.1 Routes of infection spread and of transmission of disease.

Other groups of organisms may cause disease through ingestion by the victim of substances produced during microbial growth on foods (e.g. *Clostridium botulinum*). Whether these can be regarded as pathogens for this property, for which they derive no benefit, is debatable, but they must be considered in any account of microbial pathogenicity.

The course of an infection can be considered as being a sequence of separate events (Fig. 4.1). Firstly, the potential pathogen must arrive at its portal of entry to the body or directly at its target tissue. Secondly, the organism must be able to multiply rapidly at that site in order to avoid elimination by the body. This might involve microcolonization of the tissue surface to form a biofilm. Biofilms are functional consortia of microorganisms enveloped within extensive exopolymer matrices (glycocalyx) composed either of polysaccharide or protein. As such, growth within a biofilm modulates the microenvironment of the bacteria and confers advantages upon them with respect to resistance towards phagocytosis, opsonization and antibiotics and also with respect to the acquisition of nutrients. Following this initial consolidation, the organism may expand into surrounding tissues, disperse via the circulatory systems to distant tissues and establish secondary sites of infection or consolidate further. Finally, the organism must exit the body, survive and/or re-enter another susceptible host.

2   Portals of entry

The part of the body most widely exposed to microorganisms is the skin. Intact skin is usually impervious to microorganisms. Its surface contains relatively few nutrients and is of acid pH, which is unfavourable for microbial growth. The vast majority of organisms falling onto the skin surface will die, the remainder must compete with the commensal microflora, highly adapted to this environment, and these will usually prevent the establishment of such adventitious contaminants. Infections of the skin itself, such as ringworm (*Trichophyton mentagrophytes*) rarely, if ever, involve penetration of the epidermis. Infections can, however, occur through the skin following trauma such as burns, cuts and abrasions and, in some instances, through insect or animal bites or the injection of contaminated medicines. In recent years extensive use of intravascular and extravascular medical devices and implants has led to an increase in the occurrence of hospital acquired infection. Commonly these infections involve growth of skin commensals such as *Staphylococcus epidermidis* when associated with devices which penetrate the skin barrier. The organism grows as an adhesive biofilm upon the surfaces of the device, where infection arises either from contamination of the device during implantation or growth along it from the skin. In such instances the biofilm sheds cells to the body and gives rise to bacteraemias. These readily respond to antibiotic treatment but the biofilm, which is relatively recalcitrant towards even agressive antibiotic therapy, remains and acts as a continued focus of infection. In practice the device must be removed and replaced only after successful chemotherapy.

The weak spots, or Achilles heels, of the body occur where the skin ends and mucous epithelial tissues begin (mouth, anus, eyes, ears, nose and urinogenital tract). These mucous membranes present a much more favourable environment for microbial growth than the skin, in that they are warm, moist and often rich in nutrients. They do, nevertheless, possess certain characteristics that allow them to resist infection. All, for example, possess their own highly adapted commensal microflora which must be displaced by any invading organism. These resident flora vary greatly between different sites of the body but are usually common to particular species. Each site can be additionally protected by physico-chemical barriers such as the extreme acid pH of the stomach, freely circulating non-specific antibodies and/or opsonins, and/or by the presence of macrophages and phagocytes (see Chapter 15). All infections start from contact between these tissues and the potential pathogen. This contact may be direct, from an infected individual to a healthy one; or indirect, and involve inanimate vectors (fomites) such as soil, food, drink, air, and airborne particles being ingested, inhaled or entering wounds, or via infected bedlinen and clothing. Indirect contact may also involve animal vectors (carriers).

2.1 Respiratory tract

Air contains a large amount of suspended organic matter and in enclosed spaces may hold up to 1000 microorganisms m^{-3}, almost all non-pathogenic bacteria and fungi. The average person would inhale approximately 10 000 microorganisms per day. The respiratory tract is protected against this assault by a mucociliary blanket which envelops the lower respiratory tract and nasal cavity. Particles becoming entrapped in this blanket of mucus are carried by ciliary action to the back of the throat and swallowed. The alveolar regions are protected by a lining of macrophages. To be successful, a pathogen must avoid being trapped in the mucus and swallowed, and if deposited in the alveolar sacs must avoid engulfment by macrophages or resist subsequent digestion by them.

2.2 Intestinal tract

The intestinal tract must contend with whatever it is given in terms of food and drink. The lower gut is highly populated by the commensal flora (10^{11} g^{-1} gut tissue) often associated with the intestinal wall, either embedded in layers of protective mucus or attached directly to the epithelial cells. Pathogenicity depends upon the ability of individual organisms to survive passage through the stomach and duodenum and upon their capacity for attachment to, or penetration of, the gut wall in competition with the commensal flora.

2.3 Urinogenital tract

In healthy individuals urine is sterile and constantly flushes the urinary tract. Organisms invading the urinary tract must avoid being washed out during urination. In the male, since the urethra is long (c. 20 cm), bacteria must be introduced directly into the bladder in order to initiate infection, possibly through catheterization. In the female the urethra is much shorter (c. 5 cm) and is more readily traversed by microorganisms. Bladder infections are therefore much more common in the female. Spread of the infection from the bladder to the kidneys can easily occur through reflux of urine into the ureter. As for the implantation of devices across the skin barrier (above), long-term catheterization of the bladder will promote the occurrence of bacterurias and the associated complications.

Lactic acid in the vagina gives it an acidic pH (5.0) which together with other products of metabolism inhibits colonization by most bacteria, except some lactobacilli which constitute the commensal flora. Others are unable to establish themselves unless they are extremely specialized. These tend to be associated with venereal infections.

2.4 Conjunctiva

The conjunctiva is protected by the continuous flow of secretions from lacrimal and other glands, and by frequent mechanical cleansing of its surface by the eyelid periphery during blinking. Damage to the conjunctiva, caused through mechanical abrasion or reduction in tear flow will increase microbial adhesion and allow colonization by opportunist pathogens. The likelihood of infection is thus promoted by the use of soft and hard contact lenses.

3 Consolidation

To be successful a pathogen must be able to survive at its initial portal of entry, frequently in competition with the commensal flora and subject to the attention of macrophages and wandering white blood cells. Such survival invariably requires the organism to attach itself firmly to the epithelial surface. This attachment must be highly specific in order to displace the commensal microflora and subsequently governs the course of an infection. Attachment can be mediated through provision, on the bacterial surface, of adhesive substances, such as mucopeptide and mucopolysaccharide slime layers, fimbriae (Chapter 1), pili (Chapter 1) and agglutinins (Chapter 15). These are often highly specific in their binding characteristics, differentiating, for example, between the tips and bases of villi and the epithelial cells of the upper, mid and lower gut.

The outcome of the encounter between the tissues and potential pathogens is governed by the ability of the organisms to multiply at a faster rate than they are removed from those tissues. Factors which influence this are the organism's rate of growth, the initial number of organisms arriving at the site and their ability to resist the efforts of the host tissues at killing it. The definition of virulence for pathogenic microorganisms must therefore relate to the minimum number of cells required to initiate an infection. This will vary between individuals, but will invariably be lower in compromised hosts such as diabetics, cystic fibrotics and those suffering trauma such as malnutrition, chronic infection or physical damage.

3.1 Resistance to host's defences

Most bacterial infections confine themselves to the surface of epithelial tissue (e.g. *Bordetella pertussis*, *Corynebacterium diphtheriae*, *Vibrio cholerae*). This is, to a large extent, a reflection of their inability to combat that host's deeper defences. Survival at these sites is largely due to firm attachment to the epithelial cells.

Other groups of organisms regularly establish systemic infections (e.g. *Brucella abortus*, *Salmonella typhi*, *Streptococcus pyogenes*) after traversing the epithelial surfaces. This property is associated with their ability to either gain entry into susceptible cells and thereby enjoy protection from the body's defences, or to be phagocytosed by macrophages or polymorphs yet resist their lethal action and multiply within them. Others are able to multiply and grow freely in the body's extracellular fluids. Microorganisms have evolved a number of different strategies which allow them to suppress the host's normal defences and thereby survive in the tissues. These are considered in later sections.

3.1.1 Modulation of the inflammatory response

Growth of some microorganisms releases products into the surrounding medium, many of which cause inflammation through dilation of blood vessels. This increases capillary flow and access of phagocytes. Increased lymphatic flow carries the organisms to the lymph nodes where further antimicrobial and immune forces come into play. Many of the substances released by these organisms are chemotactic towards polymorphs and tend to concentrate them at the site of infection.

Virulent strains of *Staph. aureus* produce a mucopeptide which suppresses early inflammatory oedema, a protective reaction, and therefore enhances the severity of its lesions. A related factor suppresses the chemotaxis of polymorphs, i.e. the chemical attraction of polymorphs and invading pathogens—another normal part of the body's defence.

3.1.2 Avoidance of phagocytosis

Resistance to phagocytosis is sometimes associated with specific components of the cell wall and/or with the presence of capsules surrounding the cell wall. Classic examples of these are the M-proteins of the streptococci and the polysaccharide capsules of pneumococci. The acidic polysaccharide K-antigens of *Escherichia coli* and *Sal. typhi* behave similarly, in that (i) they can mediate attachment to the intestinal epithelial cells, and (ii) they render phagocytosis more difficult. Generally, possession of an extracellular capsule will reduce the likelihood of phagocytosis.

Microorganisms are more readily phagocytosed when coated with antibody (opsonized). This is due to the presence on the white blood cells of receptors for the Fc fragment of IgM and IgG (discussed in Chapter 15). Avoidance of opsonization will clearly enhance the chances of survival of a particular pathogen. A substance called protein A can be extracted from *Staph. aureus*. This acts by non-specific interaction with IgG at the Fc region (see also Chapter 15) thereby negating the effects of opsonization through masking the Fc receptor to phagocytes. A second effect of protein A is to deplete complement from the plasma. This will react to the IgG–protein A complex in solution and will not cause damage to the organism.

3.1.3 Survival following phagocytosis

Death following phagocytosis can be avoided if the microorganisms are not exposed to the intracellular processes (killing and digestion) of the phagocyte. This is possible if fusion of the lysosomes with phagocytic vacuoles can be prevented. Such a strategy is employed by virulent

Mycobacterium tuberculosis although the precise mechanism is unknown. Other bacteria seem able to grow in the vacuoles in spite of lysosomal fusion (*Listeria monocytogenes*, *Sal. typhi*). This can be attributed to cell wall components which prevent access of the lysosomal substances (e.g. *Br. abortus*, mycobacteria) or to the production of excess catalase enzyme which neutralizes the hydrogen peroxide liberated in the vacuole (e.g. staphylococci, streptococci).

3.1.4 Killing of phagocytes

An alternative strategy is for the microorganism to kill the phagocyte. This can be achieved by the production of leucocidins (e.g. staphylococci, streptococci) which promote the discharge of lysosomal substances into the cytoplasm of the phagocyte rather than into the vacuole, thus directing the phagocyte's lethal activity towards itself.

4 Manifestation of disease

Once established, the course of a bacterial infection can proceed in a number of ways. These can be related to the ability of the organism to penetrate and invade surrounding tissues and organs. The vast majority of pathogens, being unable to combat the defences of the deeper tissues consolidate further on the epithelial surface. Others, which penetrate the epithelial layers, but no further, can be regarded as partially invasive. A small group of pathogens are fully invasive. These permeate the subepithelial tissues and are often circulated around the body to initiate secondary sites of infection remote from the initial portal of entry.

Other groups of organisms may cause disease through ingestion by the victim of substances produced during microbial growth on foods. Such disease may be regarded as an intoxication rather than as an infection and are considered further in section 5.1.1. Treatment in these cases is usually an alleviation of the harmful effects of the toxin rather than elimination of the pathogen from the body.

4.1 Non-invasive pathogens

Bord. pertussis (the aetiological agent of whooping cough) is probably the best described of these pathogens. This organism is inhaled and rapidly localizes on the mucociliary blanket of the lower respiratory tract. This localization is very selective and thought to involve agglutinins on the organism's surface. Toxins, produced by the organism, inhibit ciliary movement of the epithelial surface and thereby prevent removal of the bacterial cells to the gut. A high molecular weight exotoxin is also produced during the growth of the organism which, being of limited diffusibility, pervades the subepithelial tissues to produce inflammation and necrosis. *C. diphtheriae* (the causal organism of diphtheria) behaves similarly, attaching itself to the epithelial cells of the respiratory tract.

This organism produces a low molecular weight, diffusible toxin which enters the blood circulation and brings about a generalized toxaemia.

In the gut, many pathogens adhere to the gut wall and produce their toxic effect via toxins which pervade the surrounding gut wall or enter the systemic circulation. *V. cholerae* and some enteropathic *E. coli* strains localize on the gut wall and produce toxins which increase vascular permeability. The end result is a hypersecretion of isotonic fluids into the gut lumen, acute diarrhoea and consequent dehydration which may be fatal in juveniles. In all these instances, binding to epithelial cells is not essential but increases permeation of the toxin and prolongs the presence of the pathogen.

4.2 Partially invasive pathogens

Some bacteria are able to attach to the mucosal epithelia and then penetrate rapidly into the epithelial cells. These organisms then multiply within the protective environment of the host cell, eventually killing it and inducing disease through erosion and ulceration of the mucosal epithelium. Typically, members of the genera *Shigella* and *Salmonella* utilize this mechanism. These attach to the epithelial cells of the large and small intestines respectively and, following their entry by induced pinocytosis, multiply rapidly and penetrate laterally into adjacent epithelial cells. The mechanisms for such attachment and movement are unknown. Some species of salmonellae produce, in addition, exotoxins which induce diarrhoea (section 4.1). There are innumerable species and serotypes of *Salmonella*. These are primarily parasites of animals, but are important to humans in that they colonize farm animals such as pigs and poultry and ultimately infect such food. Salmonella food poisoning (salmonellosis), therefore, is commonly associated with inadequately cooked meats, eggs and also with cold meat products which have been incorrectly stored following contact with the uncooked product. Dependent upon the severity of the lesions induced in the gut wall by these pathogens, red blood cells and phagocytes pass into the gut lumen, along with plasma, and cause the classic 'bloody flux' of bacillary dysentery. Similar erosive lesions are produced by some enteropathic strains of *E. coli*.

4.3 Invasive pathogens

Invasive pathogens either aggressively invade the tissues surrounding the primary site of infection or are passively transported around the body in the blood, lymph, cerebrospinal fluid or pleural fluids. Some, especially aggressive organisms, do both, setting up a number of expansive secondary sites of infection in various organs.

4.3.1 *Active spread*

Active spread of microorganisms through normal subepithelial tissues is difficult in that the gel-like nature of the intracellular materials physically inhibits bacterial movement. Induced death and lysis of the tissue cells, in addition, produces a highly viscous fluid, partly due to undenatured DNA. Physical damage, such as wounds, rapidly seal with fibrin clots, thus reducing the effective routes of spread for opportunist pathogens. Organisms such as *Str. pyogenes*, *Cl. perfringens*, and to some extent the staphylococci, are able to establish themselves in tissues by virtue of their ability to produce a wide range of extracellular enzyme toxins. These are associated with killing of tissue cells, degradation of intracellular materials and mobilization of nutrients, and will be considered briefly.

Haemolysins are produced by most of the pathogenic staphylococci and streptococci. They have a lytic effect on red blood cells, releasing iron-containing nutrients.

Fibrinolysins are produced by both staphylococci (staphylokinase) and streptococci (streptokinase). These toxins dissolve fibrin clots, formed by the host around wounds and lesions to seal them, by indirect activation of plasminogen, thereby increasing the likelihood of organism spread. Streptokinase may be employed clinically in conjunction with streptodornase (Chapter 25).

Collagenases and hyaluronidases are produced by most of the aggressive invaders. These are able to dissolve collagen fibres and hyaluronic acids which function as intracellular cements. Their loss causes the tissues to break up and produce oedematous lesions.

Phospholipases are produced by organisms such as *Cl. perfringens* (a-toxin). These kill tissue cells by hydrolysing phospholipids present in cell membranes.

Amylases, peptidases and deoxyribonuclease mobilize nutrients released from lysed cells and decrease the viscosity of fluids present at the lesion by depolymerization of their biopolymer substrates.

Organisms possessing the above toxins, particularly those also possessing leucocidins, are likely to cause expanding oedematous lesions at the primary site of infection. In the case of *Cl. perfringens*, a soil microorganism which has become adapted to a saprophytic mode of life, when infection by it follows as a consequence of accidental contamination of deep wounds there ensues a process similar to that seen during the decomposition of a carcass. This organism is most likely to spread through tissues when blood circulation, and therefore oxygen tension, in the affected areas is minimal.

Abscesses formed by streptococci and staphylococci can be deep seated in soft tissues or associated with infected wounds or skin lesions. These become localized through the deposition of fibrin capsules around the infective site. Fibrin deposition is partly a response of the host tissues and partly a function of enzyme toxins such as coagulase. Phagocytic white blood cells can migrate into these abscesses in large numbers to produce significant quantities of pus; this might be digested by other phagocytes in the latter stages of the infection or discharged to the exterior or to the capillary and lymphatic network. In the latter case, blocked capillaries might serve as sites for secondary lesions. Toxins liberated from the microorganisms during their growth in such abscesses can freely diffuse to the rest of the body to set up a generalized toxaemia.

Particular strains of salmonellae (section 4.2) such as *Sal. typhi*, *Sal. paratyphi* and *Sal. typhimurium* are able not only to penetrate into intestinal epithelial cells and produce exotoxins but also penetrate beyond into subepithelial tissues. These organisms therefore produce, in addition to the usual symptoms of salmonellosis, a characteristic systemic disease (typhoid and enteric fever). Following recovery from such infection the organism is commonly found associated with the gall bladder. In this state, the recovered person will excrete the organism and form a reservoir for the infection of others.

4.3.2   *Direct passive spread*

Non-invasive pathogens may be spread passively around the body surface or target organ through direct transfer. If the infected epithelial surfaces are wet then the infective agent can often migrate to uninfected surfaces. In the respiratory tract, if ciliary movement is unaffected then the organisms will migrate upward towards the throat; if affected, then they will drain under gravity towards the alveoli. In both cases new infections will be initiated *en route*. Coughing and sneezing do much to redistribute the infective agent over fresh epithelial surfaces. Similar forms of local redistribution occur in the intestine, through the continual churning and flow of the gut contents, and also in the vagina, urethra and conjunctiva. Direct spread can also occur in internal organs and tissues. Thus, infections of the appendix can cause peritoneal inflammation or infection, and deep lung infections can, in turn, infect and inflame the pleural membranes.

4.3.3   *Indirect passive spread*

When invading microorganisms have crossed the epithelial barriers they will almost certainly be taken up in the lymphatic ducts and be delivered to the filtration and immune systems of the local lymph nodes. Sometimes this serves to spread infections further around the body. Eventually, spread may occur from local to regional lymph nodes and thence to the bloodstream. Direct entry to the bloodstream from the primary portal of entry is rare and will only occur when the organism damages the blood vessels or if it is injected directly into them. This might be the case following an insect bite. Bacteraemia such as this will often lead to secondary infections remote from the original portal of entry.

5   **Damage to tissues**

Damage caused to the host organism through infection can be direct and relate to the destructive presence of, or to the production of, toxins by microorganisms in particular target organs; or it can be indirect and relate to interactions of the antigenic components of the pathogen with the host's immune system. Effects can therefore be closely related to, or remote from, the target organ.

Symptoms of the infection can in some instances be highly specific, effecting a precise pharmacological response to a particular toxin; or they might be non-specific and relate to the usual response of the body to particular types of trauma. Damage induced by infection will therefore be considered in these categories.

5.1   Direct damage

5.1.1   *Specific effects*

The consequences of infection to the host depend to a large extent upon the tissue or organ involved. Soft tissue infections of skeletal muscle are likely to be less damaging than, for instance, infections of the heart muscle and central nervous system. Infections of the epithelial cells of small blood vessels can produce anoxia or necrosis in the tissues they supply. Cell and tissue damage is generally the result of direct local action by the microorganisms, usually concerning action at cell membranes. The target cells are usually phagocytic cells and are generally killed (e.g. by *Brucella*, *Listeria*, *Mycobacterium*). Interference with membrane function, through the action of enzymes such as phospholipase, cause the affected cells to leak. When lysosomal membranes are affected, then lysosomal enzymes disperse into the cells and tissues causing them, in turn, to autolyse. This is mediated through the vast battery of enzyme-toxins available to these organisms (section 4). If enough of these toxins are produced to enter the circulation then a generalized toxaemia might result. During their growth, other pathogens liberate toxins with precise pharmacological actions. Diseases mediated in this manner include diphtheria, tetanus and scarlet fever.

In diphtheria, the organism *C. diphtheriae* confines itself to epithelial surfaces of the nose and throat and produces a powerful toxin which affects the elongation factor involved in protein biosynthesis. The heart and peripheral nerves are particularly affected resulting in myocarditis (inflammation of the myocardium) and neuritis (inflammation of a nerve). Little damage is produced at the infective site.

Tetanus occurs when *Cl. tetani*, ubiquitous in the soil and faeces, contaminates wounds, especially deep puncture-type lesions. These might be minor trauma such as a splinter, or major ones such as battle injury. At these sites, tissue necrosis and possibly microbial growth reduce the oxygen tension to allow this anaerobe to multiply. Its growth is accompanied by the production of a highly potent toxin which passes up peripheral nerves and diffuses locally within the central nervous system. It acts like strychnine by affecting normal function at the synapses. Since the motor nerves of the brain stem are the shortest, then the cranial nerves are the first affected, with twitches of the eyes and spasms of the jaw (lockjaw).

A related organism, *Cl. botulinum*, produces a similar toxin which may contaminate food if the organism has grown in it and conditions are favourable for anaerobic growth. Meat pastes and pâtés are likely sources. This toxin interferes with acetylcholine release at cholinergic synapses and also acts at neuromuscular junctions. Death from this toxin eventually results from respiratory failure.

Many other organisms are capable of producing intoxication following their growth on foods. Most common amongst these are the staphylococci and particular strains of *Bacillus* such as *B. cereus*. Staphylococci such as *Staph. aureus* produce an enterotoxin which acts upon the vomiting centres of the brain. Nausea and vomiting therefore follow ingestion of contaminated foods, the delay between eating and vomiting varying between 1 and 6 hours, depending on the amount of toxin ingested. *B. cereus* also produces an emetic toxin but its actions are delayed and vomiting can follow up to 20 hours after ingestion. The latter organism is often associated with rice products and will propagate when the rice is cooked (spore activation) and subsequently reheated after a period of storage.

Scarlet fever is produced following infection with certain strains of *Str. pyogenes*. These produce a potent toxin which causes an erythrogenic skin rash which accompanies the more usual effects of a streptococcal infection.

5.1.2   *Non-specific effects*

If the infective agent damages an organ and affects its functioning, this can manifest itself as a series of secondary disease features. Thus, diabetes may result from an infection of the islets of Langerhans, paralysis or coma might result from infections of the central nervous system, and kidney malfunction might result from loss of tissue fluids and its associated hyperglycaemia. Exotoxins and endotoxins can also be implicated in non-specific symptoms, even when they have fairly well-defined pharmacological actions. Thus, a number of intestinal pathogens (e.g. *V. cholerae*, *E. coli*) produce potent exotoxins which affect vascular permeability. These generally act through adenyl cyclase, raising the intracellular levels of cyclic AMP (adenosine monophosphate). As a result of this the cells lose water and electrolytes to the surrounding medium, the gut lumen. A common consequence of these related, yet distinct, toxins is acute diarrhoea and haemoconcentration. Kidney malfunction might well follow and in severe cases lead to death. Symptomologically there is little difference between these conditions and food poisoning induced by ingestion of staphylococcal enterotoxin. The latter toxin is formed by the organisms during their growth on infected food substances and is absorbed actively from the gut. It acts, not at the epithelial cells of the gut, but at the vomiting centre of the central nervous system causing nausea, vomiting and diarrhoea within 6 hours.

Endotoxins form part of the cell envelopes of some bacterial species (see Chapter 1). They are shed into the surrounding medium during growth and often following autolysis of the infecting organism. They tend to be less toxic than the exotoxins and less precise in their action. Classic endotoxins are the lipopolysaccharide/protein components of Gram-negative cells, i.e. *E. coli* and the salmonellae. Various toxic effects have been attributed to these endotoxins but their role in the establishment of the infection, if any, remain unclear. The most notable effect is their pyrogenicity (Chapters 1 and 18). This relates to release by the endotoxin of endogenous pyrogen from macrophages and phagocytes. Elevation of body temperature follows within 1–2 hours.

<table>
<tr><td>5.2</td><td>

Indirect damage

</td></tr>
</table>

Inflammatory materials are released from necrotic cells and directly from the infective agent. It is not always clear to what extent these can be related to the host or pathogen. Inflammation causes swelling, pain and reddening of the tissues and sometimes loss of function of the organs affected. These reactions may sometimes be the major sign and symptom of the disease.

Many microorganisms minimize the effects against themselves by the host's defence system by mimicking the antigenic structure of the host tissue. The eventual immunological response of the host to infection then leads to the autoimmune destruction of itself. Thus infections with *Mycoplasma pneumoniae* can lead to production of antibody against normal Group O erythrocytes with concomitant haemolytic anaemia.

If antigen, released from the infective agent, is soluble then antigen–antibody complexes are produced. When antibody is present at a concentration equal to or greater than the antigen, such as in the case of an immune host, then these complexes precipitate and are removed by macrophages present in the lymph nodes. When antigen is present in excess the complexes, being small, continue to circulate in the blood and are eventually filtered off by the kidneys, becoming lodged in kidney glomeruli. A localized inflammatory response in the kidneys might then be initiated by the complement system (Chapter 15). Eventually the filtering function of the kidneys becomes impaired, producing symptoms of chronic glomerulonephritis.

6   Recovery from infection: exit of microorganisms

The primary requirement for recovery is that multiplication of the infective agent is brought under control, that it ceases to spread around the body and that the damaging consequences of its presence are arrested. Such control and recovery are brought about by the combined functioning of the phagocytic, immune and complement systems. A successful pathogen will not seriously debilitate its host; rather the continued existence of the host must be ensured in order to maximize the dissemination of the pathogen within the host population. Ideally, the organism must persist within the host for the remainder of its lifespan and be constantly released to the environment. Whilst this is the case for a number of virus infections and for some bacterial ones, it is not common. Generally, recovery from infection is accompanied by complete destruction of the organism and restoration of a sterile tissue. Alternatively, the organism might return to a commensal relationship with the host on the epithelial and skin surface.

Where the infective agent is an obligate pathogen, then a means must exist for it to infect other individuals before its eradication from the host organism. The route of exit is commonly related to the original portal of entry. Thus, pathogens of the intestinal tract are liberated in the faeces and might easily contaminate food and drinking water, whilst infective agents of the respiratory tract might be inhaled during coughing, sneezing or talking, survive in the associated water droplets and infect nearby individuals. Infective agents transmitted by insect and animal vectors may be spread through the same vectors, the insects/animals having been infected by the diseased host. For some 'fragile' organisms (e.g. *N. gonorrhoeae*, *Treponema pallidum*), direct contact transmission is the only means of transmission. In these cases intimate contact between epithelial membranes, such as occurs during sexual contact, is required for transfer to occur. For opportunist pathogens such as those associated with wound infections transfer is less important because the pathogenic role is minor. Rather, the natural habitat of the organism serves as a constant reservoir for infection.

7   Epidemiology of infectious disease

Spread of a microbial disease through a population of individuals can be considered as vertical (transferred from one generation to another) or horizontal (transfer occurring within genetically unrelated groups). The latter can be divided into common-source outbreaks, relating to infection of a number of susceptible individuals from a single reservoir of the infective agent (i.e. infected foods), or propagated-source outbreaks, where each individual provides a new source for the infection of others.

Common-source outbreaks are characterized by a sharp onset of reported cases over the course of a single incubation period and relate to a common experience of the infected individuals. Propagated-source outbreaks, on the other hand, show a gradual increase in reported cases over a number of incubation periods and eventually decline when the majority of susceptibles in the population have been affected. Factors contributing to propagated outbreaks of infectious disease are the infectivity of the agent (I), the population density (P) and the numbers of susceptible individuals in it (F). The likelihood of an epidemic is given by the product of these three factors (i.e. FIP). Changes in any one of them might initiate an outbreak of the disease in epidemic proportions. Thus, reported cases of particular diseases show periodicity, with outbreaks of epidemic proportion occurring only when FIP exceeds certain critical threshold values, related to the infectivity of the agent. Outbreaks of measles and chickenpox therefore tend to occur annually in the late summer amongst children attending school for the first time. This has the effect of concentrating all susceptible individuals in one, often confined, space at the same time. The proportion of susceptibles can be reduced through rigorous vaccination programmes (Chapter 16). Provided that the susceptible population does not exceed the threshold *FIP* value, then herd immunity against epidemic spread of the disease will be maintained.

Certain types of infectious agent (e.g. influenza virus) are able to combat herd immunity such as this through undergoing major antigenic changes. These render the majority of the population susceptible, and their occurrence is often accompanied by spread of the disease across the entire globe (pandemics).

8 Further reading

Bisno A.L. & Waldvogel F.A. (1989) *Infections Associated with Indwelling Medical Devices*. Washington: American Society for Microbiology.

Burnett F.M. & White D.O. (1972) *The Natural History of Infectious Disease*, 4th edn. Cambridge: Cambridge University Press.

Mims C.A. (1987) *The Pathogenesis of Infectious Disease*, 3rd edn. London: Academic Press.

Murray P.R., Drew W.L., Kobayashi G.S. & Thompson J.H. (1980) *Medical Microbiology*. USA: Wolf Publications.

Smith H. (1990) Pathogenicity and the microbe *in vivo*. *J Gen Microbiol*. **136**, 377–393.

Part 2 Antimicrobial agents

The theme of this section is antimicrobial agents; these are considered in three categories: first, antibiotics and *de novo* chemically synthesized chemotherapeutic agents; second, non-antibiotic antimicrobial compounds (disinfectants, antiseptics and preservatives); and third, immunological products. The subjects covered comprise the manufacture, evaluation and properties of antibiotics; antibiotic assays; the evaluation and properties of disinfectants, antiseptics and preservatives; the fundamentals of immunology; and the manufacture and quality control of immunological products. The mechanisms of action of antibiotics and non-antibiotic agents are also considered, together with an account of the ever-present problem of natural and acquired resistance. The principles involved in the clinical uses of antimicrobial drugs are discussed in Chapter 6.

Overall, the main non-clinical facets of the agents used for the treatment of microbial infection are covered, with reading material for amplification and extension of the topics.

Problems of recent years involving listeriosis, salmonellosis, giardiasis and Legionnaire's disease have received attention.

5 Types of antibiotics and synthetic antimicrobial agents

1	**Antibiotics**		**9**	**Miscellaneous antibacterial antibiotics**
1.1	Definition		9.1	Chloramphenicol
1.2	Sources		9.2	Fusidic acid
			9.3	Lincomycins
2	**β-lactam antibiotics**		9.4	Mupirocin (pseudomonic acid A)
2.1	Penicillins and mecillinams			
2.2	Cephalosporins		**10**	**Antifungal antibiotics**
2.3	Clavams		10.1	Griseofulvin
2.4	1-oxacephems		10.2	Polyenes
2.5	1-carbapenems			
2.5.1	Olivanic acids		**11**	**Synthetic antimicrobial agents**
2.5.2	Thienamycin and imipenem		11.1	Sulphonamides
2.6	Nocardicins		11.2	Diaminopyrimidine derivatives
2.7	Monobactams		11.3	Dapsone
2.8	Other β-lactams		11.4	Antitubercular compounds
2.9	Hypersensitivity		11.5	Nitrofuran compounds
			11.6	4-quinolone antibacterials
3	**Tetracylines**		11.7	Imidazole derivatives
			11.8	Flucytosine
4	**Rifamycins**		11.9	Terbinafine
			11.10	Newer antifungal drugs
5	**Aminoglycoside–aminocyclitol antibiotics**			
			12	**Antiviral drugs**
6	**Macrolides**			
			13	**Drug combinations**
7	**Polypeptide antibiotics**			
			14	**Further reading**
8	**Glycopeptide antibiotics**			
8.1	Vancomycin			
8.2	Teicoplanin			

1 Antibiotics

1.1 Definition

An antibiotic was originally defined as a substance, produced by one microorganism, which inhibited the growth of other microorganisms. The advent of synthetic methods has, however, resulted in a modification of this definition and an antibiotic now refers to a substance produced by a microorganism, or to a similar substance (produced wholly or partly by chemical synthesis), which in low concentrations inhibits the growth of other microorganisms. Chloramphenicol was an early example. Antimicrobial agents such as sulphonamides (section 11.1) and the 4-quinolones (section 11.6), produced solely by synthetic means, are often referred to as antibiotics.

99

Sources

There are three major sources from which antibiotics are obtained.

1 Microorganisms. For example, bacitracin and polymyxin are obtained from some *Bacillus* species; streptomycin, tetracyclines, etc. from *Streptomyces* species; gentamicin from *Micromonospora purpurea*; griseofulvin and some penicillins and cephalosporins from certain genera (*Penicillium*, *Acremonium*) of the family Aspergillaceae; and monobactams from *Pseudomonas acidophila* and *Gluconobacter* species. Most antibiotics in current use have been produced from *Streptomyces* spp.

2 Synthesis. Chloramphenicol is now usually produced by a synthetic process.

3 Semisynthesis. This means that part of the molecule is produced by a fermentation process using the appropriate microorganism and the product is then further modified by a chemical process. Many penicillins and cephalosporins (section 2) are produced in this way.

2 *β*-lactam antibiotics

There are several different types of *β*-lactam antibiotics that are valuable, or potentially important, antibacterial compounds. These will be considered briefly.

2.1 Penicillins and mecillinams

The penicillins (general structure, Fig. 5.1A) may be considered as being of the following types.

1 Naturally occurring. For example, produced by fermentation of moulds such as *Penicillium notatum* and *P. chrysogenum*. The most important examples are benzylpenicillin (penicillin G) and phenoxymethylpenicillin (penicillin V).

2 Semisynthetic. In 1959, scientists at Beecham Research Laboratories succeeded in isolating the penicillin 'nucleus', 6-aminopenicillanic acid (6-APA; Fig. 5.1A: R represents H). During the commercial production of benzylpenicillin, phenylacetic (phenylethanoic) acid ($C_6H_5.CH_2.COOH$) is added to the medium in which the *Penicillium* mould is growing (see

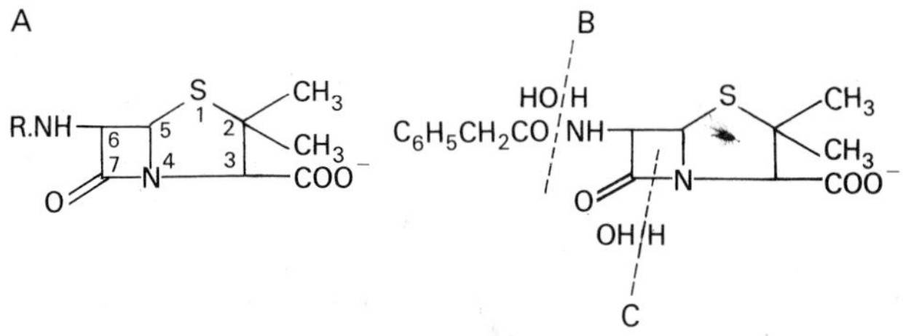

Fig. 5.1 A, General structure of penicillins; B, removal of side chain from benzylpenicillin; C, site of action of *β*-lactamases.

Table 5.1 The penicillins and mecillinams

Penicillin	Orally effective	Stability to β-lactamases from Staph. aureus	Gram −ve	Activity versus Gram −ve*	Ps. aeruginosa	Ester	Hydrolysed after absorption
1 Benzylpenicillin	−	−	−	−	−	−	−
2 Phenoxymethylpenicillin	+	−	−	−	−	−	−
3 Methicillin	−	+	+	−	−	−	−
4 Oxacillin	+	+	+	−	−	−	−
5 Cloxacillin	+	+	+	−	−	−	−
6 Flucloxacillin	+	+	+	−	−	−	−
7 Ampicillin	+	−	−	+	−	−	−
8 Amoxycillin	+	−	−	+	−	−	−
9 Carbenicillin	−	−	+	+	+	−	−
10 Ticarcillin	−	−	+	+	+	−	−
11 Temocillin	+	+	+	+	+	−	−
12 Carfecillin ⎫ Carbenicillin	+	−	+	+	+	+	+
13 Indanyl carbenicillin ⎬ esters (carindacillin) ⎭	+	−	+	+	+	+	+
14 Pivampicillin ⎫ Ampicillin	+	−	−	+	−	+	+
15 Talampicillin ⎬ esters	+	−	−	+	−	+	+
16 Bacampicillin ⎭	+	−	−	+	−	+	+
17 Piperacillin ⎫ Substituted	−	−	−	+	+	−	−
18 Azlocillin ⎬ ampicillins	−	−	−	+	+	−	−
19 Mezlocillin ⎭	−	−	−	+	+	−	−
20 Mecillinam ⎫ 6-β-amidino-	−	NA	V	+	−	−	−
21 Pivmecillinam ⎭ penicillins	+	NA	V	+	−	+	+

* Except Ps. aeruginosa. All penicillins show some degree of activity against Gram-negative cocci.
+, applicable. −, inapplicable. NA, does not apply: mecillinam and pivmecillinam have no effect on Gram-positive bacteria; V, variable.
Note: 1 Esters give high urinary levels. 2 Hydrolysis of these esters by enzyme action after absorption from the gut mucosa gives rapid and high blood levels. 3 For additional information on resistance to β-lactamase inactivation, see Chapter 10. 4 In general, all penicillins are active against Gram-positive bacteria, although this may depend on the resistance of the drug to β-lactamase (see column 3); thus, benzylpenicillin is highly active against strains of *Staphylococcus aureus* which do not produce β-lactamase, but is destroyed by β-lactamase-producing strains. 5 Temocillin (number 11) is less active against Gram-positive bacteria than ampicillin or the ureidopenicillins (substituted ampicillins).

Chapter 7). This substance is a precursor of the side chain (R; see Fig. 5.2) in benzylpenicillin. Growth of the organism in the absence of phenylacetic acid led to the isolation of 6-APA; this has a different R_F value from benzylpenicillin which allowed it to be detected chromatographically.

A second method of producing 6-APA came with the discovery that certain microorganisms produce enzymes, penicillin amidases (acylases), which catalyse the removal of the side chain from benzylpenicillin (Fig. 5.1B).

Acylation of 6-APA with appropriate substances results in new penicillins being produced which differ only in the nature of the side chain (Table 5.1; Fig. 5.2). Some of these penicillins have considerable activity against Gram-negative as well as Gram-positive bacteria, and are thus

Drug	R	Drug	R	Drug	R
1	$C_6H_5CH_2CO-$	2	$C_6H_5OCH_2CO-$	3	(2,6-dimethoxyphenyl) $-OCH_3$ / $-OCH_3$
4	isoxazolyl structure with CH_3	5	Cl-phenyl isoxazolyl with CH_3	6	Cl,F-phenyl isoxazolyl with CH_3
7	$CH.CO-$ / NH_2 (phenyl)	8	$HO-$phenyl$-CH.CO-$ / NH_2	9	phenyl$-CH.CO-$ / $COONa$
10	thienyl $-CH.CO-$ / $COONa$	11	thienyl $-CH.CO-$ / $COONa$	12	phenyl$-CH.CO-$ / CO / O-indanyl
13	phenyl$-CH.CO-$ / CO / O-indanyl	14	phenyl$-CH.CO-$ / NH_2 At 3: $COOCH_2O.C.C(CH_3)_3$ / O	15	phenyl$-CH.CO-$ / NH_2 At 3: $COO-CH$ phthalidyl
16	phenyl$-CH.CO-$ / NH_2 At 3: $COOCH_2CH_2O.COOC_2H_5$ / $COOCH(CH_3)_3O.COOC_2H_5$	17	phenyl$-CH.CO-$ / NH / CO / piperazinedione	18	phenyl$-CH.CO-$ / NH / CO / pyrazolidinone
19	phenyl$-CH.CO-$ / NH / CO / imidazolidinone $N-SO_2.CH_3$	20	$CH_2-CH_2-CH_2$ / $CH_2-CH_2-CH_2$ $N.CH=$	21	$CH_2-CH_2-CH_2$ / $CH_2-CH_2-CH_2$ $N.CH=$ At 3: $COOCH_2O.C.C(CH_3)_3$ / O

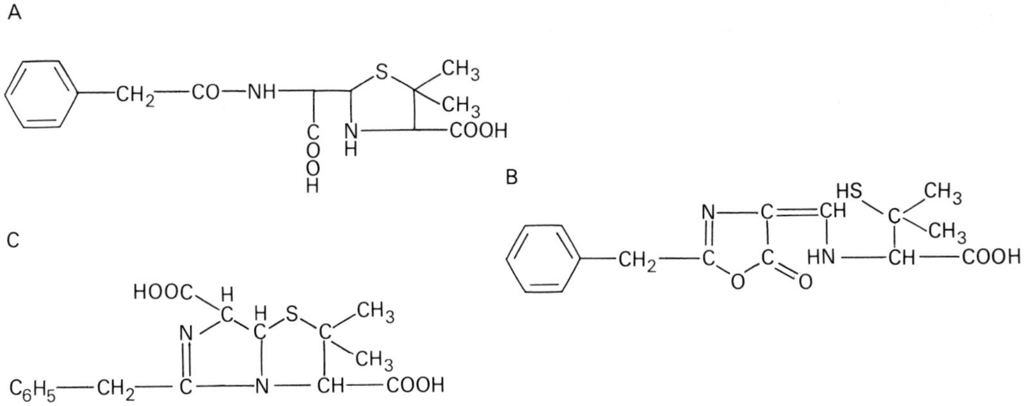

A

C

B

Fig. 5.3 Degradation products of benzylpenicillin in solution: A, penicilloic acid; B, penicillenic acid; C, penillic acid.

broad-spectrum antibiotics. Pharmacokinetic properties may also be altered.

The sodium and potassium salts are very soluble in water but they are hydrolysed in solution, at a temperature-dependent rate, to the corresponding penicilloic acid (Fig. 5.3A; see also Fig. 10.3), which is not antibacterial. Penicilloic acid is produced at alkaline pH or (via penicillenic acid; Fig. 5.3B) at neutral pH, but at acid pH a molecular rearrangement occurs, giving penillic acid (Fig. 5.3C). Instability in acid medium logically precludes oral administration, since the antibiotic may be destroyed in the stomach; for example at pH 1.3 and 35°C methicillin has a half-life of only 2–3 minutes and is therefore not administered orally, whereas ampicillin, with a half-life of 600 minutes, is obviously suitable for oral use.

Benzylpenicillin is rapidly absorbed and rapidly excreted. However, certain sparingly soluble salts of benzylpenicillin (benzathine, benethamine and procaine) slowly release penicillin into the circulation over a period of time, thus giving a continuous high concentration in the blood. Simultaneous administration of benzylpenicillin (see Fortified Procaine Penicillin, BP) may be given initially.

Pro-drugs (e.g. carbenicillin esters, ampicillin esters; Fig. 5.2, Table 5.1) are hydrolysed by enzyme action after absorption from the gut mucosa to produce high blood levels of the active antibiotic, carbenicillin and ampicillin, respectively.

Several bacteria produce an enzyme, β-lactamase (penicillinase; see Chapter 10) which may inactivate a penicillin by opening the β-lactam

Fig. 5.2 Examples of the side chain R in various penicillins (the numbers 1–19 correspond to those in Table 5.1). Numbers 20 (mecillinam) and 21 (pivmecillinam) are 6-β-amidinopenicillanic acids (mecillinams). Number 11 (temocillin) has a methoxy, OCH_3 group at position 6α: this confers high β-lactamase stability on the molecule.

103   *Types of antibiotics*

ring, as in Fig. 5.1C. However, some penicillins (Table 5.1) are considerably more resistant to this enzyme than are others, and consequently may be extremely valuable in the treatment of infections caused by β-lactamase-producing bacteria. In general, the penicillins are active against Gram-positive bacteria; some members (e.g. ampicillin) are also effective against Gram-negative bacteria though not *Ps. aeruginosa*, whereas others (e.g. carbenicillin) are active against this organism also. In particular, substituted ampicillins (piperacillin and the ureidopenicillins, azlocillin and mezlocillin) appear to combine the properties of ampicillin and carbenicillin. Temocillin is the first penicillin to be completely stable to hydrolysis by β-lactamases produced by Gram-negative bacteria.

The 6-β-amidinopenicillanic acids, mecillinam and its ester pivmecillinam, have unusual antibacterial properties, since they are active against Gram-negative but not Gram-positive organisms.

2.2	**Cephalosporins**

In the 1950s, a species of *Cephalosporium* (now known as *Acremonium*: see Chapter 7) isolated near a sewage outfall off the Sardinian coast was studied at Oxford and found to produce the following antibiotics:
1 An acidic antibiotic, cephalosporin P (subsequently found to have a steroid-like structure).
2 Another acidic antibiotic, cephalosporin N (later shown to be a penicillin, since its structure was based on 6-APA).
3 Cephalosporin C, obtained during the purification of cephalosporin N; this is a true cephalosporin, and from it has been obtained 7-aminocephalosporanic acid (7-ACA; Fig. 5.4), the starting point for new cephalosporins (see Chapter 7).

Cephalosporins consist of a six-membered dihydrothiazine ring fused to a β-lactam ring. Detailed studies of the cephalosporin molecule have been made, so that it is now possible to modify this structure to increase antibacterial activity. The activity of cephalosporins (and other β-lactams) against Gram-positive bacteria depends on antibiotic affinity for penicillin-sensitive enzymes (PSEs) detected in practice as penicillin-binding proteins (PBPs). Resistance results from altered PBPs or, more commonly, from β-lactamases. Activity against Gram-negative bacteria depends upon penetration of β-lactams through the outer membrane, resistance to β-lactamases found in the periplasmic space and binding to PBPs. (For further information on mechanisms of action and bacterial resistance, see Chapters 9 and 10.) Modifications of the cephem nucleus (Fig. 5.4) at 7α, i.e. R^3, by addition of methoxy groups increase β-lactamase stability but decrease activity against Gram-positive bacteria because of reduced affinity for PBPs. Side-chains containing a 2-aminothiazolyl group at R^1, e.g. cefotaxime, ceftizoxime, ceftriaxone and ceftazidime, yield cephalosporins with enhanced affinity for PBPs of Gram-negative bacteria and streptococci. An iminomethoxy group ($-C{=}N\cdot OCH_3$) in, for example, cefuroxime provides β-lactamase stability against common plasmid-mediated β-lactamases. A propylcarboxy group (($CH_3)_2$—C—COOH) in, for example, ceftazidime increases β-lactamase resistance and also provides activity against *Ps. aeruginosa*, whilst at the same time reducing β-lactamase induction capabilities.

Pharmacological properties may also be modified by changes in chemical structure. Some cephalosporins, for example cephalothin, cephapirin and cephacetrile, are metabolically unstable because members with a 3-acetoxymethyl group are converted by esterases *in vivo* to the antibacterially less active 3-hydroxymethyl derivatives. Cefuroxime atexil, a pro-drug, can be given orally. It is an ester of cefuroxime.

In cephalosporins susceptible to β-lactamases, opening of the β-lactam ring occurs with concomitant loss of the substituent at R^2 (except in cephalexin, where R^2 represents H; see Fig. 5.4). This is followed by fragmentation of the molecule. Provided that they are not inactivated by β-lactamases, the cephalosporins generally have a broad spectrum of activity, although there may be a wide variation. *Haemophilus influenzae*, for example, is particularly susceptible to cefuroxime.

Chemical structures of cephalosporins are provided in Fig. 5.4. Their properties are summarized in Table 5.2.

2.3 Clavams

The clavams differ from penicillins in two respects, namely the replacement of S in the penicillin thiazolidine ring (Fig. 5.1) with oxygen in the clavam oxazolidine ring (Fig. 5.5A) and the absence of the side chain at position 6. Clavulanic acid, a naturally occurring clavam isolated from *Streptomyces clavuligerus*, has poor antibacterial activity but is a potent inhibitor of staphylococcal β-lactamase and of most types of β-lactamases produced by Gram-negative bacteria, especially those with a 'penicillinase' rather than a 'cephalosporinase' type of action.

A significant development in chemotherapy has been the introduction into clinical practice of a combination of clavulanic acid with a broad-spectrum, but β-lactamase-susceptible, penicillin, amoxycillin. The spectrum of activity has been extended to include *Ps. aeruginosa* by combining clavulanic acid with the β-lactamase-susceptible penicillin, ticarcillin, under the name Timentin.

2.4 1-oxacephems

In the 1-oxacephems, for example latamoxef (moxalactam, Fig. 5.5B), the sulphur atom in the dihydrothiazine cephalosporin ring system is replaced by oxygen. This would tend to make the molecule chemically less stable and more susceptible to inactivation by β-lactamases. The introduction of the 7-α-methoxy group (as in cefoxitin, Fig. 5.4), however, stabilizes the molecule. Latamoxef is a broad-spectrum antibiotic with a high degree of stability to most types of β-lactamases, and is highly active against the anaerobe, *B. fragilis*.

R³

R¹—NH—[β-lactam/cephem core with positions 1,2,3,4,5,6,7,8, S, N5, O, COO⁻]—CH₂—R²

Cephalosporin	R¹	R²
7-ACA	H—	—O.CO.CH₃
Cephacetrile	$N\equiv C.CH_2.CO-$	—O.CO.CH₃
Cephaloridine	[thiophene]—CH₂CO—	—N⁺[pyridine]
Cephalexin	[phenyl]—CH.CO— (NH₂)	—H
Cefuroxime	[furan]—C.CO— (=N.O.CH₃)	—O.CO.NH₂
Cefoxitin	[thiophene]—CH₂CO—	—O.CO.NH₂
Cefaclor	[phenyl]—CH.CO— (NH₂)	—Cl
Cephalothin	[thiophene]—CH₂CO—	—OCOCH₃
Cephapirin	[pyridine-N]—S.CH₂CO—	—OCOCH₃
Cefsulodin	[phenyl]—CH— (SO₃Na)	—CH₂—N⁺[pyridine-C(=O)NH₂]
Cefazolin	[tetrazole N=N/N—N]—CH₂CO—	—S[1,3,4-thiadiazole]—CH₃
Cephradine	[cyclohexadienyl]—CH.CO— (NH₂)	—H
Cefamandole	[phenyl]—CH.CO— (OH)	—S[tetrazole]—N—CH₃

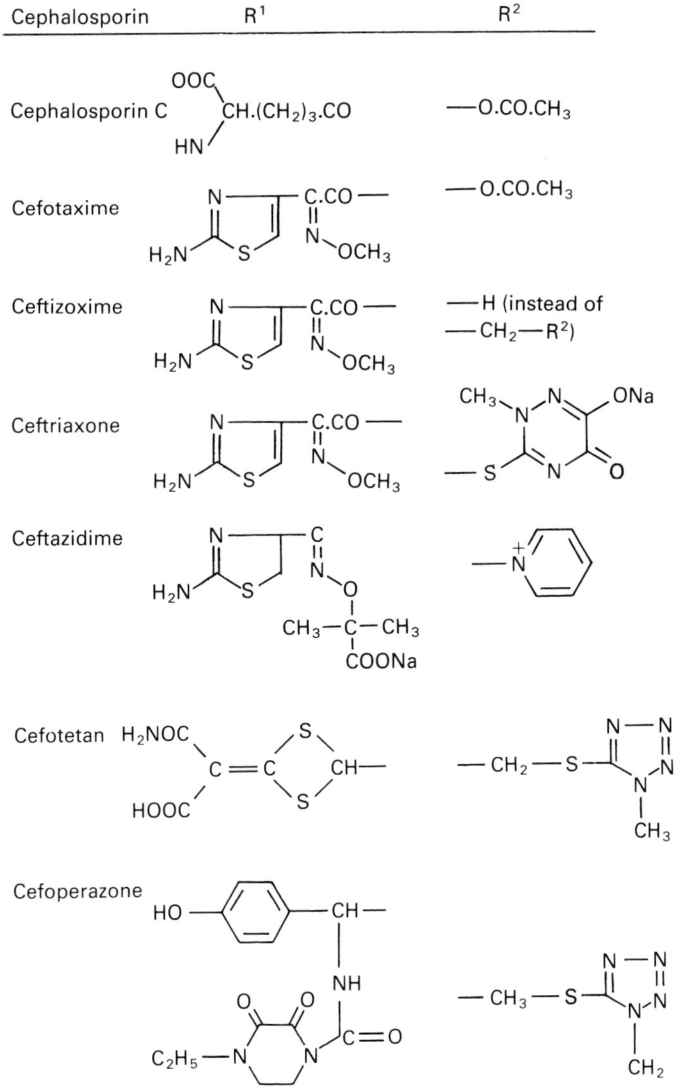

Cephalosporin R^1 R^2

Cephalosporin C

OOC$\diagdown$
$\quad$CH.(CH$_2$)$_3$.CO
HN$\diagup$

—O.CO.CH$_3$

Cefotaxime

N——C.CO—
$\quad\parallel$
H$_2$N—S—N
$\qquad\diagdown$OCH$_3$

—O.CO.CH$_3$

Ceftizoxime

N——C.CO—
$\quad\parallel$
H$_2$N—S—N
$\qquad\diagdown$OCH$_3$

—H (instead of
—CH$_2$—R^2)

Ceftriaxone

N——C.CO—
$\quad\parallel$
H$_2$N—S—N
$\qquad\diagdown$OCH$_3$

CH$_3$—N—N—ONa
—S—N—O

Ceftazidime

N——C
$\quad\parallel$
H$_2$N—S—N
$\qquad\diagdown$O
CH$_3$—C—CH$_3$
$\qquad$COONa

—N$^+$⟨pyridine⟩

Cefotetan

H$_2$NOC$\diagdown$$\qquad$S
$\qquad$C=C$\quad$CH—
HOOC$\diagup$$\qquad$S

—CH$_2$—S—⟨N—N / N—N⟩
$\qquad\qquad$CH$_3$

Cefoperazone

HO—⟨benzene⟩—CH—
$\qquad\qquad$NH
$\qquad\qquad$C=O
O$\quad$O
C$_2$H$_5$—N—N

—CH$_3$—S—⟨N—N / N—N⟩
$\qquad\qquad$CH$_2$

Fig. 5.4 (*Opposite and above*) General structure of cephalosporins and examples of side chains R^1 and R^2. (R^3 is —OCH$_3$ in cefoxitin and cefotetan and —H in other members). Cephalosporins containing an ester group at position 3 are liable to attack by esterases *in vivo*.

2.5 1-carbapenems

The 1-carbapenems (Fig. 5.5C) comprise a new family of fused β-lactam antibiotics. They are analogues of penicillins or clavams, the sulphur (penicillins) or oxygen (clavams) atom being replaced by carbon. Examples are the olivanic acids (section 2.5.1) and thienamycin and imipenem (section 2.5.2).

107 *Types of antibiotics*

Table 5.2 The cephalosporins*

Group	Examples	Properties Staphylococci†	Staphylococcal β-lactamase	Streptococci‡	Enterobacteria	Enterobacterial β-lactamases	Neisseria	Haemophilus	Ps. aeruginosa	Comment
Oral cephalosporins	Cephalexin, cephradine, cefaclor, cefadroxil	++	++	+	V	V	+	(+)	R	
Injectable cephalosporins (β-lactamase-susceptible)	Cephaloridine, cephalothin, cephacetrile, cefazolin	++	+	+	V	V	+	(+)	R	
Injectable cephalosporins (improved β-lactamase stability)	Cefuroxime, cefoxitin, cefamandole	++		++	++	++	++	++	R	Cefoxitin shows activity against *Bacteroides fragilis*
Injectable cephalosporins (still higher β-lactamase stability)	Cefotaxime, ceftazidime, ceftizoxime, ceftriaxone (also the oxacephem, latamoxef: section 2.4)	++		+++	+++	+++	+++	+++	R (ceftazidime +++)	Latamoxef has high activity against *B. fragilis*
Injectable cephalosporins (anti-pseudomonal activity)	Cefoperazone Cefsulodin	++ (+)	++ ++	+ (+)	V	V	++	++ R	++ +++	
Injectable cephalosporins (other)	Cefotetan	(+)			+++	+++			R	Inhibits *B. fragilis*

*Early cephalosporins were spelt with 'ph', more recently with 'f'. †Methicillin-resistant *Staph. aureus* (MRSA) strains are resistant to cephalosporins. ‡Enterococci are resistant to cephalosporins. +++, excellent; ++, good; +, fair; (+), poor; R, resistant; V, variable.

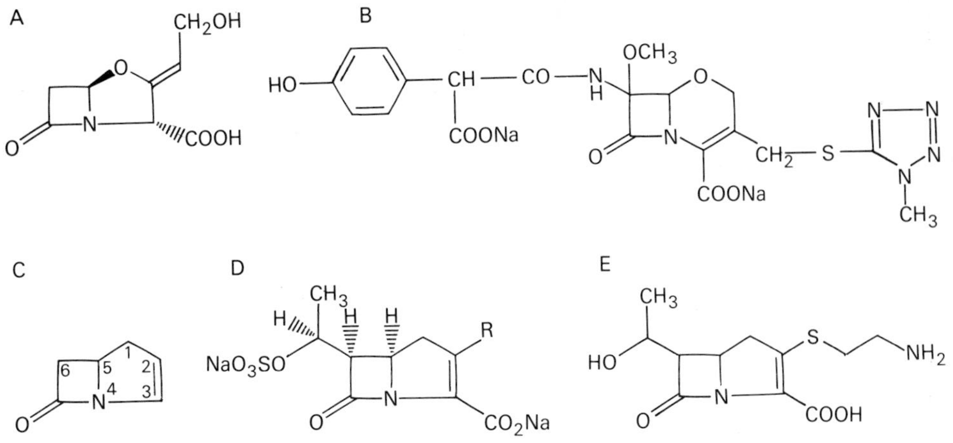

Fig. 5.5 A, Clavulanic acid; B, latamoxef; C, 1-carbapenems; D, olivanic acids (general structure); E, thienamycin.

2.5.1 Olivanic acids

The olivanic acids (general structure, Fig. 5.5D) are naturally occurring β-lactam antibiotics which have, with some difficulty, been isolated from culture fluids of *Str. olivaceus*. They are broad-spectrum antibiotics and are potent inhibitors of various types of β-lactamases.

2.5.2 Thienamycin and imipenem

Thienamycin (Fig. 5.5E) is a broad-spectrum β-lactam antibiotic with high β-lactamase resistance. Unfortunately, it is chemically unstable, although the *N*-formimidoyl derivative, imipenem, overcomes this defect. Imipenem is stable to most β-lactamases but is readily hydrolysed by kidney dehydropeptidase and is administered with a dehydropeptidase inhibitor, cilastatin, under the name Primaxin.

2.6 Nocardicins

The nocardicins (A to G) have been isolated from a strain of *Nocardia* and comprise a novel group of β-lactam antibiotics (Fig. 5.6A). Nocardicin A is the most active member, and possesses significant activity against Gram-negative but not Gram-positive bacteria.

2.7 Monobactams

The monobactams are monocyclic β-lactam antibiotics produced by various strains of bacteria. A novel nucleus, 3-aminomonobactamic acid (3-AMA, Fig. 5.6B), has been produced from naturally occurring monobactams and from 6-APA. Several monobactams have been tested and one (aztreonam, Fig. 5.6C) has been shown to be highly active

109   *Types of antibiotics*

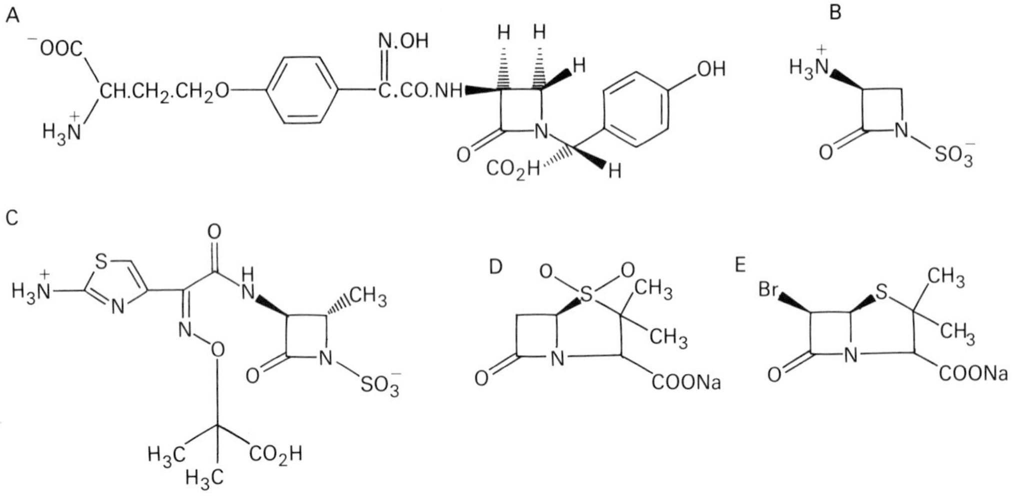

Fig. 5.6 A, Nocardicin A; B, 3-aminomonobactamic acid (3-AMA); C, aztreonam; D, penicillanic acid sulphone (sodium salt); E, β-bromopenicillanic acid (sodium salt).

against most Gram-negative bacteria and to be stable to most types of β-lactamases. It is not destroyed by staphylococcal β-lactamases but is inactive against all strains of *Staph. aureus* tested. *B. fragilis*, a Gram-negative anaerobe, is resistant to aztreonam, probably by virtue of the β-lactamase it produces, and this conclusion is supported by the finding that a combination of the monobactam with clavulanic acid (section 2.3) is ineffective against this organism.

2.8   **Other β-lactams**

Penicillanic acid derivatives are synthetically produced β-lactamase inhibitors. Examples are sulbactam, penicillanic acid sulphone (Fig. 5.6D) and β-bromopenicillanic acid (Fig. 5.6E) which inhibit some, but not all, types of β-lactamases.

2.9   **Hypersensitivity**

Some types of allergic reaction, for example immediate or delayed-type skin allergies, serum-sickness-like reactions and anaphylactic reactions, may occur in a proportion of patients given penicillin treatment. There is some, but not complete, cross-allergy with cephalosporins.

Contaminants of high molecular weight (considered to have arisen from mycelial residues from the fermentation process) may be responsible for the induction of allergy to penicillins; their removal leads to a marked reduction in the antigenicity of the penicillin. It has also been found, however, that varying amounts of a non-protein polymer (of unknown source) may also be present in penicillin and that this also may be antigenic.

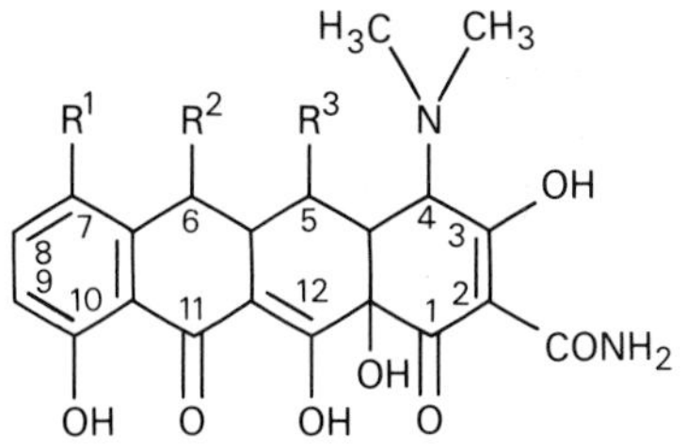

Fig. 5.7 Tetracycline antibiotics: 1, oxytetracycline; 2, chlortetracycline; 3, tetracycline; 4, demethylchlortetracycline; 5, doxycycline; 6, methacycline; 7, clomocycline; 8, minocycline; 9, thiacycline (a thiatetracycline with a sulphur atom at 6).

Drug	R^1	R^2	R^3	Drug	R^1	R^2	R^3
1	H	OH CH₃	OH	2	Cl	OH CH₃	H
3	H	OH CH₃	H	4	Cl	OH H	H
5	H	CH_3	OH	6	H	=CH_2	OH
7	Cl	OH CH₃	H	8	CH₃ CH₃ N	H_2	H
	(At 2: $CONHCH_2OH$)			9	H	—	H

The interaction of a non-enzymatic degradation product, D-benzylpenicillenic acid (formed by cleavage of the thiazolidine ring of benzylpenicillin in solution; Fig. 5.3B), with sulphydryl or amino groups in tissue proteins, to form hapten–protein conjugates, is also of importance. In particular, the reaction between D-benzylpenicillenic acid and the ε-amino group of lysine (α,ε-diamino-n-caproic acid, $NH_2(CH_2)_4 . CH(NH_2) . COOH$) residues is to be noted, because these D-benzylpenicilloyl derivatives of tissue proteins function as complete penicillin antigens.

3 **Tetracyclines**

The tetracyclines (Fig. 5.7) consist of some eight members and may be considered as a group of antibiotics obtained as by-products from the metabolism of various species of *Streptomyces*, although some members may now be thought of as being semisynthetic. Thus, tetracycline (by catalytic hydrogenation) and clomocycline are obtained from chlortetracycline, which is itself produced from *Str. aureofaciens*. Methacycline is obtained from oxytetracycline (produced from *Str. rimosus*) and hydrogenation of methacycline gives doxycycline. Demethylchlortetracycline is produced by a mutant strain of *Str. aureofaciens*. Minocycline is a derivative of tetracycline.

The tetracyclines are broad-spectrum antibiotics, i.e. they have a wide range of activity against Gram-positive and Gram-negative bacteria. *Ps. aeruginosa* is less sensitive, but is generally susceptible to tetracycline concentrations obtainable in the bladder. Resistance to the tetracyclines (see also Chapter 10) develops relatively slowly, but there is cross-resistance, i.e. an organism resistant to one member is usually resistant to all other members of this group. However, tetracycline-resistant *Staph. aureus* strains may still be sensitive to minocycline. Suprainfection ('overgrowth') with naturally tetracycline-resistant organisms, for example *Candida albicans* and other yeasts, and filamentous fungi, affecting the mouth, upper respiratory tract or gastrointestinal tract, may occur as a result of the suppression of tetracycline-susceptible microorganisms.

Thiatetracyclines contain a sulphur atom at position 6 in the molecule. One derivative, thiacycline, is more active than minocycline against tetracycline-resistant bacteria. Despite toxicity problems affecting its possible clinical use, thiacycline could be the starting point in the development of a new range of important tetracycline-type antibiotics.

4 Rifamycins

The rifamycins comprise a comparatively new antibiotic group and consist of rifamycins A to E. From rifamycin B are produced rifamide (rifamycin B diethylamide) and rifamycin SV, which is one of the most useful and least toxic of the rifamycins.

Rifampicin (Fig. 5.8), a bactericidal antibiotic, is active against Gram-positive bacteria (including *Mycobacterium tuberculosis*) and some Gram-

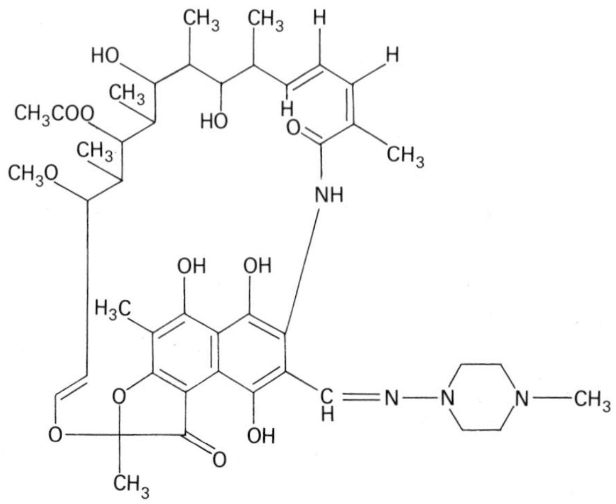

Fig. 5.8 Rifampicin.

negative bacteria (but not Enterobacteriaceae or pseudomonads). It has been found to have a greater bactericidal effect against *M. tuberculosis* than other antituberculosis drugs, is active orally, penetrates well into cerebrospinal fluid and is thus of use in the treatment of tuberculous meningitis (see also section 11.4).

Rifampicin possesses significant bactericidal activity at very low concentrations against staphylococci. Unfortunately, resistant mutants may arise very rapidly, both *in vitro* and *in vivo*. It has thus been recommended that rifampicin should be combined with another antibiotic, e.g. vancomycin, in the treatment of staphylococcal infections.

5   Aminoglycoside–aminocyclitol antibiotics

Aminoglycoside antibiotics contain amino sugars in their structure. Deoxystreptamine-containing members are neomycin, framycetin, gentamicin, kanamycin, tobramycin, amikacin, netilmicin and sisomicin. Both streptomycin and dihydrostreptomycin contain streptidine, whereas the aminocyclitol spectinomycin has no amino sugar. Examples of chemical structures are provided in Fig. 5.9.

Streptomycin was isolated by Waksman in 1944, and its activity against *M. tuberculosis* ensured its use as a primary drug in the treatment of tuberculosis. Unfortunately, its ototoxicity and the rapid development of resistance have tended to modify its usefulness, and although it still remains a front-line drug against tuberculosis it is usually used in combination with isoniazid and $p(4)$-aminosalicylic acid (section 11.4). Streptomycin also shows activity against other types of bacteria, for example against various Gram-negative bacteria and some strains of staphylococci. Dihydrostreptomycin has a similar antibacterial action but is more toxic.

Gentamicin (a mixture of three components, C_1, C_{1a} and C_2) is active against many strains of Gram-positive and Gram-negative bacteria, including some strains of *Ps. aeruginosa*. Its activity is greatly increased at pH values of about 8. It is often administered in conjunction with carbenicillin to delay the development of resistance. Gentamicin is the most important aminoglycoside antibiotic, is the aminoglycoside of choice in the UK and is widely used for treating serious infections. As with other members of this group, side-effects are dose related, dosage must be given with care, plasma levels should be monitored and treatment should not normally exceed 7 days.

Kanamycin (a complex of three antibiotics, A, B and C) is active in low concentrations against various Gram-positive (including penicillin-resistant staphylococci) and Gram-negative bacteria. It is a recognized second-line drug in the treatment of tuberculosis.

Paromomycin finds special use in the treatment of intestinal amoebiasis (it is amoebicidal against *Entamoeba histolytica*) and of acute bacillary dysentery.

Neomycin is poorly absorbed from the alimentary tract when given

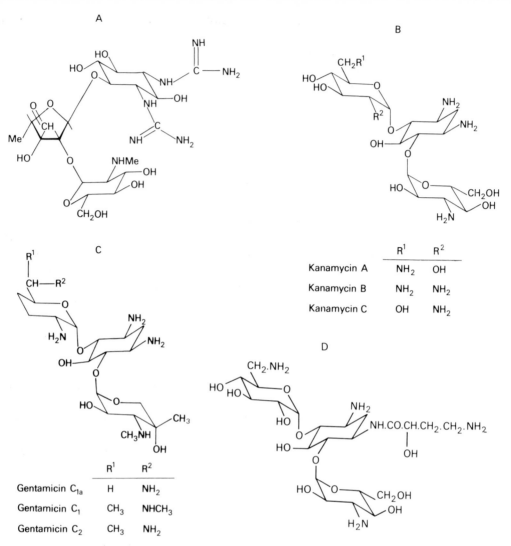

Fig. 5.9 Some aminoglycoside antibiotics: A, streptomycin; B, kanamycins; C, gentamicins; D, amikacin.

	R^1	R^2
Kanamycin A	NH$_2$	OH
Kanamycin B	NH$_2$	NH$_2$
Kanamycin C	OH	NH$_2$

	R^1	R^2
Gentamicin C$_{1a}$	H	NH$_2$
Gentamicin C$_1$	CH$_3$	NHCH$_3$
Gentamicin C$_2$	CH$_3$	NH$_2$

orally, and is usually used in the form of lotions and ointments for topical application against skin and eye infections. Framycetin consists of neomycin B with a small amount of neomycin C, and is usually employed locally.

A desirable property of newer aminoglycoside antibiotics is increased antibacterial activity against resistant strains, especially improved stability to aminoglycoside-modifying enzymes (Chapter 10). Alteration in the 3' position of kanamycin B (Fig. 5.9B) to give 3'-deoxykanamycin B (tobramycin) changes the activity spectrum. Amikacin (Fig. 5.9D) has a substituted aminobutyryl in the amino group at position 1 in the 2-deoxystreptamine ring and this enhances its resistance to some, but not all, types of aminoglycoside-modifying enzymes, as it has fewer sites of modification. Netilmicin (*N*-ethylsisomicin) is a semisynthetic derivative of sisomicin but is less susceptible than sisomicin to some types of bacterial enzymes.

The most important of these antibiotics are amikacin, tobramycin, netilmicin and especially gentamicin.

6   Macrolides

The macrolide antibiotics are characterized by possessing molecular structures that contain large lactone rings linked through glycosidic bonds with amino sugars.

The most important members of this group are erythromycin (Fig. 5.10), oleandomycin, triacetyloleandomycin and spiramycin. Erythromycin is active against most Gram-positive bacteria, *Neisseria*, *H. influenzae* and *Legionella pneumophila*, but not against the Enterobacteriaceae; its activity is pH dependent, increasing with pH up to about 8.5. Erythromycin estolate is more stable than the free base to the acid of gastric juice and is thus employed for oral use. The estolate produces higher and more prolonged blood levels and distributes into some tissues more efficiently than other dosage forms. *In vivo*, it hydrolyses to give the free base.

Staph. aureus is less sensitive to erythromycin than are pneumococci or haemolytic streptococci, and there may be a rapid development of resistance, especially of staphylococci, *in vitro*. However, *in vivo* with successful short courses of treatment, resistance is not usually a serious clinical problem. On the other hand, resistance is likely to develop when the antibiotic is used for long periods.

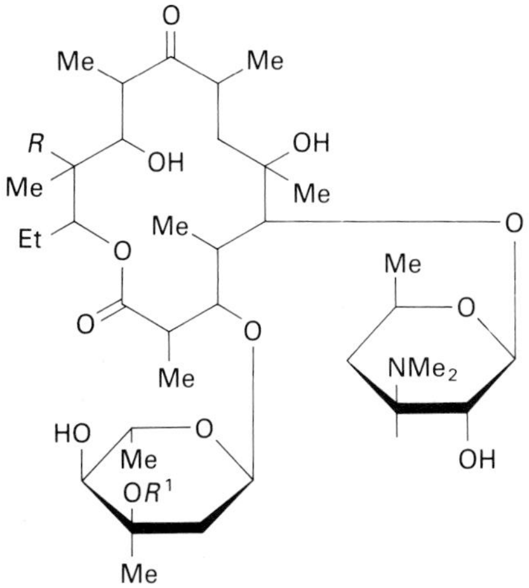

Erythromycin	R	R^1
A	OH	Me
B	H	Me
C	OH	H

Fig. 5.10 Erythromycins: erythromycin is a mixture of macrolide antibiotics consisting largely of erythronycin A.

115   *Types of antibiotics*

Oleandomycin, its ester (triacetyloleandomycin) and spiramycin have a similar range of activity as erythromycin but are less active. Resistance develops only slowly in clinical practice. However, cross-resistance may occur between all four members of this group.

Roxithromycin, a newer macrolide antibiotic, is an ether oxime derivative of erythromycin. It has similar *in vitro* activity to erythromycin, but enters macrophages and leucocytes more rapidly and achieves higher concentrations in the lysosomal component of the phagocytic cells. It is thus likely to be of particular value against *L. pneumophila* and *Chlamydia trachomatis*. Another potentially valuable new macrolide is clarithromycin.

7   Polypeptide antibiotics

The polypeptide antibiotics comprise a rather diverse group. They include (i) bacitracin, with activity against Gram-positive but not Gram-negative bacteria (except Gram-negative cocci); (ii) the polymyxins, which are active against many types of Gram-negative bacteria (including *Ps. aeruginosa* but excluding cocci, *Serratia marcescens* and *Proteus* spp.) but not Gram-positive organisms; and (iii) the two antitubercular antibiotics, capreomycin and viomycin.

Because of its highly toxic nature when administered parenterally, bacitracin is normally restricted to external usage.

The antibacterial activity of five members (A to E) of the polymyxin group is of a similar nature. However, they are all nephrotoxic although this effect is much reduced with polymyxins B and E (colistin). Colistin sulphomethate sodium is the form of colistin used for parenteral administration. Sulphomyxin sodium, a mixture of sulphomethylated polymyxin B and sodium bisulphite, has the action and uses of polymyxin B sulphate, but is less toxic.

Capreomycin and viomycin show activity against *M. tuberculosis* and may be regarded as being second-line antituberculosis drugs.

8   Glycopeptide antibiotics

Two important glycopeptide antibiotics are vancomycin and teicoplanin.

8.1   Vancomycin

Vancomycin is an antibiotic isolated from *Str. orientalis*. Vancomycin base has an empirical formula of $C_{66}H_{75}Cl_2N_9O_4$ (mol. wt 1448) and has a complex tricyclic glycopeptide structure. Modern chromatographically purified vancomycin gives rise to fewer side-effects than the antibiotic produced in the 1950s.

Vancomycin is active against most Gram-positive bacteria, including methicillin-resistant strains of *Staph. aureus* and *Staph. epidermidis*, *Enterococcus faecalis*, *Clostridium difficile* and Gram-negative cocci.

Gram-negative bacilli, mycobacteria and fungi are insusceptible. Vancomycin is bactericidal to most susceptible bacteria at concentrations near its minimum inhibitory concentration (MIC) and is an inhibitor of bacterial cell wall peptidoglycan synthesis, although at a site different from that of β-lactam antibiotics (Chapter 9).

Employed as the hydrochloride and administered by dilute intravenous injection, vancomycin is indicated in potentially life-threatening infections that cannot be treated with other effective, less toxic, antibiotics. Oral vancomycin is the drug of choice in the treatment of antibiotic-induced pseudomembranous colitis associated with the administration of antibiotics such as clindamycin and lincomycin (section 9.3).

8.2   Teicoplanin

Teicoplanin is a naturally occurring complex of five closely related tetra-cyclic molecules. Its mode of action and spectrum of activity are essentially similar to vancomycin, although it might be less active against some strains of coagulase-negative staphylococci. Teicoplanin can be administered by intramuscular injection.

9   Miscellaneous antibacterial antibiotics

Antibiotics described here (Fig. 5.11) are those which cannot logically be considered in any of the other groups above.

9.1   Chloramphenicol

Chloramphenicol has a broad spectrum of activity, but exerts a bacteriostatic effect. It has antirickettsial activity and is inhibitory to the larger viruses. Unfortunately, aplastic anaemia, which is dose-related, may result in a proportion of treated patients. It should thus not be given for minor infections and its usage should be restricted to cases where no effective alternative exists (see Chapter 6). Some bacteria (see Chapter 10) can produce an enzyme, chloramphenicol acetyltransferase, that acetylates the hydroxyl groups in the side chain of the antibiotic to produce, initially, 3-acetoxychloramphenicol and, finally, 1,3-diacetoxychloramphenicol, which lacks antibacterial activity. The design of fluorinated derivatives (Chapter 10) of chloramphenicol that are not acetylated by this enzyme could be a significant finding.

9.2   Fusidic acid

Employed as a sodium salt, fusidic acid (Fig. 5.11) is active against many types of Gram-positive bacteria, especially staphylococci, although streptococci are relatively resistant. It is employed in the treatment of staphylococcal infections, including strains resistant to other antibiotics. However, bacterial resistance may occur *in vitro* and *in vivo*.

117   *Types of antibiotics*

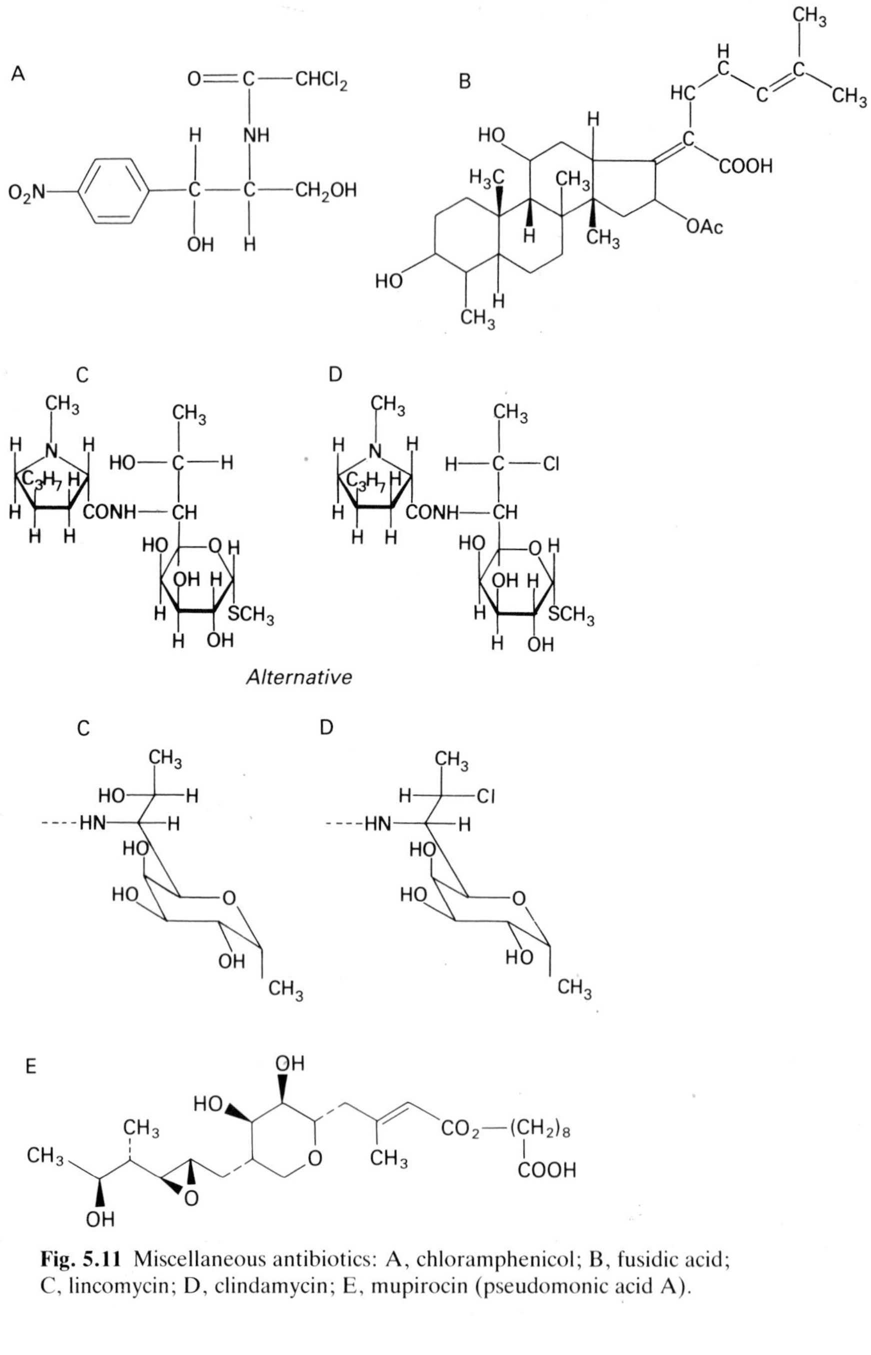

Fig. 5.11 Miscellaneous antibiotics: A, chloramphenicol; B, fusidic acid; C, lincomycin; D, clindamycin; E, mupirocin (pseudomonic acid A).

9.3 Lincomycins

Lincomycin and clindamycin (Fig. 5.11) are active against Gram-positive cocci, except *Enterococcus faecalis*. Gram-negative cocci tend to be less sensitive and enterobacteria are resistant. Cross-resistance of staphylococci may occur between lincomycins and erythromycin, but some erythromycin-resistant organisms may be sensitive to lincomycins.

<table>
<tr><td>9.4</td><td>

Mupirocin (pseudomonic acid A)

Mupirocin (Fig. 5.11) is the main fermentation product obtained from *Ps. fluorescens*. Other pseudomonic acids (B, C, D) are also produced. Mupirocin is active predominantly against staphylococci and most streptococci, but *Enterococcus faecalis* and Gram-negative bacilli are resistant. There is also evidence of plasmid-mediated mupirocin resistance in some clinical isolates of *Staph. aureus*.

Mupirocin is employed topically in eradicating nasal and skin carriage of staphylococci, including methicillin-resistant *Staph. aureus* colonization.

</td></tr>
</table>

10   Antifungal antibiotics

In contrast to the wide range of antibacterial antibiotics, there are very few antifungal antibiotics that can be used systemically.

10.1   **Griseofulvin**

This is a metabolic by-product of *P. griseofulvum*. Griseofulvin (Fig. 5.12) was first isolated in 1939, but it was not until 1958 that its antifungal activity was discovered. It is active against the dermatophytic fungi, i.e. those such as *Trichophyton* causing ringworm. It is ineffective against *Candida albicans*, the causative agent of oral thrush and intestinal candidasis, and against bacteria, and there is thus no disturbance of the normal bacterial flora of the gut.

Griseofulvin is administered orally in the form of tablets. It is not totally absorbed when given orally, and one method of increasing absorption is to reduce the particle size of the drug. Griseofulvin is deposited in the deeper layers of the skin and in hair keratin, and is therefore employed in chemotherapy of fungal infections of these areas caused by susceptible organisms.

10.2   **Polyenes**

Polyene antibiotics are characterized by possessing a large ring containing a lactone group and a hydrophobic region consisting of a sequence of four to seven conjugated double bonds. The most important polyenes are nystatin and amphotericin B (Fig. 5.12).

Nystatin has a specific action on *C. albicans* and is of no value in the treatment of any other type of infection. It is poorly absorbed from the gastrointestinal tract; even after very large doses, the blood level is insignificant. It is administered orally in the treatment of oral thrush and intestinal candidiasis infections.

Amphotericin B is particularly effective against systemic infections

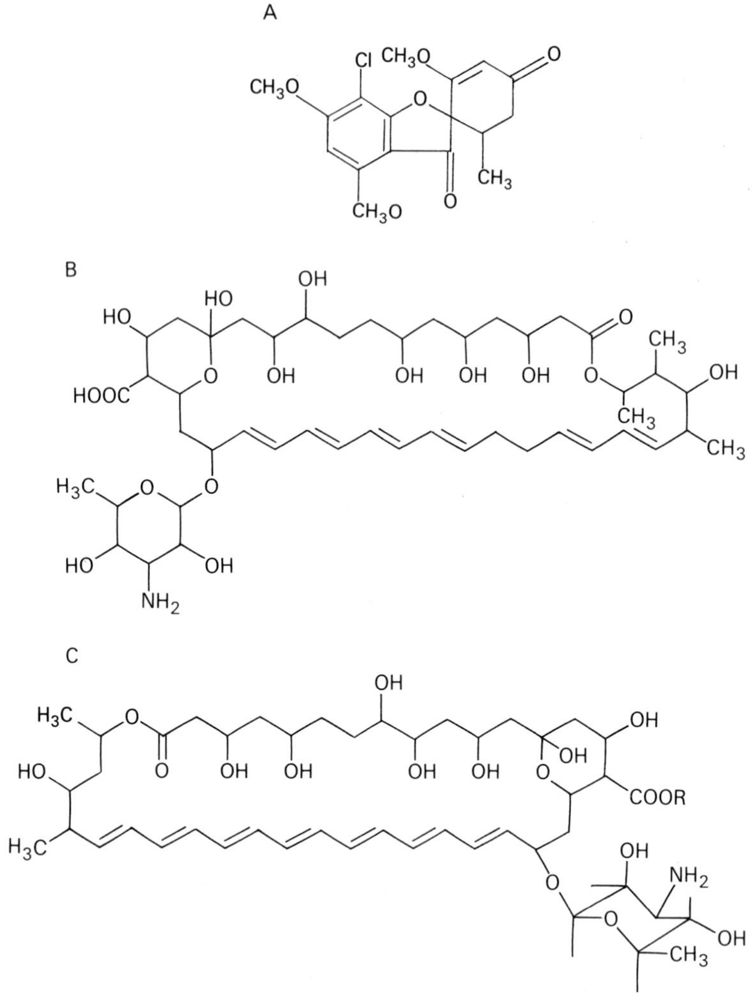

Fig. 5.12 Antifungal antibiotics: A, griseofulvin; B, nystatin; C, amphotericin (R = H) and its methyl ester (R = CH₃).

caused by *C. albicans* and *Cryptococcus neoformans*. It is poorly absorbed from the gastrointestinal tract and is thus usually administered by intravenous injection under strict medical supervision. Amphotericin methyl ester (Fig. 5.12) has equal antifungal activity *in vitro*, but much higher serum peak levels are obtained with the ester, which has been claimed to be less toxic.

11 Synthetic antimicrobial agents

11.1 Sulphonamides

Sulphonamides were introduced by Domagk in 1935. It had been shown that a red azo dye, prontosil (Fig. 5.13B), had a curative effect on mice

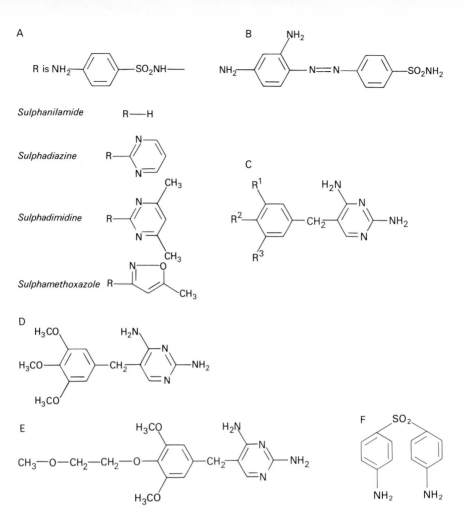

Fig. 5.13 A, Some sulphonamides; B, prontosil rubrum; C, unsubstituted diaminobenzylpyrimidines; D, trimethoprim; E, tetroxoprim; F, dapsone.

infected with β-haemolytic streptococci; it was subsequently found that, *in vivo*, prontosil was converted into sulphanilamide. Chemical modifications of the nucleus of sulphanilamide (see Fig. 5.13A) gave compounds with higher antibacterial activity, although this was often accompanied by greater toxicity. In general, it may be stated that the sulphonamides have a broadly similar antibacterial activity but differ widely in pharmacological actions.

Bacteria which are almost always sensitive to the sulphonamides include *Str. pneumoniae*, β-haemolytic streptococci, *Escherichia coli* and *Proteus mirabilis*; those almost always resistant include *Enterococcus faecalis*, *Ps. aeruginosa*, indole-positive *Proteus* and *Klebsiella*; whereas bacteria showing a marked variation in response include *Staph. aureus*, gonococci, *H. influenzae* and hospital strains of *E. coli* and *Pr. mirabilis*.

The sulphonamides show a considerable variation in the extent of their absorption into the bloodstream. Sulphadimidine and sulphadiazine are examples of rapidly absorbed ones, whereas succinylsulphathiazone and phthalylsulphathiazole are poorly absorbed and are excreted unchanged in the faeces.

From a clinical point of view, the sulphonamides are extremely useful for the treatment of uncomplicated urinary tract infection caused by *E. coli* in domiciliary practice. They have also been employed in treating meningococcal meningitis (a current problem is the number of sulphonamide-resistant meningococcal strains) and superficial eye infections. For further details on uses and toxicity, Lambert & O'Grady (1992), Greenwood (1989) and Williams (1990) should be consulted.

Co-trimoxazole is a mixture of sulphamethoxazole (five parts) and trimethoprim (one part) (section 11.2; Fig. 5.13D). The reason for using this combination is based upon the *in vitro* finding that there is a 'sequential blockade' of folic acid synthesis, in which the sulphonamide is a competitive inhibitor of dihydropteroate synthetase and trimethoprim inhibits dihydrofolate reductase (DHFR; see Chapter 9, especially Fig. 9.5). The optimum ratio of the two components may not be achieved *in vivo* and arguments continue as to the clinical value of co-trimoxazole, with many advocating the use of trimethoprim alone. Co-trimoxazole is the agent of choice in treating pneumonias caused by *Pneumocystis carinii*, a yeast (although it had been classified as protozoa). *P. carinii* is a common cause of pneumonia in patients receiving immunosuppressive therapy and in those suffering from AIDS.

11.2   Diaminopyrimidine derivatives

Small-molecule diaminopyrimidine derivatives were shown in 1948 to have an antifolate action. Subsequently, compounds were developed that were highly active against human cells (e.g. the use of methotrexate as an anticancer agent), protozoa (e.g. the use of pyrimethamine in malaria) or bacteria (e.g. trimethoprim; Fig. 5.13D). Unsubstituted diaminobenzylpyrimidines (Fig. 5.13C) bind poorly to bacterial DHFR. The introduction of one, two or especially three methoxy groups (as in trimethoprim) produces a highly selective antibacterial agent. A recent antibacterial addition is tetroxoprim (2,4-diamino-5-(3',5'-dimethoxy-4'-methoxyethoxybenzyl) pyrimidine; Fig. 5.13E) which retains methoxy groups at R^1 and R^3 and has a methoxyethoxy group at R^2. Trimethoprim and tetroxoprim have a broad spectrum of activity but resistance can arise from an insusceptible target site, i.e. an altered DHFR (see Chapter 10).

11.3   Dapsone

Dapsone (diaminodiphenylsulphone) is used specifically in the treatment of leprosy. However, because resistance to dapsone is unfortunately now well known, it is recommended that dapsone be used in conjunction with rifampicin (sections 4 and 11.4) and clofazimine. Its chemical structure is shown in Fig. 5.13F.

11.4   **Antitubercular compounds**

The three standard drugs used in the treatment of tuberculosis were streptomycin (considered above), *p*-aminosalicylic acid (PAS) and isoniazid (isonicotinylhydrazide, INH; synonym isonicotinic acid hydrazine, INAH). The tubercle bacillus rapidly becomes resistant to streptomycin, and the role of PAS was mainly that of preventing this development of resistance. The current approach is to treat tuberculosis in two phases: an *initial* phase where a combination of three drugs is used to reduce the bacterial level as rapidly as possible, and a *continuation* phase in which a combination of two drugs is employed. Front-line drugs are isoniazid, rifampicin, streptomycin and ethambutol. Pyrazinamide, which has good meningeal penetration, and is thus particularly useful in tubercular meningitis, may be used in the initial phase to produce a highly bactericidal response.

Isoniazid has no significant effect against organisms other than mycobacteria. It is given orally. Cross-resistance between it, streptomycin and rifampicin has not been found to occur.

When bacterial resistance to these primary agents exists or develops, treatment with the secondary tuberculostatic agents has to be considered. The latter group comprises capreomycin and cycloserine, and possibly kanamycin (all of which have been described briefly), ethionamide, prothionamide and thiacetazone. Like isoniazid, ethionamide is a derivative of isonicotinic acid but, strangely, tubercle bacilli do not show cross-resistance to these two substances. Prothionamide, which is as active as ethionamide and which is better tolerated, and pyrazinamide (see above), an effective tuberculocidal drug, are other examples of drugs that are derivatives of isonicotinic acid. Thiacetazone, in combination with another antitubercular drug such as isoniazid, gives highly effective therapy but its general usage is limited by its side-effects.

Figure 5.14 gives the chemical structures of the above drugs.

11.5   **Nitrofuran compounds**

The nitrofuran group of drugs (Fig. 5.15) is based on the finding over 40 years ago that a nitro group in the 5-position of 2-substituted furans endowed these compounds with antibacterial activity. Many hundreds of such compounds have been synthesized, but only a few are in current therapeutic use. In the most important nitrofurans, an azomethine group, —CH=N—, is attached at C-2 and a nitro group at C-5. Less important nitrofurans have a vinyl group, —CH=CH—, at C-2.

Biological activity is lost if: (i) the nitro ring is reduced; (ii) the —CH=N— linkage undergoes hydrolytic decomposition; or (iii) the —CH=CH— linkage is oxidized.

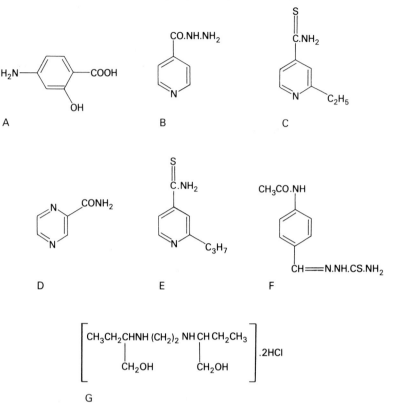

Fig. 5.14 Antitubercular compounds (see text also for details of antibiotics): A, PAS; B, isoniazid; C, ethionamide; D, pyrazinamide; E, prothionamide; F, thiacetazone; G, ethambutol.

The nitrofurans show antibacterial activity against a wide spectrum of microorganisms, but furaltadone has now been withdrawn from use because of its toxicity. Furazolidone has a very high activity against most members of the Enterobacteriaceae, and has been used in the treatment of diarrhoea and gastrointestinal disturbances of bacterial origin. Nitrofurantoin is used in the treatment of urinary tract infections; antibacterial levels are not reached in the blood and the drug is concentrated in the urine. It is most active at acid pH. Nitrofurazone is used mainly as a topical agent in the treatment of burns and wounds and also in certain types of ear infections. The nitrofurans are believed to be mutagenic.

11.6
4-quinolone antibacterials

Nalidixic acid has been used for several years as a clinically important drug in the treatment of urinary tract infections. Since its clinical introduction, other 4-quinolone antibacterials have been synthesized, some of which show considerably greater antibacterial potency. Furthermore, this means that many types of bacteria insusceptible to nalidixic acid therapy

124   *Chapter 5*

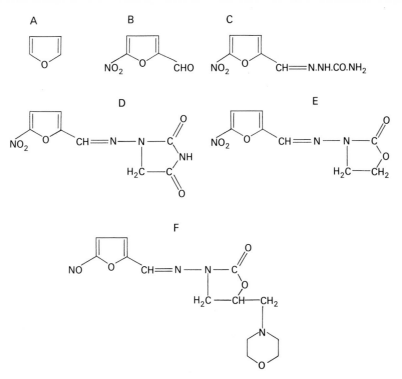

Fig. 5.15 A, furan; B, 5-nitrofurfural; C–F, nitrofuran drugs: respectively C, nitrofurazone, D, nitrofurantoin, E, furazolidone and F, furaltadone.

may be sensitive to the newer derivatives, which include ciprofloxacin, norfloxacin, oxolinic acid, ofloxacin, acrosoxacin, enoxacin and flumequine (Fig. 5.16).

Nalidixic acid is unusual in that it is active against several different types of Gram-negative bacteria, whereas Gram-positive organisms are resistant. However, the newer fluoroquinolone derivatives show superior activity against Enterobacteriaceae and *Ps. aeruginosa*, and their spectrum also includes staphylococci but not streptococci. Extensive studies with norfloxacin have demonstrated that its broad spectrum, high urine concentration and oral administration make it a drug with a promising future in the treatment of urinary infections.

The 4-quinolone antibacterials inhibit the A subunit of the essential enzyme DNA gyrase; *gyrA* mutant strains carry an alteration in the structural gene for the A subunit and are thus resistant. Resistance does not depend on drug destruction and does not appear to be plasmid-mediated (see also Chapters 9 and 10).

11.7   **Imidazole derivatives**

Metronidazole (Fig. 5.17A) inhibits the growth of pathogenic protozoa, very low concentrations being effective against the protozoa *Trichomonas*

125   *Types of antibiotics*

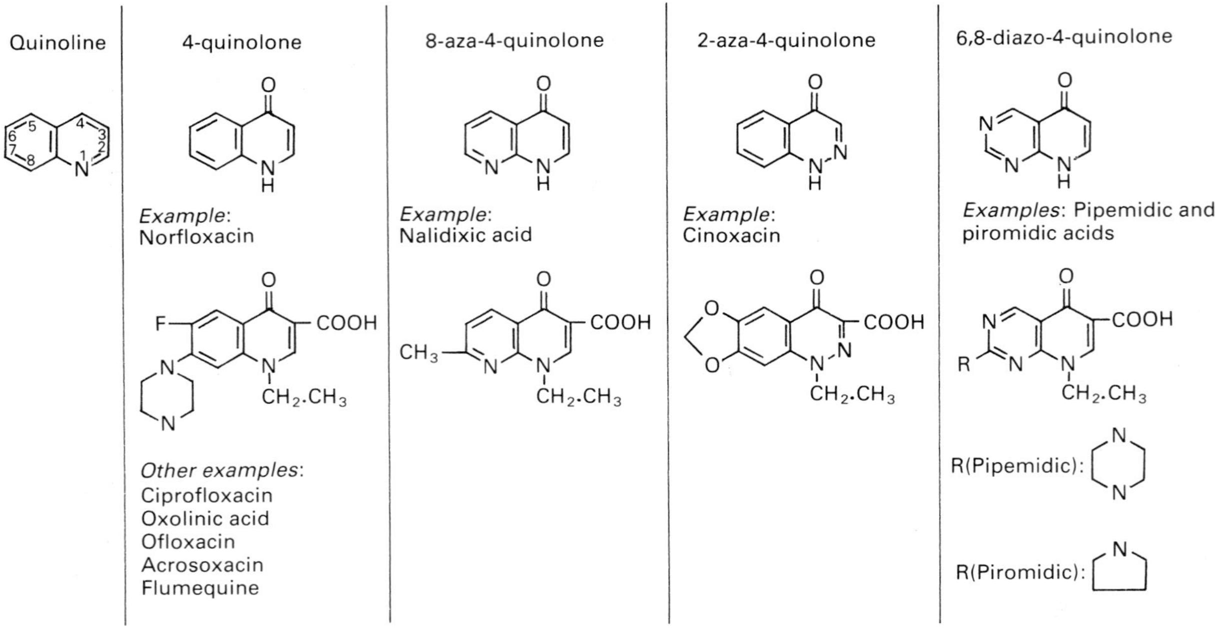

Fig. 5.16 Quinolone and antibacterial 4-quinolones. Note that the newer fluoroquine derivatives (e.g. norfloxacin, ciprofloxacin, ofloxacin) have a 6-fluoro and a 7-piperazino substituent.

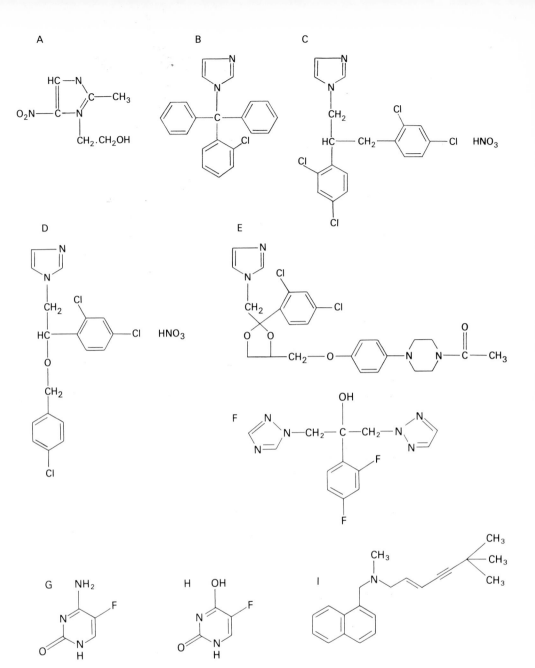

Fig. 5.17 A–F, Imidazoles; G, flucytosine; H, 5-fluorouracil and I, terbinafine. Imidazoles: A, metronidazole; B, clotrimazole; C, miconazole; D, econazole; E, ketoconazole; F, fluconazole.

vaginalis, *Ent. histolytica* and *Giardia lamblia*. It is also used to treat bacterial vaginosis caused by *Gardnerella vaginalis*. Given orally, it cures 90–100% of sexually transmitted urogenital infections caused by *T. vaginalis*. It has also been found that metronidazole is effective against anaerobic bacteria, for example *B. fragilis*, and against facultative anaerobes grown under anaerobic, but not aerobic, conditions. Metronidazole is administered orally or in the form of suppositories.

Other imidazole derivatives include clotrimazole (Fig. 5.17B), miconazole (Fig. 5.17C) and econazole (Fig. 5.17D), all of which possess a broad antimycotic spectrum with some antibacterial activity and are used topically. No development of resistance, *in vitro* or *in vivo*, has yet been reported.

These imidazole derivatives, together with the development of newer derivatives such as ketoconazole (Fig. 5.17E), which is administered orally, appear to have a promising future in the treatment of fungal infections. Fluconazole (Fig. 5.18F) is a bis-triazole derivative. Its efficacy *in vivo* is superior to what might be predicted from its poor *in vitro* activity, and sufficient to suggest that it will prove to be a useful systemically administered drug.

11.8 Flucytosine

Flucytosine (5-fluorocytosine; Fig. 5.17G) is a narrow-spectrum antifungal agent with greatest activity against yeasts such as *Candida*, *Cryptococcus* and *Torulopsis*. Evidence has been presented which shows that, once inside the fungal cell, flucytosine is deaminated to 5-fluorouracil (Fig. 5.17). This is converted by the enzyme pyrophosphorylase to 5-fluorouridine monophosphate (FUMP), diphosphate (FUDP) and triphosphate (FUTP), which inhibits RNA synthesis; 5-fluorouracil itself has poor penetration into fungi. *C. albicans* is known to convert FUMP to 5-fluorodeoxyuridine monophosphate (FdUMP), which inhibits DNA synthesis by virtue of its effect on thymidylate synthetase. Resistance can occur *in vivo* by reduced uptake into fungal cells of flucytosine or by decreased accumulation of FUTP or FdUMP.

11.9 Terbinafine

Terbinafine (Fig. 5.17I), a member of the allylamine class of antimycotics, is an inhibitor of the enzyme squalene epoxidase in fungal ergosterol biosynthesis. Terbinafine is orally active, is fungicidal and is effective against a broad range of dermatophytes and yeasts. It can also be used topically as a cream.

11.10 Newer antifungal drugs

Because there are still comparatively few antifungal agents available, research continues in attempts to develop newer, more effective, drugs.

Lipopeptides such as cilofungin are fungicidal to *Candida* and *Aspergillus* species. The pradimicins, sterol-like molecules with amino acid-containing side chains, are calcium-dependent agents that apparently form complexes with the fungal cell membrane. It is to be hoped that newer agents such as these examples will at some time in the future expand the antifungal armamentarium available to the clinician.

12   Antiviral drugs

Several compounds are known that are inhibitory to mammalian viruses in tissue culture, but only a few can be used in the treatment of human viral infections. The main problem in designing and developing antiviral agents is the lack of selective toxicity that is normally possessed by most compounds. Consequently, in comparison with antibacterial agents, very few inhibitors can be considered as being safe antiviral drugs, although the situation is improving. Possible sites of attack by antiviral agents include prevention of adsorption of a viral particle to the host cell, prevention of the intracellular penetration of the adsorbed virus, and inhibition of protein or nucleic synthesis.

Amantadine hydrochloride (Fig. 5.18A) does not prevent adsorption but inhibits viral penetration. It has a very narrow spectrum and is used prophylactically against infection with influenza A virus; it has no prophylactic value with other types of influenza virus. Methisazone (Fig. 5.18B) inhibits DNA viruses (particularly vaccinia and variola) but not RNA viruses, and has been used in the prophylaxis of smallpox.

Various nucleoside analogues have been developed that inhibit nucleic acid synthesis. Idoxuridine (2′-deoxy-5-iodouridine; IUdR; Fig. 5.18C) is a thymidine analogue which inhibits the utilization of thymidine (Fig. 5.18G) in the rapid synthesis of DNA that normally occurs in herpes-infected cells. Unfortunately, because of its toxicity, idoxuridine is unsuitable for systemic use and it is restricted to topical treatment of herpes-infected eyes. Other nucleoside analogues include the following: cytarabine (cytosine arabinoside; Ara-C; Fig. 5.18D) which has anti-neoplastic and antiviral properties and which has been employed topically to treat herpes keratitis resistant to idoxuridine; adenosine arabinoside (Ara-A; vidarabine); and ribavirin (1-β-D-ribofuranosyl-1,2,4-triazole-3, carboxamide; Fig. 5.18E) which has a broad spectrum of activity, inhibiting both RNA and DNA viruses. Vidarabine, in particular, has a high degree of selectivity against viral DNA replication and is primarily active against herpesviruses and some poxviruses. It may be used systemically or topically.

Human immunodeficiency virus (HIV) is a retrovirus, i.e. its RNA is converted in human cells by the enzyme reverse transcriptase to DNA which is incorporated into the human genome and is responsible for producing new HIV particles. Zidovudine (azidothymidine, AZT: Fig. 5.18F) is a structural analogue of thymidine (Fig. 5.18G) and is used to treat AIDS patients. Zidovudine is converted in both infected and un-

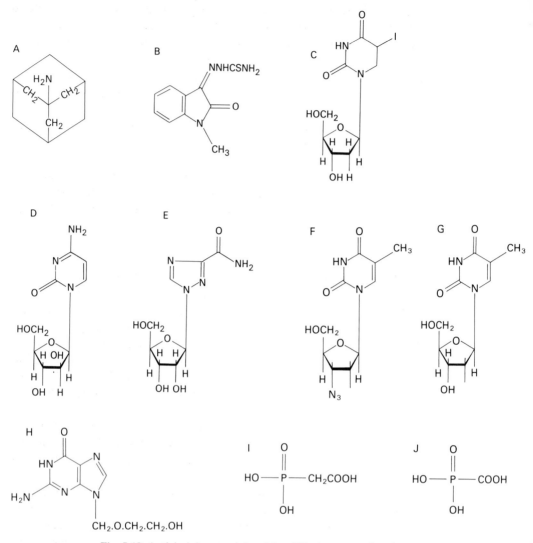

Fig. 5.18 Antiviral drugs and thymidine (G). A, amantadine (used as the hydrochloride); B, methisazone; C, idoxuridine; D, cytarabine; E, ribavirin; F, zidovudine; H, acyclovir; I, phosphonoacetic acid; J, phosphonoformic acid, used as the sodium salt.

infected cells to the mono-, di- and eventually tri-phosphate derivatives. Zidovudine triphosphate, the active form, is a potent inhibitor of HIV replication, being mistaken for thymidine by reverse transcriptase. Premature chain termination of viral DNA ensues. 2′,3′-Dideoxycytidine (DDC), a nucleoside analogue that also inhibits reverse transcriptase, is more active than zidovudine *in vitro*, and (unlike zidovudine) does not suppress erythropoiesis. DDC is not without toxicity, however, and a severe peripheral neurotoxicity, which is dose-related, has been reported.

A novel type of nucleoside analogue is acycloguanosine (acyclovir,

Fig. 5.18H) which becomes activated only in herpes-infected host cells by a herpes-specific enzyme, thymidine kinase. This enzyme initiates conversion of acyclovir initially to a monophosphate and then to the antiviral triphosphate which inhibits viral DNA polymerase. The host cell polymerase is not inhibited to the same extent, and the antiviral triphosphate is not produced in uninfected cells. Bromovinyldeoxyuridine also becomes active only in infected host cells, and this nucleoside analogue is currently under investigation. Ganciclovir, a 3'-hydroxy derivative of acyclovir, is much more active than acyclovir against human cytomegalovirus, but is also unfortunately more toxic.

Two simple molecules with potent antiviral activity are phosphonoacetic acid (Fig. 5.18I) and sodium phosphonoformate (Fig. 5.18J). The latter inhibits herpes DNA polymerase, is non-toxic when applied to the skin and, when clinical trials have been completed, could be a useful agent in treating herpes simplex labialis (cold sores). Phosphonoacetic acid has a high specificity for herpes simplex DNA synthesis, and has been shown to be non-mutagenic in experimental animals, but is highly toxic.

Finally, the properties of the antiviral agent interferon will be considered briefly. It was pointed out in Chapter 3 (section 9.1) that interferon is a low molecular weight protein, produced by virus-infected cells, that itself induces the formation of a second protein inhibiting the transcription of viral mRNA. Interferon is produced by the host cell in response to the virus particle, the viral nucleic acid and non-viral agents, including synthetic polynucleides such as polyinosinic acid:polycytidylic acid (poly I:C). There are two types of interferon.

Type I interferons These are acid-stable and comprise two major classes, leucocyte interferon (Le-IFN, IFN-α) released by stimulated leucocytes, and fibroblast interferon (F-IFN, IFN-β) released by stimulated fibroblasts.

Type II interferons These are acid-labile and are also known as 'immune' (IFN-γ) interferons because they are produced by T-lymphocytes (see Chapter 15) in the cellular immune system in response to specific antigens.

Type I interferons induce a virus-resistant state in human cells, whereas Type II are more active in inhibiting growth of tumour cells.

Disappointingly low yields of F-IFN and Le-IFN are achieved from eukaryotic cells. Recently, however, recombinant DNA technology has been employed to produce interferon in prokaryotic cells (bacteria). This aspect is considered in more detail in Chapter 24.

13 Drug combinations

A combination of two antibacterial agents may produce the following responses.

1 Synergism, where the joint effect is greater than the sum of the effects of each drug acting alone.

2 Additive effect, in which the combined effect is equal to the arithmetic sum of the effects of the two individual agents.

3 Antagonism (interference), in which there is a lesser effect of the mixture than that of the more potent drug action alone.

There are four possible justifications as to the use of antibacterial agents in combination.

1 The concept of clinical synergism, which may be extremely difficult to demonstrate convincingly. Even with trimethoprim plus sulphamethoxazole, where true synergism occurs *in vitro*, the optimum ratio of the two components may not always be present *in vivo*, i.e. at the site of infection in a particular tissue.

2 A wider spectrum of cover may be obtained, which may be (i) desirable as an emergency measure in life-threatening situations; or (ii) of use in treating mixed infections.

3 The emergence of resistant organisms may be prevented. A classical example here occurs in combined antitubercular therapy (see earlier).

4 A possible reduction in dosage of a toxic drug may be achieved.

Indications for combined therapy are now considered to be much fewer than was originally thought. There is also the problem of a chemical or physical incompatibility between two drugs. However, one combination with an important role to play in chemotherapy is amoxycillin plus the β-lactamase inhibitor clavulanic acid. A more recent combination is the broad-spectrum penicillin, ticarcillin, with clavulanic acid and another of ampicillin with sulbactam (a β-lactamase inhibitor). It must also be noted that a combination of two β-lactams does not necessarily produce a synergistic effect. Some antibiotics are excellent inducers of β-lactamase, and consequently a reduced response (antagonism) may be produced.

14   **Further reading**

Andriole V.T. (Ed.) (1988) *The Quinolones*. London: Academic Press.

Bean B. (1992) Antiviral therapy: current concepts and practices. *Clin Mic Rev.* **5**, 146–182.

British National Formulary. London: British Medical Association & The Pharmaceutical Press. (The chapter on drugs used in the treatment of infections is a particularly useful section. New editions of the BNF appear at regular intervals.)

Brown A.G. (1981) New naturally occurring β-lactam antibiotics and related compounds. *J Antimicrob Chemother.* **7**, 15–48.

Chopra I., Hawkey P.M. & Hinton M. (1992) Tetracyclines, molecular and clinical aspects. *J Antimicrob Chemother.* **29**, 245–277.

D'Arcy P.F. & Scott E.M. (1978) Antifungal agents. In *Progress in Drug Research* (Ed. E. Jucker), vol. 22, pp. 93–147. Basel & Stuttgart: Birkhäuser Verlag.

Greenwood D. (Ed.) (1989) *Antimicrobial Chemotherapy*, 2nd edn. London: Bailliere Tindall.

Hamilton-Miller J.M.T. (1991) From foreign pharmacopoeias: 'new' antibiotics from old? *J Antimicrob Chemother.* **27**, 702–705.

Lambert H.P. & O'Grady F. (1992) *Antibiotic and Chemotherapy*, 6th edn. London & Edinburgh: Churchill Livingstone.

Neu H.C. (1985) Relation of structural properties of β-lactam antibiotics to antibacterial activity. *Am J Med.* **79** (Suppl. 2A), 2–13.

Norrby S.R. (Ed.) (1988) *New Antiviral Strategies*. Edinburgh: Churchill Livingstone.

Oxford J.S. (1979) Inhibition of herpes virus by a new compound—acyclic guanosine. *J Antimicrob Chemother.* **5**, 333–334.

Reeves D.S. & Howard A.J. (Eds) (1991) New macrolides—the respiratory antibiotics for the 1990s. *J Hosp Infect.* **19** (Suppl. A).

Russell A.D. (1988) Design of antimicrobial chemotherapeutic agents. In *Introduction to Principles of Drug Design* (Ed. H.J. Smith), 2nd edn, pp. 264–308. Bristol: John Wright.

Russell A.D. & Chopra I. (1990) *Understanding Antibacteral Action and Resistance.* Chichester: Ellis Horwood.

Sammes P.G. (Ed.) *Topics in Antibiotic Chemistry*, vols 1–5. Chichester: Ellis Horwood.

Shanson D.C. (1989) *Microbiology in Clinical Practice*, 2nd edn. London: Wright.

Smith J.T. (1984) Awakening the slumbering potential of the 4-quinolone antibacterials. *Pharm J.* **233**, 299–305.

Stokes E.J. & Ridgway G.L. (1987) *Clinical Microbiology*, 6th edn. London: Edward Arnold.

Tyrrell D.A.J., Phillips I., Goodwin C.S. & Blowers R. (1979) *Microbial Disease: The Use of the Laboratory in Diagnosis. Therapy and Control.* London: Edward Arnold.

Williams J.D. (1990) Antimicrobial substances. In *Topley & Wilson's Principles of Bacteriology, Virology and Immunity, vol. 1, General Microbiology and Immunity* (Eds A.H. Linton & H.M. Dick), 8th edn, pp. 105–151. London: Edward Arnold.

Wise R. & Reeves D.S. (Eds) (1979) Proceedings of a symposium: advances in therapy with antibacterial folate inhibitors. *J Antimicrob Chemother.* **5** (Suppl. B).

Wood, M.J. (1991) More macrolides. *Br Med J.* **303**, 594–595.

6   Clinical uses of antimicrobial drugs

1   Introduction

2   Principles of use of antimicrobial
     drugs
2.1   Susceptibility of infecting
       organisms
2.2   Host factors
2.3   Pharmacological factors
2.4   Drug resistance
2.5   Drug combinations
2.6   Adverse reactions
2.7   Superinfection
2.8   Chemoprophylaxis

3   Clinical use
3.1   Respiratory tract infections
3.1.1   Upper respiratory tract infections

3.1.2   Lower respiratory tract infections
3.2   Urinary tract infections
3.2.1   Pathogenesis
3.2.2   Drug therapy
3.3   Gastrointestinal infections
3.4   Skin and soft tissue infections
3.5   Central nervous system infections

4   Antibiotic policies
4.1   Rationale
4.2   Types of antibiotic policies
4.2.1   Free prescribing policy
4.2.2   Restricted reporting
4.2.3   Restricted dispensing

5   Further reading

1   Introduction

The worldwide use of antimicrobial drugs continues to rise; in 1989 these agents accounted for an expenditure of £11 700 million. In the UK, general practice use accounts for approximately 90% of all antibiotic prescribing and largely involves oral and topical agents. Hospital use accounts for the remaining 10% of antibiotic prescribing with a much heavier use of injectable agents. Although this chapter is concerned with the clinical use of antimicrobial drugs, it should be remembered that these agents are also extensively used in veterinary practice as well as in animal husbandry as growth promoters. In humans the therapeutic use of anti-infectives has revolutionized the management of most bacterial infections, many parasitic and fungal diseases and, with the availability of acyclovir and zidovudine (azidothymidine, AZT) (see Chapters 3 and 5), selected herpes virus infections and human immunodeficiency virus (HIV) infection respectively. Although originally used for the treatment of established bacterial infections, antibiotics have proved useful in the prevention of infection in various high-risk circumstances; this applies especially to patients undergoing various surgical procedures where perioperative antibiotics have significantly reduced postoperative infectious complications.

The advantages of effective antimicrobial chemotherapy are self-evident, but this has led to a significant problem in ensuring that they are always appropriately used. Surveys of antibiotic use have demonstrated that more than 50% of antibiotic prescribing is actually inappropriate; this may reflect prescribing in situations where antibiotics are either ineffective, such as viral infections, or that the selected agent, its dose, route of administration or duration of use are inappropriate. Of particular concern is the prolonged use of antibiotics for surgical prophylaxis. Apart from being wasteful of health resources, prolonged use encourages super-infection by drug-resistant organisms and unnecessarily increases the risk of adverse drug reactions. Thus, it is essential that the clinical use of these agents be based on a clear understanding of the principles that have evolved to ensure safe, yet effective, prescribing.

Further information about the properties of antimicrobial agents described in this chapter can be found in Chapter 5.

2   Principles of use of antimicrobial drugs

2.1   Susceptibility of infecting organisms

Drug selection should be based on knowledge of its activity against infecting microorganisms. Selected organisms may be predictably susceptible to a particular agent, and laboratory testing is therefore rarely performed. For example, *Streptococcus pyogenes* is uniformly sensitive to penicillin. In contrast, the susceptibility of many Gram-negative enteric bacteria is less predictable and laboratory guidance is essential for safe prescribing. The susceptibility of common bacterial pathogens and widely prescribed antibiotics is summarized in Table 6.1. It can be seen that although certain bacteria are susceptible *in vitro* to a particular agent, use of this drug may be inappropriate, either on pharmacological grounds or because other less toxic agents are preferred.

2.2   Host factors

In vitro susceptibility testing does not always predict clinical outcome. Host factors play an important part in determining outcome and this applies particularly to circulating and tissue phagocytic activity. Infections can progress rapidly in patients suffering from either an absolute or functional deficiency of phagocytic cells. This applies particularly to those suffering from various haematological malignancies, such as the acute leukaemias, where phagocyte function is impaired both by the disease and also by the use of potent cytotoxic drugs which destroy healthy, as well as malignant, white cells. Under these circumstances it is essential to select agents which are bactericidal, since bacteristatic drugs, such as the tetracyclines, sulphonamides or chloramphenicol, rely on host phagocytic activity to clear bacteria. Widely used bactericidal agents include the aminoglycosides, broad-spectrum penicillins, the cephalosporins and quinolones (see Chapter 5).

In some infections the pathogenic organisms are located intracellularly within phagocytic cells and, therefore, remain relatively protected from drugs which penetrate cells poorly, such as the penicillins and cephalosporins. In contrast, erythromycin, rifampicin and chloramphenicol

Table 6.1 Sensitivity of selected bacteria to common antibacterial agents

	Staph. aureus (pen. sensitive)	Staph. aureus (pen. resistant)	Strep. pyogenes and Strep. pneumoniae	Enterococcus	Cl. perfringens	N. gonorrhoeae	N. meningitidis	H. influenzae	E. coli	Klebsiella spp.	Proteus spp. (indole-negative)	Proteus spp. (indole-positive)	Serratia spp.	Salmonella spp.	Shigella spp.	Pseudomonas spp.	B. fragilis	Other Bacteroides spp.	Chlamydia spp.	M. pneumoniae	Rickettsia spp.
Penicillin V/G	+	R	+*	+	+	+*	+	±	R	R	R	R	R	R	R	R	R	+	R	R	R
Methicillin, oxacillin, cloxacillin, dicloxacillin, nafcillin	+	+*	+	R	(+)	(±)	(±)	R	R	R	R	R	R	R	R	R	R	(±)	R	R	R
Ampicillin, amoxycillin	(+)	R	+*	+	+	+*	+	±	±	R	+	R	R	±	±	R*	R	+	R	R	R
Carbenicillin, ticarcillin	+	R*	(+)	R	(±)	(±)	(+)	(+)	±	R	+	±	R	(+)	(+)	+*	±	±	R	R	R
Cefazolin	+	+	+	R	+	(+)	(+)	±	+	±	+	R	R	(+)	(+)	R	R	R	R	R	R
Cefamandole, cefuroxime	+	+	+	R	+	(+)	(+)	+	+	+	+	+	±	(+)	(+)	R	R/	R/	R	R	R
Cefoxitin	+	+	+	R	+	(+)	(+)	+	+	+	+	+	±	(+)	(+)	R	+	±	R	R	R
Cefotaxime, latamoxef	+	+	+	R	+	+	+	+	+	+	+	+	+	(+)	(+)	±	R/	+	R	R	R
Ceftazidime	+	+	+	R	+	+	+	+	+	+	+	+	+	(+)	(+)	+	R	±	R	R	R
Erythromycin	+	+	+	R	(+)	+	(+)	±	R	R	R	R	R	R	R	R	±	±	+	+	R
Clindamycin	+*	+*	+*	R	+	R	R	R	R	R	R	R	R	R	R	R	+	+	R	R	R
Tetracyclines	+*	+*	±	+	+	+	(+)	+	+	±	±	±	±	(±)	(±)	R	+	+	R	R	+
Chloramphenicol	±	+	+	+	+	+	+	+	+	±	±	+	+	+	+	R	±	±	+	+	+
Ciprofloxacin	+	+	R	±	R	R	R	+*	+	+	+	+	+	+*	+	(+)	+	+	+	+	+
Gentamicin, tobramycin, amikacin, netilmicin	±	±	R	R	R	R	R	+*	+	+*	R	R	+*	+*	(+)	+*	R	R	+	+	+
Sulphonamides	+	+	±	±	(±)	±	±	±	±	±	±	±	R	±	±	R	±	±	±	±	R
Trimethoprim-sulphamethoxazole	+	+	+	+	R	+	+	+	+	+	+	+	R	+	+	R	R	R	+	R	R

+, Sensitive; R, resistant; ±, some strains resistant; (), not appropriate therapy; *, rare strains resistant.

readily penetrate phagocytic cells. Legionnaires' disease is an example of an intracellular infection and is treated with rifampicin and/or erythromycin.

2.3 Pharmacological factors

Clinical efficacy is also dependent on achieving satisfactory drug concentrations at the site of the infection; this is influenced by the standard pharmacological factors of absorption, distribution, metabolism and excretion. If an oral agent is selected, gastrointestinal absorption should be satisfactory. However, it may be impaired by factors such as the presence of food, drug interactions (including chelation), or impaired gastrointestinal function either as a result of surgical resection or malabsorptive states. Although effective, oral absorption may be inappropriate in patients who are vomiting or have undergone recent surgery; under these circumstances a parenteral agent will be required and has the advantage of providing rapidly effective drug concentrations.

Antibiotic selection also varies according to the anatomical site of infection. Lipid solubility is of importance in relation to drug distribution. For example, the aminoglycosides are poorly lipid-soluble and although achieving therapeutic concentrations within the extracellular fluid compartment, penetrate the cerebrospinal fluid (CSF) poorly. Likewise the presence of inflammation may affect drug penetration into the tissues. In the presence of meningeal inflammation, β-lactam agents achieve satisfactory concentrations within the CSF, but as the inflammatory response subsides drug concentrations fall. Hence it is essential to maintain sufficient dosaging throughout the treatment of bacterial meningitis. Other agents such as chloramphenicol are little affected by the presence or absence of meningeal inflammation.

Therapeutic drug concentrations within the bile duct and gall bladder are dependent upon biliary excretion. In the presence of biliary disease, such as gallstones or chronic inflammation, the drug concentration may fail to reach therapeutic levels. In contrast, drugs which are excreted primarily via the liver or kidneys may require reduced dosaging in the presence of impaired renal or hepatic function. The malfunction of excretory organs may not only risk toxicity from drug accumulation, but will also reduce urinary concentration of drugs excreted primarily by glomerular filtration. This applies to the aminoglycosides and the urinary antiseptics, nalidixic acid and nitrofurantoin, where therapeutic failure of urinary tract infections may complicate severe renal failure.

2.4 Drug resistance

Drug resistance may be a natural or an acquired characteristic of a microorganism. This may result from impaired cell wall or cell envelope penetration, enzymatic inactivation or altered binding sites. Acquired drug resistance may result from mutation, adaptation or gene transfer. Spontaneous mutations occur at low frequency, as in the case of

Mycobacterium tuberculosis where a minority population of organisms are resistant to isoniazid. In this situation the use of isoniazid alone will eventually result in overgrowth by this subpopulation of resistant organisms.

Genetic resistance may be chromosomally- or plasmid-mediated. Plasmid-mediated resistance has been increasingly recognized among Gram-negative enteric pathogens. By the process of conjugation (Chapter 10), resistance plasmids may be transferred both between bacteria of the same and different species and also different genera. Such resistance can code for multiple antibiotic resistance. For example, the penicillins, cephalosporins, chloramphenicol and the aminoglycosides are all subject to enzymatic inactivation which may be plasmid-mediated. Knowledge of the local epidemiology of resistant pathogens within a hospital, and especially within high dependency areas such as intensive care units, is invaluable in guiding appropriate drug selection.

The underlying mechanisms of resistance are considered in Chapter 10.

2.5   Drug combinations

Antibiotics are generally used alone, but may on occasion be prescribed in combination. Combining two antibiotics may result in synergism, indifference or antagonism. In the case of synergism, microbial inhibition is achieved at concentrations below that for each agent alone and may prove advantageous in treating relatively insusceptible infections such as enterococcal endocarditis, where a combination of penicillin and gentamicin is synergistically active. Another advantage of synergistic combinations is that it may enable the use of toxic agents where dose reductions are possible. For example, meningitis caused by the fungus *Cryptococcus neoformans* responds to an abbreviated course of amphotericin B when it is combined with 5-flucytosine, thereby reducing the risk of toxicity from amphotericin B.

Combined drug use is occasionally recommended to prevent resistance emerging during treatment. For example, treatment may fail when fusidic acid is used alone to treat *Staphylococcus aureus* infections, because resistant strains develop rapidly; this is prevented by combining fusidic acid with flucloxacillin. Likewise, tuberculosis is treated with a minimum of two agents, such as rifampicin and isoniazid; again drug resistance is prevented which may result if either agent is used alone.

The commonest reason for using combined therapy is in the treatment of confirmed or suspected mixed infections where a single agent alone will fail to cover all pathogenic organisms. This is the case in serious abdominal sepsis where mixed aerobic and anaerobic infections are common and the use of metronidazole in combination with either an aminoglycosider or a broad-spectrum cephalosporin is essential. Finally, drugs are used in combination in patients who are seriously ill and in whom uncertainty exists concerning the microbiological nature of their infection. This initial 'blind therapy' frequently includes a broad-spectrum penicillin or cephalosporin in combination with an aminoglycoside. The regimen should be modified in the light of subsequent microbiological information.

2.6   **Adverse reactions**

Regrettably, all chemotherapeutic agents have the potential to produce adverse reactions with varying degrees of frequency and severity, and these include hypersensitivity reactions and toxic effects. These may be dose-related and predictable in a patient with a history of hypersensitivity or a previous toxic reaction to a drug or its chemical analogues. However, many adverse events are idiosyncratic and therefore unpredictable.

Hypersensitivity reactions range in severity from fatal anaphylaxis, in which there is widespread tissue oedema, airway obstruction and cardio-vascular collapse, to minor and reversible hypersensitivity reactions such as skin eruptions and drug fever. Such reactions are more likely in those with a history of hypersensitivity to the drug, and are more frequent in patients with previous allergic diseases such as childhood eczema or asthma. It is important to question patients closely concerning hyper-sensitivity reactions before prescribing, since it precludes the use of all compounds within a class, such as the sulphonamides or tetracyclines, while cephalosporins should be used with caution in patients allergic to penicillin since these agents are structurally related. They should be avoided entirely in those who have had a previous severe hypersensitivity reaction to penicillin.

Drug toxicity is often dose-related and may affect a variety of organs or tissues. For example, the aminoglycosides are both nephrotoxic and ototoxic to varying degrees; therefore, dosaging should be individualized and the serum assayed, especially where renal function is abnormal, to avoid toxic effects and non-therapeutic drug concentrations. An example of dose-related toxicity is chloramphenicol-induced bone marrow sup-pression. Chloramphenicol interferes with the normal maturation of bone marrow stem cells and high concentrations may result in a steady fall in circulating red and white cells and also platelets. This effect is generally reversible with dose reduction or drug withdrawal. This dose-related toxic reaction of chloramphenicol should be contrasted with idiosyncratic bone marrow toxicity which is unrelated to dose and occurs at a much lower frequency of approximately 1:40000 and is frequently irreversible, end-ing fatally. Toxic effects may also be genetically determined. For example, peripheral neuropathy may occur in those who are slow acetylators of isoniazid, while haemolysis occurs in those deficient in the red cell enzyme glucose-6-phosphate dehydrogenase, when treated with sulphonamides or primaquine.

2.7 Superinfection

Anti-infective drugs not only affect the invading organism undergoing treatment but also have an impact on the normal bacterial flora, especially of the skin and mucous membranes. This may result in microbial overgrowth of resistant organisms with subsequent superinfection. One example is the common occurrence of oral or vaginal candidiasis in patients treated with broad-spectrum agents such as ampicillin or tetracycline. A more serious example is the development of pseudo-membranous colitis from the overgrowth of toxin-producing strains of *Clostridium difficile* present in the bowel flora following the use of clindamycin and other broad-spectrum antibiotics. This condition is managed by drug withdrawal and oral vancomycin. Rarely, colectomy (excision of part or whole of the colon) may be necessary for severe cases.

2.8 Chemoprophylaxis

An increasingly important use of antimicrobial agents is that of infection prevention, especially in relationship to surgery. Infection remains one of the most important complications of many surgical procedures, and the recognition that peri-operative antibiotics are effective and safe in preventing this complication has proved a major advance in surgery. The principles that underly the chemoprophylactic use of antibacterials relate to the predictability of infection for a particular surgical procedure, both in terms of its occurrence, microbial aetiology and susceptibility to antibiotics. Therapeutic drug concentrations present at the operative site at the time of surgery rapidly reduce the number of potentially infectious organisms and prevents wound sepsis. If prophylaxis is delayed to the postoperative period then efficacy is markedly impaired. It is important that chemoprophylaxis be limited to the peri-operative period, the first dose being administered approximately 1 hour before surgery for injectable agents and repeated for two to three repeat doses postoperatively. Prolonging chemoprophylaxis beyond this period is not cost-effective and increases the risk of adverse drug reactions and superinfection. One of the best examples of the efficacy of surgical prophylaxis is in the area of large bowel surgery. Before the widespread use of chemoprophylaxis, postoperative infection rates for colectomy were often 30% or higher; these have now been reduced to around 5–10%.

Chemoprophylaxis has been extended to other surgical procedures where the risk of infection may be low but its occurrence has serious consequences. This is especially true for the implantation of prosthetic joints or heart valves. These are major surgical procedures and although infection may be infrequent its consequences are serious and on balance the use of chemoprophylaxis is cost-effective.

Examples of chemoprophylaxis in the non-surgical arena include the prevention of endocarditis with amoxycillin in patients with valvular heart disease undergoing dental surgery, and the prevention of secondary cases of meningococcal meningitis with rifampicin among household contacts of an index case.

3 Clinical use

The choice of antimicrobial chemotherapy is initially dependent upon the clinical diagnosis. Under some circumstances the clinical diagnosis implies a microbiological diagnosis which may dictate specific therapy. For example, typhoid fever is caused by *Salmonella typhi* which is generally sensitive to chloramphenicol or co-trimoxazole. However, for many infections, establishing a clinical diagnosis implies a range of possible microbiological causes and requires laboratory confirmation from samples collected, preferably before antibiotic therapy is begun. Laboratory isolation and susceptibility testing of the causative agent establish the diagnosis with certainty and make drug selection more rational. However, in many circumstances, especially in general practice, microbiological documentation of an infection is not possible. Hence knowledge of the usual microbiological cause of a particular infection and its susceptibility to antimicrobial agents is essential for effective drug prescribing. The following section explores a selection of the problems associated with antimicrobial drug prescribing for a range of clinical problems.

3.1 Respiratory tract infections

Infections of the respiratory tract are among the commonest of infections, and account for much consultation in general practice and a high percentage of acute hospital admissions. They are divided into infections of the upper respiratory tract, involving the ears, throat, nasal sinuses and the trachea, and the lower respiratory tract, where they affect the airways, lungs and pleura.

3.1.1 Upper respiratory tract infections

Acute pharyngitis presents a diagnostic and therapeutic dilemma. The majority of sore throats are caused by a variety of viruses; fewer than 20% are bacterial and hence potentially responsive to antibiotic therapy. However, antibiotics are widely prescribed and this reflects the difficulty in discriminating streptococcal from non-streptococcal infections clinically in the absence of microbiological documentation. Nonetheless, *Strep. pyogenes* is the most important bacterial pathogen and this responds to oral penicillin. However, up to 10 days treatment is required for its eradication from the throat. This requirement causes problems with compliance since symptomatic improvement generally occurs within 2–3 days.

Although viral infections are important causes of both otitis media and sinusitis, they are generally self-limiting. Bacterial infections may complicate viral illnesses, and are also primary causes of ear and sinus infections. *Strep. pneumoniae* and *Haemophilus influenzae* are the commonest bacterial pathogens. Amoxycillin is widely prescribed for these infections since it is microbiologically active, penetrates the middle ear and sinuses, is well tolerated and has proved effective.

3.1.2   *Lower respiratory tract infections*

Infections of the lower respiratory tract (LRT) include pneumonia, lung abscess, bronchitis, bronchiectasis and infective complications of cystic fibrosis. Each presents a specific diagnostic and therapeutic challenge which reflects the variety of pathogens involved and the frequent difficulties in establishing an accurate microbial diagnosis. The laboratory diagnosis of LRT infections is largely dependent upon culturing sputum. Unfortunately this may be contaminated with the normal bacterial flora of the upper respiratory tract during expectoration. In hospitalized patients, the empirical use of antibiotics before admission substantially diminishes the value of sputum culture and may result in overgrowth by non-pathogenic microbes, thus causing difficulty with the interpretation of sputum culture results. Alternative diagnostic samples include needle aspiration of sputum directly from the trachea or of fluid within the pleural cavity. Blood may also be cultured and serum examined for antibody responses or microbial antigens. In the community, few patients will have their LRT infection diagnosed microbiologically and the choice of antibiotic is based on clinical diagnosis.

Pneumonia.   The range of pathogens causing acute pneumonia includes viruses, bacteria and, in the immunocompromised host, parasites and fungi. Table 6.2 summarizes these pathogens and indicates drugs appropriate for their treatment. Clinical assessment includes details of the

Table 6.2 Microorganisms responsible for pneumonia and the therapeutic agent of choice

Pathogen	Drug(s) of choice
Streptococcus pneumoniae	Penicillin
Staphylococcus aureus	Flucloxacillin ± fusidic acid
Haemophilus influenzae	Amoxycillin or cefuroxime
Klebsiella pneumoniae	Cefotaxime ± gentamicin
Pseudomonas aeruginosa	Gentamicin ± azlocillin
Mycoplasma pneumoniae	Erythromycin
Legionella pneumophila	Erythromycin ± rifampicin
Chlamydia psittaci	Tetracycline
Mycobacterium tuberculosis	Rifampicin + isoniazid + ethambutol + pyrazinamide*
Herpes simplex, varicella/zoster	Acyclovir
Candida or *Aspergillus* spp.	Amphotericin B
Anaerobic bacteria	Penicillin or metronidazole

* Reduce to two drugs after 6–8 weeks.

evolution of the infection, any evidence of a recent viral infection, the age of the patient and risk factors such as corticosteroid therapy or pre-existing lung disease. The extent of the pneumonia, as assessed clinically or by X-ray, is also important.

Strep. pneumoniae remains the commonest cause of pneumonia and responds well to penicillin. In addition, a number of atypical infections may cause pneumonia and include *Mycoplasma pneumoniae*, *Legionella pneumophila*, psittacosis and occasionally Q fever. With psittacosis there may be a history of contact with parrots or budgerigars; while Legionnaires' disease has often been acquired during hotel holidays in the Mediterranean area. The atypical pneumonias, unlike pneumococcal pneumonia, do not respond to penicillin. Legionnaires' disease is treated with erythromycin and in the presence of severe pneumonia, rifampicin is added to the regimen. Mycoplasma infections are best treated with either erythromycin or tetracycline, while the latter drug is indicated for both psittacosis and Q fever.

Lung abscess. Destruction of lung tissue may lead to abscess formation and is a feature of aerobic Gram-negative bacillary and *Staph. aureus* infections. In addition, aspiration of oropharyngeal secretion can lead to chronic low-grade sepsis with abscess formation and the expectoration of foul-smelling sputum which characterizes anaerobic sepsis. The latter condition responds to high-dose penicillin, which is active against most of the normal oropharyngeal flora, while metronidazole may be appropriate for strictly anaerobic infections. In the case of aerobic Gram-negative bacillary sepsis, aminoglycosides, with or without a broad-spectrum cephalosporin, are the agents of choice. Acute staphylococcal pneumonia is an extremely serious infection and requires treatment with high-dose flucloxacillin alone or in combination with fusidic acid.

Cystic fibrosis. Cystic fibrosis is a multisystem, congenital abnormality which often affects the lungs and results in recurrent infections, initially with *Staph. aureus*, subsequently with *H. influenzae* and eventually leads on to recurrent *Ps. aeruginosa* infection. The latter organism is associated with copious quantities of purulent sputum which is extremely difficult to expectorate. *Ps. aeruginosa* is a cofactor in the progressive lung damage which is eventually fatal in these patients. Repeated courses of antibiotics are prescribed and although they have improved the quality and longevity of life, infections caused by *Ps. aeruginosa* are difficult to treat and require repeated hospitalization and administration of parenteral anti-biotics such as an aminoglycoside, either alone or in combination with an antipseudomonal penicillin. The dose of aminoglycosides tolerated by these patients is often higher than in normal individuals and is associated with larger volumes of distribution for these and other agents. Some benefit may also be obtained from inhaled aerosolized antibiotics. Unfortunately drug resistance may emerge and makes drug selection more dependent upon laboratory guidance.

3.2   **Urinary tract infections**

Urinary tract infection is a common problem in both community and hospital practice. Although occurring throughout life, infections are more common in pre-school girls and women during their childbearing years, although in the elderly the sex distribution is similar. Infection is predisposed by factors which impair urine flow. These include congenital abnormalities, reflux of urine from the bladder into the ureters, kidney stones and tumours and, in males, enlargement of the prostate gland. Bladder catheterization is an important cause of urinary tract infection in hospitalized patients.

3.2.1   *Pathogenesis*

In those with structural or drainage problems the risk exists of ascending infection to involve the kidney and occasionally the bloodstream. Although structural abnormalities may be absent in women of childbearing years, infection can become recurrent, symptomatic and extremely distressing. Of greater concern is the occurrence of infection in the preschool child since normal maturation of the kidney may be impaired and result in progressive damage which presents as renal failure in later life.

From a therapeutic point of view, it is essential to confirm the presence of bacteriuria (a condition in which there are bacteria in the urine) since symptoms alone are not a reliable method of documenting infection. This applies particularly to bladder infection where the symptoms of burning micturition (dysuria) and frequency can be associated with a variety of non-bacteriuric conditions. Patients with symptomatic bacteriuria should always be treated. However, the necessity to treat asymptomatic bacteriuric patients varies with age and the presence or absence of underlying urinary tract abnormalities. In the pre-school child it is essential to treat all urinary tract infections and maintain the urine in a sterile state so that normal kidney maturation can proceed. Likewise in pregnancy there is a risk of infection ascending from the bladder to involve the kidney. This is a serious complication and may result in premature labour. Other indications for treating asymptomatic bacteriuria include the presence of underlying renal abnormalities such as stones which may be associated with repeated infections caused by *Proteus* spp.

3.2.2   *Drug therapy*

The antimicrobial treatment of urinary tract infection presents a number of interesting challenges. Drugs must be selected for their ability to achieve high urinary concentrations and, if the kidney is involved, adequate tissue concentrations. Safety in childhood or pregnancy is important since repeated or prolonged medication may be necessary. The choice of agent will be dictated by the microbial aetiology and suscepti-

Table 6.3 Urinary tract infection—distribution of pathogenic bacteria in the community and hospitalized patients

Organism	Community (%)	Hospital (%)
Escherichia coli	75	55
Proteus mirabilis	10	13
Klebsiella or *Enterobacter* spp.	4	18
Enterococci	6	5
Staphylococcus epidermidis	5	4
Pseudomonas aeruginosa	–	5

bility findings, since the latter can vary widely among Gram-negative enteric bacilli, especially in those who are hospitalized. Table 6.3 shows the distribution of bacteria causing urinary tract infection in the community and in hospitalized patients. The greater tendency towards infections caused by *Klebsiella* spp. and *Ps. aeruginosa* should be noted since antibiotic sensitivity is more variable for these pathogens. Drug resistance has increased substantially in recent years and has reduced the value of formerly widely prescribed agents such as the sulphonamides and ampicillin.

Uncomplicated community-acquired urinary tract infection presents few problems with management. Drugs such as trimethoprim, co-trimoxazole, ciprofloxacin and ampicillin are widely used. Cure rates are high for ciprofloxacin and the trimethoprim-containing regimens, although drug resistance to ampicillin has increased. Treatment for 3 days is generally satisfactory and is usually accompanied by prompt control of symptoms. Single-dose therapy with amoxycillin 3 g or co-trimoxazole 1920 mg (4 tablets) has also been shown to be effective in selected individuals. Alternative agents include nitrofurantoin and nalidixic acid, although these are not as well tolerated.

It is important to demonstrate the cure of bacteriuria with a repeat urine sample collected 4–6 weeks after treatment, or sooner should symptoms fail to subside. Recurrent urinary tract infection is an indication for further investigation of the urinary tract to detect underlying pathology which may be surgically correctable. Under these circumstances it also is important to maintain the urine in a sterile state. This can be achieved with repeated courses of antibiotics, guided by laboratory sensitivity data. Alternatively, long-term chemoprophylaxis for periods of 6–12 months to control infection by either prevention or suppression is widely used. Trimethoprim is the most commonly prescribed chemoprophylactic agent and is given as a single nightly dose. This achieves high urinary concentrations throughout the night and generally ensures a sterile urine. Nitrofurantoin is an alternative agent.

Infection of the kidney demands the use of agents which achieve adequate tissue as well as urinary concentrations. Since bacteraemia (a condition in which there are bacteria circulating in the blood) may complicate infection of the kidney it is generally recommended that antibiotics be administered parenterally. Although ampicillin was formerly widely used, drug resistance is now common and agents such as cefuroxime or cefotaxime are often preferred, since the aminoglycosides, although highly effective and preferentially concentrated within the renal cortex, carry the risk of nephrotoxicity.

Infections of the prostate tend to be persistent, recurrent and difficult to treat. This is in part due to the more acid environment of the prostate gland which inhibits drug penetration by many of the antibiotics used to treat urinary tract infection. Agents which are basic in nature, such as erythromycin, achieve therapeutic concentrations within the gland but unfortunately are not active against the pathogens responsible for bacterial prostatitis. Trimethoprim, however, is a useful agent since it is preferentially concentrated within the prostate and active against many of the causative pathogens. It is important that treatment be prolonged for several weeks, since relapse is common.

3.3   Gastrointestinal infections

The gut is vulnerable to infection by viruses, bacteria, parasites and occasionally fungi. Virus infections are the most prevalent but are not subject to chemotherapeutic intervention. Bacterial infections are more readily recognized and raise questions concerning the role of antibiotic management. Parasitic infections of the gut are beyond the scope of this chapter.

Bacteria cause disease of the gut as a result of either mucosal invasion or toxin production or a combination of the two mechanisms as summarized in Table 6.4. Treatment is largely directed at replacing and maintaining an adequate intake of fluid and electrolytes. Antibiotics are generally not recommended for infective gastroenteritis, but deserve

Table 6.4 Bacterial gut infections—pathogenic mechanisms

Origin	Site of infection	Mechanism
Campylobacter jejuni	Small and large bowel	Invasion
Salmonella spp.	Small and large bowel	Invasion
Shigella spp.	Large bowel	Invasion ± toxin
Escherichia coli		
enteroinvasive	Large bowel	Invasion
enterotoxigenic	Small bowel	Toxin
Clostridium difficile	Large bowel	Toxin
Staphylococcus aureus	Small bowel	Toxin
Vibrio cholerae	Small bowel	Toxin
Clostridium perfringens	Small bowel	Toxin
Yersinia spp.	Small and large bowel	Invasion
Bacillus cereus	Small bowel	Invasion ± toxin
Vibrio parahaemolyticus	Small bowel	Invasion ± toxin

consideration where they have been demonstrated to abbreviate the acute disease or to prevent complications including prolonged gastrointestinal excretion of the pathogen where this poses a public health hazard.

It should be emphasized that most gut infections are self-limiting. However, attacks can be severe and may result in hospitalization. Antibiotics are used to treat severe campylobacter and shigella infections; erythromycin and co-trimoxazole, respectively, are the preferred agents. Such treatment abbreviates the disease and eliminates gut excretion in shigella infection. However, in severe campylobacter infection the data are currently equivocal, although the clinical impression favours the use of erythromycin for severe infections. The role of antibiotics for campylobacter and shigella infections should be contrasted with gastrointestinal salmonellosis, for which antibiotics are contraindicated since they do not abbreviate symptoms and are associated with more prolonged gut excretion and introduce the risk of adverse drug reactions. However, in severe salmonellosis, especially at extremes of age, systemic toxaemia and bloodstream infection can occur and under these circumstances treatment with either chloramphenicol or co-trimoxazole is appropriate.

Typhoid and paratyphoid fevers (known as enteric fevers), although acquired by ingestion of salmonellae, *Sal. typhi* and *Sal. paratyphi*, respectively, are largely systemic infections and antibiotic therapy is mandatory; chloramphenicol remains the drug of choice although co-trimoxazole is a satisfactory alternative. Prolonged gut excretion of *Sal. typhi* is a well-known complication of typhoid fever and is a major public health hazard in developing countries. Treatment with high-dose ampicillin can eliminate the gall-bladder excretion which is the major site of persistent infection in carriers. However, the presence of gallstones reduces the chance of cure.

Cholera is a serious infection causing epidemics throughout Asia. Although a toxin-mediated disease, largely controlled with replacement of fluid and electrolyte losses, tetracycline has proved effective in eliminating the causative vibrio from the bowel, thereby abbreviating the course of the illness and reducing the total fluid and electrolyte losses.

Traveller's diarrhoea may be caused by one of many gastrointestinal pathogens (Table 6.4). However, enterotoxigenic *E. coli* is the commonest pathogen. Whilst it is generally short lived, traveller's diarrhoea can seriously mar a brief period abroad, be it for holiday or business purposes. Although not universally accepted, the use of short-course co-trimoxazole can abbreviate an attack and may also be used prophylactically when travelling to developing countries where standards of food hygiene and sanitation are unreliable.

3.4   **Skin and soft tissue infections**

Infections of the skin and soft tissue commonly follow traumatic injury to the epithelium but occasionally may be blood-borne. Interruption of the integrity of the skin allows ingress of microorganisms to produce superficial, localized infections which on occasion may become more deep-seated and spread rapidly through tissues. Skin trauma complicates surgical incisions and accidents, including burns. Similarly, prolonged immobilization can result in pressure damage to skin from impaired blood flow. It is most commonly seen in patients who are unconscious.

Microbes responsible for skin infection often arise from the normal skin flora which includes *Staph. aureus*. In addition *Strep. pyogenes*, *Ps. aeruginosa* and anaerobic bacteria are other recognized pathogens. Viruses also affect the skin and mucosal surfaces, either as a result of generalized infection or localized disease as in the case of herpes simplex. The latter is amenable to antiviral therapy in selected patients, although for the majority of patients, virus infections of the skin are self-limiting.

Strep. pyogenes is responsible for a range of skin infections: impetigo is a superficial infection of the epidermis which is common in childhood and is highly contagious; cellulitis is a more deep-seated infection which spreads rapidly through the tissues to involve the lymphatics and occasionally the bloodstream; erysipelas is a rapidly spreading cellulitis commonly involving the face, which characteristically has a raised leading edge due to lymphatic involvement. Penicillin is the drug of choice for all these infections although in severe instances parenteral administration is appropriate. The use of topical agents, such as tetracycline, to treat impetigo may fail since drug resistance is now recognized.

Staph. aureus is responsible for a variety of skin infections which require therapeutic approaches different from those of streptococcal infections. Staphylococcal cellulitis is indistinguishable clinically from streptococcal cellulitis and responds to cloxacillin or flucloxacillin, but generally fails to respond to penicillin owing to penicillinase (β-lactamase) production. *Staph. aureus* is an important cause of superficial, localized skin sepsis which varies from small pustules to boils and occasionally to a more deeply invasive, suppurative skin abscess known as a carbuncle. Antibiotics are generally not indicated for these conditions. Pustules and boils settle with antiseptic soaps or creams and often discharge spontaneously, whereas carbuncles frequently require surgical drainage. *Staph. aureus* may also cause postoperative wound infections, sometimes associated with retained suture material, and settles once the stitch is removed. Antibiotics are only appropriate in this situation if there is extensive accompanying soft tissue invasion.

Anaerobic bacteria are characteristically associated with foul-smelling wounds. They are found in association with surgical incisions following intra-abdominal procedures and pressure sores which are usually located over the buttocks and hips where they become infected with faecal flora. These infections are frequently mixed and include Gram-negative enteric bacilli which may mask the presence of underlying anaerobic bacteria. The principles of treating anaerobic soft tissue infection again emphasize the need for removal of all foreign and devitalized material. Antibiotics such as metronidazole or clindamycin should be considered where tissue invasion has occurred.

The treatment of infected burn wounds presents a number of peculiar facets. Burns are initially sterile, especially when they involve all layers of the skin. However, they rapidly become colonized with bacteria whose growth is supported by the protein-rich exudate. Staphylococci, *Strep. pyogenes* and, particularly, *Ps. aeruginosa* frequently colonize burns and may jeopardize survival of skin grafts and occasionally, and more seriously, result in bloodstream invasion. Treatment of invasive *Ps. aeruginosa* infections requires combined therapy with an aminoglycoside, such as gentamicin or tobramycin, and an antipseudomonal agent, such as azlocillin, ticarcillin or ceftazidime. This produces high therapeutic concentrations which generally act in a synergistic manner. The use of aminoglycosides in patients with serious burns requires careful monitoring of serum concentrations to ensure that they are therapeutic yet non-toxic, since renal function is often impaired in the days immediately following a serious burn. Excessive sodium loading may complicate the use of large doses of antipseudomonal penicillins such as carbenicillin and to a lesser extent ticarcillin.

3.5	**Central nervous system infections**

The brain and spinal cord and its surrounding covering of meninges is subject to infection, which is generally blood-borne but may also complicate neurosurgery, penetrating injuries or direct spread from infection in the middle ear or nasal sinuses. Viral meningitis is the most common infection but is generally self-limiting. Occasionally destructive forms of encephalitis occur; an example is herpes simplex encephalitis. Bacterial infections include meningitis and brain abscess and carry a high risk of mortality, while, in those who recover, residual neurological damage or impairment of intellectual function may follow. This occurs despite the availability of antibiotics active against the responsible bacterial pathogens. Fungal infections of the brain, although rare, are increasing in frequency, particularly among immunocompromised patients who either have underlying malignant conditions or are on potent cytotoxic drugs.

The treatment of bacterial infections of the central nervous system highlights a number of important therapeutic considerations. Bacterial meningitis is caused by a variety of bacteria although their incidence varies with age. In the neonate, *E. coli* and group B streptococci account for the majority of infections, while in the pre-school child *H. influenzae* is the commonest pathogen. *Neisseria meningitidis* has a peak incidence between 5 and 15 years of age, while pneumococcal meningitis is predominantly a disease of adults.

Penicillin is the drug of choice for the treatment of group B streptococcal, meningococcal and pneumococcal infections but, as discussed earlier, CSF concentrations of penicillin are significantly influenced by the intensity of the inflammatory response. To achieve therapeutic concentrations within the CSF, high dosages are required, and in the case of pneumococcal meningitis should be continued for 10–14 days.

Resistance of *H. influenzae* to ampicillin has increased in the past decade and varies geographically. Thus it can no longer be prescribed with confidence as initial therapy, and chloramphenicol, despite its toxic potential, is recommended. However, once laboratory evidence for β-lactamase activity is excluded, ampicillin can be safely substituted.

E. coli meningitis carries a mortality of greater than 40% and reflects both the virulence of this organism and the pharmacokinetic problems of achieving adequate CSF antibiotic levels. Gentamicin is widely used but rarely achieves a CSF concentration greater than $1\,mg\,l^{-1}$ owing to its poor lipid solubility. Attempts to improve gentamicin concentrations by directly injecting the drug into the spinal fluid have not improved survival. More recently, potent broad-spectrum cephalosporins such as cefotaxime and ceftazidime have been shown to achieve better therapeutic levels and are now widely used to treat Gram-negative bacillary meningitis. Treatment again must be prolonged for periods ranging from 2 to 4 weeks.

Brain abscess presents a different therapeutic challenge. An abscess is locally destructive to the brain and causes further damage by increasing intracranial pressure. The infecting organisms are varied but those arising from middle ear or nasal sinus infection are often polymicrobial and include anaerobic bacteria, microaerophilic species and Gram-negative enteric bacilli. Less commonly, a pure *Staph. aureus* abscess may complicate blood-borne spread. Brain abscess is a neurosurgical emergency and requires drainage. However, antibiotics are an important adjunct to treatment. The polymicrobial nature of many infections demands prompt and careful laboratory examination to determine optimum therapy. Drugs are selected not only on their ability to penetrate the blood–brain barrier and enter the CSF but also on their ability to penetrate the brain substance. Chloramphenicol has been widely used owing to its favourable pharmacokinetics. However, metronidazole has proved a valuable alternative agent in such infections, although it is not active against microaerophilic streptococci which must be treated with high-dose benzylpenicillin.

4   Antibiotic policies

4.1   Rationale

The plethora of available antimicrobial agents presents both an increasing problem of selection to the prescriber and difficulties to the diagnostic laboratory as to which agents should be tested for susceptibility. Differences in antimicrobial activity among related compounds are often of minor importance but can occasionally be of greater significance and may be a source of confusion to the non-specialist. This applies particularly to large classes of drugs, such as the penicillins and cephalosporins, where there has been an explosion in the availability of new agents in recent years. Guidance, in the form of an antibiotic policy, has a major role to play in providing the prescriber with a range of agents appropriate to his/her needs and should be supported by laboratory evidence of susceptibility to these agents.

In recent years, increased awareness of the cost of medical care has led to a major review of various aspects of health costs. The pharmacy budget has often attracted attention since, unlike many other hospital expenses, it is readily identifiable in terms of cost and prescriber. Thus an antibiotic policy is also seen as a means whereby the economic burden of drug prescribing can be reduced or contained. There can be little argument with the recommendation that the cheaper of two compounds should be selected where agents are similar in terms of efficacy and adverse reactions. Likewise generic substitution is also desirable provided there is bioequivalence. It has become increasingly impractical for pharmacists to stock all the formulations of every antibiotic currently available, and here again an antibiotic policy can produce significant savings by limiting the amount of stock held. A policy based on a restricted number of agents also enables price reduction on purchasing costs through competitive tendering. The above activities have had a major influence on containing or reducing drug costs, although these savings have often been lost as new and often expensive preparations become available, particularly in the field of biological and anticancer therapy.

Another argument in favour of an antibiotic policy is the occurrence of drug-resistant bacteria within an institution. The presence of sick patients and the opportunities for the spread of microorganisms can produce outbreaks of hospital infection. The excessive use of selected agents has been associated with the emergence of drug-resistant bacteria which have often caused serious problems within high dependency areas, such as intensive care units or burns units where antibiotic use is often high. One oft-quoted example is the occurrence of a multiple-antibiotic resistant *K. aerogenes* within a neurosurgical intensive care unit in which the organism became resistant to all currently available antibiotics and was associated with the widespread use of ampicillin. By prohibiting the use of all antibiotics, and in particular ampicillin, the resistant organism rapidly disappeared and the problem was resolved.

In formulating an antibiotic policy, it is important that the susceptibility of microorganisms be monitored and reviewed at regular intervals. This applies not only to the hospital as a whole, but to specific high dependency units in particular. Likewise general practitioner samples should also be monitored. This will provide accurate information on drug susceptibility to guide the prescriber as to the most effective agent.

4.2 Types of antibiotic policies

There are a number of different approaches to the organization of an antibiotic policy. These range from a deliberate absence of any restriction on prescribing to a strict policy whereby all anti-infective agents must have expert approval before they are administered. Restrictive policies vary according to whether they are mainly laboratory controlled, by employing restrictive reporting, or whether they are mainly pharmacy controlled, by restrictive dispensing. In many institutions it is common practice to combine the two approaches.

4.2.1   *Free prescribing policy*

The advocates of a free prescribing policy argue that strict antibiotic policies are both impractical and limit clinical freedom to prescribe. It is also argued that the greater the number of agents in use the less likely it is that drug resistance will emerge to any one agent or class of agents. However, few would support such an approach, which is generally an argument for mayhem.

4.2.2   *Restricted reporting*

Another approach that is widely practised in the UK is that of restricted reporting. The laboratory, largely for practical reasons, tests only a limited range of agents against bacterial isolates. The agents may be selected primarily by microbiological staff or following consultation with their clinical colleagues. The antibiotics tested will vary according to the site of infection, since drugs used to treat urinary tract infections often differ from those used to treat systemic disease.

There are specific problems regarding the testing of certain agents such as the cephalosporins where the many different preparations have varying activity against bacteria. The practice of testing a single agent to represent first generation, second generation or third generation compounds is questionable, and with the new compounds susceptibility should be tested specifically to that agent. By selecting a limited range of compounds for use, sensitivity testing becomes a practical consideration and allows the clinician to use such agents with greater confidence.

4.2.3   *Restricted dispensing*

As mentioned above, the most Draconian of all antibiotic policies is the absolute restriction of drug dispensing pending expert approval. The expert opinion may be provided by either a microbiologist or infectious disease specialist. Such a system can only be effective in large institutions where staff are available 24 hours a day. This approach is often cumbersome, generates hostility and does not necessarily create the best educational forum for learning effective antibiotic prescribing.

A more widely used approach is to divide agents into those approved for unrestricted use and those for restricted use. Agents on the unrestricted list are appropriate for the majority of common clinical situations. The restricted list may include agents where microbiological sensitivity information is essential, such as for vancomycin and certain aminoglycosides. In addition, agents which are used infrequently but for specific indications, such as parenteral amphotericin B, are also restricted in use. Other compounds which may be expensive and used for specific indications, such as broad-spectrum β-lactams in the treatment of *Ps. aeruginosa* infections, may also be justifiably included on the restricted list. Items omitted from the restricted or unrestriced list are generally not stocked, although they can be obtained at short notice as necessary.

Such a policy should have a mechanism whereby desirable new agents are added as they become available and is most appropriately decided at a therapeutics committee. Policing such a policy is best effected as a joint arrangement between senior pharmacists and microbiologists. This combined approach of both restricted reporting and restricted prescribing is extremely effective and provides a powerful educational tool for medical staff and students faced with learning the complexities of modern antibiotic prescribing.

5　Further reading

Bennett W.M., Aronoff W.M., Morrison G., Golper T.A., Pulliam J., Wolfson M. & Singer I. (1983) Drug prescribing in renal failure—dosing guidelines for adults. *Am J Kidney Dis.* **3**, 155–193.

Finch R.G. (1988) Antibacterial chemotherapy. *Med Int.* **5**, 2146–2154.

Greenwood D. (1989) *Antimicrobial Chemotherapy*, 2nd edn. Oxford, New York & Tokyo: Oxford University Press.

Kucers A. & Bennett N.McK. (1987) *The Use of Antibiotics*, 4th edn. London: William Heinemann.

Lambert H.P. & O'Grady F. (1992) *Antibiotic and Chemotherapy*, 6th edn. Edinburgh and London: Churchill Livingstone.

Mandell G.L., Douglas R.G. & Bennett J.E. (Eds) (1990) *Principles and Practice of Infectious Diseases*, 3rd edn. New York: John Wiley.

7 Manufacture of antibiotics

1	**Introduction**	3.4.2	Fed nutrients
		3.4.3	Stimulation by PAA
2	**Choice of examples**	3.4.4	Termination
		3.5	Extraction
3	**The production of benzylpenicillin**	3.5.1	Removal of cells
3.1	The organism	3.5.2	Isolation of penicillin G
3.2	Inoculum preparation	3.5.3	Treatment of crude extract
3.3	The fermenter		
3.3.1	Oxygen supply	**4**	**The production of penicillin V**
3.3.2	Temperature control		
3.3.3	Defoaming agents and	**5**	**The production of cephalosporin C**
	instrumentation		
3.3.4	Media additions	**6**	**Acknowledgements**
3.3.5	Transfer and sampling systems		
3.4	Control of the fermentation	**7**	**Further reading**
3.4.1	Batched medium		

1 Introduction

The science of antibiotic fermentations is still imperfectly developed, despite the ever-increasing use of complex instrumentation, the application of feedback control techniques and the use of computers. This technology is involved with a living cell population which is changing both quantitatively and qualitatively throughout the productive cycle; no two ostensibly identical 'batches' are ever wholly alike. Dealing with this variation accounts for the difficulty, the attraction, the challenge and the excitement for those who, like the author, practise in this field.

2 Choice of examples

For the model antibiotic production process there can be really only one choice—the manufacture of benzylpenicillin (penicillin G, originally just 'penicillin'). This, the most renowned of antibiotics and the first to have been manufactured in bulk, is still universally prescribed. There is also a large and increasing demand for it as input material for semisynthetic antibiotics (Chapter 5). Although originally made by surface liquid culture, penicillin G is now produced by deep fermentation under stirred and aerated conditions, in vessels as large as $250\,m^3$ capacity.

No single product can exemplify all the important features of antibiotic manufacture. The discussion of penicillin G production leads naturally into the field of β-lactams manufacture generally; brief accounts of penicillin V (phenoxymethylpenicillin) and cephalosporin C give opportunities to mention further key aspects.

Two cautionary points must be made right at the outset. Firstly, important though the β-lactams are, they are but one of many families of antibiotics (Chapter 5). Secondly, most of the industrial microorganisms used to make β-lactams are fungi; this is atypical of antibiotics as a whole where *Streptomyces* spp. predominate. Chapter 5 and some of the Further Reading at the end of this chapter provide the broad perspective, including information on those antibiotics made by total or partial chemical synthesis, against which this present account with its necessarily selective subject matter should be read.

All the examples are of 'batched' fermentations, that is, of processes in which a volume of sterile medium in a vessel is inoculated, the broth is fermented for a defined period, the tank is then emptied and the proceeds extracted to yield the antibiotic. The vessel is then rebatched with medium and the sequence repeated, as often as required. In related fermentation industries, such as brewing, 'continuous' fermentation, as well as the older batch fermentation, is now employed. The production of microorganisms as food—the so-called 'single cell protein'—is carried out entirely by continuous process, sterile medium being added without interruption to the fermentation system with a balancing withdrawal (or 'harvesting') of broth for product extraction. The length of a 'continuous' fermentation is measured in weeks or months, rather than in hours or days as in the batch process. As yet, however, continuous fermentation is not common practice in the antibiotics industry.

Little more than simple recognition is given to those operations that occur after the fermentation stage, i.e. the recovery, purification, quality testing and sterile packaging of the products, even though these usually account for most of the total manufacturing costs. The quality of the fermented material can markedly affect the efficiency of all the succeeding operations, for at the end of a typical fermentation, the antibiotic concentration will rarely exceed $20\,\mathrm{g\,l^{-1}}$ and may be as low as $0.5\,\mathrm{g\,l^{-1}}$.

Details of the manufacture of streptomycin and griseofulvin are to be found in the previous editions of this book.

3   The production of benzylpenicillin

3.1   The organism

The original organism for the production of penicillin, *Penicillium notatum*, was isolated by Fleming in 1926 as a chance contaminant. However, all the penicillin-producing strains used in deep culture stem from a common ancestor that was isolated and purified from an infected Cantaloupe melon obtained at a market in Peoria, Illinois, USA. The infecting organism was *P. chrysogenum*. From this one ancestral fungus each penicillin manufacturer has evolved a particular production strain by a series of mutagenic treatments, each followed by the selection of improved variants. These selected variants have proved capable of producing amounts of penicillin far greater than those produced by the 'wild' strain, especially when fermented on media under particular control regimens developed in parallel with the strains. The original, 'wild' strains

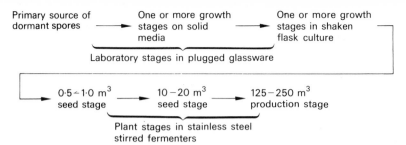

Fig. 7.1 Stages in the preparation of inoculum for the penicillin G fermentation.

of *Penicillium* produced a yellow pigment devoid of antibiotic properties which coloured the final product. Production mutants do not produce the pigment and yield a colourless end-product.

Production strains are stored in a dormant form by any of the standard culture preservation techniques. Thus, a spore suspension may be mixed with a sterile, finely divided, inert support and desiccated. Alternatively, spore suspensions in appropriate media can be lyophilized or stored in a liquid nitrogen biostat.

All laboratory manipulations are carried out in laminar flow cabinets in rooms in which filtered air is maintained at a slight positive pressure relative to their outer environment. Operators wear sterilized clothing and work aseptically. Antibiotic fermentations are, of strict necessity, pure culture processes.

3.2   Inoculum preparation

The aim is to develop for the production stage fermenter, a pure inoculum in sufficient volume and in the fast-growing (logarithmic) phase so that a high population density is soon obtained. Figure 7.1 shows a typical route by which the inoculum is produced. The time taken for each stage is measured in days and decreases as the sequence progresses.

The media are designed to provide the organism with all the nutrients that it requires. Adequate oxygen is provided in the form of sterile air and the temperature is controlled at the desired level. Principal criteria for transfer to the next stage in the progression are freedom from contamination and growth to a predetermined cell density.

The organism grows as branching filaments (hyphae) and by the time that the culture has progressed to the production stage it has a soup-like consistency.

3.3   The fermenter

A typical fermenter is a closed, vertical, cylindrical, stainless steel vessel with convexly dished ends and of $25-250\,m^3$ capacity. Its height is usually

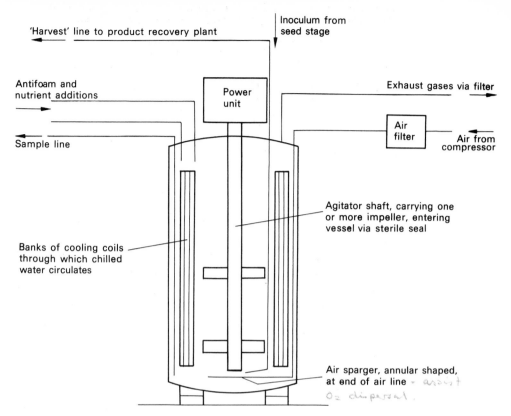

Fig. 7.2 Diagram of a typical fermenter.

two to three times its diameter. Figure 7.2 shows such a vessel diagrammatically, and Fig. 7.3 gives a view inside an actual vessel.

3.3.1   Oxygen supply

The penicillin fermentation needs oxygen, which is supplied as filter-sterilized air from a compressor. As oxygen is poorly soluble in water, steps are taken to assist its passage into the liquid phase and from aqueous solution into the mould. In a conventional fermenter, air is introduced at the bottom of the vessel via a ring 'sparger' that breaks the flow into a myriad of bubbles to increase the transfer area. These bubbles lose oxygen as they rise up the tank and, at the same time, carbon dioxide diffuses into them. The transfer of oxygen is further assisted by impellers mounted on a rotating vertical shaft driven by a powerful electric motor. Baffles are also included to achieve the correct blend of shear and of bulk circulation from the power supplied, and generally to promote intimate contact of cells and nutrients. Aeration is a major expense as very large amounts of energy are consumed. This has led to considerable research into novel, energetically more efficient methods of aeration and the next

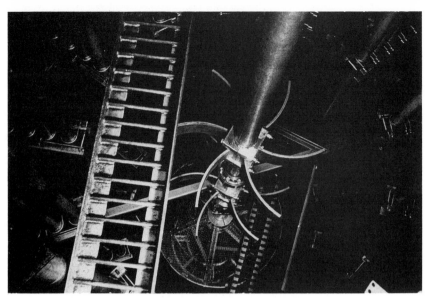

Fig. 7.3 View looking down into a 125-m³ stainless steel fermenter. (Courtesy of Glaxochem Ltd.)

generation of fermenters may include some that are radically different in design.

3.3.2 Temperature control

The production of penicillin G is very sensitive to temperature, the tolerance being less than 1°C. A lot of metabolic heat is generated and the fermentation temperature has to be reduced by controlled cooling. This heat transfer is achieved by circulating chilled water through banks of pipes inside the vessel or through external 'limpet' coils, continuous lengths of pipe welded in a shallow spiral round the vessel, providing a cooling jacket.

3.3.3 Defoaming agents and instrumentation

Microbial cultures foam when they are subjected to vigorous mechanical stirring and aeration. If this foaming is not controlled, culture is lost by entrainment in the exhaust gases and so there are systems, often automatic, for detecting incipient foaming, for temporarily applying back-pressure to contain the culture within the vessel and for the aseptic addition of defoaming agents.

Instrumentation is also fitted to provide a continuous display of important variables such as temperature and pH, the power used by the electric motor, the airflow and the amount of carbon dioxide gas in the exhaust gas stream. Manual or computer feedback control can be based either directly on the signals provided by the probes and sensors or on derived variables calculated from those signals, such as the respiratory coefficient or the rate of change of pH.

3.3.4 Media additions

Not all the nutrients required during fermentation are initially provided in the culture medium. Provision is therefore made to add these while the fermentation is in progress, usually via automatic systems that allow a present programme of continuous or discrete aseptic additions.

3.3.5 Transfer and sampling systems

Aseptic systems are provided to transfer the inoculum to the vessel, to allow the taking of routine samples during fermentation, and to transfer the final contents to the extraction plant when fermentation is complete. Sampling is essential to monitor the amount of growth, the running levels of key nutrients and the penicillin concentration. It is necessary also to check that there has been no contamination by unwanted microorganisms.

3.4 Control of the fermentation

Should oxygen availability fall below a critical level, penicillin G biosynthesis is greatly reduced although culture growth continues. Thus, if growth in the fermenter proceeds unchecked at the rate prevailing in the seed stages, the culture would become very dense and the available aeration would no longer be sufficient to maintain penicillin production. Accordingly, conditions are so adjusted that fast growth is achieved only until the cell population has reached the maximum density that the vessel can support. Further net growth is constrained by deliberately limiting the supply of a key nutrient (in practice, a sugar). The cells can then be stimulated to an 'overproduction' of penicillin G while restricting the amount of growth and a stable, highly productive cell population can be sustained.

3.4.1 Batched medium

The medium initially placed in the fermenter is a complete one but designed only to support the desired amount of early growth. The principal nitrogen source is corn steep liquor (CSL), a by-product of the maize starch-producing industry. This material was originally found to be specifically useful for the penicillin fermentation, but is now recognized as valuable in many fungal antibiotic fermentation media. Apart from its primary purpose in supplying cheap and readily available nitrogen, CSL also contains a useful range of carbon compounds, such as acids and sugars, inorganic ions and growth factors—in short, it is virtually a complete growth medium in itself.

The medium contains subsidiary nitrogen sources and additional essential nutrients such as calcium (added in the form of chalk to counter the natural acidity of the CSL), magnesium, sulphate, phosphate, potassium and trace metals. The medium is sterilized with steam at 120°C either in the fermenter itself or in ancillary plant, which may be worked continuously.

3.4.2 *Fed nutrients*

The sterile medium is stirred and aerated and its pH and temperature are set to the correct values on the process control monitors. It is then inoculated and the growth phase begins. The initial carbon source is sufficient in quantity to maintain early growth but not sufficient to provide the energy that penicillin production and maintenance of the cell population need during the rest of the fermentation. Carbon for these subsequent stages is 'fed' continuously in such a way as to limit net growth. Either sucrose or glucose is used, possibly as cheaper, impure forms, such as molasses or starch hydrolysate. As the concentration of residual sugar in the broth is too low to measure, the rate of feeding has to be learnt by experience and modified on the basis of systematic observation. An alternative way of attaining carbon limitation without the complication of a carefully monitored carbon feed is to supply all the carbohydrate at the outset as lactose. The rate-limiting hydrolysis of lactose to hexose is then relied upon to give a steady, slow feed of assimilable carbohydrate. Originally, all penicillin G was manufactured using lactose in this way and some manufacturers still prefer this technique.

Calcium, magnesium, phosphate and trace metals added initially are usually sufficient to last throughout the fermentation, but the microorganism needs further supplies of nitrogen and sulphur to balance the carbon feed. Nitrogen is often supplied as ammonia gas. The word 'balance' is used quite deliberately; the whole system is a balanced one. Thus, the carbon and nitrogen feeds not only satisfy the organism's requirements for these elements in the correct molar ratio, they also maintain an adequate reserve of ammonium ion and contribute to pH control, the carbon metabolism being acidogenic and balanced by the alkalinity of the ammonia. Sulphate is usually supplied in common with the sugar feed and, by getting the ratio correct, there is a balanced presentation of sulphate with an adequate pool of intermediates.

All feeds are sterilized before they are metered into the fermenter. Contaminants resistant to the antibiotic rarely find their way into the fermenter, but when they do, their effects are so damaging that prevention is of paramount importance. A resistant, β-lactamase-producing, fast-growing bacterial contaminant can destroy the penicillin already made, as well as consuming nutrients intended for the fungus, causing loss of pH control and causing interference with the subsequent extraction process.

The growth phase passes rapidly into the antibiotic production phase. The optimum pH and temperature for growth are not those for penicillin production and there may be changes in the control of these parameters. The only other event that marks the onset of the production phase is the addition of phenylacetic acid (PAA) by continuous feed.

Stimulation by PAA

Phenylacetic acid supplies the side chain of penicillin G (see also Chapter 5); without PAA, the organism synthesizes only small quantities of this penicillin. Indeed, it was the chance presence of phenylacetyl compounds in CSL (formed from phenylalanine in the grain by the natural bacterial flora during processing) that caused it to be established in early experiments as the best of the cheap complex nitrogen sources and led to the use of PAA. Not only does PAA stimulate penicillin G biosynthesis but it also suppresses the formation of other (unwanted) penicillins. High levels of PAA are, however, toxic to the organism and so it cannot be added indiscriminately. PAA is expensive. The feed provides an adequate standing level of PAA without approaching the toxic limit; the feed is reduced just before the end of the process so that the amount of unused (irrecoverable) precursor in the final culture is not excessive.

The building blocks for the biosynthesis of penicillin G are three amino acids, α-aminoadipic acid, cysteine and valine, and PAA. The amino acids condense to a tripeptide, ring closure of which gives the penicillin ring structure with an α-aminoadipyl side chain, isopenicillin N. The side chain is then displaced by a phenylacetyl group from PAA to give penicillin G.

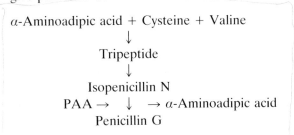

α-Aminoadipic acid + Cysteine + Valine
$\downarrow$
Tripeptide
$\downarrow$
Isopenicillin N
PAA $\rightarrow$ $\downarrow$ $\rightarrow$ α-Aminoadipic acid
Penicillin G

There comes a time when sequential improvements in penicillin productivity obtained by standard strain improvement techniques (physical and chemical mutagenesis in conjunction with a variety of selection techniques that apply pressure for high yielding variants) become subject to rate-limiting returns. At first, it is easy to double the 'titre' with each campaign; later in the genealogy even a 5% improvement would be regarded as excellent.

Recent developments by academic and industrial geneticists may well prove to have transformed this situation. Tremendous progress has been made since the mid-1980s both in the isolation and manipulation of the biosynthetic genes in this pathway and in the related routes to the cephalosporins (via the cephalosporin C-producing fungus *Acremonium chrysogenum*) and the cephamycins (via the cephamycin C-producing bacterium *Streptomyces clavuligerus*). Antibiotic manufacturers can now apply recombinant DNA technology to the industrial strains of filamentous microorganisms used to produce *β*-lactams and there are exciting prospects of making genetic changes that will very significantly increase productivity. These are discussed further, later in this chapter. There is plenty of scope for improvement: the best current industrial strains and processes convert little more than 10% of all elemental carbon into penicillin.

3.4.4   *Termination*

When to stop a fermentation is a very complex decision and several factors have to be taken into account. Quite often a manufacturer will find it appropriate to harvest shortly after the first signs of a faltering in the efficiency of conversion of the most costly raw material (sugar) into penicillin.

3.5   **Extraction**

3.5.1   *Removal of cells*

At harvest, the penicillin G is in solution exocellularly, together with a range of other metabolites and medium constituents. The first step is to remove the cells by filtration. This stage is carried out under conditions that avoid contamination of filtrate with *β*-lactamase-producing microorganisms which could lead to serious or total loss of product.

3.5.2   *Isolation of penicillin G*

The next stage is to isolate the penicillin G. Solvent extraction is the generally accepted process. In aqueous solution at pH 2–2.5 there is a high partition coefficient in favour of certain organic solvents such as amyl acetate, butyl acetate and methyl isobutyl ketone. The extraction has to be carried out quickly, for penicillin G is very unstable at these low pH values. The penicillin is then extracted back into an aqueous buffer at pH 7.5, the partition coefficient now being strongly in favour of the aqueous phase. The solvent is recovered by distillation for reuse.

3.5.3   *Treatment of crude extract*

Penicillin G is produced as various salts according to its intended use, whether as an input to semisynthetic *β*-lactam antibiotics manfacture or for clinical use in its own right.

The treatment of the crude penicillin extract varies according to the objective but involves formation of an appropriate salt, probably followed by charcoal treatment to remove pyrogens, and by sterilization. This last is usually achieved by filtration but pure metal salts of penicillin G can be safely sterilized by dry heat if desired.

For parenteral use, the antibiotic is packed in sterile vials as a powder (for reconstitution before use) or suspension. For oral use it is prepared in any of the standard presentations, such as film-coated tablets. Searching tests are carried out on an appreciable number of random samples of the finished product to ensure that it satisfies the stringent requirements for potency, purity, freedom from pyrogens and sterility.

4   The production of penicillin V

From penicillin G it is natural to move on to the other β-lactam antibiotics, beginning with penicillin V. This antibiotic is made by a very similar process to that used for penicillin G, but with phenoxyacetic acid as the precursor instead of PAA. In the biosynthetic pathway, the α-aminoadipyl side chain of isopenicillin N is replaced by a phenoxyacetyl group.

The microorganism is again *P. chrysogenum*. A manufacturer may use the same mutant strain to make both products or may have different mutants for the two penicillins.

Parallel situations of a single organism producing more than one natural product occur with other types of antibiotics; for example, strains of *Streptomyces aureofaciens* are used both for chlortetracycline and demethylchlortetracycline fermentations.

Just like penicillin G, penicillin V still finds extensive clinical use as a valuable antibiotic in its own right and it too can be used as a starting material for the manufacture of the semisynthetic penicillins, none of which can be made by direct fermentation.

5   The production of cephalosporin C

It is possible to convert penicillin V or G to cephalosporins by chemical ring expansion. The first generation cephalosporin cephalexin, for example, can be made in this way. Most cephalosporins used in clinical practice, however, are semisynthetics produced from the fermentation product cephalosporin C.

The ancestral strain of *Acremonium chrysogenum* (at that time called *Cephalosporium acremonium*) was isolated on the Sardinian coast following an observation that the local sewage outflow into the sea cleared at a quite remarkable rate. The original water samples were taken in 1945, roughly at the start of the penicillin production era, and it was some 15 years later that the first of the clinically useful cephalosporins became available.

The biosynthetic route to cephalosporin C is identical to that of the penicillins as far as isopenicillin N (section 3.4.3). The further route to cephalosporin C is shown below. Note the branch into a third series of β-lactam drugs, the cephamycins.

Isopenicillin N
↓
Penicillin N
↓
Desacetoxy cephalosporin C
↓
Desacetyl cephalosporin C → Cephamycin C (in certain *Streptomyces*)
↓
Cephalosporin C

The similarities in the routes to the three classes of antibiotics have facilitated progress in understanding the underlying molecular genetics; the results from studies of individual enzymes or enzyme systems have rapidly built up into a solid body of knowledge of β-lactam biosynthesis generally.

Most of the genes coding for the relevant enzymes have been isolated. Methods have been developed for increasing gene copy number in the hope that it will then prove possible to promote the expression of the replicate genes, relieve putative enzyme bottle-necks and produce more antibiotic. One example of success in this field has already been published for cephalosporin C production; others are attempting to do the same for penicillin. β-lactam synthetic genes have been shown to occur in clusters on the chromosome, raising the possibility of increasing the gene dosage for the whole set of biosynthetic genes. Whether this will be effective, or whether the limitation to biosynthesis lies in the supply of primary metabolite remains to be seen.

Given the similarity in the biosynthetic pathways to cephalosporin C and penicillin G, it is not surprising that the manufacturing processes are broadly similar. In common with many other antibiotic fermentations, no specific precursor feed is necessary for cephalosporin C. There is sufficient acetyl group substrate for the terminal acetyltransferase reaction available from the organism's metabolic pool.

The product is extracted from the culture fluid by adsorption on to carbon or resins rather than by solvent. This illustrates an important general point that antibiotic manufacturing processes differ from one another much more in their product recovery stages, than in their fermentation stages.

Cephalosporin C has no clinically useful antibiotic activity and it is all used to make semisynthetic cephalosporins.

6   Acknowledgements

The author wishes to record his appreciation to his colleague, Mr N. J. Maishman, for his advice on microbial genetics.

Further reading

Bu'Lock J.D., Nisbet L.J. & Winstanley D.J. (Eds) (1983) *Bioactive Microbial Products, vol. II, Development and Production*. London: Academic Press.

Calam C.T. (1987) *Process Development in Antibiotic Fermentations*. Cambridge Studies in Biotechnology, 4 (Eds Sir James Baddiley, N.H. Carey, J.F. Davidson, I.J. Higgins & W.G. Potter). Cambridge: Cambridge University Press.

Hugo W.B. & Mol H. (1972) Antibiotics and chemotherapeutic agents. In *Materials and Technology* (Eds L.W. Codd., K. Dijkoff, J.H. Fearon, C.J. van Oss., H.G. Roeberson & E.G. Stanford). London & Amsterdam: Longman & de Bussy.

Peberdy J.F. (Ed.) (1987) *Penicillium and Acremonium*. Biotechnology Handbooks, 1 (Series eds T. Atkinson & R.F. Sherwood). New York and London: Plenum Press. (See, in particular, Chapters 2 and 5.)

Queener S.W. (1990) Molecular biology of penicillin and cephalosporin biosynthesis. *Antimicrob Agents Chemother.* **34**, 943–948.

Smith J.E. (1985) *Biotechnology Principles*. Aspects of Microbiology Series No. 11. Wokingham: Van Nostrand Reinhold.

Stowell J.D., Bailey P.J. & Winstanley D.J. (Eds) (1986) *Bioactive Microbial Products, vol. III, Downstream Proessing*. London: Academic Press.

Van Damme E.J. (1984) *Biotechnology of Industrial Antibiotics*. New York: Marcel Dekker.

Verrall M.S. (Ed.) (1985) *Discovery and Isolation of Microbial Products*. Society of Chemical Industry Series in Biological Chemistry and Biotechnology. Chichester. Ellis Horwood.

A good source of articles on individual antibiotics, groups of antibiotics, fermentation plant and related topics is the series *Progress in Industrial Microbiology* edited originally by D.J.D. Hockenhull and published by Heywood Books, London. These articles normally carry extensive references to the original literature.

8 Principles of methods of assaying antibiotics

1	**Introduction**		4.1.1	Pump
1.1	Assay calibration		4.1.2	Injector
1.2	Assay precision		4.1.3	Detector
1.3	Accuracy		4.2	Stationary phases
1.4	Determination of assay performance		4.3	Reverse phase chromatography (RPLC)
			4.4	Paired-ion chromatography
2	**Microbiological assays**		4.5	Mobile phase formulation
2.1	Agar diffusion assays		4.6	Sample preparation
2.1.1	One-dimensional assay		4.7	Calibration
2.1.2	Two/three-dimensional assay		4.7.1	Internal standardization
2.1.3	Dynamics of zone formation		4.8	Controlling precision
2.1.4	Controlling reproducibility			
2.1.5	Zone measurement		**5**	**Immunoassay**
2.1.6	Calibration		5.1	Principles of immunoassay
2.2	Rapid microbiological assay methods		5.2	Preparation of antibody
			5.3	Drug label for preparing drug tracer
2.2.1	Urease		5.4	Measurement of bound label
2.2.2	Luciferase assay		5.4.1	Heterogeneous immunoassay
			5.4.2	Homogeneous immunoassay
3	**Radioenzymatic (transferase) assays**		5.5	Calibration
3.1	Calibration		5.5.1	Heterogeneous immunoassay
3.2	Non-isotopic modification		5.5.2	Homogeneous immunoassay
			5.6	Controlling precision
4	**HPLC**			
4.1	LC equipment		**6**	**Further reading**

1 Introduction

The need to assay antimicrobial agents (the term here is used to cover antibiotics and other antimicrobial drugs) arises in three major situations: (i) during manufacture for determining potency and quality control; (ii) for determining the pharmacokinetics of a drug in animals or humans; and (iii) for monitoring and controlling antimicrobial chemotherapy.

The first involves assaying relatively high concentrations of pure drug in an uncomplicated solution such as water or buffer. The latter two involve measuring low concentrations of drug in biologial fluids such as serum or urine, which by their nature contain many extraneous materials that may interfere with the antibiotic determination. The fluid in which the drug to be assayed is dissolved is usually referred to as the *matrix*.

Traditionally, antimicrobial agents have been assayed by microbiological assay. However, in recent years a greater awareness of the problems of poor assay specificity where drugs are partially metabolized or when other antibiotics may also be present, together with a requirement for more rapid techniques, has encouraged the investigation of other methods. These include enzymatic, immunological and chromatographic assays. This chapter will briefly cover the principles of microbiological assay, enzymatic assay, immunoassay and high-performance liquid chromatography (HPLC).

1.1 Assay calibration

Whichever technique is used for the assay, proper calibration is essential if the result is to be expressed in absolute units (e.g. $mg\,l^{-1}$). Thus a pure sample of the drug to be assayed or a sample of known potency is essential for the preparation of calibrator solutions. Some drugs are hygroscopic, and potency may be expressed as 'as-is' potency or 'dried' potency. As-is potency is the potency of the powder without drying, assuming it has been properly stored such that water is not lost or gained. Dried potency is the potency after drying to constant weight under defined conditions. If as-is potency is used, the drug must be stored such that it cannot absorb or lose water; if dried potency is used, then the drug must be dried before weighing. Once suitable standard material is obtained, calibrator solutions (working standards) covering an appropriate range of concentrations must be prepared. The number and concentration range of the calibrators will depend on the type of assay being performed and this is discussed more fully under individual methods. The matrix in which the calibrators are dissolved is also important and, unless it can be shown otherwise, should be similar to the matrix of the samples. This is particularly important when assaying drugs in serum, since protein binding can often profoundly influence the results of microbiological assay. No assay can produce accurate results without the appropriate calibrators accurately prepared in a suitable matrix.

1.2 Assay precision

Precision is a measure of reproducibility and is determined by replicating a single sample a number of times and determining the mean result ($\bar{X}$), the standard deviation (SD) and the coefficient of variation (SD/$\bar{X}$ × 100). Precision within a single run (intra-assay precision) and between runs (inter-assay precision) should both be determined. The degree of precision required in a particular situation will determine such factors as the number of replicates required for each calibrator and sample solution, and the number and concentration range of calibrators. The precision of many assays varies with concentration and should be determined with low, medium and high concentration samples.

1.3 Accuracy

Assuming that the calibrator solutions were correctly prepared from the appropriate drug, the accuracy of a particular result will depend on assay precision and assay specificity. Poor specificity may result if the samples contain endogenous interfering substances, or other antimicrobial agents, or active metabolites of the drug being assayed. If other drugs or drug metabolites are present, inaccuracy will be expressed as a positive bias (i.e. results higher than expected). A negative bias will result if there are antagonists present. Inaccuracy due to poor precision will usually show no bias, and that due to over- or under-potent calibrators will show negative and positive bias respectively.

1.4   Determination of assay performance

In assessing the performance characteristic of a newly developed assay, intra-assay and inter-assay precision over the whole range of expected concentrations should be determined. Accuracy should be checked with spiked samples (i.e. of known concentration) over the whole range of concentrations and also, when measuring drugs in biological fluids, with samples from individuals who have received the drug enterally or parenterally since *in vivo* metabolites may only be apparent in these people. Substances which might interfere must be checked for interference both alone and in the presence of the drug being assayed. Ideally, a large number of samples should be assayed by the new method and a reference method, the results compared by linear regression, and the correlation coefficient of the two methods determined. When a method is used routinely, internal controls (samples of known value) should be included in every run and laboratories assaying clinical specimens should, where possible, participate in an external quality control scheme.

2   Microbiological assays

In microbiological assays the response of a growing population of micro-organisms to the antimicrobial agent is measured. An exellent and detailed account of microbiological assays is given by Hewitt & Vincent (1989).

2.1   Agar diffusion assays

In these, the drug diffuses into agar seeded with a susceptible microbial population and produces a zone of growth inhibition. The assay may be one-, two- or three-dimensional.

2.1.1   *One-dimensional assay*

In this type of assay, capillary tubes containing agar seeded with indicator organism are overlaid with sample. The antibiotic diffuses downwards into the agar and a zone of inhibition is formed. This technique is now little used. It is difficult to set up and standardize but, nevertheless, may be applicable for assaying antibiotics anaerobically or when only very small samples are available. This technique shows good precision.

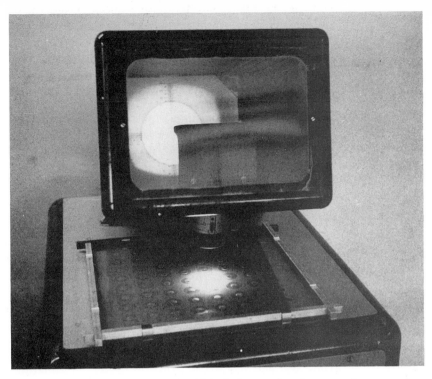

Fig. 8.1 Photograph of a typical microbiological assay plate demonstrating zones of inhibition.

2.1.2   *Two/three-dimensional assay*

This is the commonest form of microbiological assay used today. Samples to be assayed are applied in some form of reservoir (porcelain cup, paper disc or well) to a thin layer of agar seeded with indicator organism. The drug diffuses into the medium and, as with the one-dimensional assay, after incubation a zone of growth inhibition forms, in this case as a circle around the reservoir (Fig. 8.1). All other factors being constant, the diameter of the zone of inhibition is, within limits, related to the concentration of antibiotic in the reservoir.

2.1.3   *Dynamics of zone formation*

During incubation the antibiotic diffuses from the reservoir, and that part of the microbial population away from the influence of the antibiotic increases by cell division. The edge of a zone is formed when the minimum concentration of antibiotic which will inhibit the growth of the organism on the plate (critical concentration) reaches, for the first time, a population density too great for it to inhibit. The position of the zone edge is thus determined by the initial population density, growth rate of the organism and the rate of diffusion of the antibiotic.

169   *Principles of methods of assaying antibiotics*

This critical concentration (C') at the edge of a zone of inhibition, when it is formed, can be calculated from the formula:

$$\ln C' = \frac{\ln Cd^2}{4DT_0}$$

where C is the concentration of drug in the reservoir; d, the distance from the reservoir to the zone edge; D, the diffusion coefficient (constant for a given antibiotic in a given matrix diffusing into a given medium at a given temperature); and T_0, the critical time at which the position of the zone edge was determined.

In any assay where T_0 and D are constant, a plot of $\ln C$ against d^2 for a range of concentrations will, within limits, give a straight line which can be extrapolated to determine C'. C' represents the minimum value of C which will give a zone of inhibition and is independent of D and T_0. D and T_0 can be manipulated to decrease or increase the zone sizes for concentrations of C greater than C'. Pre-incubation will increase the number of microorganisms on the plate, and thus the critical population density will be achieved more rapidly (i.e. T_0 is reduced) so zones will be smaller. Reducing the microbial growth rate will result in larger zones. Increasing the sample size or reducing the agar thickness will increase zone size and vice versa. When designing an assay, indicator organism, medium, sample size and incubation temperature must be optimized to give a suitably large range of zone sizes over the required range of antibiotic concentrations.

2.1.4   *Controlling reproducibility*

Since zone sizes depend on so many variables (sample size, agar thickness, indicator organism, population density, organism growth rate, drug diffusion rate) greater care must be taken to standardize them to achieve good precision. A large $25 \times 25\,$cm or $30 \times 30\,$cm flat-bottomed plate should be used and should be carefully levelled before the agar is poured. The effects of variations in agar composition are minimized by preparing and using aliquots of large batches. Variations in inoculum size of the indicator organism can be reduced by growing a large volume, dispensing it into aliquots sufficient for a single plate, and storing them in liquid nitrogen to preserve viability. If spore inocula are used these may be stored for long periods under conditions inhibiting germination and preserving viability. Calibrators and samples should be dosed together on a single plate and variation in thickness of the agar, edge-effects, and possible variations in incubation temperature due to uneven warming in the incubator should be minimized by using some kind of predetermined 'random layout'. A Latin square arrangement (Fig. 8.2A), where the number of replicates equals the number of specimens, ensures maximum precision. However, less demanding methods using fewer replicates are usually acceptable for pharmacokinetic or clinical assays (Fig. 8.2B,C).

A

3	4	8	1	7	6	5	2
2	7	3	8	5	1	4	6
8	1	5	2	4	3	6	7
4	2	1	6	3	5	7	8
7	8	2	5	6	4	3	1
6	5	7	3	8	2	1	4
1	3	6	4	2	7	8	5
5	6	4	7	1	8	2	3

Latin square

B

5	1	3	7	14	10	12	16
8	4	2	6	9	13	15	11
14	7	5	16	12	1	3	10
2	11	9	4	15	6	8	13
3	10	12	6	5	2	14	7
15	6	8	13	2	11	9	4
12	16	14	10	3	7	5	1
9	13	15	11	8	4	2	6

16 doses (samples or calibrators in quadruplicate)

C

C1	S2	C2	C4	C3
S1	C5	S3	S4	S5
S3	C3	C1	S2	C4
S4	S1	S5	C5	C2
C2	C4	C3	S4	C1
S2	C5	S1	S5	S3

5 samples and 5 calibrators in triplicate

Fig. 8.2 Example 'random' patterns for use in microbiological plate assay.

2.1.5   *Zone measurement*

For maximum precision, a magnifying zone-reader should be used. Also, to avoid subjective bias, the person reading the plate should be unaware whether he/she is reading a calibrator or a test zone. 'Random' arrangements like those in Fig. 8.2 can help to ensure this, the pattern being decoded after the plate is read.

2.1.6   *Calibration*

Standard curves.   At least two, and up to seven, calibrators covering the required range of concentrations may be used and these should be equally

171   *Principles of methods of assaying antibiotics*

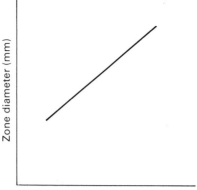

Fig. 8.3 Standard curve for an agar plate diffusion assay.

spaced on a logarithmic scale (e.g. 0.5, 1, 2, 4, 8, 16, 32 mg l^{-1}). The number of replicates of each calibrator should be the minimum necessary to give the desired precision. A manual plot of either zone size against log$_{10}$ concentration or (zone size)^2 against log$_{10}$ concentration will often give a near straight line (Fig. 8.3). Alternatively, if the calibrator concentrations cover more than a fourfold concentration range, such plots give a slight curve. Results can be obtained by manual interpolation from the curve or computed. Bennett's method for calculating a calibration curve with the general formula

$$y = a + bx + cx^2$$

where y is zone size, x is the log (drug concentration) and a, b and c are constants, is given below using Perkins' simplified method of giving coded values to the calibrators. Thus converting the equation to

$$y = A + BX + CX^2$$

where A, B and C are new constants. The coded values may be found on p. 159 of *Laboratory Methods in Antimicrobial Chemotherapy* (Reeves *et al*. 1980). If $y_1, y_2 \ldots y_n$ are the mean zone sizes for the coded calibrators $X_1, X_2 \ldots X_n$, then

$$\Sigma y = y_1 + y_2 + y_3 \ldots y_n$$
$$\Sigma Xy = X_1 y_1 + X_2 y_2 + X_3 y_3 \ldots X_n y_n$$
$$\Sigma X^2 y = X_1^2 y_1 + X_2^2 y_2 + X_3^2 y_3 \ldots X_n^2 y_n$$

$$A = \frac{D\Sigma y - E\Sigma X^2 y}{F}$$

$$B = \frac{\Sigma Xy}{E}$$

$$C = \frac{\Sigma y - nA}{E}$$

Then to calculate the concentration of drug in a sample giving a mean zone size y_t

$$X_t = \frac{\sqrt{(B^2 - 4AC + 4Cy_t)} - B}{2C}$$

and

$$x_t = x_1 + \frac{\log d}{2}(X_t + n - 1)$$

where d is the dilution factor for the calibrators and x_1 is the log of the concentration of the lowest calibrator. The solution is then antilog x_t.

A microcomputer can readily be programmed to derandomize the zone pattern, take the mean of the zone sizes, compute the standard curve and calculate the results for the tests, thus allowing zone sizes to be read directly from the plate into the computer.

Two-by-two assay. In situations where the likely concentration range of the tests will lie within a relatively narrow range (e.g. in determining potency of pharmaceutical preparations) and high precision is sought, then a Latin square design with tests and calibrators at two or three levels of concentration may be used. For example an 8×8 Latin square can be used to assay three samples and one calibrator, or two samples and two calibrators at two concentrations each (over a two or fourfold range), with a coefficient of variation of around 3%. Using this technique, parallel dose–response lines should be obtained for the calibrators and the tests at the two dilutions (Fig. 8.4). Using such a method, potency can be computed or determined from carefully prepared nomograms.

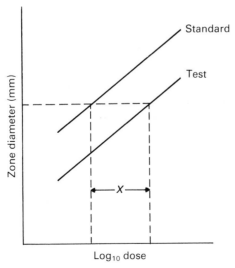

Fig. 8.4 Graphical representation of a two-by-two assay response. X is the horizontal distance between the two lines. The antilog of X gives the relative potency of the standard and test.

Since conventional plate assays require several hours' incubation and are consequently rather slow, rapid microbiological methods have been developed by exploiting techniques which measure variations in growth rate after a short incubation. These rapid methods suffer from the same problems of specificity as conventional microbiological methods.

2.2.1 *Urease*

When *Proteus mirabilis* grows in a urea-containing medium it hydrolyses the urea to ammonia and consequently raises the pH of the medium. This production of urease is inhibited by aminoglycoside antibiotics (inhibitors of protein synthesis; Chapter 9). The assay is run with two series of tubes of urea-containing medium to which a range of calibrator solutions is added. To one series of tubes is added a volume of sample equal to the volume of calibrator, to the other is added a half volume of sample. The tubes are then inoculated with *Pr. mirabilis* and incubated for 60–75 minutes after which the pH is accurately measured to 0.01 pH units. By plotting pH against $\log_{10}$ (calibrator concentration) for each series, two calibration curves are obtained. The vertical distance between the two curves is equal to the logarithm of half the concentration of drug in the sample. A sensitive and reliable pH meter is essential for this assay. In practice, it is difficult to obtain reliable results by this method.

2.2.2 *Luciferase assay*

In this technique, firefly luciferase is used to measure small amounts of adenosine triphosphate (ATP) in a bacterial culture, ATP levels being reduced by the inhibitory action of aminoglycoside antibiotics. Calibrators and test solutions (after preliminary heating if the matrix is serum) are added to tubes of medium containing a growing bacterial culture. After incubation (*c.* 90 minutes), the cultures are treated with apyrase (to destroy extracellular ATP), extracted with sulphuric acid/ethylenediamine tetra-acetic acid (EDTA) and the intracellular ATP determined with the firefly enzyme in a luminometer. A calibration curve is constructed by plotting intracellular ATP content against $\log_{10}$ (calibrator concentration). This technology has been little used to date but may find more application in the future as more active and reliable luciferase preparations become available.

3 **Radioenzymatic (transferase) assays**

These have been used for assaying aminoglycosides and chloramphenicol, and depend on the fact that bacterial resistance to aminoglycosides (Chapter 5), such as gentamicin, tobramycin, amikacin, netilmicin, streptomycin, spectinomycin, etc., and chloramphenicol is frequently associated with the presence of specific enzymes (often coded for by transmissible plasmids) which either acetylate, adenylylate or phosphorylate the antibiotics, thereby rendering them inactive (Chapter 10). Aminoglycosides may be susceptible to attack by aminoglycoside acetyltransferases (AAC), aminoglycoside adenylyltransferases (AAD), or aminoglycoside phosphotransferases (APH). Chloramphenicol is attacked by chloramphenicol acetyltransferases (CAT). Acetyltransferases attack susceptible amino groups and require acetyl coenzyme A, while AAD or APH enzymes attack susceptible hydroxyl groups and require ATP (or another nucleotide triphosphate). In the case of the aminoglycoside-modifying enzymes, current nomenclature uses the three-letter prefix AAC, AAD (or ANT) or APH followed by the number of the carbon atom which bears the susceptible group. Isoenzymes acting at the same site are numbered sequentially. For example there are three AAC enzymes, AAC(3), AAC(2′) and AAC(6′), and AAC(3) has several isoenzymes, AAC(3)-1, AAC(3)-2, etc.

Several AAC and AAD enzymes have been used for assays. The enzyme and the appropriate radiolabelled cofactor ([1-^{14}C] acetyl coenzyme A, or [2-^{3}H] ATP) are used to radiolabel the drug being assayed. Enzyme is simply prepared by breaking the cells of a suitable bacterial culture by high-frequency (ultrasonic) sound waves, or by submitting them to a change of solution strength such that a change in osmotic pressure (osmotic shock) breaks open the cells, no purification being necessary. All the assays have a similar protocol. The radiolabelled drug is separated from the reaction mixture after the reaction has been allowed to go to completion; the amount of radioactivity extracted is directly proportional to the amount of drug present. Aminoglycosides are usually separated by binding them to phosphocellulose paper, whereas chloramphenicol is usually extracted using an organic solvent.

These types of assay are rapid (taking approx. 2 hours), show good precision and are much more specific than microbiological methods. The aminoglycoside enzyme-mediated procedures are, however, less specific for individual aminoglycosides than immunoassays. AAC(6′) has the widest spectrum and can be used to assay amikacin, netilmicin, tobramycin and three of the four major C components of gentamicin. AAC(3) or AAC(2′) can be used to assay gentamicin, tobramycin and netilmicin but not amikacin; one AAC(3) isoenzyme will also acetylate apramycin. If AAD enzymes are used and the drug is present in serum, preheating to destroy endogenous ATPase is necessary.

Assay specificity is generally no problem unless an aminoglycoside is assayed in the presence of chloramphenicol or vice versa. In this case it is necessary to use enzyme preparations which do not contain both AAC and CAT enzymes, otherwise they will compete for the available acetyl coenzyme A and errors may result. Cationic radionucleotides in the sample are a potential source of error in samples from patients. If such interference is possible a reagent-free sample should be processed to determine the amount of radioactivity it contains.

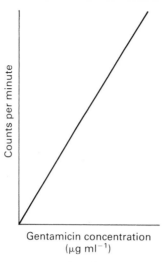

Fig. 8.5 Relationship between concentration of aminoglycoside antibiotic and the transfer of radioactivity from ATP to phosphocellulose.

3.1   **Calibration**

When reactants are present in sufficient quantities and the reaction goes to completion, a plot of counts per minute against calibrator concentration is linear (Fig. 8.5). Deviations from this would suggest that one of the reagents is at fault or that the enzyme reaction is not going to completion. Because the response is linear, results are simple to plot or compute.

3.2   **Non-isotopic modification**

Disadvantages of the above method are the need for expensive scintillation-counting equipment and the usual problems associated with the use of radioactive chemicals. To overcome these problems, a photometric variation of the AAC aminoglycoside assay has been described. The sulphydryl reagent 5,5'-dithiobis(2-nitrobenzoic acid) (DTNB) is introduced into the assay system, reacting with the reduced coenzyme A as it is produced, to yield a yellow product which can be measured in a spectrophotometer.

Aminoglycoside + acetyl CoA → acetyl-aminoglycoside + CoASH
CoASH + DTNB → yellow product.

Semi-purified enzyme is required for use in the photometric adaptation.

4   **HPLC**

HPLC (high-performance or high-pressure liquid chromatography) is an advanced form of liquid chromatography and current practice is to refer

176   *Chapter 8*

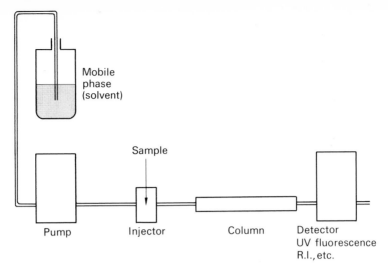

Fig. 8.6 Diagrammatic representation of a high-performance liquid chromatograph.

to HPLC simply as LC (to avoid the confusion between pressure and performance, and because pressures are no longer necessarily high). In LC (Fig. 8.6) a liquid mobile phase (a mixture of solvents with or without additives) is pumped over a stationary phase. The sample to be assayed is introduced into the mobile phase and the chromatographic separation takes place at the stationary phase. After the stationary phase, the mobile phase next passes through a detector which enables the separated substances to be identified (by measuring the characteristic time, known as retention time, that they take to be eluted) and quantified (by measuring either the height or the area of a peak on a chart-recorder trace or integrator). The technique is nondestructive and consequently it is also possible to collect the eluate and test fractions for antimicrobial activity.

LC can be used to separate and quantify a vast range of compounds, including most antibiotics. Its particular advantages are as follows.
1   Speed, a result usually being obtained in a matter of minutes.
2   High precision (for quality control work).
3   High sensitivity (for assaying drugs in serum).
4   High specificity and selectivity. Very closely related compounds such as the gentamicin C components, epimers (e.g. of latamoxef or cefotetan), and pro-drugs (e.g. chloramphenicol and chloramphenicol succinate) can be separated. It is especially useful for the assay of those antibiotics metabolized to microbiologically active and inactive metabolites (e.g. cefotaxime (Fig. 8.7), metronidazole, 4-quinolones), and for difficult mixtures (e.g. benzylpenicillin with other β-lactams and other antibiotics).
5   It is easily automated.

The major disadvantage of LC is that, since samples must be assayed

177   *Principles of methods of assaying antibiotics*

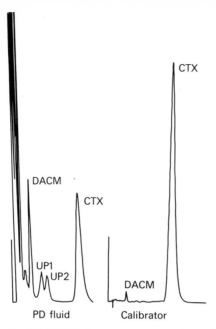

Fig. 8.7 HPLC chromatogram of peritoneal fluid from a patient with impaired renal function to whom cefotaxime was administered intraperitoneally. CTX, cefotaxime; UP1 and UP2, two microbiologically inactive metabolites; DACM, microbiologically active desacetyl cefotaxime.

sequentially, a single chromatograph can only deal with a modest workload (low sample capacity).

4.1   LC equipment

Before dealing with the principles, some consideration of the necessary equipment is in order (see Fig. 8.6).

4.1.1   Pump

An LC pump must deliver a pulse-free flow at up to $10\,\text{ml}\,\text{min}^{-1}$ (or greater) against back-pressures of $20\,700-27\,600\,\text{kPa}$ ($3000-4000\,\text{p.s.i.}$) (or greater). Irregularities in flow rate will lead to imprecision, especially if peak areas are being measured (see below).

4.1.2   Injector

This is the device for introducing the sample into the solvent flow. Injectors are available for handling sample sizes from 0.5 to $2000\,\mu\text{l}$. Most injectors will handle the range $10-200\,\mu\text{l}$. Autoinjectors and autosamplers capable of unattended handling of numbers of samples are available. Whatever injector is used, the injection volume must be reproducible or imprecision will result unless an internal standard is included in every sample (see below).

Detector

A variety of LC detectors are available, including those measuring ultra-violet (UV) absorbance, fluorescence, refractive index, electrochemical activity, etc. UV absorbance detectors are commonly employed and may be fixed (selectable) wavelength or continuously variable wavelength detectors. Theoretically the latter are preferable since they can be tuned to the absorbance maximum of the drug being assayed. However, in the past, baseline instability and electronic noise when used at maximum sensitivity have made them inferior to fixed wavelength detectors. Fixed wavelength detectors covering a wide range of wavelengths are available (e.g. 313, 280, 254, 229, 214 nm), which are very stable and sensitive and thus well suited to the assay of many antibiotics in serum. Variable wavelength detector design is, however, improving rapidly.

4.2   **Stationary phases**

The stationary phase is the heart of the LC system and is usually con-tained in a stainless steel column of approx. 4 mm internal diameter and 10–25 cm in length. For rapid, high-efficiency chromatography, micro-particulate stationary phases (5–10 μm particle size) are used. Conven-tional LC columns are expensive and difficult to pack; an alternative approach uses a plastic column which is radially compressed during use (e.g. Waters Z-module). Such columns are, like conventional columns, highly efficient, but are less expensive and produce much lower back-pressures. When assaying biological samples the analytical column should be protected by a short guard column to prolong its life by preventing irreversibly absorbed molecules reaching it. It should contain a low-efficiency packing of similar selectivity to that in the analytical column.

Any type of high-performance chromatography is possible with this general system (ion exchange, absorption, partition, gel permeation) and an extensive range of commercial packings are available, but for most antibiotic assays a technique known as reverse phase chromatography (RPLC) has become very popular as it enables a large variety of mol-ecular species to be assayed on a single column.

4.3   **Reverse phase chromatography (RPLC)**

Traditional chromatography uses a polar stationary phase (paper, silica, etc.) and an organic solvent system. RPLC is merely the reverse of this situation. The stationary phase is non-polar and the mobile phase is polar (watery). The stationary phase used for RPLC is usually a hydro-carbon bonded surface, most commonly silica with chemically bonded octadecylsilyl groups ($C_{18}H_{37}Si$—). Such bonded packing materials are commercially available and referred to as C_{18} or ODS packings (e.g. MicroBondapak C_{18}, Hypersil-5-ODS).

RPLC is a better alternative to ion-exchange chromatography for the assay of ionizable substances (like many antibiotics). Acidic drugs can be chromatographed by buffering the mobile phase at a low pH and thus suppressing their ionization. C_{18} columns cannot be used above pH 8 since they break down. Positively charged drugs are therefore best assayed by RPLC using an ion-pairing technique.

Paired-ion chromatography

In ion-pair (IP) chromatography a hydrophobic counter-ion (e.g. heptane sulphonic acid) with affinity for the stationary phase is added to the mobile phase. Concentration- and pH-dependent equilibria occur as shown below.

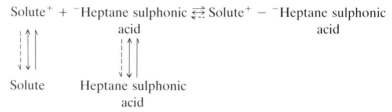

The pH is kept low enough to ensure that both solute and counter-ion remain ionized. Acids may also be separated by IP-RPLC, if necessary, using a basic IP reagent such as tetrabutylammonium phosphate. Mixtures of basic and acidic drugs (e.g. trimethoprim and sulphonamides) may be separated and assayed simultaneously by IP; the one drug pairs with the counter-ion and the other, having its ionization suppressed, is separated by conventional RP interactions.

Mobile phase formulation

For a given stationary phase, such as an ODS packing, the desired separation is achieved by adjusting the composition of the mobile phase. For RPLC of polar drugs, the mobile phase is usually a mixture of water and organic solvent (e.g. methanol) with an additive (buffer, IP reagent, etc.) to give the required selectivity. Antibiotics within a single class (e.g. β-lactams) will vary in the proportion of organic solvent required in the mobile phase to give a reasonable retention time. For example, in 25% methanol on MicroBondapak C_{18}, cefsulodin had a retention time of 1.8 minutes and cafamandole nafate one of 26 minutes at a flow rate of $2\,\mathrm{ml\,min}^{-1}$. Generally speaking, the greater the proportion of organic solvent in the mobile phase the shorter the retention time of any given solute. When dealing with complex samples such as serum, the mobile phase composition must be formulated to separate the drug(s) to be assayed from the other compounds in the sample in the shortest possible time.

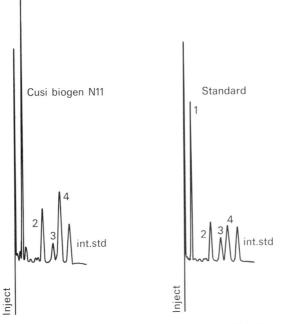

Fig. 8.8 HPLC chromatograms of orthophthalaldehyde derivatized gentamicin sulphate illustrating: (i) the separation of the four gentamicin components (1, C1; 2, C1a; 3, C2a; 4, C2); (ii) the use of an internal standard (int. std), in this case nonylamine; and (iii) the variation in percentage composition between different preparations of gentamicin, here a standard preparation (standard) and a commercial batch (Cusi biogen N11).

4.6   **Sample preparation**

Whatever sample is being assayed, all particulate matter must be removed and the sample should not precipitate when mixed with mobile phase. Serum proteins must first be removed since they may precipitate in the mobile phase and/or combine irreversibly with the stationary phase. Standard methods of protein precipitation may be used; the organic solvent acetonitrile is useful as mixing it with serum (ratio 1:1 or 2:1) produces a coarse, easily centrifuged protein precipitate.

Some antibiotics, notably the aminoglycosides, must be derivatized since they do not absorb UV light and are therefore difficult to detect. A number of standard techniques have been applied to gentamicin to give UV-absorbing or fluorescent derivatives. Figure 8.8 shows a chromatogram of an orthophthalaldehyde derivative of gentamicin, and demonstrates the separation of the four major C components.

4.7   **Calibration**

As with other assays, calibration with appropriate calibrator solutions is required. Normally a linear standard curve with an intercept of zero will

181   *Principles of methods of assaying antibiotics*

result if calibrator concentration is plotted against detector response (peak height or peak area). Once linearity is established it is possible to use single-point calibration for clinical assays although this time-saving procedure is not normally acceptable for pharmacokinetic studies requiring greater accuracy. The concentration c of a test is determined from

$$c = \frac{C \times R_t}{R_{cal}}$$

where C is the calibrator concentration, R_{cal} the detector response to the calibrator and R_t the detector response to the test.

Where drugs are present in serum or some other fluid where extraction/precipitation is necessary, then percentage recovery of the drug must be determined over a range of concentrations. If recovery is less than 100%, then as long as it is reproducible, proper calibration is possible. An extraction/precipitation method giving irreproducible recoveries is not suitable for accurate assays.

4.7.1 *Internal standardization*

An internal standard is a chemical entity which is added to the sample to improve the precision of quantification. For any LC separation an internal standard must be:
1   Chemically stable.
2   Capable of detection and able to give an acceptable peak shape.
3   Separated from other components in the sample and elute close to the test peak.
4   Reproducibly recovered in any extraction procedure.

The chromatogram in Fig. 8.8 has an internal standard. With internal standardization the equation above becomes

$$c = \frac{C}{R_{cal}/R_{ic}} \times \frac{R_t}{R_{it}}$$

where R_{ic} is the detector response of the internal standard in the calibrator chromatogram and R_{it} that in the test chromatogram.

4.8   Controlling precision

If peak heights are measured then imprecision will result if the retention time alters as subsequent samples are assayed. Peaks will become higher if the retention time is reduced and lower if the retention time increases. Measuring areas instead of heights eliminates this problem since area is independent of retention time. If, however, the solvent flow rate is not uniform, imprecision will result. Peak areas will seem greater if the flow rate slowed when the peak eluted, and smaller if the flow rate increased. Which of these is the greater source of error in any particular system must be determined experimentally.

Imprecision will also result if, during a long run, column efficiency drops (loss of efficiency makes peaks wider and shorter). Monitoring efficiency can guard against this, a simple equation of efficiency being

$$N = 5.54\left(\frac{Rt}{W_{hh}}\right)^2$$

where N (number of theoretical plates) is efficiency, Rt is the retention time, and W_{hh} is the peak width at one-half height, in the same units as Rt.

5   Immunoassay

An immunoassay is a quantitative technique which uses specific antibody as a reagent. Immunoassays have been developed for a number of antibiotics. Their special features are as follows.

1 Very high specificity, the degree of which is determined by antibody specificity. Thus immunoassay can eliminate the problems of poor specificity associated with microbiological assay. It is possible, for example, to assay gentamicin immunologically in the presence of virtually any other antibiotic, including amikacin and tobramycin. Any consideration of immunoassays must, however, bear in mind that if a drug is converted to metabolites which cross-react with the antibody, poor specificity will result. This has caused problems in the development of an immunoassay for chloramphenicol.

2 Very high sensitivity. Most immunoassay techniques can measure antibiotic concentrations far below that usually encountered. Consequently, samples are often diluted and only small volumes are required.

3 Long, costly development time. As specific antibody is a first requirement, a development time of at least several months is needed. Costs are high because the antibody will usually be raised in animals. For this reason most immunoassays are developed commercially and sold as kits.

4 Speed and simplicity. Some immunoassay techniques are technically simple, easily automated, and very fast. For example, it is possible to assay 20 samples in 10 minutes by automated polarization fluoroimmunoassay.

Immunoassays for antibiotics are, at the moment, limited to the aminoglycosides (gentamicin, tobramycin, amikacin, netilmicin, sisomicin, kanamycin, etc.), vancomycin and chloramphenicol, although others are being developed.

5.1   Principles of immunoassay

Most immunoassays depend on the principle of competitive binding. Drug molecules (D) and labelled drug molecules (D*) compete for binding sites on a limited amount of specific antidrug antibody (Ab) according to the equations:

$$Ab + D \rightarrow AbD \tag{8.1}$$

$$Ab + D^* \rightarrow AbD^* \tag{8.2}$$

$Ab + D + D^* \rightarrow$ competition between D and D* for a
limited amount of Ab. (The greater the amount of D,
the lower the amount of D* which binds to Ab.) $\tag{8.3}$

Thus in equation (8.1), antidrug antibody (Ab) and drug (D) bind to form an antigen–antibody complex. In equation (8.2) the same reaction occurs between Ab and the drug which has been chemically labelled in some way with an atom or molecule referred to as a tracer (D* or drug tracer). Equation (8.3) describes the immunoassay situation. D* (a fixed amount) and D (from the sample being assayed) compete for a limited amount of Ab. The proportion of D* which binds to Ab is dependent on the concentration of D, which can therefore be estimated by measuring the amount of D* which binds to Ab (the bound fraction) or the amount which fails to bind (the free fraction) (see Fig. 8.9). Equation (8.2) can be likened to an immunoassay situation where the sample contains no D, i.e. maximum binding of D* occurs.

5.2 Preparation of antibody

Most drugs are not immunogenic (i.e. will not themselves give rise to an antibody response when injected into an animal). To induce a drug to act as an immunogen it must be chemically coupled to a carrier protein such as bovine serum albumin. A number of chemical methods of achieving this are available, most of which involve coupling via free —NH$_2$ or —COOH groups. The antibody raised by such a conjugate must bind specifically and avidly with the native (uncoupled) drug if it is to be suitable for immunoassay purposes. For quenching fluoroimmunoassay, the antibody must have suitable quenching properties as well (see below).

5.3 Drug label for preparing drug tracer

The second essential ingredient of the immunoassay is the labelled drug reagent (D*). Radioimmunoassay, the prototype immunoassay, uses a radioactive tracer as the label. However, in theory, any label enabling the simple measurement of 'bound' or 'free' could be used. Some labels which have been used include radioisotopes (radioimmunoassay), fluorochromes (fluoroimmunoassay), enzymes (enzyme immunoassay), enzyme substrates (substrate-labelled enzyme immunoassay), and luminescent compounds (luminescent immunoassay).

5.4 Measurement of bound label

This can be accomplished either by the separation of bound from free labels (heterogeneous or separation immunoassay), or by choosing a label which is modified in some way when the drug is bound to antibody and can therefore be measured directly without a separation step (homogeneous or non-separation immunoassay).

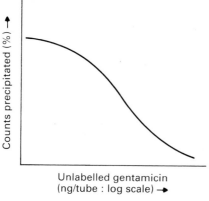

Fig. 8.9 Standard curve for the radioimmunoassay of a typical aminoglycoside antibiotic, gentamicin.

5.4.1 *Heterogeneous immunoassay*

There are a number of heterogeneous methods using a number of labels. For example, radioimmunoassays using either iodine-125 or tritium labels, and separation fluoroimmunoassays using fluorescein label and separation enzyme immunoassays (e.g. SOPHEA).

Bound label is separated by either (i) precipitation using any reagent suitable for antibody precipitation, for example second antibody (i.e. a specific antibody recognizing the antidrug antibody as an antigen, e.g. if the antidrug antibody was raised in rabbits then goat anti-rabbit IgG could be used), ammonium sulphate, etc.; or (ii) the use of solid-phase antidrug antibody. This can be antibody-coated tubes or antibody bound to particles such as dextran. A novel system, commercially available, uses magnetic antibody particles and the separation is achieved with a magnet.

Clearly a possible source of error in these assays is poor separation of bound and free labels. In general solid-phase techniques are more reliable than precipitation techniques.

5.4.2 *Homogeneous immunoassay*

These methods have no separation step and are therefore potentially simpler and faster, and may be reproduced more exactly. There is no homogeneous radioimmunoassay since radioactive emission is not affected by antibody binding.

Homogeneous fluoroimmunoassays. These can be quenching fluoro-immunoassay or polarization fluoroimmunoassay.
1 *Quenching fluoroimmunoassay.* When fluorescein-labelled drug binds to a suitable antibody the fluorescent emission of the label is reduced (quenched). Therefore if fluorescence is measured before (F1) and after (F2) addition of antibody, quenching (F1−F2) can be determined. Not all

185 *Principles of methods of assaying antibiotics*

antibody, even that of high titre and high avidity, is necessarily good 'quenching' antibody and therefore its production for quenching assays may be a problem.

2 *Polarization fluoroimmunoassay* (Abbott TDX). If fluorescein-labelled drug is excited with polarized light, the polarization of its fluorescent emission is reduced due to the Brownian motion of the molecule. If the labelled drug is bound to antibody, polarization is retained. By measuring the degree of polarization of the emitted light the amount of antibody binding can be determined.

EMIT (Syva).   The drug is labelled with an enzyme in such a way as to preserve enzymic activity. Binding to antidrug antibody inhibits the enzyme either by steric hindrance or by inducing conformational changes. The enzyme commonly used is bacterial glucose-6-phosphate dehydrogenase, as its activity can be measured in a spectrophotometer. When glucose-6-phosphate is dehydrogenated, the cofactor NAD is reduced to $NADH_2$ which is more absorbent at 340 nm than NAD.

Substrate-labelled assay (Ames TDA).   The drug is labelled with a fluorigenic enzyme substrate (umbelliferyl-β-D-galactoside), which is hydrolysed by β-galactosidase to a fluorescent product. Antibody binding inhibits the enzymic degradation, thus only free labelled drug is broken down by the enzyme.

5.5   **Calibration**

As with microbiological assays, the relationship between drug concentration and response is not mathematically simple and, in general, five or more calibrators, covering a suitable concentration range, are used. In addition, immunoassays require a zero calibrator (maximum binding, see equation (8.2) above) and sometimes an antibody-free test (NSB or non-specific binding).

5.5.1   *Heterogeneous immunoassay*

Plotting percentage binding against log (calibrator concentration) gives a sigmoid curve. This can be mathematically straightened by, for example, a logit transformation.
      For each calibrator

Percentage binding, $B = (R_c - R_{nsb}) \times 100/(R_{mb} - R_{nsb})$

Logit $B = \log(B/100 - B)$

where R_c is the reading (radioactive counts, fluorescence, etc.) given by the calibrator concentration c, R_{nsb} the reading of the NSB tube and R_{mb} the reading of the maximum binding tube. A plot of logit B against log c will be linear within the normal range of calibrators for many assays, although some may demand more complex mathematical manipulation.

Here the type of calibration depends on the type of label used. Most can, however, be mathematically straightened for computing purposes (often fluorimeters or spectrophotometers are programmed to do this, i.e. are designed and built specifically for running commercial drug assay kits).

5.6   Controlling precision

With immunoassays precision depends on four factors.
1 Instrument performance. Most radioactivity counters or spectrophotometers will perform satisfactorily. Fluorimeters used to be notoriously poor but newer, purpose-built machines usually perform well.
2 Efficiency of separation (heterogeneous assays only). As stated above, solid-phase assays are probably more reproducible than precipitation methods. In both, care and technical skill are required.
3 Time and temperature of incubation. This is most important with enzyme immunoassays. EMIT reactions are read kinetically over 45 seconds in a cuvette heated to 30°C, and electronic timing and temperature control are essential. Ames TDA assays are read after a fixed incubation period (6–20 minutes) and precisely timed additions of the fluorigenic reagent are essential. In most other immunoassays, incubation is continued until equilibrium is reached and timing is not as critical, provided the incubation period is long enough (usually 10–30 minutes).
4 Precise pipetting. Poor pipetting is a major source of error since only microlitre quantities of sample and reagent are used. Operator error can be minimized by extensive use of pipettor-diluters (EMIT, Ames TDA) or by automation (EMIT Autolab system, Abbott TDX polarization fluoroimmunoassay, Ames Optimate).

6   Further reading

PLATE ASSAY, IMMUNOASSAY, TRANSFERASE
Lorian V. (1980) *Antibiotics in Laboratory Medicine*. Baltimore: Williams & Wilkins.
Reeves D.S., Phillips I., Williams J.D. & Wise R. (Eds) (1980) *Laboratory Methods in Antimicrobial Chemotherapy*. Edinburgh: Churchill Livingstone.

PLATE ASSAY, HPLC, IMMUNOASSAY, TRANSFERASE
Reid E. & Wilson D. (Eds) (1990) *Analysis for Drugs and Metabolites including Anti-infective Agents*. Methodological Surveys in Biochemistry and Analysis, vol. 20. Royal Society of Chemistry.
Richens A. & Marks V. (Eds) (1981) *Therapeutic Drug Monitoring*. Edinburgh: Churchill Livingstone.

PLATE ASSAY
Hewitt W. & Vincent S. (1989) *Theory and Application of Microbiological Assay*. London: Academic Press.
Kavanagh F. (1963, 1972) *Analytical Microbiology*, vols I & II. London: Academic Press.
Reeves D.S. & Bywater M.J. (1976) Assay of antimicrobial agents. In *Selected Topics in Clinical Bacteriology* (Ed. J. de Louvois), pp. 21–78. London: Baillière Tindall.

UREASE

Board R.G. & Lovelock D.W. (Eds) (1975) *Some Methods for Microbiological Assay.* Society for Applied Bacteriology Technical Series No. 8. London: Academic Press.

LUCIFERASE

Nilsson L. (1982) Rapid bioluminescent assay of serum amikacin. *J Antimicrob Chemother.* **10**, 125–130.

TRANSFERASE

White L.O. & Reeves D.S. (1983) Enzymatic assay of aminoglycoside antibiotics. In *Antibiotics: Assessment of Antimicrobial Activity and Resistance* (Eds A.D. Russell & L.B. Quesnel), pp. 199–210. Society for Applied Bacteriology Technical Series No. 18. London: Academic Press.

HPLC

Hawks G.L. (1982) *Biological–Biomedical Applications of Liquid Chromatography,* vol. V. New York: Marcel Dekker.

Reeves D.S. & Ullmann U. (Eds) (1986) *High Performance Liquid Chromatography in Medical Microbiology.* Stuttgart: Gustav Fischer Verlag.

Warnock D.W., Richardson M.D. & Turner A. (1982) High performance liquid chromatographic (HPLC) and other non-biological methods for quantitation of antifungal drugs. *J Antimicrob Chemother.* **10**, 467–478.

White L.O. (1981) HPLC in clinical microbiology laboratories. *J Antimicrob Chemother.* **8**, 1–3.

Yoshikawa T.T., Maitra S.K., Schotz M.C. & Guze L.B. (1980) High pressure liquid chromatography for quantitation of antimicrobial agents. *Rev Infect Dis.* **2**, 169–181.

9 Mechanisms of action of antibiotics

1	**Introduction**	**4**	**Chromosome function**
		4.1	The basis of selective inhibition
2	**The bacterial cell wall**	4.2	Quinolone carboxylic acids
2.1	The nature of peptidoglycan	4.3	Metronidazole and nitrofurantoin
2.2	Biosynthesis of peptidoglycan	4.4	Rifampicin
2.3	Inhibitors of peptidoglycan		
	biosynthesis	**5**	**Folate antagonists**
2.3.1	β-lactams	5.1	Folate metabolism in bacteria and
2.3.2	Vancomycin and teicoplanin		mammalian cells
2.3.3	Daptomycin	5.2	Sulphonamides
2.3.4	D-cycloserine	5.3	Trimethoprim
3	**Protein synthesis**	**6**	**The cytoplasmic membrane**
3.1	The basis of selective inhibition	6.1	Composition and susceptibility to
3.2	Aminoglycosides		disruption
3.3	Tetracyclines	6.2	Polyenes
3.4	Chloramphenicol	6.3	Imidazoles
3.5	Erythromycin	6.4	Naftidine
3.6	Lincomycin and clindamycin		
3.7	Fusidic acid	**7**	**Further reading**
3.8	Mupirocin (pseudomonic acid)		

1 Introduction

Unlike disinfectants and antiseptics (Chapters 11 and 13), antibiotics exert *selective* toxicity towards microbial cells. That is, they kill or inhibit microbes at low concentrations which have little or no effect upon mammalian cells. This chapter will describe the molecular basis of the selective toxicity of the major groups of antibiotics, concentrating upon those used to treat microbial infections in humans. Table 9.1 lists their general target areas and summarizes the basis of their selective toxicity. Note that some antibiotics inhibit targets which are unique to bacteria and essential for their normal growth (e.g. β-lactams which inhibit peptidoglycan synthesis). Others (e.g. tetracyclines) owe their selective toxicity to active uptake by bacteria but exclusion from mammalian cells; the intracellular target is equally sensitive in both kinds of cell.

Properties of antibiotics described in this chapter are presented in Chapter 5, and mechanisms of resistance in Chapter 10.

2 The bacterial cell wall

2.1 The nature of peptidoglycan

Peptidoglycan (see Fig. 1.2, Chapter 1) is a vital component of virtually all bacterial cell walls, accounting for approximately 50% of the weight of

Table 9.1 Microbial targets for antibiotics

Target	Antibiotics	Basis of selective toxicity
Cell wall (peptidoglycan synthesis)	β-lactams Vancomycin Teicoplanin Daptomycin Cycloserine	Peptidoglycan, an essential component of bacterial walls, is absent from mammalian cells; D-alanyl-D-alanine is unique to peptidoglycan
Ribosome function (protein synthesis)	Aminoglycosides	Selective for bacterial 30S subunit, active uptake
	Tetracyclines	Active uptake by bacteria, exclusion from mammalian cells
	Chloramphenicol	Selective for bacterial 50S subunit
	Erythromycin	Selective for bacterial 50S subunit
	Lincomycin, clindamycin	Selective for bacterial 50S subunit
	Fusidic acid	Active uptake by bacteria, exclusion from mammalian cells
	Mupirocin (pseudomonic acid)	Selective for bacterial isoleucyl tRNA synthetase
Chromosome function (replication and transcription of DNA)	Quinolone carboxylates	DNA gyrase unique to bacteria
	Metronidazole, nitrofurantoin	Converted to active form in bacteria
	Rifampicin	Selective for bacterial RNA polymerase
Folate metabolism	Sulphonamides	Dihydropteroic acid synthetase unique to bacteria
	Trimethoprim	Selective inhibition of bacterial DHFR
Cytoplasmic membrane	Polyenes	Higher affinity for ergosterol than cholesterol
	Imidazoles	Selective inhibition of ergosterol biosynthesis
	Naftidine	Selective inhibition of ergosterol biosynthesis by blocking squalene epoxidase

the Gram-positive cell wall and 10–20% of the Gram-negative envelope. It is responsible for both the shape and integrity of the cells. Since no similar component exists in mammalian cells, interference with its biosynthesis provides an excellent basis for selective toxicity towards bacteria. The linear glycan strands of peptidoglycan (comprising alternating units of *N*-acetylmuramic acid and *N*-acetylglucosamine) are cross-linked by short peptide bridges which give the polymer its mechanical strength. If the biosynthesis and assembly of cross-linked peptidoglycan is inhibited by antibiotic action, the peptidoglycan is unable to support the cell wall and as a consequence the cells take on abnormal shapes and eventually lyse and die.

Biosynthesis of peptidoglycan

The precursors of peptidoglycan are synthesized in the cytoplasm, assembled on a lipid carrier molecule in the cytoplasmic membrane and inserted into the cell wall forming uncross-linked glycan strands. The strands are finally cross-linked with pre-existing peptidoglycan by enzymes located on the outer face of the cytoplasmic membranes (Fig. 9.1). These enzymes (known as penicillin-sensitive enzymes, PSEs, and visualized by gel electrophoresis as the penicillin-binding proteins, PBPs) are the targets of β-lactam action. This overall sequence of events is common to all bacteria but there are minor variations in the details, particularly in the nature of the amino acids forming the peptide cross-links of different species of bacteria. Biosynthesis of *Escherichia coli* peptidoglycan will be used as an example throughout since it has a composition common to many species and has been studied extensively with regard to β-lactam action.

In the cytoplasm the two subunits of peptidoglycan, *N*-acetylmuramyl-pentapeptide and *N*-acetylglucosamine are synthesized, each linked to a carrier molecule, uridine diphosphate (UDP). This is a common feature of macromolecular synthesis whereby energy for the coupling of the precursors is provided by release of the nucleotides which play no further part in the biosynthetic pathway. The pentapeptide sequence attached to *N*-acetylmuramic acid comprises L-alanine, D-glutamic acid, *meso*-diaminopimelic acid and two D-alanine residues which occupy the free carboxyl end. The D-forms of amino acids are rare in nature and diaminopimelic acid, which plays a vital role later in the cross-linking step, is unique to peptidoglycan. The lipid carrier (or lipid intermediate) is a highly lipophilic molecule, undecaprenyl phosphate, which stays in the cytoplasmic membrane and functions as an acceptor of the peptidoglycan precursors on the cytoplasmic face of the membrane. *N*-acetylmuramylpentapeptide and *N*-acetylglucosamine are sequentially transferred from their respective sugar nucleotide precursors to the lipid carrier with release of uridine monophosphate (UMP) and UDP, respectively. The precursors, joined by a β-1,4-glycosidic linkage, are attached to the lipid carrier via a pyrophosphate bond. In this form they are rendered lipid-soluble and are translocated across the cytoplasmic membrane and inserted into the existing peptidoglycan of the cell wall in a region where the wall is to be extended during cell growth. The point and manner of insertion must be under considerable and subtle control for the organism to maintain its characteristic shape.

A linear glycan strand attached to another molecule of the lipid carrier is thought to be the acceptor site in the wall for the newly translocated disaccharide pentapeptide. The linear glycan is transferred to the disaccharide pentapeptide lipid carrier and joined to it by formation

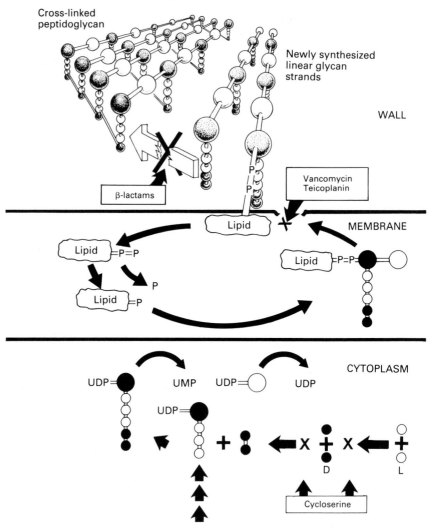

Fig. 9.1 Biosynthesis of peptidoglycan. The large open circles represent *N*-acetylglucosamine; the large filled circles represent *N*-acetylmuramic acid to which is linked initially a pentapeptide chain comprising L-alanine, D-glutamic acid and *meso*-diaminopimelic acid (small open circles) terminating in two D-alanine residues (small filled circles). The lipid molecule is undecaprenyl phosphate.

of a glycosidic bond between the *N*-acetylmuramic acid residue on the glycan strand and the *N*-acetylglucosamine of the newly translocated disaccharide pentapeptide. This transglycosylation reaction extends the linear glycan strand by one disaccharide unit and releases one molecule of lipid carrier pyrophosphate. The carrier loses one phosphate by the action of a pyrophosphorylase and is then available to participate in another cycle of translocating a disaccharide pentapeptide from the cytoplasm into the wall.

Meanwhile the final cross-linking stage of peptidoglycan assembly takes

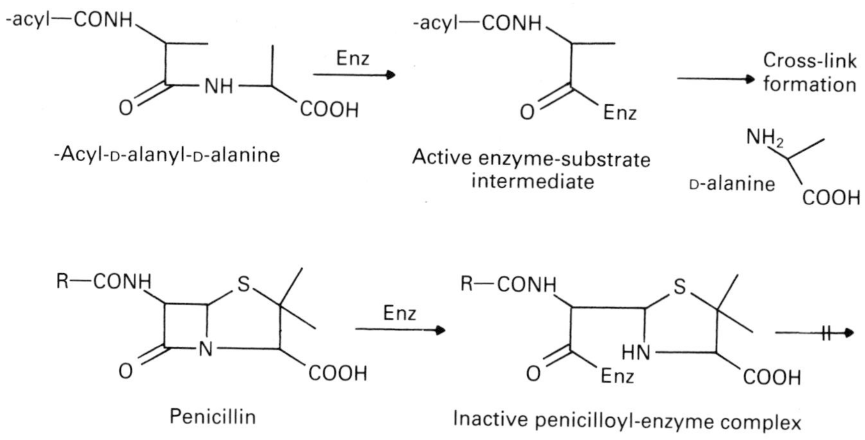

Fig. 9.2 Interaction of transpeptidase (Enz) with its natural substrate, acyl-D-alanyl-D-alanine in the first stage of the transpeptidation reaction to form an acyl–enzyme intermediate. A similar reaction with a penicillin results in the formation of an inactive penicilloyl–enzyme complex.

place in the wall. The newly assembled linear glycan strand is joined to the existing peptidoglycan in the wall by formation of peptide bonds. Transpeptidases (TPases) located on the outer face of the cytoplasmic membrane carry out this function. First, the TPases bind to the terminal D-alanyl-D-alanine residues of the pentapeptide substituents on the N-acetylmuramic units of the linear glycan strand. Peptidase action cleaves the peptide bond releasing the terminal D-alanine which plays no further part in the process. At this stage the TPase is transiently linked to the carboxyl terminus of the remaining D-alanine, forming a reactive enzyme-substrate intermediate (Fig. 9.2). The second stage of the transpeptidation reaction occurs when the enzyme transfers the acyl group of the terminal D-alanine residue to an amino acceptor group situated on the diaminopimelic acid residue of a nearby glycan strand of peptidoglycan. A new peptide bond is formed between the carboxyl group of D-alanine on the newly synthesized glycan strand and the amino group of diaminopimelic acid on existing peptidoglycan, and the TPase enzyme is released. Similar reactions occur at other points along the glycan strand, which thus becomes firmly incorporated into the peptidoglycan network of the wall. Since the cross-linking reaction occurs on the outer face of the cytoplasmic membrane it is remote from conventional sources of energy, such as adenosine triphosphate (ATP), to drive the reaction. It is thought that the energy released by breaking the D-alanyl-D-alanine linkage is used to form the peptide cross-link.

The peptidoglycan of *E. coli* and many bacilli is only 20–30% cross-linked yet all the peptide substituents on each N-acetylmuramic acid unit are tetrapeptides, lacking the fifth D-alanine. Another enzyme, D,D-carboxypeptidase (CPase) is responsible for removing the terminal D-alanine from some of the pentapeptides of newly formed glycan strands.

Its action is the same as that of the TPase, except that after cleavage of the D-alanyl-D-alanine bond and release of the terminal D-alanine, the enzyme releases the tetrapeptide without forming a cross-link. The degree of cross-linking is therefore controlled by the relative activities of the TPases and CPases.

Still other enzymes are involved in peptidoglycan assembly. An endo-peptidase is capable of breaking the cross-links, i.e. it reverses the action of the TPase. This possibly creates sites in the peptidoglycan into which new glycan strands can be inserted. Transglycosylase enzymes also extend the glycan chain length in peptidoglycan by joining on glycan strands which are subsequently cross-linked by TPases.

2.3 Inhibitors of peptidoglycan biosynthesis

Penicillins and cephalosporins are the most important antibiotics which interfere with peptidoglycan biosynthesis. Other non-classical β-lactams (see Chapter 5) such as cefoxitin (a cephamycin), latamoxef (an oxacephem), imipenem (N-formimidoylthienamycin, a carbapanem) and aztreonam (a monobactam) also act in this area with the same general mechanism, i.e. inhibition of cross-linking. Various antibiotics are known to inhibit peptidoglycan biosynthesis at other stages but most have little use as therapeutic antimicrobial agents. Vancomycin is an exception because it has an important place in the treatment of multiple-resistant staphylococcal infections and pseudomembranous colitis. Teicoplanin acts in a similar manner to vancomycin and is currently undergoing clinical evaluation. Daptomycin acts at an earlier stage and is also undergoing clinical evaluation for use in β-lactam-resistant Gram-positive bacterial infections. Cycloserine has limited use as an antituberculous drug.

2.3.1 β-lactams

These antibiotics inhibit TPases and CPases by binding covalently to the active sites of the enzymes. As structural analogues of D-alanyl-D-alanine, the β-lactams are mistakenly recognized by the enzymes as the natural substrates and the β-lactam bond is attacked. The cleaved β-lactam becomes covalently linked to the enzyme forming an acyl–enzyme complex which is much more stable than that of the enzyme and natural substrate (the half-life is of the order of 10 minutes). The enzymes are therefore trapped as inactive complexes by the β-lactams and are unable to participate in peptidoglycan assembly (Fig. 9.2). A close structural similarity between the β-lactams and D-alanyl-D-alanine might not be immediately obvious. However, examination of bond lengths and angles shows that the reactive β-lactam bond lies in a similar position to the peptide bond of D-alanyl-D-alanine (Figs 9.2 and 9.3), a feature first recognized by Tipper and Strominger in 1965.

Although both TPase and CPase activities are inhibited by β-lactams, the TPase is the lethal target. Concentrations of β-lactams required to

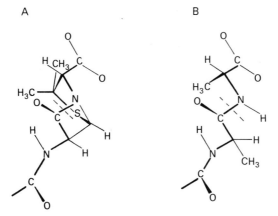

A   B

Fig. 9.3 A, comparison of the structure of the nucleus of the penicillin molecule with B, the D-alanyl-D-alanine end group of the precursor of bacterial peptidoglycan. The broken lines show the correspondence in position between the labile bond of penicillin and the bond broken during the transpeptidation reaction associated with the cross-linking in peptidoglycan.

inhibit bacterial growth closely match those required to inhibit TPase, which fulfils a vital function in the cells, whereas CPase activity is not essential for cell viability. Interestingly, nearly 90% of the penicillin G (benzylpenicillin) that reacts with the enzymes of *E. coli* binds to the non-essential CPase, the remaining 10% being responsible for the lethal action. This information has come from studying the enzymes which become covalently linked to radiolabelled penicillin G, taking advantage of the relatively stable complexes formed. These PBPs can be separated by electrophoresis and individual levels measured.

Studies in mutants with defective PBPs, competition assays with un-labelled β-lactams and labelled penicillin G, and observations of the morphological effects upon whole cells have revealed some useful infor-mation on the function of the PBPs and the mechanisms of action of the β-lactams. The first point to note is the multiplicity of PBPs; at least seven distinct forms occur in *E. coli* (Table 9.2) and these are denoted PBP 1A–6 with 1Bs denoting a group of PBPs of several enzymes. The functions and enzymic activities of all the PBPs have not been deter-mined, TPase activity being particularly difficult to retain and assay in purified form. The 1Bs group of PBPs have TPase and transglycosylase activity. The second point is that β-lactams with any selective activity towards PBP 1Bs, 2 or 3 should exhibit distinct morphological effects of lysis, formation of oval-shaped cells or formation of filaments, respec-tively. Some β-lactams do show these selective morphological effects: cephaloridine causes lysis; mecillinam is highly selective for PBP 2 and produces oval-shaped cells; whilst many of the 'third generation' cephalosporins cause filamentation. Most β-lactams produce a combina-tion of effects depending on the concentration employed. The ultimate result in any case is cell death, although a precise explanation of how this

Table 9.2 Penicillin-binding proteins of *E. coli*: some properties and functions

PBP	Molecular weight	% of total PBP	Morphological effects of inactivation	Enzymic activity	Possible function
1A	90 000	6	?	?	Compensates for 1Bs, minor TPase?
1Bs	87 000	2	Lysis	TPase and transglycosylase	Major TPase, involved in integrity of cells
2	66 000	0.7	Oval cells	?	Maintains rod shape
3	60 000	2	Filaments	?	Septum formation, cell division
4	49 000	4	None	CPase, TPase, endopeptidase ⎫	
5	42 000	65	None	CPase ⎬	Regulation of cross-linkage
6	40 000	21	None	CPase ⎭	

occurs is still not entirely clear. Lysis is a lethal event and autolytic enzymes present in the cell wall may be activated by β-lactam action. An important feature of β-lactams is their broad spectrum of action, which reflects the essential nature and common function of PBPs in a wide range of bacteria.

2.3.2 Vancomycin and teicoplanin

These complex glycopeptide molecules, like the β-lactams, have a specific affinity for the D-alanyl-D-alanine portion of peptidoglycan precursors. They bind to this region and inhibit the transfer of the linear glycan acceptor in the wall to the *N*-acetylmuramylpentapeptide-*N*-acetylglucosamine on its lipid carrier. Peptidoglycan assembly is thereby halted at an earlier stage than transpeptidation. Due to their high molecular weights these compounds show poor penetration into Gram-negative cells and only have useful activity against Gram-positive organisms.

2.3.3 Daptomycin

This cyclic lipopeptide antibiotic appears to act upon peptidoglycan biosynthesis at an earlier stage than vancomycin and teicoplanin because it prevents them causing the accumulation of the peptidoglycan precursor, UDP-*N*-acetylmuramylpentapeptide. Daptomycin does not enter the cytoplasm and probably exerts its action at the cytoplasmic membrane. Although it interferes with peptidoglycan synthesis, causing changes in the shape of sensitive organisms, it also affects the synthesis of lipids and possibly other components of the Gram-positive cell wall, including teichoic acids.

2.3.4 D-cycloserine

This antibiotic, which is particularly active against mycobacteria, acts at an early cytoplasmic stage of peptidoglycan synthesis in which the

D-alanyl-D-alanine unit is formed. Two enzymes are inhibited by cycloserine; the first is alanine racemase, which converts L-alanine to D-alanine, and the second is a synthetase that forms the dipeptide, D-alanyl-D-alanine. Cycloserine bears a structural resemblance to one of the possible conformations of D-alanine, the enzymes bind it and are inhibited.

3   Protein synthesis

3.1   The basis of selective inhibition

Figure 9.4 gives a basic outline of the mechanism of protein synthesis involving the ribosome, mRNA and a series of aminoacyl transfer RNA (tRNA) molecules (one for each of the amino acids found in proteins).

The mechanism of protein synthesis is essentially similar in bacteria and mammalian cells but there are several important differences in ribosome structure and accessory factors which form the targets for selective inhibition by antibiotics. A wide range of valuable antibiotics act by inhibiting protein synthesis, ranging from the complex macrolides (e.g. erythromycin) and aminoglycosides (e.g. gentamicin), to the simple structure of chloramphenicol. Bacterial ribosomes are smaller than their mammalian counterparts; those from bacteria have a sedimentation coefficient (S) of 70S, comprising two subunits of 50S and 30S, those from mammalian cells of 80S, made up from 60S and 40S subunits. S is a measure of the rate of deposition in an ultracentrifuge and is measured in reciprocal seconds.

In many cases selective inhibition of bacteria results from selective binding of the antibiotic to the bacterial ribosome, either to the 50S or 30S subunit. In the case of the tetracyclines, selectivity derives from active uptake of the antibiotic by bacteria and virtual exclusion from mammalian cells, isolated 70S and 80S ribosomes being equally sensitive to inhibitors.

3.2   Aminoglycosides

Aminoglycosides interfere selectively with the 30S subunit of bacterial ribosomes. Most studies have been made with streptomycin which has been shown to bind tightly to one of the 21 distinct protein components of the subunit. This information comes from studies with radiolabelled streptomycin and mutants which contain ribosomes resistant to inhibition by streptomycin. Binding of the antibiotic to the protein, which is the binding site for initiation factor IF-3, prevents initiation of protein synthesis. It also distorts the A site of the 30S subunit and interferes with the correct positioning of the aminoacyl-tRNA. This results either in inhibition of protein synthesis or 'misreading' of the genetic code, i.e. incorporation of wrong amino acids into the polypeptide chain. It has been suggested that misreading leads to an accumulation of toxic, non-

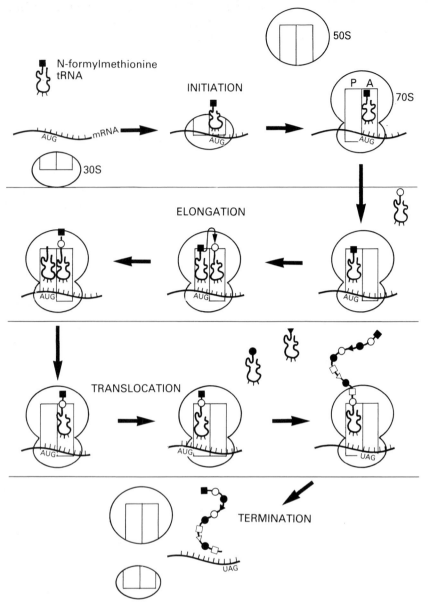

Fig. 9.4 Outline of the main events in protein synthesis: initiation, elongation, translocation and termination. AUG is an initiation codon on the mRNA; it codes for *N*-formylmethionine and initiates the formation of the 70S ribosome. UAG is a termination codon; it does not code for any amino acid and brings about termination of protein synthesis.

functional proteins which would eventually prove fatal to the bacteria. However, addition of an aminoglycoside to a growing culture of bacteria causes a complete cessation of protein synthesis within a few minutes and it seems unlikely that sufficient toxic, non-functional proteins could be produced in this time. The bactericidal action therefore seems to result from tight binding of the antibiotics to the 30S subunit and irreversible inactivation of the ribosome.

Other aminoglycosides also bind to the 30S subunit at a site which is close to, but not identical with, that of streptomycin. They are assumed to inhibit protein synthesis in the same way. The effectiveness of the aminoglycosides is enhanced by their active uptake by bacteria. Uptake proceeds in three phases. Firstly, a rapid initial uptake occurs within a few seconds of contact which represents binding to negatively charged surface components of bacteria. This is usually referred to as the energy-independent phase (EIP) of uptake. In the case of Gram-negative bacteria, the aminoglycosides damage the outer membrane causing release of some lipopolysaccharide, protein and phospholipid. Secondly, there follows an energy-dependent phase of uptake (termed EDP I) lasting about 10 minutes, in which the aminoglycoside is transported across the cytoplasmic membrane by the proton motive force or on respiratory quinones. A second energy-dependent phase of uptake (EDP II), which leads to further intracellular accumulation, follows after the aminoglycoside has bound to the ribosomes. The precise nature of the transport mechanisms involved in EDP I and II is still uncertain. They are essential to the antimicrobial action and do not occur in anaerobic organisms, which are consequently resistant to aminoglycosides.

3.3 Tetracyclines

This group of antibiotics bind to the 30S subunit of bacterial ribosomes and prevent attachment of aminoacyl-tRNA to the A site. The equivalent 40S subunits of mammalian ribosomes are also sensitive to tetracyclines; the basis of the selective toxicity towards bacteria resides in a differential permeability. Bacteria actively transport tetracyclines into the cytoplasm, achieving a 50-fold concentration gradient over the extracellular concentration, whereas no similar transport process occurs in mammalian cells. Tetracyclines presumably make illicit use of a normal bacterial transport process. The importance of the uptake process in their mode of action is shown by the fact that resistance to tetracyclines is associated either with failure to accumulate the antibiotic or to an active efflux system which removes the drug from the cells.

3.4 Chloramphenicol

This antibiotic selectively inhibits bacterial protein synthesis by binding to the 50S subunit in the region of the A site. In so doing, it inhibits peptidyl transferase, a component of the ribosome. This enzyme is responsible for forming the new peptide bond between the growing peptide in the P site and the amino acid on the aminoacyl-tRNA in the A site. Very little is known about the penetration of bacterial cells by chloramphenicol. It also penetrates mammalian cells but is not toxic because it does not affect protein synthesis by 80S ribosomes. The ability to penetrate mammalian cells is of value in treating infections caused by intracellular bacteria such as *Salmonella typhi*.

3.5   Erythromycin

This antibiotic is a member of the macrolide group; it selectively inhibits bacterial protein synthesis by binding to the 50S subunit. The binding site is close to that of chloramphenicol because binding of one antibiotic to the ribosome prevents binding of the other. It is the translocation process, rather than peptidyl transferase, that is inhibited by erythromycin. Translocation is the process which occurs after the peptide chain elongation reaction of peptidyl transferase. The ribosome moves along the mRNA by one codon (the set of three bases which specify an amino acid), displacing tRNA from the P site, moving the extended peptidyl-tRNA into the P site and leaving the A site free for the next aminoacyl-tRNA specified by the codon in the A site. Erythromycin does not prevent the peptidyl-tRNA leaving the A site, but it inhibits release of tRNA from the P site and obstructs occupation of the P site by peptidyl-tRNA.

3.6   Lincomycin and clindamycin

These agents bind to a region of the 50S subunit close to that of chloramphenicol and erythromycin and block elongation of the peptide chain by inhibition of peptidyl transferase.

3.7   Fusidic acid

This steroidal antibiotic does not act upon the ribosome itself but upon one of the factors associated with it, elongation factor EF-G. It also acts upon the equivalent factor in mammalian cells, EF-2, and owes its selective toxicity, like the tetracyclines, to active uptake by bacteria and exclusion from mammalian cells. EF-G has guanosine triphosphatase (GTPase) activity, hydrolysis of GTP being used to drive the translocation process. Fusidic acid binds to EF-G, inhibiting the GTPase function and blocking translocation by forming a stable complex between EF-G, guanosine diphosphate (GDP) and the ribosome.

3.8   Mupirocin (pseudomonic acid)

The target of mupirocin is one of a group of enzymes which couple amino acids to tRNA for delivery to the ribosome and incorporation into protein. Part of the mupirocin structure resembles isoleucine and this is thought to be the basis for its specific inhibition of bacterial isoleucyl-tRNA synthetase.

4 Chromosome function

4.1 The basis of selective inhibition

As with protein synthesis, the mechanisms of chromosome replication and transcription in bacteria are essentially similar to those of mammalian cells. There are, however, important differences in the detailed functioning and properties in the enzymes involved and these differences are exploited by a number of antibiotics as the basis of selective toxicity. In *E. coli* the basic sequence of events in chromosome replication is as follows.

1 Nucleotide and deoxynucleotide precursors are synthesized in the cytoplasm.

2 The two strands of DNA double helix are unfolded by the action of topoisomerases. One of these, the gyrase, is unique to bacteria and is the target for selective inhibition by the quinolone carboxylic acids. The unwound strands are kept unfolded by binding of a protein called Albert's protein.

3 An RNA polymerase synthesizes short primer strands of RNA on each strand at specific initiator sites.

4 DNA polymerase III then joins short DNA strands on to the RNA primers. The DNA strands are called Okasaki fragments.

5 DNA polymerase I (which has nucleotidase activity) removes the RNA primers and replaces them with DNA strands.

6 A DNA ligase joins the DNA fragments together.

Transcription, the copying of a gene sequence from DNA to a single strand of mRNA, is carried out by RNA polymerase. This is a complex of units comprising the core enzyme, made up of four proteins and another protein, termed the σ-factor. The σ-protein is responsible for recognizing the initiation signal for transcription, and the core enzyme possesses the activity to incorporate and join together the nucleotides in the sequence specified by the DNA. Mammalian cells possess an analogous RNA polymerase but there are sufficient differences in structure to permit selective inhibition of the microbial enzyme by the antibiotic rifampicin.

Numerous antibiotics inhibit chromosome function, many at the level of nucleotide synthesis, others by binding to the double helix and preventing strand separation and therefore replication and transcription. Many of these are toxic to mammalian cells, and some are used as anticancer agents. Very few have genuine selective toxicity towards bacteria.

4.2 Quinolone carboxylic acids

This general chemical name is used to describe a group of synthetic agents originated by nalidixic acid and recently joined by many others, chemically synthesized, with greatly improved activity, such as norfloxacin and ciprofloxacin (Chapter 5). They inhibit specifically the microbial

201   *Mechanisms of action of antibiotics*

enzyme, DNA gyrase, and thus block chromosomal replication. DNA gyrase (bacterial topoisomerase II) catalyses the ATP-dependent negative supercoiling of DNA. It acts together with DNA topoisomerase I to promote supercoiling and relaxation of the bacterial chromosome. These enzymes are vital in DNA replication, expression and repair. The gyrase acts by introducing a transient double-stranded break in a segment of DNA; a second segment is passed through the break which is then resealed. The gyrase is a tetrameric complex of two A and two B subunits which are all required for activity. The A subunits are involved in breaking and resealing DNA and the B subunits hydrolyse ATP to generate energy. It is thought that the quinolone carboxylic acids stabilize (trap) transiently broken DNA in a complex with the gyrase. The reason for subsequent death of the cells remains unclear.

4.3   Metronidazole and nitrofurantoin

The antimicrobial action of these agents is thought to be due to unstable metabolites produced by reduction inside bacteria after uptake. The unidentified products appear to cause strand breakage of DNA. Whatever the basis of the mechanism of action might be, metronidazole is specifically active against anaerobic organisms, in which the metabolite is produced under the low redox conditions. Resistance to metronidazole is rare.

4.4   Rifampicin

Rifampicin binds to one of the proteins, the B subunit of the core enzyme of RNA polymerase, and blocks the action of the enzyme at the initiation stage of transcription. Although the σ-factor is involved in recognizing the initiation signal on the DNA, rifampicin does not interact with it. Rifampicin does not inhibit transcription once it has been initiated. One of the problems with this antibiotic is that resistance develops quickly. The basis of resistance is an alteration in the RNA polymerase so that it no longer binds rifampicin.

5   Folate antagonists

5.1   Folate metabolism in bacteria and mammalian cells

There is a basic difference in folate metabolism between bacteria and mammalian cells. Bacteria are unable to take up and utilize exogenous folate and must synthesize it themselves. This is carried out in a series of reactions involving first the synthesis of dihydropteroic acid from one molecule each of pteridine and p-aminobenzoic (4-aminobenzoic) acid. Glutamic acid is then added to dihydropteroic acid to form dihydrofolate (DHF). DHF is reduced by dihydrofolate reductase (DHFR) to tetrahydrofolate (THF) using $NADPH_2$ as a cofactor (Fig. 9.5). THF is the

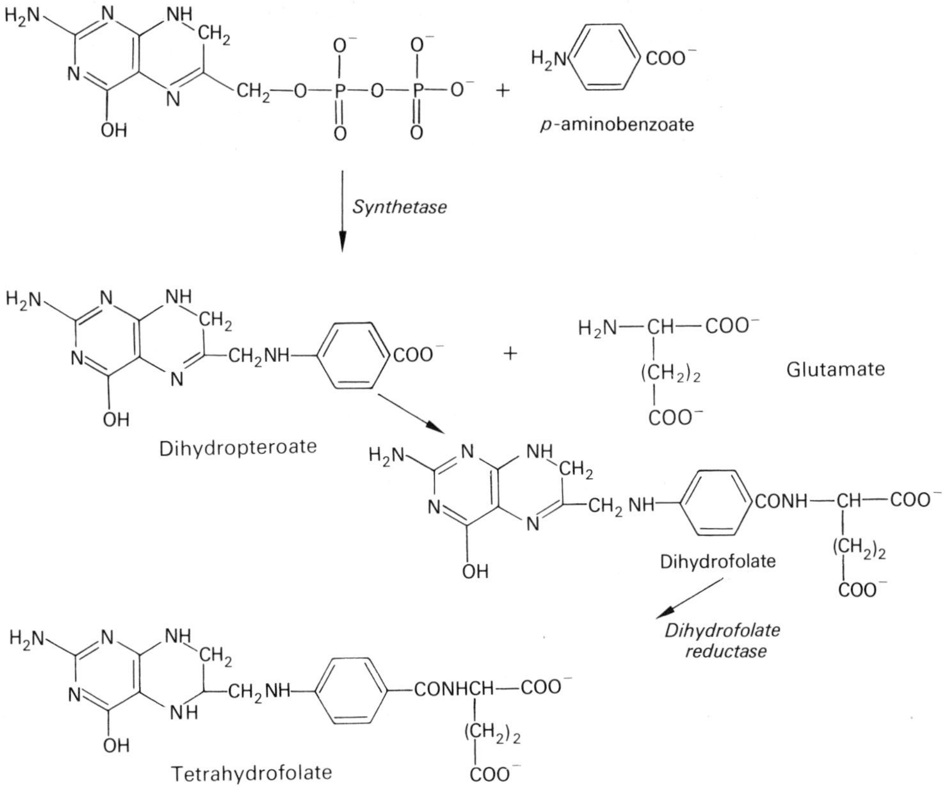

Fig. 9.5 Final steps in the biosynthesis of tetrahydrofolate by bacteria.

carrier of single carbon fragments (— CHO and — CH$_3$) used in the biosynthesis of adenine, guanine, thymine and methionine. Thus interference with any stage of THF synthesis will affect the cell's ability to synthesize DNA, RNA and protein and inhibit growth. Mammalian cells take up preformed DHF from dietary nutrients, and convert it to THF by mammalian DHFR using THF as a single carbon carrier in the same way as bacteria. Both the sulphonamides and trimethoprim inhibit selectively bacterial folate metabolism, the former by inhibiting dihydropteroic acid synthesis, which does not occur in mammalian cells, and the latter by a selective affinity for bacterial DHFR (sections 5.2 and 5.3, Fig. 9.5). Both of these antibacterial agents only inhibit bacterial growth in the absence of metabolites such as adenine, guanine, thymine and methionine. If these are available in the environment as nutrients, bacteria will use them preferentially and overcome the drug inhibition.

5.2   **Sulphonamides**

Sulphonamides competitively inhibit the incorporation of *p*-aminobenzoic acid into dihydropteroic acid and thence into DHF and THF. They are structural analogues of *p*-aminobenzoic acid, and there is evidence that

203   *Mechanisms of action of antibiotics*

they are incorporated in place of the natural substrate into folic acid analogues that are also inhibitors of folate metabolism. Excess of *p*-aminobenzoic acid will reverse the inhibitory action of sulphonamide, but like thymine, adenine, guanine and methionine, this nutrient is not normally available at the site of infections for which they are used.

5.3   Trimethoprim

There are many analogues of DHF that inhibit DHFR. Most are toxic to mammalian cells as well as to bacteria. Trimethoprim is an exception because it exhibits a highly specific inhibitory action upon bacterial DHFR. The bacterial enzyme is several thousand times more sensitive than the mammalian enzyme, and trimethoprim exerts no effective mammalian toxicity at the concentrations used to treat bacterial infections. The basis for the selective inhibition of the microbial DHFR by trimethoprim is associated with the subtle differences in the size and shape of the active site of the enzyme compared with its mammalian counterpart.

Trimethoprim is used alone to treat infections but, frequently, it is used in combination with a sulphonamide, usually with sulphamethoxazole as co-trimoxazole. The two stage interference upon folate metabolism is synergistic, i.e. the combination is much more effective than either drug used alone. It was hoped that resistance would not develop readily to a combination of agents, but this has occurred.

6   The cytoplasmic membrane

6.1   Composition and susceptibility to disruption

All cells are surrounded by a membrane (the cytoplasmic membrane) which separates them from the environment and controls the passage of substances in and out of the cells. The integrity of the cytoplasmic membrane is vital to the normal function of the cells. Bacterial membranes differ in one important respect from those of fungi and mammalian cells—bacterial membranes do not contain sterols; fungi, on the other hand, contain predominantly ergosterol as the sterol component and mammalian cells contain cholesterol. In basic structure, the cell membranes are similar. In bacteria they comprise a phospholipid bilayer made up from several different phospholipids (e.g. phosphatidylethanolamine, phosphatidylglycerol and diphosphatidylglycerol) which account for 20–30% of the weight of the membrane. Proteins (50–70% of the weight) are distributed throughout the phospholipid bilayer, some on either side of the membrane and some protruding through and exposed on both sides. The membrane structure is held together non-covalently by a combination of ionic, hydrophobic and hydrogen bonding between the phospholipid and protein constituents. This balance of interactions can be disturbed by the intrusion of molecules, membrane-active agents, which destroy the integrity of the membrane causing leakage of cytoplasmic contents or impairment of metabolic functions associated with the membrane.

Most of the membrane-active agents which function in this way, e.g. alcohols, phenols, quaternary ammonium compounds (QACs) and bisbiguanides (considered in Chapters 11 and 14) have very poor selectivity, i.e. they have similar effects upon mammalian cells. Consequently, they cannot be used systemically to treat bacterial infections as they would lyse blood cells and damage tissue membranes. Instead they are used as skin antiseptics, disinfectants and preservatives. Polymyxin is one of the few membrane-active agents which can be used systemically, although it is toxic and its major use as an antipseudomonal agent has been virtually discontinued in favour of β-lactams and aminoglycosides. Two other groups of agents have an important role in the chemotherapy of fungal infections, namely the polyenes and the imidazoles. The basis of their selective toxicity towards fungi depends on the more subtle differences in composition and biosynthesis of fungal and mammalian cell membranes.

6.2   **Polyenes**

This group of antibiotics, of which amphotericin B and nystatin are the most frequently used, owe their selective antifungal activity to their strong affinity for ergosterol. The hydrophobic polyene region of the antibiotic binds to the sterol component in fungal membranes. In so doing, the hydroxylated hydrophilic portion of the polyene antibiotic is pulled into the membrane interior. Possibly the association together of several polyene molecules in the membrane forms an aqueous channel through which cytoplasmic constituents leak from the cells. The first components released from the fungi are K^+ ions, followed by amino acids and nucleotides. As the pH of the cytoplasm falls, macromolecules are degraded and the cells are killed. The affinity of polyenes for cholesterol is lower than for ergosterol. Polyenes are therefore more active against fungi than mammalian cells, although irreversible kidney damage is a major problem when they are used intravenously to treat severe systemic infections. The problem can be reduced by administration of amphotericin as a liposome complex with cholesterol and phosphatidylglycerol.

6.3   **Imidazoles**

Miconazole and ketoconazole are the most widely used of the antifungal imidazoles (see also Chapter 5). They have a broad spectrum of activity which includes dermatophytes, dimorphic fungi and yeasts; they also exhibit some activity against Gram-positive bacteria but are not used to treat bacterial infections. Several possible mechanisms have been implicated in their antifungal action. The first is membrane damage due to binding of the imidazoles to unsaturated fatty acids present in the phospholipid components of the membrane. Imidazoles induce some leakage of cytoplasmic constituents from treated cells. It is difficult to explain the excellent selective toxicity of the imidazoles entirely in terms of membrane damage and leakage, since mammalian cell membranes also contain unsaturated fatty acids. The second mechanism of action concerns selective inhibition of a step in ergosterol biosynthesis. In particular, the antifungal imidazoles inhibit the demethylation of the 14-α-methyl group from lanosterol, leading to a toxic accumulation of lanosterol. Another reported effect is an inhibition of triglyceride and phospholipid synthesis. Finally, ketoconazole inhibits electron transport in the respiratory chain of fungi growing under aerobic conditions, succinate and NADH oxidases being inhibited. The importance of this effect to the overall antifungal activity is possibly reflected in the minimum inhibitory concentration, which is much higher under anaerobic conditions.

6.4   **Naftidine**

This synthetic allylamine derivative inhibits squalene epoxidase, one of the enzymes involved in the complex pathway of fungal sterol biosynthesis. Probably acting as a structural analogue of squalene, naftidine causes the accumulation of this unsaturated hydrocarbon, and a decrease in ergosterol in the fungal cell membrane.

7   **Further reading**

Actor P., Daneo-Moore L., Higgins M.L., Salton M.R.J. & Shockman G.D. (1988) *Antibiotic Inhibition of Bacterial Cell Surface Assembly and Function.* Washington: American Society for Microbiology.

Brajtburg J., Powderly W.G., Kobayashi G.S. & Medoff G. (1990) Amphotericin B: current understanding of mechanisms of action. *Antimicrob Agents Chemother.* **34**, 183–188.

Chopra I. & Linton A. (1986) The antibacterial effects of low concentrations of antibiotics. *Adv Microb Physiol.* **28**, 212–259.

Franklin T.J. & Snow G.A. (1989) *Biochemistry of Antimicrobial Action*, 4th edn. London: Chapman & Hall.

Greenwood D. (1989) *Antimicrobial Chemotherapy*, 2nd edn. Oxford: Oxford University Press.

Hlavka J.J. & Boothe J.H. (1985) *The Tetracyclines. Handbook of Experimental Pharmacology 78.* Berlin: Springer-Verlag.

Kerridge D. (1986) Mode of action of clinically important antifungal drugs. *Adv Microb Physiol.* **27**, 1–72.

Nagarajan R. (1991) Antibacterial activities and modes of action of vancomycin and related glycopeptides. *Antimicrob Agents Chemother.* **35**, 605–609.

Peterson P.K. & Verhoef J. (Eds) (1986) *The Antimicrobial Agents Annual 1.* Amsterdam: Elsevier. (See also Numbers 2 (1987) and 3 (1988) in the same series.)

Ristuccia A.M. & Cunha B.A. (1984) *Antimicrobial Therapy.* New York: Raven Press.

Rubinstein E., Adam D., Moellering R. & Waldvogel F. (Eds) (1989) Second international symposium on new quinolones. *Rev Infect Dis.* **11** (Suppl. 5), S897–S1431.

Russell A.D. & Chopra I. (1990) *Understanding Antibacterial Action and Resistance.* New York: Ellis Horwood.

St Georgiev V. (Ed.) (1988) *Antifungal Drugs*. Annals of the New York Academy of Sciences, vol. 544. New York: New York Academy of Sciences.

Tipper D.J. (1988) *Antibiotic Inhibitors of Bacterial Cell Wall Biosynthesis*, 2nd edn. Oxford: Pergamon.

Williams A.H. & Gruneberg R.N. (1988) Teicoplanin. *J Antimicrob Chemother*. **22**, 397–401.

Wolfson J.S. & Hooper D.C. (1989) Fluoroquinolone antimicrobial agents. *Clin Microbiol Rev*. **2**, 378–424.

10   Bacterial resistance to antibiotics

1   **Inherent and acquired resistance**

2   **The genetic basis of acquired antibiotic resistance**
2.1   Spontaneous mutations
2.2   Acquisition of antibiotic resistance by the transfer of genetic information.
2.2.1   Conjugation
2.2.2   Transference of resistance-determining genes in Gram-negative bacteria
2.2.3   Transposons
2.2.4   Clinical importance of drug resistance caused by R-plasmids
2.2.5   Conjugative transfer of drug-resistance plasmids in Gram-positive bacteria
2.3   Transduction
2.4   Transformation

3   **Biochemical mechanisms of resistance**
3.1   Conversion of an active drug to a non-toxic derivative
3.1.1   Inactivation of penicillins and cephalosporins

3.1.2   Genetics and physiology of β-lactamase synthesis
3.1.3   Penicillins and cephalosporins stable to β-lactamases
3.1.4   Inactivation of chloramphenicol by acetylation
3.1.5   Inactivation of aminoglycoside antibiotics
3.2   Changes in the target site resulting in resistance
3.2.1   Streptomycin
3.2.2   Methicillin
3.2.3   Erythromycin
3.2.4   Trimethoprim
3.2.5   Sulphonamides
3.2.6   Quinolones
3.3   Reduction in cellular permeability to antibiotics
3.3.1   Permeability of the outer membrane
3.3.2   Antagonism of antibiotic transport processes

4   **What can be done to combat the problem of bacterial resistance to antibiotics?**

5   **Further reading**

1   Inherent and acquired resistance

A distinction must be made between *inherent* and *acquired* resistance of bacteria to antibiotics. Certain bacteria are, and as far as we know always have been, more or less resistant to some antibiotics. For example, Gram-negative bacteria are inherently resistant to a number of important antibiotics that are very effective against Gram-positive organisms (Table 10.1). Within the Gram-negative group, *Pseudomonas aeruginosa* presents an especially intractable chemotherapeutic problem as it has unusually high intrinsic resistance to many antibiotics. This inherent resistance of Gram-negative bacteria seems to be associated with the impermeability of the complex outer layers of the cell envelope to some drugs, which prevents the attainment of an inhibitory concentration within the cell. The non-specific resistance of Gram-negative bacteria is recognized as a limitation in the treatment of infections of these organisms. However, the general pattern of resistance is well known and stable, so that drugs are prescribed to which the infecting organism are not inherently resistant.

Table 10.1 Sensitivity of Gram-positive and Gram-negative bacteria to common antibacterial drugs*

Drugs active against Gram-positive and Gram-negative bacteria	Drugs with poor activity against Gram-negative bacteria because of permeability barriers
Tetracyclines	Benzylpenicillin (Penicillin G)
Aminoglycosides	Methicillin
Carbapenems	Macrolides
Sulphonamides	Lincomycin
Fluoroquinolones	Rifamycins
Chloramphenicol	Fusidic acid
Phosphonomycin	Vancomycin
Nitrofurans	Novobiocin
Ampicillin and carbenicillin	
Certain cephalosporins	

* See also Chapters 5 and 14. The difference in response of Gram-positive and Gram-negative bacteria to chlorhexidine is considered in Chapter 14 (Table 14.1).

The problem of acquired resistance to antibiotics is quite different. Because of the rapid multiplication rates and large sizes of bacterial populations, the ability of these populations to respond to environmental changes is remarkable. The adaptation of bacteria to the toxicity of antibiotics is due, therefore, to the flexibility of bacterial populations. When a new antibiotic is introduced into clinical practice for the treatment of infections caused by bacteria that are not inherently resistant to the drug, the majority of infections respond to the new drug, but after months or years of continuous use, reports often appear describing treatment failures due to infections by strains of bacteria that are resistant to the drug.

The time of emergence and rate of spread of the resistant organisms is unpredictable and the ease of acquisition of resistance by bacteria under laboratory conditions may be misleading with regard to the likelihood of resistance appearing in clinical practice. Despite the common occurrence of acquired bacterial resistance to many antibiotics it is not necessarily inevitable. Thus *Streptococcus haemolyticus* has remained sensitive to benzylpenicillin after more than 40 years' exposure to the drug. The causative organism of syphilis, *Treponema pallidum*, has also remained sensitive to penicillin. Resistance may be extremely slow to emerge. For example, resistance of staphylococci to neomycin did not appear for fully 9 years after the introduction of the antibiotic; staphylococcal resistance to the penicillin derivative, methicillin, was relatively uncommon for many years.

These examples are, however, exceptional and the development of antibiotic resistance must usually be expected. Clinical practice should therefore be planned to maximize the period before resistant organisms emerge and to limit the subsequent spread of resistant organisms.

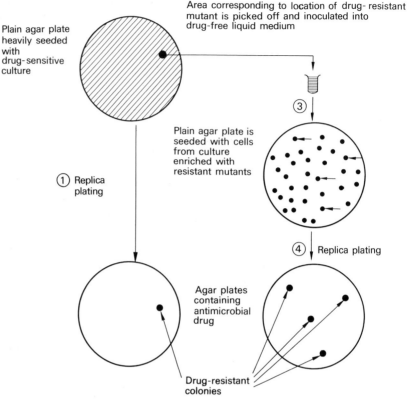

Plain agar plate heavily seeded with drug-sensitive culture

② Area corresponding to location of drug-resistant mutant is picked off and inoculated into drug-free liquid medium

① Replica plating

Plain agar plate is seeded with cells from culture enriched with resistant mutants

③

④ Replica plating

Agar plates containing antimicrobial drug

Drug-resistant colonies

Fig. 10.1 The technique of replica plating reveals the presence of drug-resistant bacteria in a population which is overall drug sensitive. (Redrawn with permission from *Biochemistry of Antimicrobial Action* by T.J. Franklin & G.A. Snow, 1989.)

2 The genetic basis of acquired antibiotic resistance

The acquired ability of bacteria to grow and multiply in the presence of an antibiotic reflects a difference in the genetic make-up between these and sensitive organisms. When a bacterial population adapts to the presence of an antibiotic, sensitive cells are gradually replaced by resistant cells that carry the genes for resistance. How does this genetic change from drug sensitivity to resistance come about?

2.1 Spontaneous mutations

The simple technique of replica plating of bacterial colonies (Fig. 10.1) shows that drug-resistant cells constantly arise in bacterial populations although usually with low frequency. Replica plating reveals that drug-resistant cells appear *in the absence of the antibiotic*, i.e. the drug does not induce the development of resistance. In the presence of antibiotic, resistant cells grow at the expense of sensitive cells and a resistant population eventually supersedes the previously sensitive one.

The appearance of resistant cells can be explained by the relatively infrequent occurrence (approximately 1 per 10^7 cells per cell division) of spontaneous gene mutations which confer drug resistance. Despite the low mutation rate, the enormous numbers of cells in bacterial populations ensure the eventual emergence of bacteria resistant to most antibiotics. In addition, recent evidence suggests that the rate of spontaneous mutations may actually increase when bacteria are subjected to a selective environmental pressure. When that pressure is due to the presence of an antibiotic the chances that a mutation causing resistance will arise increase, even though, as previously stated, the antibiotic does not specifically induce resistance. The rate of occurrence of resistant cells can also be increased by the exposure of bacterial cultures to mutagens. A single mutation at an appropriate genetic locus usually produces only a small increase in resistance and the level of resistance (as opposed to the frequency) gradually builds up with successive mutations at other sites, each conferring a small increment in resistance. Occasionally, though, a single mutation may result in a dramatic leap from sensitivity to high-level resistance, for example ribosomal resistance to streptomycin (see section 3.2.1).

However, many of the resistant bacteria arising from spontaneous mutations only proliferate under carefully controlled laboratory conditions and are unlikely to be encountered in clinical practice. The common procedure of subjecting important pathogens to increasing concentrations of new antibacterial agents in the hope of predicting the likely occurrence of resistant organisms in clinical practice can be misleading. Thus, whilst it is likely that the origins of antibiotic resistance lie in mutations affecting genes responsible for normal cellular activities, the contribution of novel or recurring mutations to the emergence of resistant organisms in the clinical setting may not be very large. Rather it is the selection of pre-existing mutants and the remarkable ability of many bacteria to acquire resistance genes by the processes of conjugation, transduction and possibly transformation that account for most examples of clinically important antibiotic resistance.

2.2 Acquisition of antibiotic resistance by the transfer of genetic information

2.2.1 *Conjugation*

The ability to transfer and receive genetic material by cellular conjugation occurs widely in Gram-negative bacteria. Bacterial conjugation was discovered by J. Lederberg in the late 1940s. During the 1950s it was found that copies of the genes for antibiotic resistance could be transferred in this way from resistant to sensitive organisms, so spreading resistance rapidly through bacterial populations. In Japan at that time it was observed that a strain of *Shigella*, isolated during an epidemic of dysentery, was resistant to several drugs including tetracyclines, streptomycin, chloramphenicol and the sulphonamides. Since the frequency of mutation providing resistance to one drug is about 10^{-7} per cell division, the frequency of mutation to four drugs is 10^{-28}! Therefore, it was extremely unlikely that this resistant strain of *Shigella* had arisen by spontaneous mutations during the dysentery epidemic. Even more startling was the isolation of both sensitive and resistant *Shigella* and also of other multiple-resistant species such as *Escherichia coli* from the same patient. Japanese microbiologists suggested that sets of genes conferring resistance to several antibiotics simultaneously were being transferred between bacteria of the same and different species. It was soon established that the mechanism of resistance transfer closely resembled the transfer of other genetic determinants by cell-to-cell conjugation. Within a few years of the discovery of *transferable* or *infectious multiple drug resistance* in Gram-negative bacteria in Japan, the same phenomenon was found to be widespread in many other parts of the world, including the UK, Europe and USA. A striking increase in drug resistance in *Salmonella typhimurium* was charted by Anderson and his colleagues at the Public Health Laboratories in London. In 1963–64, 21% of the cultures of *Sal. typhimurium* were found to be resistant to several antibiotics, including streptomycin, sulphonamides and tetracyclines. By 1965–66, 50–60% of *Sal. typhimurium* examined were drug resistant and many cultures were simultaneously resistant to streptomycin, sulphonamides, tetracyclines, ampicillin, neomycin, kanamycin and chloramphenicol. All of these resistance markers were transferable to other Gram-negative bacteria by conjugation.

This marked increase in the resistance of *Sal. typhimurium* with the concomitant risk of resistance spreading to other Gram-negative bacteria sharing the same environment has been attributed to the widespread use of antibiotics as growth promoters in poultry and pigs, and for the treatment of bacterial infections in a wider range of animals. Animals exposed to antibiotics act as a reservoir for resistant bacteria from which transference to human hosts by cross-infection is a common occurrence. Fortunately, an increasing awareness of this problem led to restrictions in certain countries on the use in animal husbandry of antibacterial drugs of importance in human medicine. This resulted in an encouraging decline in the incidence of resistant Gram-negative bacteria in farm animals. A disturbing exception was the persistence of bacteria resistant to tetracyclines long after the withdrawal of these antibiotics from the farmyard in the UK. Subsequent thinking has tended to minimize the contribution to human disease of the reservoir of drug-resistant bacteria in farm animals and has, instead, emphasized the potential of the inappropriate use of antibiotics in human patients for encouraging the spread of resistant bacterial pathogens (Richmond & Petrocheilou, 1981). For example, when the treatment regime is insufficient to ensure complete elimination of the infecting bacteria, this allows the emergence of resistant cells. This, coupled with inadequate procedures to contain resistant organisms and minimize cross-infections amongst patients and staff, has often been responsible for the contamination of the hospital environment with these dangerous bacteria.

Transference of resistance-determining genes in Gram-negative bacteria

The ability to transfer genes that confer drug resistance by cellular conjugation is due to the presence in the bacterial cell of DNA elements known as *R-plasmids* that replicate separately from, but usually under the control of, the bacterial chromosome. A typical plasmid found in resistant Gram-negative bacteria consists of two distinct but frequently linked elements: (i) a suite of genes that enables the cell to conjugate with a sensitive bacterium and to transfer to it a copy of the entire plasmid; (ii) one or more linked genes each conferring resistance to a specific antibacterial drug. The linked complex of conjugation and resistance determinants takes the form of a double-stranded circular molecule of DNA. The resistance determinants associated with R-plasmids confer resistance to many drugs, including the tetracyclines, β-lactams, the sulphonamides, the aminoglycosides, chloramphenicol and trimethoprim. The presence of an R-plasmid in a bacterial cell does not exclude the possibility of chromosomally determined resistance markers occurring in the same cell. For example, resistance to the quinolones, which are useful in the treatment of urinary tract infections, is determined by chromosomal mutations. Quinolone resistance can, therefore, occur in the presence of a battery of R-plasmid resistance markers.

The replication of the R-plasmid, although it occurs separately from that of the bacterial chromosome, is nevertheless normally coordinated with chromosomal replication. Usually there is a fixed number of R-plasmids per chromosome, for example one to four in *Ecscherichia coli* depending on the particular R-plasmid. Under some circumstances, especially in *Proteus mirabilis*, the number of R-plasmid copies may increase. This occurs in *Pr. mirabilis* during stationary phase and also in cells exposed to very high concentrations of certain antibiotics. The presence of multiple copies of the resistance genes leads to enhanced levels of resistance to certain antibiotics. 'Gene dosage' effects result in increased production of specific enzymes that inactivate antibiotics such as β-lactams, chloramphenicol and streptomycin (see section 3). At present, however, it is not known whether resistance is influenced by gene dosage effects in clinical situations.

Some R-plasmid bearing cells (R^+ cells) possess hair-like structures known as pili that extend out from the bacterial surface (Fig. 10.2). The pili, whose synthesis is under the control of the conjugation component of the R-plasmid, are essential to the conjugation phenomenon with R^- bacteria and the transfer of an R-plasmid. The synthesis of the pilus is repressed except for a short period following the acquisition of the R-plasmid so that not all cells in an R^+ population possess pili, some cells having lost their pili in mechanical 'accidents'. When R^- cells are

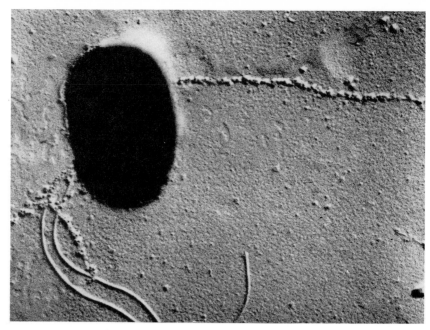

Fig. 10.2 Electron micrograph of an *E. coli* cell carrying pili coded by an R-plasmid. The pili are visualized by the attachment along their lengths of specific phage particles. The smooth filaments extending from the bacterial surface are flagellae. Original magnification ×48 000. (Courtesy of Dr T.D. Hennessey and M. Bentley.)

introduced into the culture, Brownian motion ensures random contacts between R^+ and R^- cells. Such contacts result in the attachment of the pili of the R^+ cells to R^- cells. This signals the R^+ cell to commence transfer of a copy of its R-plasmid to the R^- cell. The pili serve to hold the R^+ and R^- cells in close contact and the transfer of plasmid DNA is believed to occur through a transmembrane pore formed by the localized fusion of the cytoplasmic membranes of the two cells. Transfer of plasmid DNA is initiated at a specific site on the DNA called *oriT*.

A nick is introduced into the strand of DNA destined for transfer at *oriT* and a cellular signal resulting from the formation of the mating pair then initiates unwinding of the two DNA strands to permit the transmission of the nicked strand. Transfer of the single strand of DNA is accompanied by replication of this strand in the receipient cells and of the retained strand in the donor cells so that both partners are left with double-stranded circular DNA R-plasmids. Expression of drug resistance in the newly R^+ cell immediately follows the formation of a double-stranded circularized R-plasmid. For a period of about an hour after the acquisition of an R-plasmid, pili are synthesized and the cell enjoys a brief fertile spell when it is able to transfer copies of its R-plasmid to R^- cells that it contacts. In this way drug resistance spreads rapidly through a bacterial population. Pilus synthesis eventually ceases, however, due to the progressive accumulation of a substance within the cells which suppresses the activity of the pilus-determining genes.

2.2.3   *Transposons*

The discovery of the remarkable 'jumping genes' or transposons has provided an explanation for the frequent occurrence of drug-resistance genes in bacteria. At one time it was believed that the exchange of genes among plasmids, phages and bacterial chromosomes could only occur by classical recombination mediated by the product of the bacterial *recA* gene and involving the exchange of DNA in domains of extensive genetic homology. This relatively restricted mechanism permits DNA exchange only between closely related genomes and it did not satisfactorily explain the very diverse shuffling of resistance genes in bacterial genomes. It is now known, however, that the acquisition and exchange of genetic material are not limited by the *recA* mechanism but can also occur as a result of the promiscuous activity of transposons, which are able to insert themselves into many different genomic sites that have no homology with the transposons. Some very simple transposons only carry information concerned with the insertion function. More complex transposons, however, have other genes including those for resistance to β-lactams, aminoglycosides, sulphonamides, tetracyclines, chloramphenicol and trimethoprim. The integrated transposon DNA is replicated along with the host DNA. Almost certainly it is the activities of such transposons which have resulted in R-plasmids with resistance markers to many antibiotics.

2.2.4   *Clinical importance of drug resistance caused by R-plasmids*

The dramatic rapidity with which R-plasmid-mediated resistance spreads through bacterial populations and between different species and genera under laboratory conditions, coupled with the frequent detection of R^+ bacteria associated with clinical infections, aroused serious fears for the future of antibiotic treatment of diseases caused by Gram-negative bacteria. There is no doubt that the presence of R^+ bacteria in microbial populations associated with infections can result in the emergence of a drug-resistant population of cells during antibiotic therapy. However, the current view is that R-plasmid transfer from R^+ to R^- cells in the tissues or in bowel contents or urine may be relatively infrequent, perhaps because of interfering chemicals, for example bile salts, or inappropriate conditions. The emergence of an R^+ population of bacteria during antibiotic therapy is more likely the result of a selection process for resistant cells rather than of extensive conjugative transfer of resistance.

Following antibiotic therapy the number of R^+ bacteria in the faeces declines and there is evidence that in the absence of antibiotic therapy the colonizing ability of R^+ *E. coli* in the gut is significantly poorer than that of the R^- strains of *E. coli* that normally inhabit the gut. Apparently the R-plasmids themselves do not impair colonizing ability but rather the strains of organism which normally harbour R-plasmids are poor colonizers.

In summary, R-plasmid-mediated drug resistance continues to pose a threat to the successful treatment of Gram-negative infections with anti-biotics. Fortunately, however, because of a number of peculiarities of the natural history of R-plasmids and their host bacteria the spread of drug resistance due to the conjugative activity of R^+ bacteria is not as rapid as was previously feared.

| 2.2.5 | *Conjugative transfer of drug-resistance plasmids in Gram-positive bacteria* |

It is now clear that drug-resistance genes can also be transferred by conjugation in various species of Gram-positive bacteria including *Streptococcus*, *Staphylococcus*, *Bacillus* and *Clostridium*. The genes conferring conjugative activity are found on plasmids, prophages and transposons. Amongst the streptococci and probably also in staphylococci, potential recipients secrete pheromone peptides which stimulate the production of a protein called adhesin by the donor cells. This protein causes donor and recipients to aggregate and gene transfer takes place within the cellular aggregates. The alarming increase in the incidence of resistant staphlyococci in recent years may be associated with plasmid-mediated conjugative transfer of resistance genes. There is some indication that conjugative activity in staphylococci may be favoured by the absorbent materials in wound dressings, bedding and clothing.

| 2.3 | **Transduction** |

The transfer of genetic information between bacteria by bacteriophage (bacterial virus) transduction occurs in Gram-positive and Gram-negative bacteria. Essentially the process begins when a phage infects a susceptible cell. The intracellular virus may replicate immediately, resulting in cell lysis and the release of many new viruses, or it may enter the *lysogenic* state (Chapter 3) by becoming integrated into the host genome. This may be chromosomal or plasmid DNA. Sooner or later the latent phage is induced to replicate by a change in environmental conditions and so cause cell lysis and the release of virus particles. Transduction occurs when the replicating phage DNA picks up a piece of adjacent host DNA and incorporates it into the complete phage particle. In generalized transduction a wide range of genetic material from the host DNA is recombined with phage DNA and, upon induction of phage replication, is packaged inside the phage envelope. Subsequent infection of other bacteria by the phage particles leads to the introduction of both phage and bacterial DNA into the new host. However, the phage is disabled by the acquisition of bacterial DNA since this leads to the exclusion of a corresponding length of phage DNA from the phage particle. The infecting phage now only replicates within the new host as a prophage integrated into host DNA. In the P1 phage of *E. coli* all of the phage DNA may be replaced by host DNA during transduction. Generalized transducing phages capable of transmitting drug resistance genes are well recognized in both Gram-negative and Gram-positive bacteria.

However, in the clinical setting transduction may be more important in spreading resistance among Gram-positive bacteria than among Gram-negative cells. Transduction of drug-resistance plasmids in *Staphylococcus aureus* has been observed in infected mice and on human skin. However, the clinical evidence for drug-resistance transfer by transduction is epidemiological. The sudden appearance of neomycin resistance in *Staph. aureus* after some 9 years' use of this antibiotic may have been due to transduction. As in the case of R-plasmid transfer, however, it seems likely that phage transduction of resistance is of less importance in the emergence of resistant populations than the selection of bacteria harbouring drug-resistance plasmids during therapy. Cross-infection of patients with resistant bacteria is also undoubtedly a major factor in the spread of resistance.

| 2.4 | **Transformation** |

A third means by which genetic information can be acquired by bacterial cells is by *transformation* (Chapter 1). Transformation is confined essentially to Gram-positive bacteria and works in this way: DNA released upon cell lysis may be absorbed by 'competent' cells and integrated into their genomes. Usually only fragments of the transforming DNA are successfully incorporated. Rather surprisingly, in view of the widespread occurrence of deoxyribonucleases (DNases) in the tissues, there is evidence that transformation does occur in experimental animals infected with pneumococcus. The genetic markers chosen for this experiment were the smooth and rough colonial forms of the pneumococcus but in principle there is no reason why drug-resistance markers could not be acquired by transformation. However, it is perhaps unlikely that the clinical problem of resistance among Gram-positive pathogens owes much to transformation.

| 3 | **Biochemical mechanisms of resistance** |

Among the clinically important strains of drug-resistant bacteria, certain mechanisms of resistance are commonly found.

1 Conversion of an active drug to an inert product by an enzyme found only in resistant microorganisms.

2 A change in the antibiotic target site resulting in drug resistance.

3 Acquisition, via gene transfer, of a resistant form of the target enzyme.

4 A reduction in cellular permeability to the antibiotic resulting in its exclusion from the bacterium.

Bacterial resistance to a particular antibiotic or related series of drugs

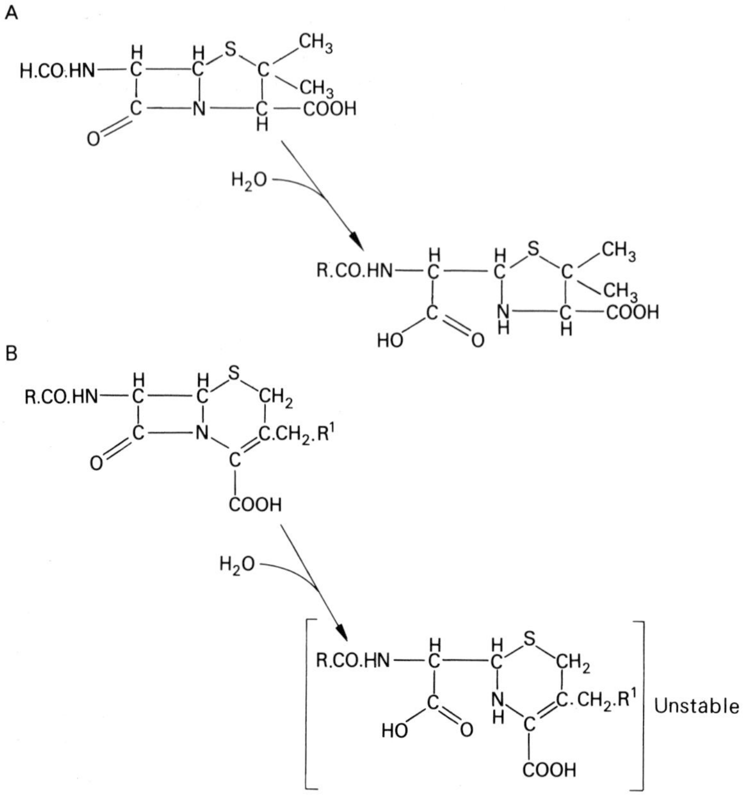

Fig. 10.3 A, degradation of penicillins and B, cephalosporins by β-lactamases. The degradation product of penicillin, penicilloic acid, is relatively stable but the corresponding cephalosporin product decomposes spontaneously to a complex mixture.

usually depends on a single mechanism, although occasionally a combination of mechanisms may be found in the same organism. Other modes of resistance are encountered among resistant strains of bacteria developed in the laboratory but these are largely of academic interest. A number of examples of the clinically important mechanisms of resistance will now be considered under the various headings.

3.1 **Conversion of an active drug to a non-toxic derivative**

3.1.1 *Inactivation of penicillins and cephalosporins*

Quite soon after the introduction of penicillin into medical practice penicillin-resistant *Staph. aureus* were isolated from patients who failed to respond to therapy. It was found that these bacteria converted the drug to an inactive product, penicilloic acid (Fig. 10.3; see also Chapter 5). The structurally related cephalosporins undergo a similar fate (Fig. 10.3). In both cases, the reaction is catalysed by a family of related enzymes, the β-lactamases. These enzymes are produced by many Gram-positive and Gram-negative bacteria and undoubtedly account for the high levels of penicillin and cephalosporin resistance. While penicillin-resistant bacteria are often also resistant to cephalosporins, and vice versa, the level of cross-resistance depends very much on the relative susceptibilities of the two groups of antibiotics to particular β-lactamases.

3.1.2

Genetics and physiology of β-lactamase synthesis

In many clinically important resistant strains of bacteria, the production of β-lactamase is controlled by plasmid genes. There is considerable homology amongst the amino acid sequences of the β-lactamases of Gram-positive bacteria, suggesting a common evolutionary origin for these enzymes. The numerous Gram-negative β-lactamases, which to judge from their amino acid composition appear to be related to each other, are quite different from the Gram-positive enzymes and there is little to suggest that Gram-negative and Gram-positive β-lactamases evolved from a common ancestor.

In Gram-positive bacteria the most important β-lactamase is probably that produced by *Staph. aureus*, of which four minor variants are known. The enzyme is inducible, i.e. the very low level of the enzyme observed in the absence of a β-lactam is dramatically increased when cells carrying the gene for β-lactamase encounter the antibiotic. The drug binds to and inactivates an intracellular repressor protein which normally inhibits the transcription of the β-lactamase gene. The enzyme, which preferentially hydrolyses members of the penicillin family, is released into the extra-cellular environment in considerable amounts. This destroys the antibiotic giving rise to the 'inoculum effect': the β-lactamase produced by the relatively few cells of a small inoculum can be overwhelmed by the drug, whereas a larger inoculum can dispose of correspondingly more drug, permitting the bacterial population to expand.

Although inducible β-lactamases are found in some Gram-negative species such as *Enterobacter*, *Ps. aeruginosa*, *Citrobacter*, *Serratia* and indole-positive *Proteus*, many Gram-negative cells synthesize β-lactamases constitutively, i.e. continuously but usually in much smaller quantities than the Gram-positive cells. Furthermore, the Gram-negative β-lactamases are not released into the external environment. The complex outer envelope of Gram-negative bacteria, which confers considerable intrinsic resistance to β-lactams, obviates the need for the high-level synthesis of β-lactamase seen in Gram-positive bacteria.

In most Gram-negative bacteria, β-lactam resistance results from plasmid-mediated β-lactamases of which more than 30 forms have been defined by various biochemical criteria such as substrate profile and isoelectric point. The most important of the plasmid enzymes is known as TEM-1, which inactivates both penicillins and cephalosporins. Recently described variants of TEM-1 and its close relative, TEM-2, also confer resistance to the so-called 'third generation' cephalosporins such as

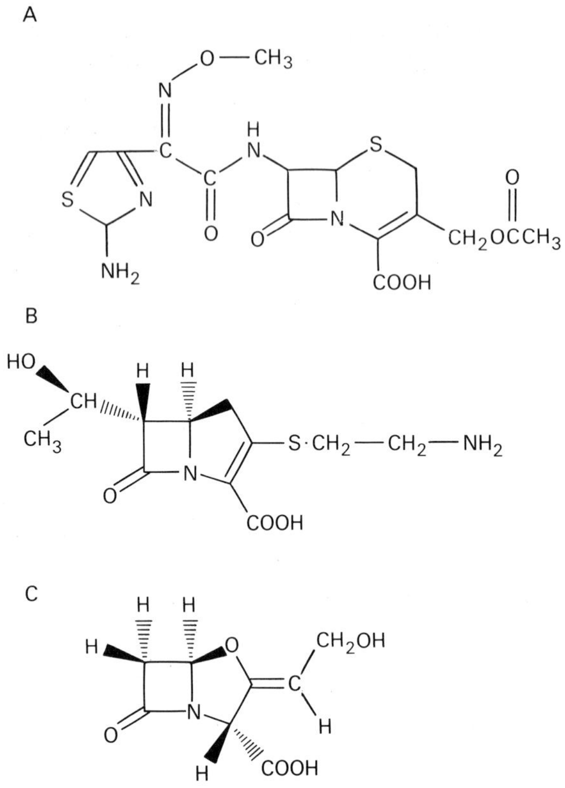

Fig. 10.4 Some recent drugs which counter β-lactamase-dependent resistance. A, cefotaxime, a semisynthetic cephalosporin stable to many important β-lactamases; B, thienamycin, a naturally occurring antibiotic which has broad β-lactamase stability; C, clavulanic acid, a naturally occurring inhibitor of β-lactamase which can be combined with β-lactamase-sensitive drugs.

cefotaxime (see Fig. 10.4A) which are relatively unaffected by the original TEM-1.

Clinically important mutations have appeared which affect those Gram-negative species mentioned above that normally only produce β-lactamases after antibiotic induction. These single-step mutations in chromosomal genes result in high-level constitutive enzyme synthesis. Although the enzyme may be poorly active against recently developed drugs, sufficient is produced to give rise to clinically significant resistance.

3.1.3   *Penicillins and cephalosporins stable to β-lactamases*

The availability of the penicillin and cephalosporin 'nuclei', 6-amino-penicillanic and 6-aminocephalosporanic acids, permitted the substitution of many different chemical groups at the amino groups (Chapter 5). Additionally, in the case of the cephalosporins, substitution is possible at the 3-position. These chemical modifications have produced new derivatives with enhanced stability to β-lactamases and a marked improvement in clinical performance against pathogens that synthesize β-lactamase. Cefotaxime (Fig. 10.4A) is a good example of a semisynthetic cephalosporin which is highly active against many previously resistant bacteria. However, as already mentioned, strains of bacteria have now emerged with β-lactamases which degrade even these powerful antibiotics. These include the recent TEM variants and also the constitutive mutants of the normally inducible β-lactamases of some Gram-negative species.

Thienamycin (imipenem), a naturally occurring β-lactam antibiotic which is neither a penicillin nor a cephalosporin (Fig. 10.4B), shows excellent stability towards most β-lactamases, including all forms of the TEM enzymes.

Another approach to the problem of pathogens which are resistant to β-lactams has been the development of inhibitors of β-lactamase which can be combined with β-lactamase-susceptible antibiotics. A notable success has been the use of the naturally occurring inhibitor clavulanic acid (Fig. 10.4C) in combination with the long-established penicillin derivative amoxycillin, a molecule that is otherwise readily hydrolysed by β-lactamase. Resistance to combinations is emerging only slowly and probably depends on changes in the β-lactam targets in the cell and possibly on diminished cellular permeability to the drugs. So far this type of resistance has not posed a significant clinical threat.

3.1.4   *Inactivation of chloramphenicol by acetylation*

Chloramphenicol is reserved largely for the treatment of typhoid fever and for other life-threatening infections, particularly those caused by *Haemophilus influenzae*. The discovery that resistance to chloramphenicol could be disseminated amongst Gram-negative bacteria by R-plasmid transfer aroused fears that the treatment of typhoid might be seriously prejudiced. These fears were realized during a major typhoid epidemic in Mexico when extensive resistance of *Sal. typhi* to chloramphenicol was encountered.

Chloramphenicol resistance mediated by R-plasmids in Gram-negative bacteria by transmissible plasmids in *Staph. aureus* is due to the presence of an enzyme, chloramphenicol acetyltransferase (CAT), which acetylates the hydroxyl groups in the side chain (Fig. 10.5). The resulting 1,3-diacetoxychloramphenicol is inactive. In Gram-positive cells the synthesis of CAT is induced by chloramphenicol. Since chloramphenicol is an inhibitor of protein synthesis, the induction process is characterized by an initial lag period during which chloramphenicol is 'calling for' the synthesis of CAT and at the same time inhibiting protein synthesis. Nevertheless, small amounts of CAT are formed which gradually depress the concentration of chloramphenicol below the inhibitory level. The synthesis of CAT then rapidly accelerates. In Gram-negative bacteria the enzyme is constitutively synthesized and is a tetrameric protein with identical subunits. In Gram-negative bacteria there is a family of CATs whose

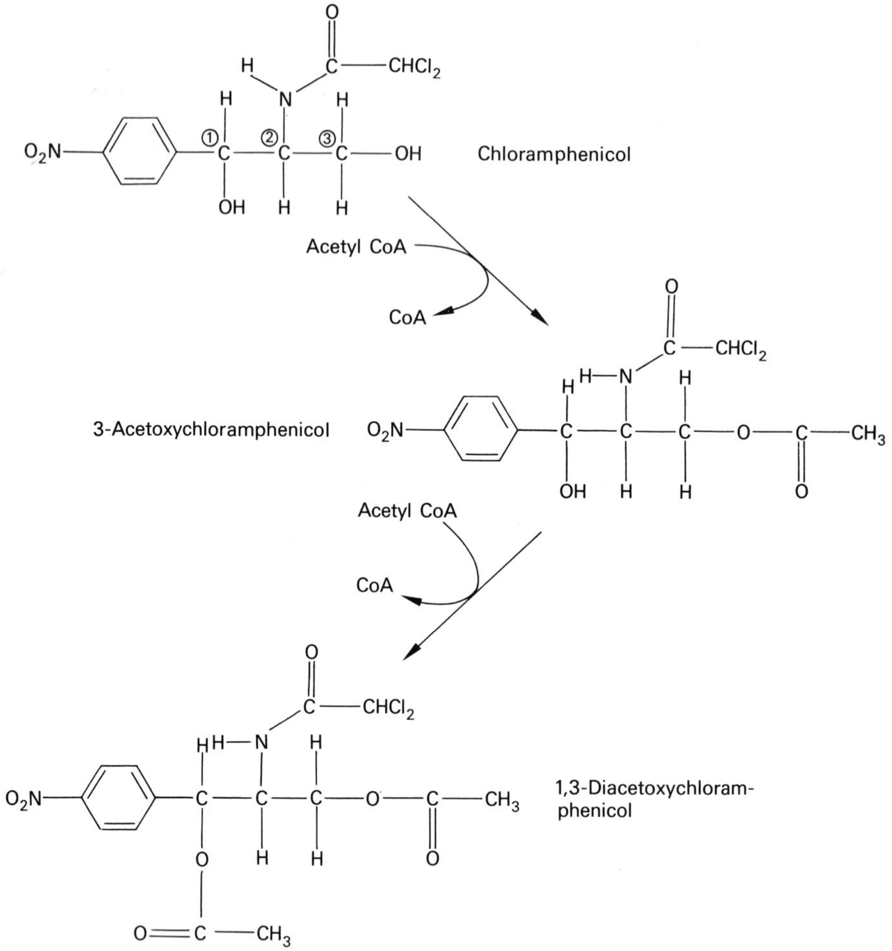

Fig. 10.5 Inactivation of chloramphenicol by the bacterial enzyme chloramphenicol acetyltransferase which successively acetylates the hydroxyl groups at positions 3 and 1.

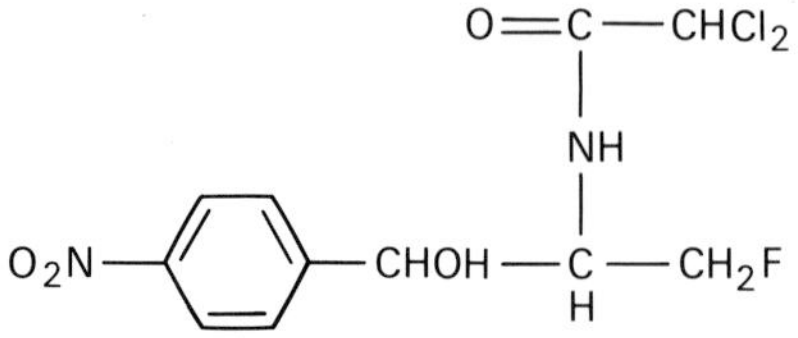

Fig. 10.6 A synthetic derivative of chloramphenicol which is not a substrate for chloramphenicol acetyltransferase. Note that the terminal hydroxyl group of the side chain is replaced by a fluorine atom.

members have individual characteristics determined by different classes of R-plasmid. Organic chemists have devised derivatives of chloramphenicol which are not substrates for CAT and therefore retain antibacterial activity in the presence of CAT-producing cells. An example is given in

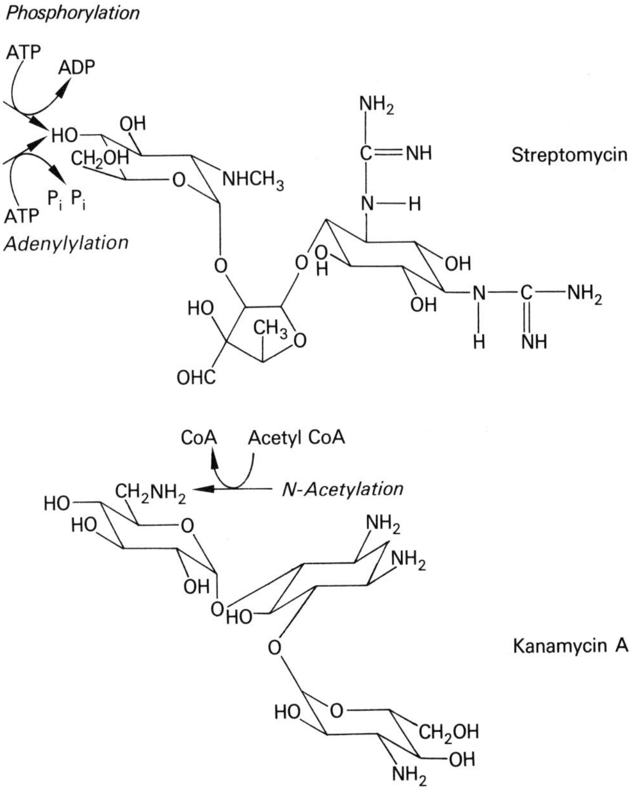

Fig. 10.7 Three examples of enzymic inactivation of aminoglycosides. Streptomycin is not subject to *N*-acetylation while kanamycin can also be inactivated by *O*-adenylylation and *O*-phosphorylation.

Fig. 10.6 in which the terminal hydroxyl group of the side chain, which is normally the first to be acetylated, is replaced by an inert fluorine atom. Despite their apparent advantage, these CAT-resistant chloramphenicol derivatives have not found a significant place in clinical medicine possibly because of the risk that the occasional but devastating bone marrow toxicity associated with the use of chloramphenicol might also occur with the CAT-resistant derivatives.

3.1.5   *Inactivation of aminoglycoside antibiotics*

The aminoglycosides are inactivated by a considerable number of enzymes, mostly coded by R-plasmids. Despite the variety of enzymes that confer full or partial resistance to the aminoglycosides, there are only three types of inactivation reaction: (i) *N*-acetylation of susceptible amino groups; (ii) adenylylation (also referred to as nucleotidylation); or (iii) phosphorylation of certain hydroxyl groups. The source of the acetyl group is acetyl CoA and the adenyl and phosphoryl groups are derived from adenosine triphosphate (ATP). In Fig. 10.7 three typical reactions are

223   *Bacterial resistance to antibiotics*

Table 10.2 Bacterial enzymes which inactivate aminoglycoside antibiotics

Enzymes	Substrates
Streptomycin–spectinomycin adenylyltransferase	Streptomycin, spectinomycin
Streptomycin phosphotransferase	Streptomycin
Kanamycin acetyltransferase	Kanamycin, neomycins, gentamicin, amikacin
Neomycin–kanamycin phosphotransferases I & II	Kanamycin, neomycins, gentamicin
Gentamicin adenylyltransferase	Kanamycin, gentamicin
Gentamicin acetyltransferases I & II	Gentamicin
Lividomycin phosphotransferase	Neomycins, paromomycin, lividomycin B
Bifunctional *N*-acetyltransferase–*O*-phosphotransferase	Amikacin, gentamicin

illustrated. Streptomycin is subject to adenylylation and phosphorylation, but not *N*-acetylation whereas kanamycin is subject to all three reactions. Table 10.2 lists the enzymes discovered so far which inactivate aminoglycoside antibiotics. Most aminoglycosides are subject to more than one inactivation reaction.

There is a process of intracellular accumulation of active aminoglycosides which depends upon the induction of an energy-dependent process by the drugs themselves. However, the enzymically modified aminoglycosides are unable to induce this process and are also inactive against ribosomal function. The semisynthetic aminoglycoside, amikacin (Fig. 5.9, Chapter 5), was designed to withstand the attack of many of the inactivating enzymes. However, with extended clinical use amikacin resistance has emerged, caused by all three mechanisms of aminoglycoside inactivation. In Gram-negative bacteria, *N*-acetylation is the most commonly encountered mechanism. An intriguing enzyme found in some strains of *Staphylococcus* is able to inactivate aminoglycosides either by *N*-acetylation or by *O*-phosphorylation. The enzyme inactivates amikacin by *N*-acetylation. In contrast, the most widely used aminoglycoside, gentamicin, is inactivated by the bifunctional enzyme by *O*-phosphorylation.

3.2 Changes in the target site resulting in resistance

3.2.1 Streptomycin

Perhaps the best known example of mutations causing changes in the target site of an antibiotic are those which affect the 30S ribosomal subunit (Chapter 9). In *E. coli* a single amino acid replacement in either one or two specific positions of protein S12 of the 30S subunit results in ribosomes resistant to streptomycin; lysine 42 may be replaced by asparagine, threonine or arginine, while lysine 87 may be replaced by arginine.

Enzymic inactivation of streptomycin is the most frequently encountered mechanism of resistance in clinical medicine but there are clinically important strains of *Strep. faecalis* and *Staph. aureus* which have chromosomally determined streptomycin resistance affecting the ribosome.

3.2.2 Methicillin

A type of *Staph. aureus* that has given rise to increasing concern in hospitals around the world is resistant to methicillin. Resistance appears to be due to the presence of a penicillin binding protein termed PBP2a or PBP2′ which has a reduced affinity for methicillin, most third generation cephalosporins and the monobactams (Chapter 5). The mutant organisms are therefore resistant to a broad range of important β-lactam antibiotics. The gene responsible for this type of resistance, *mecA*, appears to have been acquired and integrated into a chromosomal location. If the cells harbour plasmid genes for inducible β-lactamase the synthesis of PBP2a is also inducible by β-lactams. The similarity of the promoter regions of the *mecA* and β-lactamase genes permits the coordinate regulation of both genes. The level of methicillin resistance is not, however, directly proportional to the amount of PBP2a produced and other genetic regulators influence the expression of methicillin resistance that are incompletely understood.

3.2.3 Erythromycin

Resistant bacteria of several species have altered 50S ribosomal subunits with reduced affinity for this antibiotic. Mutations affecting proteins L4 or L12 may be involved. However, in *Staph. aureus* and *Streptococcus* spp., which are important therapeutic targets for erythromycin, resistance depends on an enzyme-mediated change in the 23S RNA of the ribosome. In the resistant cells a plasmid-determined RNA methylase catalyses dimethylation of a specific adenine residue of the RNA molecule. This reaction confers cross-resistance to lincomycin. The methylases occur in both constitutive and inducible forms.

3.2.4 Trimethoprim

This drug, which is generally used in combination with the sulphonamide sulphamethoxazole, is an inhibitor of bacterial dihydrofolate reductase (DHFR; Chapter 9), a key enzyme in folic acid metabolism. R-plasmid-mediated resistance to trimethoprim is well characterized. R-plasmid genes code for four types of DHFR that are resistant to trimethoprim. Type I enzymes are several thousand times more resistant to trimethoprim than the normal target enzyme of the bacterial cell, while Type II DHFRs are virtually *totally* resistant to the inhibitor. The Type III enzyme, in contrast, is much more sensitive to trimethoprim. As might be expected, bacteria harbouring Type I and II enzymes are considerably more resistant to trimethoprim than strains harbouring the Type III enzyme. Recently, yet another trimethoprim resistance plasmid, Type IV, has been described. The enzyme coded by the Type IV plasmid is only slightly less susceptible to trimethoprim than the chromosomal enzyme but its synthesis is increased up to 600 times by the presence of the drug. The organism thereby escapes the inhibitory action of the drug by overwhelming it with excess DHFR.

3.2.5   Sulphonamides

R-plasmid-determined resistance to trimethoprim often occurs in combination with resistance to sulphonamides. In addition to coding for trimethoprim-resistant DHFR, the plasmids also carry a gene for one of two variants of the sulphonamide target enzyme, dihydropteroate synthetase, which are very resistant to sulphonamides. Enzymic resistance to sulphathiazole, for example, is increased by about 1000-fold while the K_m for the substrate, p-aminobenzoic acid, is unchanged. Organisms with these enzymes grow quite happily in the presence of combinations of trimethoprim and sulphonamides.

3.2.6   Quinolones

This group of drugs has become increasingly important as the narrow spectrum of the original members of the group, nalidixic and oxolinic acids, has been extended to include Gram-positive pathogens by the introduction of fluoroquinolone derivatives such as ciprofloxacin and norfloxacin (Fig. 5.16, Chapter 5). Chromosomally determined resistance to nalidixic and oxolinic acids occurs fairly readily and causes changes in either the A or B subunits of the quinolone target enzyme, DNA gyrase, so that the enzyme is unaffected at normally antibacterial drug concentrations. Resistance of this type of ciprofloxacin and norfloxacin has so far been uncommon and in any case appears to confer a level of resistance which is lower than clinically achievable blood concentrations of the drugs.

3.3   Reduction in cellular permeability to antibiotics

In order to suppress the growth of bacteria, a drug must achieve an inhibitory concentration at its target site. The concentration at the target is determined by the external concentration of the drug and by the speed with which it penetrates the various permeability barriers which separate the drug from its target. Therefore, a decrease in cellular permeability to the drug may depress the drug concentration at the target site below the inhibitory level. A decrease in the rate of penetration of an antibiotic into a bacterium also enables enzymic inactivation mechanisms to deal more effectively with incoming antibiotics. A reduction in permeability can arise in several ways.

Permeability of the outer membrane

The complex structure of the outer envelope of Gram-negative bacteria ensures that they are intrinsically less sensitive than Gram-positive bacteria to a variety of antibiotics (Table 10.1). Some antibiotics active against Gram-negative bacteria cross their outer membrane via water-filled channels created by pore-forming proteins in the membrane known as *porins*. These pores permit the diffusion of hydrophilic molecules of molecular weight up to 600–700. In general, therefore, Gram-negative bacteria are more resistant to hydrophobic antibiotics whose passage across the outer membrane is hindered by their inability to enter the water pores. *Ps. aeruginosa* is also resistant to many hydrophilic drugs probably because most of the pores are closed. Mutations affecting the structure and function of the porins may decrease the uptake of water-soluble antibiotics. The lipopolysaccharide (LPS) molecules of the outer leaflet of the outer membrane also affect its permeability. Strains of *E. coli* and *Ps. aeruginosa* which are hypersensitive to antibiotics apparently produce abnormal LPS which markedly decreases the effectiveness of the outer membrane as a penetration barrier.

Certain nutrients cross the outer membrane of Gram-negative bacteria by mechanisms involving specific receptors in the membrane. The sideromycin antibiotics, which resemble certain iron-sequestering growth factors known as siderophores, cross the outer membrane by using the siderophore receptors. The clinical use of the sideromycins against Gram-negative bacteria was cut short because of the extraordinary ease with which bacteria acquired resistance by deleting sideromycin transport across the outer membrane. The iron requirements of the resistant cells were presumably satisfied by the activities of other iron-transporting systems in the cell envelope.

Antagonism of antibiotic transport processes

Tetracyclines. The antibacterial action of the tetracycline antibiotics is facilitated by the ability of susceptible bacteria to accumulate these drugs within the cytoplasm by a transport process located in the cytoplasmic membrane. This results in intracellular concentrations of tetracycline which inhibit ribosomal function. Bacterial resistance to tetracyclines is common amongst Gram-negative and Gram-positive species and constitutes a serious threat to the clinical use of these drugs. There seems little doubt that resistance to the tetracyclines is caused principally by a plasmid-mediated specific antagonism of the tetracycline accumulation process.

In both Gram-positive and Gram-negative bacteria which harbour plasmids with a tetracycline-resistance marker, the antagonism of tetracycline accumulation is induced by subinhibitory concentrations of tetracyclines and is associated with the synthesis of at least one protein which is incorporated into the cytoplasmic membrane.

Among the Enterobacteriaceae there are four groups of tetracycline-resistance determinants, *tet A–D*. Each determinant codes for several proteins, which include a membrane-located protein that actively promotes the efflux of tetracycline antibiotics from the cell. The proteins coded by *tetA*, *tetB* and *tetC* have been studied in detail and are related proteins of molecular weights between 34 000 and 36 000. An analogous protein, molecular weight 32 000, has been characterized in the Gram-positive bacterium *Bacillus subtilis*. The efflux process appears to be energized by the normal proton gradient across the cytoplasmic membrane. Not all tetracyclines are extruded with equal efficiency by each marker. Tetracycline itself is extruded by each of the *tet* proteins but the semisynthetic tetracycline, minocycline (Chapter 5), is only susceptible to *tetB*.

Quinolones. A form of resistance to these important drugs in *E. coli* is associated with decreased uptake which appears to result from a combination of two changes: a diminution in the amount of an outer membrane porin protein, *OmpF* and possibly an energy-dependent quinolone efflux system in the cytoplasmic membrane analogous to the *tet* system.

4   What can be done to combat the problem of bacterial resistance to antibiotics?

The immense efforts made over the last 50 years to understand bacterial resistance in terms of its genetics, biochemistry, clinical and epidemiological aspects have been successful to a degree unimaginable at the start of the antibiotic era in the 1940s. Despite this progress in defining the problem, attempts to control resistance and limit its impact upon medical treatment have been only partially successful. The usefulness of virtually every new antibacterial drug introduced into medicine has, sooner or later, been threatened by the emergence of resistant strains. However, important lessons have been learned concerning the use of antibiotics which can help to limit the rate of emergence and spread of resistant strains. First, the use of antibacterial drugs must be confined to the minimum needed to control serious disease caused by bacterial infections. There is no doubt that the indiscriminate use of antibiotics for trivial illness, for ill-defined prophylactic purposes and in animal feedstuffs has been a major contributory factor to the problem of resistance. Dramatic results in reducing the incidence of resistant bacteria in hospitals have been obtained by imposing tight restrictions on the administration of antibiotics. Hospital experience has shown that careful attention to procedures to prevent cross-infection of patients and staff is essential if foci of resistant organisms are to be contained. Serious problems with methicillin-resistant *Staph. aureus* (MRSA) have emphasized the importance of good hospital practice in limiting the spread of resistant pathogens. Where antibiotics with important human applications continue to be used in veterinary medicine, care is needed to prevent the transmission of resistant bacteria, whether or not they are human pathogens, from animals to man.

The contribution of laboratory research to the control of resistance has been highly significant. The development of semisynthetic β-lactams which are resistant to the action of β-lactamases has had an enormous impact on the treatment of infections caused by previously resistant pathogens. It is encouraging that bacterial adaptation to these important drugs has been relatively slow and only partially successful. More modest advances have been made in preparing derivatives of tetracycline (minocycline) and of aminoglycosides (amikacin) which show improved activity against some resistant strains.

Despite the excellent properties of many antibacterial drugs, the continuing threat posed by the emergence of resistant bacteria emphasizes the need for drugs with novel or extensively modified chemical constitutions. The introduction of a new antibiotic presents bacteria with fresh problems of survival and adaptation and usually provides a period of 'grace', of unpredictable length, before resistance appears. The practice of co-administering two drugs, each of which blocks a separate step in metabolism in bacteria is another interesting approach to countering resistance. It was thought that bacteria would be much less able to develop resistance to two differently acting drugs. However, the discovery that bacteria can accept and transfer 'packages' of genes which confer resistance simultaneously to several different drugs has limited the impact of this form of combination therapy. The best hopes for containing the resistance problem rest with the ingenuity of medicinal chemists to design new drugs combined with sensible medical practice in hospitals and in the community.

5   Further reading

Anderson E.S. (1968) Ecology of transferable drug resistance in enterobacteria. *Annu Rev Micobiol.* **22**, 131–180.

Brown M.R.W. (Ed.) (1975) *Resistance of Pseudomonas aeruginosa.* London: John Wiley & Sons.

Chambers H.F., Hartman B.J. & Tomasz A. (1985) Increased amounts of a novel penicillin-binding protein in a strain of methicillin-resistant *Staphylococcus aureus* exposed to nafcillin. *J Clin Invest.* **76**, 325–331.

Clewell D.B. & Gawron-Burke C. (1986) Conjugative transposons and the dissemination of antibiotic resistance in the Streptococci. *Annu Rev Microbiol.* **40**, 635–659.

Datta N. (Ed.) (1984) Antibiotic resistance in bacteria. *Br Med Bull.* **40**, 1–106.

Franklin T.J. & Snow G.A. (1989) *Biochemistry of Antimicrobial Action*, 4th edn. London: Chapman & Hall.

Hall B.G. (1991) Increased rates of advantageous mutations in response to environmental challenges. *Am Soc Microbiol News,* **57**, 82–86.

Jacoby G.A. (1985) Genetics and epidemiology of resistance. In *Scientific Basis of Antimicrobial Therapy*, (Eds D. Greenwood & F. O'Grady) 38th Symposium of The Society for General Microbiology, pp. 185–218. Cambridge: Cambridge University Press.

Jacoby G.A. & Archer G.L. (1991) New mechanisms of bacterial resistance to anti-microbial agents. *N Engl J Med*. **324**, 601–612.

Lyon B.R. & Skurray R. (1987) Antimicrobial resistance of *Staphylococcus aureus*: genetic basis. *Microbiol Rev*. **51**, 88–134.

Marples R.R. & Cooke E.M. (1985) Workshop on methicillin-resistant *Staphylococcus aureus*. *J Hosp Infect*. **6**, 342–348.

Nikaido H. & Vaara M. (1985) Molecular basis of bacterial outer membrane permeability. *Microbiol Rev*. **49**, 1–32.

Novick R.P. (1990) Staphylococcal plasmids and their replication. *Annu Rev Microbiol*. **43**, 537–565.

Richmond M.H. & Petrocheilou V. (1981) The ecology of transferable antibiotic resistance. In *The Future of Antibiotherapy and Antibiotic Research* (Eds L. Ninet, P.E. Bost, D.H. Bouanchand & J. Florent), pp. 59–71. London: Academic Press.

Russell A.D. & Chopra I. (1990) *Understanding Antibacterial Action and Resistance*. Chichester: Ellis Horwood.

Sanders C.C. (1987) Chromosomal cephalosporinases responsible for multiple resistance to newer β-lactam antibiotics. *Annu Rev Microbiol*. **41**, 573–593.

Sanders C.C. & Wiedemann B. (1988) New developments in resistance to β-lactam antibiotics among non-fastidious Gram-negative organisms. *Rev Infect Dis*. **10**, 677–912.

Young H.-K. & Amyes S.G.B. (1986) A new mechanism of plasmid trimethoprim resistance. *J Biol Chem*. **261**, 2503–2505.

11 Chemical disinfectants, antiseptics and preservatives

1 Introduction

2 Factors affecting choice of antimicrobial agent
2.1 Properties of the chemical agent
2.2 Microbiological challenge
2.2.1 Vegetative bacteria
2.2.2 *Mycobacterium tuberculosis*
2.2.3 Bacterial spores
2.2.4 Fungi
2.2.5 Viruses
2.3 Intended application
2.4 Environmental factors
2.5 Toxicity of the agent

3 Types of compound
3.1 Acids and esters
3.1.1 Benzoic acid
3.1.2 Sorbic acid
3.1.3 Sulphur dioxide, sulphites and metabisulphites
3.1.4 Esters of *p*(4)-hydroxybenzoic acid (parabens)
3.2 Alcohols
3.2.1 Alcohols used for disinfection and antisepsis
3.2.2 Alcohols as preservatives
3.3 Aldehydes
3.3.1 Glutaraldehyde
3.3.2 Formaldehyde
3.3.3 Formaldehyde-releasing agents

3.4 Biguanides
3.4.1 Chlorhexidine and alexidine
3.4.2 Polyhexamethylene biguanides
3.5 Halogens
3.5.1 Chlorine
3.5.2 Hypochlorites
3.5.3 Organic chlorine compounds
3.5.4 Chloroform
3.5.5 Iodine
3.5.6 Iodophors (iodophores)
3.6 Heavy metals
3.6.1 Mercurials
3.7 Hydrogen peroxide and peracid compounds
3.8 Phenols
3.8.1 Phenol (carbolic acid)
3.8.2 Tar acids
3.8.3 Non-coal tar phenols (chloroxylenol and chlorocresol)
3.8.4 Bisphenols
3.9 Surface-active agents
3.9.1 Cationic surface-active agents
3.10 Other antimicrobials
3.10.1 Diamidines
3.10.2 Dyes
3.10.3 Quinoline derivatives

4 Disinfection policies

5 Further reading

1 Introduction

Disinfectants, antiseptics and preservatives are chemicals which have the ability to destroy or inhibit the growth of microorganisms and which are used for this purpose.

Disinfectants. Disinfection is the process of removing microorganisms, including potentially pathogenic ones, from the surfaces of inanimate objects. The British Standards Institution further defines disinfection as not necessarily killing all microorganisms, but reducing them to a level acceptable for a defined purpose, for example a level which is harmful neither to health nor to the quality of perishable goods. Chemical disinfectants are capable of different levels of action. The term 'high-level disinfection' indicates destruction of all microorganisms but not necessarily bacterial spores; 'intermediate level disinfection' indicates destruction of all vegetative bacteria including *Mycobacterium tuberculosis* but may

231

exclude some viruses and fungi and have little or no sporicidal activity; 'low-level disinfection' can destroy most vegetative bacteria, fungi and viruses, but this will not include spores and some of the more resistant microorganisms. Some high-level disinfectants have good sporicidal activity and have been ascribed the name 'liquid chemical sterilant' or 'chemosterilant' to indicate that they can effect a complete kill of all microorganisms, as in sterilization.

Antiseptics. Antisepsis is defined as destruction or inhibition of micro-organsims on living tissues, having the effect of limiting or preventing the harmful results of infection. It is *not* a synonym for disinfection (British Standards Institution). The chemicals used are applied to skin and mucous membranes; therefore as well as having adequate antimicrobial activity, they must not be toxic or irritating for skin. Antiseptics are mostly used to reduce the microbial population on the skin prior to surgery or on the hands to help prevent spread of infection by this route. Antiseptics are often lower concentrations of the agents used for disinfection.

Preservatives. These are included in pharmaceutical preparations to prevent microbial spoilage of the product and to minimize the risk of the consumer acquiring an infection when the preparation is administered. Preservatives must be able to limit proliferation of microorganisms that may be introduced unavoidably during manufacture and use of non-sterile products such as oral and topical medications. In sterile products such as eye-drops and multidose injections, preservatives should kill any microbial contaminants introduced inadvertently during use. It is essential that a preservative is not toxic in relation to the intended route of administration of the preserved preparation. Preservatives therefore tend to be employed at low concentrations and consequently levels of anti-microbial action also tend to be of a lower order than for disinfectants or antiseptics. This is illustrated by the *British Pharmacopoeia* requirements for preservative efficacy where a degree of bactericidal activity is necessary, although this should be obtained within a few hours or over several days of microbial challenge depending on the type of product to be preserved. In oral liquid preparations, a fungistatic action only is required by the *British Pharmacopoeia* (1988).

Other terms are considered in Chapter 12 (see section 1.1 and Fig. 12.1 in that chapter).

There are many antimicrobial agents used in pharmacy for the above functions. The aim of this chapter is to introduce the range of chemicals in current use and to indicate their activities and applications.

2 Factors affecting choice of antimicrobial agent

Choice of the most appropriate antimicrobial compound for a particular purpose depends on:

1 properties of the chemical agent,
2 microbiological challenge,
3 intended application,
4 environmental factors,
5 toxicity of the agent.

2.1 Properties of the chemical agent

The process of killing or inhibiting the growth of microorganisms using an antimicrobial agent is basically that of a chemical reaction and the rate and extent of this reaction will be influenced by the factors of concentration of chemical, temperature, pH and formulation. The influence of these factors on activity is discussed fully in Chapter 12 and is referred to in discussing the individual agents. Tissue toxicity influences whether a chemical can be used as an antiseptic or preservative and this unfortunately limits the range of chemicals for these applications or necessitates the use of lower concentrations of the chemical.

2.2 Microbiological challenge

The types of microorganism present and the levels of microbial contamination (the bioburden) both have a significant effect on the outcome of chemical treatment. If the bioburden is high, long exposure times or higher concentrations of antimicrobial may be required. Microorganisms vary in their sensitivity to the action of chemical agents. Some organisms, either because of their resistance to disinfection (for further discussion see Chapter 14) or because of their significance in cross-infection or nosocomial (hospital-acquired) infections, merit attention.

The efficacy of an antimicrobial agent must be investigated by appropriate capacity, challenge and in-use tests to ensure that a standard is obtained which is appropriate to the intended use (Chapter 12). In practice, it is not usually possible to know which organisms are present on the articles being treated. Thus, it is necessary to categorize chemicals according to their antimicrobial capabilities and for the user to have an awareness of what level of antimicrobial action is required in a particular situation.

2.2.1 Vegetative bacteria

At in-use concentrations, chemicals used for disinfection should be capable of killing most vegetative bacteria within a reasonable contact period. Antiseptics and preservatives are also expected to have a broad spectrum of antimicrobial activity but at the in-use concentrations, after exerting an initial biocidal effect, their main function may be biostatic. Gram-negative bacilli, which are the main causes of nosocomial infections, are often more resistant than Gram-positive species. *Pseudomonas aeruginosa*, an opportunistic pathogen (i.e. is pathogenic if the oppor-

Table 11.1 Antibacterial activity of commonly used disinfectants and antiseptics

	Activity against		General level of antibacterial activity*
Class of compound	M. tuberculosis	Spores	
Alcohols	+	−	Intermediate
Ethanol/isopropyl			
Aldehydes			
Glutaraldehyde	+	+	High
Formaldehyde	+	+	High
Biguanides			
Chlorhexidine	−	−	Intermediate
Halogens			
Hypochlorite/chloramines	+	+	High
Iodine/iodophors	+	+	Intermediate, problems with *Ps. aeruginosa*
Hydrogen peroxide	+	+	High
Phenolics			
Clear soluble fluids	+	+	High
Chloroxylenol	−	−	Low
Bisphenols	−	−	Low, poor against *Ps. aeruginosa*
Quaternary ammonium compounds			
Benzalkonium chloride	−	−	Intermediate
Cetrimide	−	−	Intermediate

* Activity will depend on concentration, time of contact, temperature, etc. (see Chapter 12) but these are the activities expected if in-use concentrations were being employed.

tunity arises; see also Chapter 4), has gained a reputation as the most resistant of the Gram-negative organisms. However, problems mainly arise when a number of additional factors such as heavily soiled articles or diluted or degraded solutions are involved.

2.2.2 Mycobacterium tuberculosis

M. tuberculosis (the tubercle bacillus) and other acid-fast bacilli are resistant to many aqueous bactericides. Tuberculosis remains an important public health hazard, and the greatest risk of acquiring infection is from the undiagnosed patient. Equipment used for respiratory investigations can become contaminated with mycobacteria if the patient is a carrier of this organism. It is important to be able to disinfect the equipment to a safe level to prevent transmission of infection to other patients.

2.2.3 Bacterial spores

Bacterial spores are the most resistant of all microbial forms to chemical treatment. The majority of antimicrobial agents have no useful sporicidal action, with the exception of the aldehydes, hypochlorites and hydrogen

Table 11.2 Antifungal activity of disinfectants and antiseptics (adapted from Scott *et al.* 1986)

Antimicrobial agent	Time (min) to give >99.99% kill* of		
	Aspergillus niger	*Trichophyton mentagrophytes*	*Candida albicans*
Phenolic (0.36%)	<2	<2	<2
Chlorhexidine gluconate (0.02%, alcoholic)	<2	<2	<2
Iodine (1%, alcoholic)	<2	<2	<2
Povidone-iodine (10%, alcoholic and aqueous)	10	<2	<2
Hypochlorite (0.2%)	10	<2	5
Cetrimide (1%)	<2	20	<2
Chlorhexidine gluconate (0.05%) + cetrimide (0.5%)	20	>20	>2
Chlorhexidine gluconate (0.5%, aqueous)	20	>20	>2

* Initial viable counts were *c.* 1×10^6.

peroxide. Such chemicals are sometimes used as an alternative to physical methods for sterilization of heat-sensitive equipment. In these circumstances, correct usage of the agent is of paramount importance since safety margins are lower in comparison with physical methods of sterilization (Chapter 20).

The antibacterial activity of disinfectants and antiseptics is summarized in Table 11.1.

2.2.4   *Fungi*

The vegetative fungal form is often as sensitive as vegetative bacteria to antimicrobial agents. Fungal spores (conidia and chlamydospores; see Chapter 2) may be more resistant but this resistance is of much lesser magnitude than for bacterial spores. The ability to rapidly destroy pathogenic fungi such as the opportunistic yeast, *Candida albicans*, and filamentous fungi such as *Trichophyton mentagrophytes*, and spores of common spoilage moulds such as *Aspergillus niger* is put to advantage in many applications of use. Many disinfectants have good activity against these fungi (Table 11.2).

2.2.5   *Viruses*

Susceptibility of viruses to antimicrobial agents may depend on whether the viruses possess a lipid envelope. Non-lipid viruses are frequently more resistant to disinfectants and it is also likely that such viruses cannot be readily categorized with respect to their sensitivities to antimicrobial agents. These viruses are responsible for many nosocomial infections, e.g. rotaviruses, picornaviruses and adenoviruses (see Chapter 3) and it may

Table 11.3 Chemical disinfection of HIV (adapted from *HIV—The Causative Agent of AIDS and Related Conditions*, ACDP, 1990)

Agent	Application	Comment
Chlorine-releasing preparations e.g. Hypochlorite 10 000 p.p.m. available chlorine	Spillage of HIV-contaminated blood and body fluid	Use fresh solution. Deteriorates on storage and may be adversely affected by organic matter
Hypochlorite 1000 p.p.m. available chlorine	Minor contamination of inanimate surfaces	Corrosive to metals. Bleaches fabrics
Aldehydes e.g. Glutaraldehyde 2% (w/v)	Reserved for non-corrosive treatment of delicate items	Must be freshly activated
Alcohol 70%	Minor surface contamination	Confirmatory evidence still required in presence of protein so use alternative if possible

be necessary to select an antiseptic or disinfectant to suit specific circumstances. There is much concern for the safety of personnel handling articles contaminated with pathogenic viruses such as hepatitis B and human immunodeficiency virus (HIV), which causes AIDS. HIV is an enveloped virus and fortunately tests have shown that, like other enveloped viruses, it is sensitive to many of the commonly used antiseptics and disinfectants. Some agents have been recommended by the Advisory Committee on Dangerous Pathogens for use where such viruses are thought to be present depending on the circumstances and level of contamination; these are listed in Table 11.3.

The virucidal activity of chemicals is difficult to determine in the laboratory. Tissue culture techniques are the most common methods for growing and estimating viruses; however, antimicrobial agents may also adversely affect the tissue culture. Further information is provided in Chapter 12.

2.3 Intended application

The intended application of the antimicrobial agent, whether for preservation, antisepsis or disinfection, will influence its selection and also affect its performance. For example, in medicinal preparations the ingredients in the formulation may antagonize preservative activity.

In disinfection of instruments, the chemicals used must not adversely affect the instruments, e.g. cause corrosion of metals, affect clarity or integrity of lenses, or change texture of synthetic polymers. Many materials such as fabrics, rubber and plastics are capable of adsorbing certain disinfectants, e.g. quaternary ammonium compounds (QACs; section 3.9.1) are adsorbed by fabrics, while phenolics (section 3.8) are adsorbed by rubber, the consequence of this being a reduction in concentration of active compound. A disinfectant can only exert its effect if it is in contact with the item being treated. Therefore access to all parts of an instrument or piece of equipment is essential. For small items, total immersion in the disinfectant must also be ensured.

2.4　　**Environmental factors**

Organic matter can have a drastic effect on antimicrobial capacity either by adsorption or chemical inactivation, thus reducing the concentration of active agent in solution or by acting as a barrier to the penetration of the disinfectant. Blood, body fluids, pus-, milk, food residues or colloidal proteins, even present in small amounts, all reduce the effectiveness of antimicrobial agents to varying degrees and some are seriously affected. In their normal habitats, microorganisms have a tendency to adhere to surfaces and are thus less accessible to the chemical agent. Some organisms are specific to certain environments and their destruction will be of paramount importance in the selection of a suitable agent, e.g. *Legionella* in cooling towers and non-potable water supply systems, *Listeria* in the dairy and food industry and hepatitis in blood-contaminated articles.

Dried organic deposits may inhibit penetration of the chemical agent. Where possible, *objects to be disinfected should be thoroughly cleaned.* The presence of ions in water can also affect activity of antimicrobial agents; thus water for testing biocidal activity can be made artificially 'hard' by addition of ions.

These factors can have very significant effects on activity and are summarized in Table 11.4.

2.5　　**Toxicity of the agent**

In choosing an antimicrobial agent for a particular application some consideration must be given to its toxicity. Recently introduced legislation, the Control of Substances Hazardous to Health (COSHH) Regulations, is directed towards the health and safety precautions required in handling toxic or potentially toxic agents. In respect of disinfectant use these regulations affect, particularly, the use of phenolics, formaldehyde and glutaraldehyde. Toxic volatile substances, in general, should be kept in covered containers to reduce the level of exposure to irritant vapour and they should be used with an extractor facility. Limits governing the exposure of individuals to such substances are now listed, e.g. $0.7\,\mathrm{mg\,m}^{-3}$ (0.2 p.p.m.) glutaraldehyde for both short- and long-term exposure. The aldehydes, glutaraldehyde less so than formaldehyde, may affect the eyes, skin (causing contact dermatitis) and induce respiratory distress. Face protection and impermeable nitrile rubber gloves should be worn when using these agents. Table 11.4 lists the toxicity of many of the disinfectants in use and other concerns of toxicity are described for individual agents below.

Table 11.4 Properties of commonly used disinfectants and antiseptics

Class of compound	Effect of organic matter	pH Optimum	Toxicity	Other factors
Alcohols: ethanol/ isopropanol	Slight		Avoid broken skin, eyes	Poor penetration, good cleansing properties
Aldehydes: glutaraldehyde	Slight	pH 8	Respiratory complaints and contact dermatitis reported. Eye sensitivity	Non-corrosive, useful for heat-sensitive instruments
Biguanides: chlorhexidine	Severe	pH 7–8	Avoid contact with eyes and mucous membranes. Sensitivity may develop	Incompatible with soap and anionic detergents. Inactivated by hard water, some materials and plastic
Chlorine compounds: hypochlorite	Severe	Acid/neutral pH	Irritation of skin, eyes and lungs	Corrosive to metals
Iodine preparations: iodophors	Severe	Acid pH	Eye irritation	May corrode metals
Phenolics: clear soluble fluids	Slight	Acid pH	Protect skin and eyes	Adsorbed by rubber/plastic
black/white fluids	Moderate/severe		Very irritant	Greatly reduced by dilution
chloroxylenol	Severe		Sensitivity. May irritate skin	Adsorbed by rubber/plastic
QACs: cetrimide and benzalkonium chloride	Severe	Alkaline pH	Avoid contact with eyes	Incompatible with soap and anionic detergents. Adsorbed by fabrics

Where the atmosphere of a workplace is likely to be contaminated, sampling and analysis of the atmosphere may need to be carried out on a periodic basis with a frequency determined by conditions.

3 Types of compound

The following section presents in alphabetical order by chemical grouping the agents most often employed for disinfection, antisepsis and preservation. This information is summarized in Table 11.5.

3.1 Acids and esters

Antimicrobial activity, within a pharmaceutical context, is generally found only in the organic acids. These are weak acids and will therefore dissociate incompletely to give the three entities HA, H⁺ and A⁻ in solution. As the undissociated form, HA, is the active antimicrobial agent, the ionization constant, K_a, is important and the pK_a of the acid must be considered especially in formulation of the agent.

3.1.1 Benzoic acid

This is an organic acid, C_6H_5COOH, which is included, alone or in combination with other preservatives, in many pharmaceuticals. Although the compound is often used as the sodium salt, the non-ionized acid is the active substance. A limitation on its use is imposed by the pH of the final product as the pK_a of benzoic acid is 4.2, at which pH 50% of the acid is ionized. It is advisable to limit use of the acid to preservation of pharmaceuticals having a maximum final pH of 5.0 and if possible less than 4.0. Concentrations of 0.05–0.1% are suitable for oral preparations. A disadvantage of the compound is the development of resistance by some organisms, involving in some cases metabolism of the acid resulting in complete loss of activity. Benzoic acid also has some use in combination with other agents, salicylic acid among others, in the treatment of superficial fungal infections.

3.1.2 Sorbic acid

This compound is a widely used preservative as the acid or its potassium salt. The pK_a is 4.8 and, as with benzoic acid, activity decreases with increasing pH and ionization. It is most effective at pH 4 or below. Pharmaceutical products such as gums, mucilages and syrups are usefully preserved with this agent.

3.1.3 Sulphur dioxide, sulphites and metabisulphites

Sulphur dioxide has extensive use as a preservative in the food and beverage industries. In a pharmaceutical context, sodium sulphite and metabisulphite or bisulphite have a dual role acting as preservatives and antioxidants.

3.1.4 Esters of p(4)-hydroxybenzoic acid (parabens)

A series of alkyl esters (Fig. 11.1) of p(4)-hydroxybenzoic acid was originally prepared to overcome the marked pH-dependent activity of the acid.

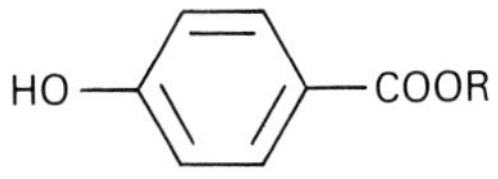

HO —⟨benzene ring⟩— COOR

Fig. 11.1 p-Hydroxybenzoates (R is methyl, ethyl, propyl, butyl or benzyl).

Table 11.5 Examples of the main antimicrobial groups in disinfection and preservation

Antimicrobial agent	Antiseptic activity		Disinfectant activity		Preservative activity	
	Concentration	Typical formulation/application	Concentration	Typical formulation/application	Concentration	Typical formulation/application
Acids and esters e.g. benzoic acid parabens					0.05–0.1% 0.25%	For oral and topical formulations
Alcohols*, e.g. ethyl or isopropyl	50–90% in water	Skin preparation	50–90% in water	Clean surface prep., thermometers	25%	Vaccine
Aldehydes, e.g. glutaraldehyde	10%	Gel for warts	2.0%	Solution for instruments	<0.1%	Cosmetics
Biguanides, e.g. chlorhexidine† (gluconate, acetate, etc.)	0.01% 0.05% 0.2% 0.5% (in 70% alcohol) 1.0% 4.0%	Bladder irrigation Lozenge Mouthwash Skin prep. Dusting powder, cream Pre-op. scrub in surfactant	0.05% 0.5%	Storage of instruments, clean instrument disinfection (30 min) Emergency instrument disinfection (2 min)	0.0025% 0.01%	Contact lens solutions Eye-drops
Chlorine, e.g. hypochlorite	≤0.5% available chlorine	Solution for skin and wounds	1–10%	Solution for surfaces and instruments		
Hydrogen peroxide	6%	Solution for wounds and ulcers, mouthwash				

Agent	Conc.	Use	Conc.	Use	Conc.	Use
Iodine compounds						
e.g. free iodine povidone-iodine	1.0%	Aqueous or alcoholic (70%) solution				
	0.5%	Dry powder spray				
	1.0%	Mouthwash				
	4.0%	Shampoo				
	10.0%	Pre-op. scrub (aqueous or alcoholic)	10.0%	Aqueous or alcoholic solution		
Mercurials						
e.g. thiomersal					0.001%	Contact lens solutions
					0.01%	Vaccines
e.g. phenylmercuric nitrate					0.001–0.002%	Eye-drops
Phenolics						
e.g. tar acids (clear soluble phenolics)			1–2%	Solution		
e.g. non-coal tar (chloroxylenol)	0.5%	Dusting powder				
	1.3%	Solution				
e.g. bisphenol (hexachlorophane)	0.5%	Lotion for prevention of bedsores and cross-infection				
	3.0%	Pre-op. scrub				
QACs,						
e.g. cetyltri methylammonium bromide (cetrimide)	0.1%	Solution for wounds and burns	0.1%	Storage or sterile instruments	0.01%	Eye-drops
	0.5%	Cream	1.0%	Instruments (1 hour)		
	1.0%	Skin solution				
	10.0%	Shampoo, psoriasis				

* Also used in combination with other agents, e.g. chlorhexidine, iodine.
† Several formulations available having x% chlorhexidine and 10x% cetrimide.

These parabens, the methyl, ethyl, propyl and butyl esters, are less readily ionized having pK_a values in the range 8–8.5 and exhibit good preservative activity even at pH levels of 7–8 though optimum activity is again displayed in acidic solutions. This broader pH range allows extensive and successful use of the parabens as pharmaceutical preservatives. The agents are active against a wide range of fungi but are less active against bacteria, especially the pseudomonads which may utilize the parabens as a carbon source. They are frequently used as preservatives of emulsions, creams and lotions where two phases exist. Combinations of esters are most successful for this type of product in that the more water-soluble methyl ester (0.25%) protects the aqueous phase whereas the propyl or butyl esters (0.02%) give protection to the oil phase. Such combinations are also considered to extend the range of activity. As inactivation of parabens occurs with non-ionic surfactants due care should be taken in formulation with these.

3.2 Alcohols

3.2.1 Alcohols used for disinfection and antisepsis

The aliphatic alcohols, notably ethanol and isopropanol, which are used for disinfection and antisepsis, are bactericidal against vegetative forms, including *Mycobacterium* species, but are not sporicidal. Alcohols have poor penetration of organic matter and their use is therefore restricted to clean conditions. They possess properties such as a cleansing action and volatility and are able to achieve a rapid and large reduction in skin flora, which makes them suitable for skin preparation prior to injection or other surgical procedures.

Ethanol (CH_3CH_2OH) is widely used as a disinfectant and antiseptic. The presence of water is essential for activity, hence 100% ethanol is ineffective. Concentrations between 60 and 95% are bactericidal but it is usually employed as a 70% solution for the disinfection of skin and clean instruments or surfaces. A 70% solution is also a powerful virucide and

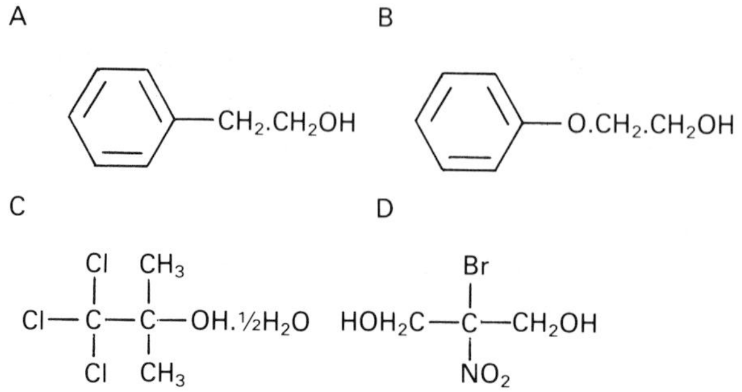

Fig. 11.2 Structural formulae of alcohols used in preserving and disinfection: A, 2-phenylethanol; B, 2-phenoxyethanol; C, chlorbutol (trichloro-*tert*-butanol); D, bronopol (2-bromo-2-nitropropan-1,3-diol).

mycobactericidal agent. It has also been effectively combined with other agents such as chlorhexidine and iodine to produce more active preparations. Ethanol is a popular choice in pharmaceutical preparations and cosmetic products as a solvent and preservative.

Isopropyl alcohol (isopropanol, $CH_3 . CHOH . CH_3$) has slightly greater bactericidal activity than that of ethanol but is also about twice as toxic. It is less active against viruses, particularly non-enveloped viruses, and should be considered a limited-spectrum virucide. Used at concentrations of 70% and above, it is an acceptable alternative to ethanol for pre-operative skin treatment and is also employed as a preservative for cosmetics.

3.2.2   *Alcohols as preservatives*

The aralkyl alcohols and more highly substituted aliphatic alcohols (Fig. 11.2) are used mostly as preservatives. These include the following.
1 Benzyl alcohol ($C_6H_5CH_2OH$). This has antibacterial and weak local anaesthetic properties and is used as an antimicrobial preservative at a concentration of 1%, although its use in cosmetics is restricted.
2 Chlorbutol (trichlorobutanol; trichloro-*tert*-butanol; trichlorobutanol; 1,1,1-trichloro-2-methylpropan-2-ol). Typical in-use concentration is 0.5%. It has been used as a preservative in injections and eye-drops. It is unstable, decomposition occurring at acid pH during autoclaving, while alkaline solutions are unstable at room temperature.
3 Phenylethanol (phenylethyl alcohol; 2-phenylethanol). Typical in-use concentration is 0.25–0.5%. It is reported to have greater activity against Gram-negative organisms and is usually employed in conjunction with another agent.
4 Phenoxyethanol (2-phenoxyethanol). Typical in-use concentration is 1%. It is more active against *Ps. aeruginosa* than against other bacteria and is usually combined with other preservatives such as the hydroxy-benzoates to broaden the spectrum of antimicrobial activity.
5 Bronopol (2-bromo-2-nitropropan-1,3-diol). Typical in-use concentration is 0.01–0.1%. It has a broad spectrum of antibacterial activity, including activity against *Pseudomonas* species. The main limitation on the use of bronopol is that when exposed to light at alkaline pH, especially if accompanied by an increase in temperature, solutions decompose, turning yellow or brown. A number of decomposition products including formaldehyde are produced. In addition, nitrite ions may be produced and react with any secondary and tertiary amines present forming nitrosamines, which are potentially carcinogenic.

3.3   **Aldehydes**

A number of aldehydes possess antimicrobial properties, including sporicidal activity; however, only two, formaldehyde and glutaraldehyde,

243   *Chemical disinfectants, antiseptics and preservatives*

are used for disinfection. Both these aldehydes are highly effective biocides and their use as 'chemosterilants' reflects this.

3.3.1   Glutaraldehyde

Glutaraldehyde (pentanedial; $CHO(CH_2)_3CHO$) has a broad spectrum of antimicrobial activity and rapid rate of kill, most vegetative bacteria being killed within a minute of exposure, although bacterial spores may require 3 hours or more. The latter depends on the intrinsic resistance of spores which may vary widely. It has the further advantage of not being affected significantly by organic matter. The glutaraldehyde molecule possesses two aldehyde groupings which are highly reactive and their presence is an important component of biocidal activity. The monomeric molecule is in equilibrium with polymeric forms, and the physical conditions of temperature and pH have a significant effect on this equilibrium. At a pH of 8, biocidal activity is greatest but stability is poor due to polymerization. In contrast, acid solutions are stable but considerably less active, although as temperature is increased, there is a breakdown in the polymeric forms which exist in acid solutions and a concomitant increase in free active dialdehyde, resulting in better activity. In practice, glutaraldehyde is generally supplied as an acidic 2% aqueous solution, which is stable on prolonged storage. This is then 'activated' prior to use by addition of a suitable alkylating agent to bring the pH of the solution to its optimum for activity. The activated solution will have a limited shelf-life, in the order of 2 weeks, although more stable formulations are available. Glutaraldehyde is employed mainly for the cold, liquid chemical sterilization of medical and surgical materials that cannot be sterilized by other methods.

3.3.2   Formaldehyde

Formaldehyde (HCHO) can be used in either the liquid or gaseous state for disinfection purposes. In the vapour phase it has been used for decontamination of safety cabinets and rooms; however, recent trends have been to combine formaldehyde vapour with low temperature steam (LTSF) for the sterilization of heat-sensitive items (Chapter 20). Formaldehyde vapour is highly toxic and potentially carcinogenic if inhaled; thus its use must be carefully controlled. It is not very active at temperatures below 20°C and requires a relative humidity of at least 70%. The agent is not supplied as a gas but as either a solid polymer, paraformaldehyde, or a liquid, formalin, which is a 34–38% aqueous solution. The gas is liberated by heating or mixing the solid or liquid with potassium permanganate and water. Formalin, diluted 1:10 to give a 3.4–3.8% formaldehyde solution, may be used for disinfecting surfaces. In general, however, solutions of either aqueous or alcoholic formaldehyde are too irritant for routine application to skin, while poor penetration and a tendency to polymerize on surfaces limit its use as a disinfectant.

Formaldehyde-releasing agents

Various formaldehyde condensates have been developed to reduce the irritancy associated with formaldehyde while maintaining activity and these are described as formaldehyde-releasing agents or masked-formaldehyde compounds.

Of these, noxythiolin (*N*-hydroxy-*N*-methylthiourea) has the greatest pharmaceutical use as an antimicrobial agent. The compound is supplied as a dry powder and on aqueous reconstitution slowly releases formaldehyde and *N*-methylthiourea. Antimicrobial activity is considered to be due to both the noxythiolin molecule and the released formaldehyde. Noxythiolin is used both topically and in accessible body cavities as an irrigation solution and in the treatment of peritonitis. The compound has extensive antibacterial and antifungal properties.

Polynoxylin (poly[methylenebis(hydroxymethyl)urea]) is a similar compound available in gel and lozenge formulations.

Taurolidine (bis-[1,1-dioxoperhydro-1,2,4-thiadiazinyl-4]methane) is a condensate of two molecules of the amino acid taurine and three molecules of formaldehyde. It is more stable than noxythiolin in solution and has similar uses. The activity of taurolidine is stated to be greater than that of formaldehyde.

3.4   Biguanides

3.4.1   *Chlorhexidine and alexidine*

Chlorhexidine is an antimicrobial agent first synthesized at Imperial Chemical Industries in 1954 in a research programme to produce compounds related to the biguanide antimalarial, proguanil. Compounds containing the biguanide structure could be expected to have good antibacterial effect and thus the major part of the proguanil structure is found in chlorhexidine. The chlorhexidine molecule, a bisbiguanide, is symmetrical. A hexamethylene chain links two biguanide groups to each of which a 4-chlorophenyl radical is bound (Fig. 11.3B). A related compound is the bisbiguanide, alexidine, which has use as an oral antiseptic and antiplaque agent. Alexidine differs from chlorhexidine in that it possesses ethylhexyl end-groups (Fig. 11.3A).

Chlorhexidine base is not readily soluble in water and therefore the freely soluble salts, acetate, gluconate and hydrochloride are used in formulation. Chlorhexidine exhibits the greatest antibacterial activity at pH 7–8 where it exists exclusively as a di-cation. The cationic nature of the compound results in activity being reduced by anionic compounds including soap and many anions, due to the formation of insoluble salts. Anions to be wary of include bicarbonate, borate, carbonate, chloride, citrate and phosphate with due attention being paid to the presence of hard water. Deionized or distilled water should preferably be used for dilution purposes. Reduction in activity will also occur in the presence of blood, pus and other organic matter.

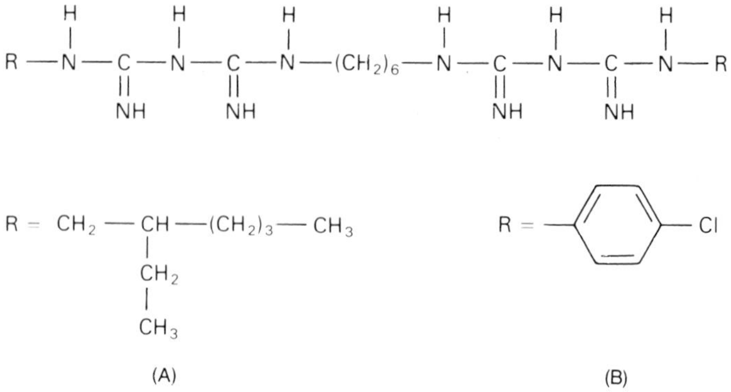

Fig. 11.3 Bisbiguanides. A, alexidine; B, chlorhexidine.

Chlorhexidine has widespread use, in particular as an antiseptic. It has significant antibacterial activity though Gram-negative bacteria are less sensitive than Gram-positive. A concentration of 1:2 000 000 prevents growth of, for example, *Staphylococcus aureus* whereas a 1:50 000 dilution prevents growth of *Ps. aeruginosa*. Reports of pseudomonad contamination of aqueous chlorhexidine solutions have prompted the inclusion of small amounts of ethanol or isopropanol. Chlorhexidine is ineffective at ambient temperatures against bacterial spores and *M. tuberculosis*. A limited antifungal activity has been demonstrated which unfortunately restricts its use as a general preservative. Skin sensitivity has occasionally been reported though, in general, chlorhexidine is well tolerated and non-toxic when applied to skin or mucous membranes and is an important pre-operative antiseptic.

3.4.2   *Polyhexamethylene biguanides*

The antimicrobial activity of chlorhexidine, a bisbiguanide, exceeds that of monomeric biguanides. This has stimulated the development of polymeric biguanides containing repeating biguanide groups linked by hexamethylene chains. One such compound is a commercially available heterodisperse mixture of polyhexamethylene biguanides (PHMB) having the general formula:

where n varies with a mean value of 5.5. The compound has a broad spectrum of activity against Gram-positive and Gram-negative bacteria and has low toxicity. PHMB is generally used as an environmental biocide though, recently, it has been employed in the USA as an antimicrobial in various ophthalmic products.

246   *Chapter 11*

Chlorine and iodine have been used extensively since their introduction as disinfecting agents in the late eighteenth and early nineteenth centuries. Preparations containing these halogens, such as Dakin's solution and tincture of iodine, were early inclusions in many pharmacopoeias and national formularies. More recent formulations of these elements have improved activity, stability and ease of use.

3.5.1   *Chlorine*

A large number of antimicrobially active chlorine compounds are commercially available, one of the most important being liquid chlorine. This is supplied as an amber liquid, made by compressing and cooling gaseous chlorine. The terms liquid and gaseous chlorine refer to elemental chlorine whereas the word 'chlorine' is normally used to signify a mixture of OCl^-, Cl_2, HOCl and other active chlorine compounds in aqueous solution. The potency of chlorine disinfectants is usually expressed in terms of parts per million (p.p.m.) or percentage of available chlorine (avCl).

3.5.2   *Hypochlorites*

Hypochlorites are the oldest and remain the most useful of the chlorine disinfectants being readily available and inexpensive. They exhibit a rapid kill against a wide spectrum of microorganisms including fungi and viruses. However, it is unfortunate that, at concentrations which are permitted in water, the oocysts of the water-borne protozoan *Cryptosporidium parvum*, emerging as the fourth most frequent cause of diarrhoea, are not killed. High levels of available chlorine will enable eradication of acid-fast bacilli and bacterial spores. The compounds are compatible with most anionic and cationic surface-active agents and are relatively inexpensive to use. To their disadvantage they are corrosive, suffer inactivation by organic matter and can become unstable. Hypochlorites are available as powders or liquids, most frequently as the sodium or potassium salts of hypochlorous acid (HOCl). Sodium hypochlorite exists in solution as follows:

$$NaOCl + H_2O \rightleftharpoons HOCl + NaOH.$$

Undissociated hypochlorous acid is a strong oxidizing agent and its potent antimicrobial activity is dependent on pH as shown:

$$HOCl \rightleftharpoons H^+ + OCl^-.$$

At low pH the existence of HOCl is favoured over OCl^- (hypochlorite ion). The relative microbiocidal effectiveness of these forms is of the order of 100:1. By lowering the pH of hypochlorite solutions the antimicrobial activity increases to an optimum at about pH 5; however this is

247   *Chemical disinfectants, antiseptics and preservatives*

concurrent with a decrease in stability of the solutions. This problem may be alleviated by addition of NaOH (see above equation) in order to maintain a high pH during storage for stability. The absence of buffer allows the pH to be lowered sufficiently for activity on dilution to use-strength. It is preferable to prepare use-dilutions of hypochlorite on a daily basis.

3.5.3 Organic chlorine compounds

A number of organic chlorine, or chloramine, compounds are now available for disinfection and antisepsis. These are the N-chloro ($=N-Cl$) derivatives of, for example, sulphonamides giving compounds such as chloramine-T and dichloramine-T and halazone (Fig. 11.4), which may be used for the disinfection of contaminated drinking water.

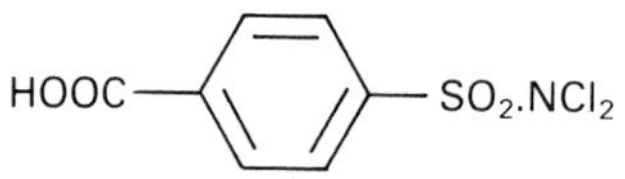

Fig. 11.4 Halazone.

A second group of compounds, formed by N-chloro derivatization of heterocyclic compounds containing a nitrogen in the ring, includes the sodium and potassium salts of dichloroisocyanuric acid (e.g. NaDCC). These are available in powder or tablet form and, in contrast to hypochlorite, are very stable on storage, if protected from moisture. In water they will give a known chlorine concentration. The antimicrobial activity of the compounds is similar to that of the hypochlorites when acidic conditions of use are maintained.

3.5.4 Chloroform

Chloroform ($CHCl_3$) has a narrow spectrum of activity. It has been used extensively as a preservative of pharmaceuticals since the last century though recently limitations have been placed on its use. Marked reductions in concentration may occur through volatilization from products resulting in the possibility of microbial growth.

3.5.5 Iodine

Iodine has a wide spectrum of antimicrobial activity. Gram-negative and Gram-positive organisms, bacterial spores (on extended exposure), mycobacteria, fungi and viruses are all susceptible. The active agent is the elemental iodine molecule, I_2. As elemental iodine is only slightly soluble in water, iodide ions are required to prepare aqueous solutions such as Aqueous Iodine Solution BP 1988 (Lugol's Solution) containing 5% iodine in 10% potassium iodide solution. Iodine (2.5%) may also be dissolved in ethanol (90%) and potassium iodide (2.5%) solution to give Weak Iodine Solution BP 1988 (Iodine Tincture).

The antimicrobial activity of iodine is less dependent than chlorine on temperature and pH, though alkaline pH should be avoided. Iodine is also less susceptible to inactivation by organic matter. Disadvantages in the use of iodine in skin antisepsis are staining of skin and fabrics coupled with possible sensitizing of skin and mucous membranes.

3.5.6   *Iodophors (iodophores)*

In the 1950s iodophors (*iodo* meaning iodine and *phor* meaning carrier) were developed to eliminate the side-effects of iodine while retaining its antimicrobial activity. These allowed slow release of iodine on demand from the complex formed. Essentially, four generic compounds may be used as the carrier molecule or complexing agent. These give polyoxymer iodophors (i.e. with propylene or ethyene oxide polymers), cationic (quaternary ammonium) surfactant iodophors, non-ionic (ethoxylated) surfactant iodophors and polyvinylpyrrolidone iodophors (PVP-I or povidone-iodine). The non-ionic or cationic surface-active agents act as solubilizers and carriers, combining detergency with antimicrobial activity. The former type of surfactant, especially, produces a stable efficient formulation, the activity of which is further enhanced by the addition of phosphoric or citric acid to give a pH below 5 on use-dilution. The iodine is present in the form of micellar aggregates which disperse on dilution, especially below the critical micelle concentration (c.m.c.) of the surfactant, to liberate free iodine.

When iodine and povidone are combined, a chemical reaction takes place forming a complex between the two entities. Some of the iodine becomes organically linked to povidone though the major portion of the complexed iodine is in the form of tri-iodide. Dilution of this iodophor results in a weakening of the iodine linkage to the carrier polymer with concomitant increases in elemental iodine in solution and antimicrobial activity.

The amount of free iodine the solution can generate is termed the 'available iodine'. This acts as a reservoir for active iodine releasing it when required and therefore largely avoiding the harmful side-effects of high iodine concentration. Consequently, when used for antisepsis, iodophors should be allowed to remain on the skin for 2 minutes to obtain full advantage of the sustained-release iodine.

3.6   **Heavy metals**

Mercury and silver have long been known to have antibacterial properties and preparations of these metals were among the earliest used antiseptics. However, they are no longer in regular use today, having been replaced by less toxic compounds. Other metals such as zinc, copper, aluminium and tin have weak antibacterial properties but are used in medicine for other functions, e.g. aluminium acetate and zinc sulphate are employed as astringents.

3.6.1   Mercurials

Plasmid-mediated resistance has resulted in many pathogenic micro-organisms becoming insensitive to the antimicrobial effects of mercurials. The organomercurial derivatives which are still in regular use in pharmacy are thiomersal and phenlymercuric nitrate or acetate (PMN or PMA) (Fig. 11.5).

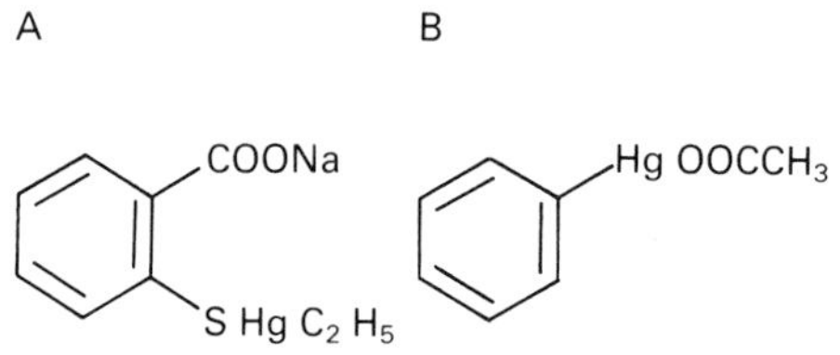

A   B

Fig. 11.5 Some organomercurials: A, thiomersal (sodium ethylmercurithiosalicylate); B, phenylmercuric acetate.

The phenylmercuric salts (0.002%) are recommended by the *British Pharmacopoeia* for preservation of eye-drops and injections. Thiomersal is also employed as a preservative for eye-drops and in lower concentration, 0.001–0.004%, as a preservative for contact lens solutions. Both mercurials are absorbed from solution by rubber closures and plastic containers to a significant extent.

3.7   Hydrogen peroxide and peracid compounds

Hydrogen peroxide (H_2O_2) and peracid compounds are powerful anti-microbial agents and this includes good sporicidal activity. The germicidal properties of hydrogen peroxide have been known for more than a century, but use of low concentrations of unstable solutions did little for its reputation. Stabilized solutions in concentrations of 3–6% are effective for general disinfection purposes and for cleansing wounds and ulcers, while higher concentrations (up to 25%) have application for liquid chemical sterilization. Peracetic acid (CH_3COOOH) is also an excellent bactericide, fungicide and sporicide. Its disadvantages are that it is corrosive to some metals, is toxic and may be inactivated by organic matter.

3.8   Phenols

Phenols (Fig. 11.6) are widely used as disinfectants and preservatives. The phenolics for disinfectant use have good antimicrobial activity and are rapidly bactericidal but generally are not sporicidal. Their activity is markedly diminished by dilution and is also reduced by organic matter.

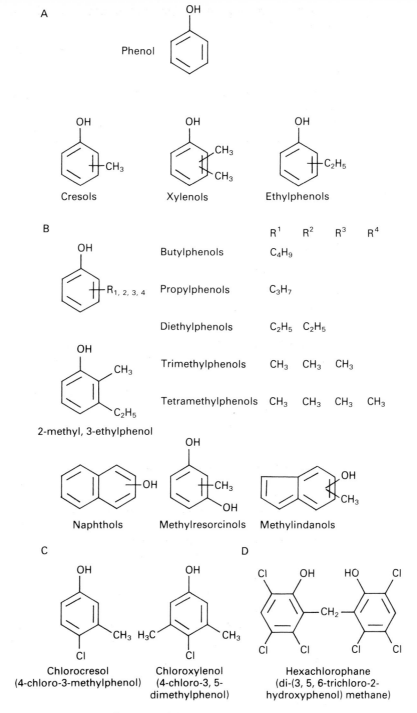

Fig. 11.6 Structural formulae of phenolic disinfectants: A, clear soluble fluids; B, black and white fluids; C, chlorinated phenols; D, bisphenols.

They are more active at acid pH. The main disadvantages of phenols are their caustic effect on skin and tissues and their systemic toxicity. The more highly substituted phenols are less toxic and can be used as preservatives and antiseptics; however, they are also less active than the simple phenolics, especially against Gram-negative organisms.

3.8.1   *Phenol (carbolic acid)*

Phenol no longer plays any significant role as an antibacterial agent. It is of historical interest, since it was introduced by Lister in 1867 as an antiseptic and has been used as a standard for comparison with other disinfectants, which are then given a phenol coefficient in tests such as the Rideal–Walker Test (Chapter 12).

3.8.2   *Tar acids*

Many of the phenols which are used in household and other commercial disinfectant products are produced from the tar obtained by distillation of coal or more recently petroleum. They are known as the tar acids. These phenols are separated by fractional distillation according to their boiling range into phenols, cresols, xylenols and high boiling point tar acids. As the boiling point increases the properties of the products alter as shown:

Phenols	Boiling point increases
Cresols	Bactericidal activity increases
Xylenols	Inactivation by organic matter increases
High boiling point	Water solubility decreases
tar acids	↓ Tissue toxicity decreases

The phenols from the higher boiling point fractions have greater antimicrobial activity but must be formulated so as to overcome their poor solubility. A range of solubilized and emulsified phenolic disinfectants are available including the *clear soluble fluids, black fluids* and *white fluids*.

Clear soluble fluids.   Cresol is a mixture of *o*-, *m*- and *p*-methyl phenol (Fig. 11.6A). Because of its poor solubility, it is solubilized with a soap prepared from linseed oil and potassium hydroxide. It forms a clear solution on dilution. This preparation, known as Lysol (Cresol and Soap Solution BP 1968) has been widely used as a general purpose disinfectant but has largely been superseded by less irritant phenolics.

By using a higher boiling point fraction than cresols, consisting of xylenols and ethylphenols (Fig. 11.6A), a more active, less corrosive product which retains activity in the presence of organic matter, is obtained. It is also solubilized with a soap to give a clear soluble fluid. A variety of proprietary products for general disinfection purposes are available with these phenols as active ingredients. They possess rapid bactericidal activity, including mycobacteria, but are only slowly sporicidal.

Black fluids and white fluids.   Black fluids and white fluids are prepared by solubilizing the high boiling point tar acids (Fig. 11.6B). Black fluids are homogeneous solutions, which form an emulsion on dilution with water. White fluids are finely dispersed emulsions of tar acids, which on dilution with water produce more stable emulsions than do black fluids. Both types of fluid have good bactericidal activity. Preparations are very irritant and corrosive to the skin and are strong smelling; however, they are relatively inexpensive and are useful for household and general disinfection purposes. They must be used in adequate concentrations as activity is reduced by organic matter and is markedly affected by dilution.

3.8.3   *Non-coal tar phenols (chloroxylenol and chlorocresol)*

Many derivatives of phenol are now made by a synthetic process. Homologous series of substituted derivatives have been prepared and tested for antimicrobial activity. A combination of alkyl substitution and halogenation has produced useful derivatives including chlorinated phenols which are constituents of a number of proprietary disinfectants. Two of the most widely used derivatives are 4-chloro-3-methylphenol (*p*-chloro-*m*-cresol, chlorocresol, Fig. 11.6C) which is mostly employed as a preservative at a concentration of 0.1%, and 4-chloro-3,5-dimethylphenol (*p*-chloro-*m*-xylenol, chloroxylenol, Fig. 11.6C) which is used for skin disinfection, although less than formerly. Chloroxylenol is sparingly soluble in water and must be solubilized, for example in a suitable soap solution in conjunction with terpineol or pine oil. Its antimicrobial capacity is weak and is reduced by the presence of organic matter.

3.8.4   *Bisphenols*

Bisphenols are composed of two phenolic groups connected by various linkages. Hydroxy halogenated derivatives, such as hexachlorophane (Fig. 11.6D) and triclosan, are the most active microbiologically, but are bacteriostatic at use-concentrations and have little antipseudomonal activity. The use of hexachlorophane is also limited by its serious toxicity. Both hexachlorophane and triclosan have limited application in medicated soaps and washing creams.

3.9   **Surface-active agents**

Surface-active agents or surfactants are classified as anionic, cationic, non-ionic or ampholytic according to the ionization of the hydrophilic group in the molecule. A hydrophobic, water-repellent, group is also present. Within the various classes a range of detergent and disinfectant activity is found. The anionic and non-ionic surface-active agents, for example, have strong detergent properties but exhibit little or no antimicrobial activity. They can, however, render certain bacterial species more sensitive to some antimicrobial agents, possibly by altering the

253   *Chemical disinfectants, antiseptics and preservatives*

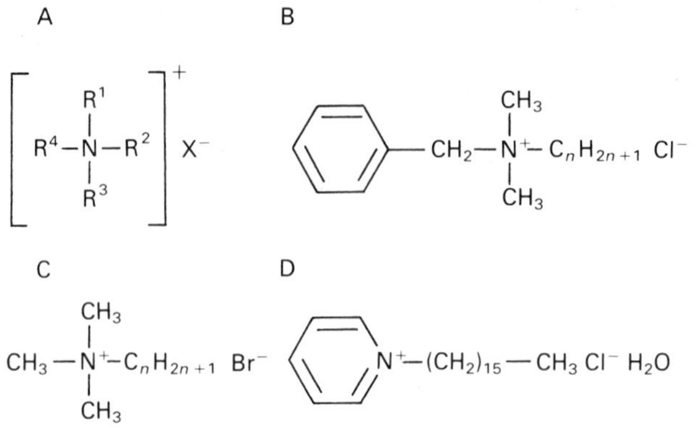

A   B

$$\left[\begin{array}{c} R^1 \\ | \\ R^4-N-R^2 \\ | \\ R^3 \end{array}\right]^+ X^-$$

(benzene ring)—CH_2—$\overset{+}{N}$—C_nH_{2n+1} Cl^- with two CH_3 groups on N

C   D

CH_3—$\overset{+}{N}$—C_nH_{2n+1} Br^- with two CH_3 groups on N

(pyridine ring)$\overset{+}{N}$—$(CH_2)_{15}$—CH_3 Cl^- H_2O

Fig. 11.7 Quaternary ammonium compounds: A, general structure of QACs; B, benzalkonium chloride ($n = 8–18$); C, cetrimide ($n = 12$, 14 or 16); D, cetylpyridinium chloride.

permeability of the outer envelope. Ampholytic or amphoteric agents can ionize to give anionic, cationic and zwitterionic (positively and negatively charged ions in the same molecule) activity. Consequently, they display both the detergent properties of the anionic surface-active agents and the antimicrobial activity of the cationic agents. They are used quite extensively in Europe for pre-surgical hand scrubbing, medical instrument disinfection and floor disinfection in hospitals.

Of the four classes of surface-active agents, however, the cationic compounds arguably play the most important role in an antimicrobial context.

3.9.1   *Cationic surface-active agents*

The cationic agents used for their antimicrobial activity all fall within the group known as the quaternary ammonium compounds, which are variously described as QACs, quats or onium ions. These are organically substituted ammonium compounds as shown in Fig. 11.7A where the R substituents are alkyl or heterocyclic radicals to give compounds such as cetyltrimethylammonium bromide (cetrimide), cetylpyridinium chloride and benzalkonium chloride. Inspection of the structures of these compounds (Fig. 11.7B) indicates the requirement for good antimicrobial activity of having a chain length in the range C_8 to C_{18} in at least one of the R substituents. In the pyridinium compounds (Fig. 11.7C) three of the four covalent links may be satisfied by the nitrogen in a pyridine ring.

The QACs are most effective against microorganisms at neutral or slightly alkaline pH and become virtually inactive below pH 3.5. Not surprisingly, anionic agents greatly reduce the activity of these cationic agents. Incompatibilities have also been recorded with non-ionic agents, possibly due to the formation of micelles. The presence of organic matter such as serum, faeces and milk will also seriously affect activity.

QACs exhibit greatest activity against Gram-positive bacteria with a lethal effect observed using concentrations as low as 1:200000. Gram-negative bacteria are more resistant requiring a level of 1:30000 or higher still if *Ps. aeruginosa* is present. Bacteriostasis is obtained at higher dilutions. A limited antifungal activity, more in the form of a bacterio-static than a bactericidal effect, is exhibited. The QACs have not been shown to possess any useful sporicidal activity. This narrow spectrum of activity therefore limits the usefulness of the compounds. Since they are generally well tolerated and non-toxic when applied to skin and mucous membranes, the compounds have considerable use in treatment of wounds and abrasions and they are used as preservatives in certain preparations. Benzalkonium chloride and cetrimide are employed exten-sively in surgery, urology and gynaecology as aqueous and alcoholic solutions and as creams. In many instances they are used in conjunction with a biguanide disinfectant such as chlorhexidine. The detergent proper-ties of the QACs also provide a useful activity, especially in hospitals, for general environmental sanitation.

3.10 Other antimicrobials

The range of chemicals which can be shown to have antimicrobial pro-perties is beyond the scope of this chapter. The agents included in this section have limited use or are of historic interest.

3.10.1 *Diamidines*

The activity of diamidines is reduced by acid pH and in the presence of blood and serum. Microorganisms may acquire resistance by serial subculture in the presence of increasing doses of the compounds. Propamidine, as the isethionate salt in the form of eye-drops (0.1%), is the major diamidine derivative employed as an antimicrobial agent.

3.10.2 *Dyes*

Crystal violet (gentian violet), brilliant green and malachite green are triphenylmethane dyes widely used to stain bacteria for microscopic examination. They also have bacteriostatic and fungistatic activity and have been applied topically for the treatment of infections. Staining of skin and clothes is a disadvantage of these agents. Due to concern about possible carcinogenicity, they are now rarely used.

The acridine dyes, including proflavine, acriflavine and aminacrine, have also been employed for skin disinfection and treatment of infected wounds or burns. They are slow-acting and mainly bacteriostatic in effect, with no useful fungicidal or sporicidal activity.

3.10.3 *Quinoline derivatives*

The quinoline derivatives of pharmaceutical interest are little used now. The antimicrobial activity of the derivatives is generally good against the

Gram-positive bacteria though less so against Gram-negative species. The compound most frequently used in a pharmaceutical context is dequalinium chloride, a bisquaternary ammonium derivative of 4-aminoquinaldinium. As it is a cationic surface-active agent it is incompatible with anionic agents. It is formulated as a lozenge for the treatment of oropharyngeal infections.

4   Disinfection policies

The aim of a disinfection policy is to control the use of chemicals for disinfection and antisepsis and give guidelines on their use. The preceding descriptions within this chapter of the activities, advantages and disadvantages of the many disinfectants available allow considerable scope for choice and inclusion of agents in a policy to be applied to such areas as industrial plant, walls, ceilings, floors, air, cleaning equipment and laundries and to the extensive range of equipment in contact with hospital patients.

The control of microorganisms is of prime importance in hospital and industrial environments. Where pharmaceutical products (either sterile or non-sterile) are manufactured, contamination of the product may lead to its deterioration and to infection in the user. In hospital there is the additional consideration of patient care, and therefore protection from nosocomial infection and prevention of cross-infection must also be covered. Hospitals generally have a disinfection policy, though the degree of adherence to, and implementation of, the policy content can vary. A specialized committee, involving the pharmacist, should formulate a policy and regularly reassess its efficiency. Reference to Tables 11.1–11.3 indicates the susceptibility of various microorganisms to the range of agents available. Table 11.5 presents examples of the range of formulations and uses of the agents available.

Although scope exists for choice of disinfectant in many of the areas covered by a policy, in certain instances specific recommendations are made as to the type, concentration and usage of disinfectant in particular circumstances. An example of this concerns the cleaning and disinfection of endoscopes which was the subject of a 1988 report by the Working Party of the British Society of Gastroenterology. The report recommended aldehyde preparations (e.g. 2%) as the first line antibacterial and antiviral disinfectant with a 4-minute soak of the instruments sufficient for inactivation of hepatitis B virus and HIV. Similarly, the area of use of hypochlorite solutions will dictate the strength of solution (avCl) required. Where blood and body fluid spill occurs, a 1% avCl (10 000 p.p.m.) solution is required. Lower strengths, 0.1% and 0.0125% avCl, are recommended for disinfection of general working surfaces and baby feeding bottles respectively.

Categories of risk (to patients) may be assigned to equipment coming into contact with a patient, dictating the level of decontamination required and degree of concern. *High-risk items* have close contact with broken skin or mucous membrane or are those introduced into a sterile area of the body and should therefore be sterile. These include sterile instruments, gloves, catheters, syringes and needles. Liquid chemical disinfectants should only be used if heat or other methods of sterilization are unsuitable. *Intermediate-risk items* are in close contact with skin or mucous membranes and disinfection will normally be applied. Endoscopes, respiratory and anaesthetic equipment, wash bowls, bed-pans and similar items are included in this category. *Low-risk items* or areas include those detailed earlier such as walls, floors, etc., which are not in close contact with the patient. Cleaning is obviously important with disinfection being required, for example, in the event of contaminated spillage.

5 Further reading

Advisory Committee on Dangerous Pathogens (1990) HIV—the causative agent of AIDS and related conditions. Second revision of guidelines. London: Health and Safety Executive.

Anon. (1988) Cleaning and disinfection of equipment for gastrointestinal flexible endoscopy: interim recommendations of a Working Party of the British Society of Gastroenterology. *Gut*, **29**, 1134–1151.

Anon. (1992) *Glutaraldehyde and You*. London: Health and Safety Executive.

Block S.S. (Ed.) (1991) *Disinfection, Sterilisation and Preservation*, 4th edn. Philadelphia: Lea & Febiger.

British Medical Association (1989) A code of practice for sterilisation of instruments and control of cross infection. London: BMA (Board of Science and Education).

British Standards Institution (1986) *Terms relating to disinfectants*. BS 5283: 1986 Glossary. Section one. London: British Standards Institution.

British Standards Institution (1991) *Guide to choice of chemical disinfectants*. BS 7152: 1991. London: British Standards Institution.

Control of Substances Hazardous to Health (COSHH) Regulations (1988). Statutory Instrument No. 1657.

Eggers H.J. (1990) Experiments on antiviral activity of hand disinfectants. Some theoretical and practical considerations. *Zentralblatt Bakt.* **273**, 36–51.

La Rocca R, La Rocca M.A.K. & Ansell J.M. (1983) Microbiology of povidone-iodine—an overview. In *Proceedings of the International Symposium on Povidone* (Eds G.A. Digenis & J.M. Ansell), pp. 101–109. Lexington, Kentucky: University of Kentucky.

Russell A.D. (1990) Bacterial spores and chemical sporicidal agents. *Clin Microbiol Rev.* **3**, 99–119.

Russell A.D., Hugo W.B. & Ayliffe G.A.J. (Eds) (1992) *Principles and Practice of Disinfection, Preservation and Sterilization*, 2nd edn. Oxford: Blackwell Scientific Publications.

Scott E.M., Gorman S.P. & McGrath S.J. (1986) An assessment of the fungicidal activity of antimicrobial agents used for hard-surface and skin disinfection. *J Clin Hosp Pharm.* **11**, 199–205.

12 Evaluation of non-antibiotic antimicrobial agents

1 **Introduction**
1.1 Definition of terms
1.2 Dynamics of disinfection

2 **Factors affecting the disinfection process**
2.1 Effect of temperature
2.1.1 Practical meaning of the temperature coefficient
2.2 Effect of dilution
2.2.1 Practical meaning of the concentration exponent
2.3 Effect of pH
2.3.1 Rate of growth of the inoculum
2.3.2 Potency of the antibacterial agent
2.3.3 Effect on the cell surface
2.4 Effect of surface activity
2.5 Presence of interfering substances
2.6 Effect of inoculum size

3 **Evaluation of liquid disinfectants**
3.1 Suspension tests
3.1.1 Phenol coefficient tests
3.1.2 Capacity use-dilution test
3.2 Quantitative suspension tests
3.3 Mycobactericidal activity
3.4 Sporicidal activity
3.5 *In vivo* tests
3.5.1 Skin tests
3.5.2 Other *in vivo* tests
3.5.3 Toxicity tests
3.6 Estimation of bacteriostasis
3.6.1 Serial dilution
3.6.2 Ditch-plate technique
3.6.3 Cup-plate technique
3.6.4 Solid dilution method
3.6.5 Gradient-plate technique
3.7 Tests for antifungal activity

3.7.1 Fungicidal activity
3.7.2 Fungistatic activity
3.7.3 Choice of test organism
3.8 Virucidal activity
3.8.1 Tissue culture or egg inoculation
3.8.2 Plaque assays
3.8.3 'Acceptable' animal model
3.8.4 Duck hepatitis B virus: a possible model of infectivity of human hepatitis B virus
3.8.5 Immune reaction
3.8.6 Virus morphology
3.8.7 Endogenous reverse transcriptase

4 **Semi-solid antibacterial preparations**
4.1 Tests for bacteriostatic activity
4.2 Tests for bactericidal activity
4.3 Tests on skin
4.4 General conclusions

5 **Solid disinfectants**

6 **Evaluation of air disinfectants**
6.1 Determination of viable airborne microorganisms
6.2 Experimental evaluation

7 **Preservatives**
7.1 Evaluation of preservatives
7.2 Preservative combinations
7.2.1 Synergy in preservative combinations
7.2.2 Evaluation of synergy
7.2.3 Rapid methods

8 **Appendix: British Standards**

9 **Further reading**

1   Introduction

1.1   Definition of terms

- *Bactericide*. An agency which kills bacteria.
- *Sporicide*. An agency which kills spores.
- *Bacteriostat*. An agency which prevents the reproduction and multiplication of bacteria.
- *Virucide (viricide)*. An agency which skills viruses.

258

- *Fungicide.* An agency which kills fungi.
- *Fungistat.* An agency which prevents fungal proliferation.

The foregoing terms are unequivocal and are the terms of choice in scientific writing; however, other terms are also in common use.

Disinfectant. This term implies a substance with bactericidal action. Clearly, if an environment is to be made free from the ability to reinfect, its bacterial population must be destroyed. A detailed description of the meaning of the terms 'disinfectant' and 'disinfection' was provided in Chapter 11 (section 1).

Sanitizer. This term, sometimes used in the public health context, refers to an agent that reduces the number of bacterial contaminants to a safe level.

Antiseptic. This term means 'against sepsis', which in general means wound infection. A bacteriostatic agent may prevent sepsis developing in the body especially if the normal body defences against sepsis are operative. For further details, see Chapter 11.

Another common usage of the terms disinfectant and antiseptic is to use the former for preparations to be applied to inert surfaces and the latter to preparations for application to living tissues.

Many of the standard works include only the word 'disinfection' in their title yet deal with all classes of compounds and with a wide range of application. It is unrewarding to be too dogmatic about these terms; many substances can function in both capacities depending upon their concentration and time of contact. A more general term, *biocide*, is now widely used to denote a chemical agent that, literally, kills microorganisms.

It is doubtful if there is a difference other than degree between bacteriostatic and bactericidal action. The three situations, growth, bacteriostasis and killing are represented graphically in Fig. 12.1. The question posed by this notion, to which often there is no precise answer, is: How long will a culture of bacteria remain viable when prevented from reproducing?

1.2 **Dynamics of disinfection**

Changes in the population of viable bacteria in an environment are determined by means of a viable count, and a plot of this count against time gives a dynamic picture of any pattern of change (see Fig. 12.1 curve A). The typical growth curve of a bacterial culture is constructed from data obtained in this way. The pattern of bacterial death in a lethal environment may be obtained by the same technique, when a death or mortality curve is obtained (Fig. 12.1, curve C).

Inspection of the death curves obtained from viable count data had early elicited the idea that because there was usually an approximate, and under some circumstances a quite excellent, linear relationship between

259   *Evaluation of non-antibiotic antimicrobial agents*

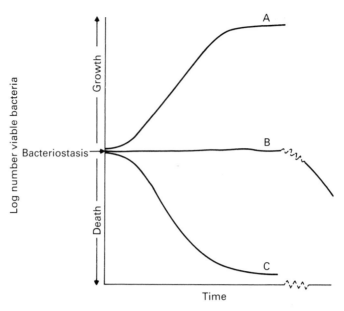

Fig. 12.1 The fate of a bacterial population when inoculated into: A, nutrient medium, normal growth curve. B, bacteriostatic environment. No change in viable population; after a prolonged time-interval the viable population will probably begin to fall. C, bactericidal environment. A sigmoid death curve is shown.

the logarithm of the number of survivors and time, then the disinfection process was comparable to a unimolecular reaction. This implied that the rate of killing was a function of the amount of one of the participants in the reaction only, i.e. in the case of the disinfection process the number of viable cells. From this observation there followed the notion that the principles of first-order kinetics could be applied to the disinfection process and that a rate or velocity constant in an equation of the type shown below could be used as a measure of the efficiency of a disinfectant:

$$K = \frac{1}{t} \log \frac{N_0}{N} \tag{12.1}$$

where K is the rate or velocity constant, N_0 is the initial number of organisms, N is the final number of organisms, and t is the time for the viable count to fall from N_0 to N.

This may be understood more fully by reference to Fig. 12.2. Curve A shows the type of response which would be obtained if the lethal process followed precisely the pattern of a first-order reaction. Some experimental curves do, in fact, follow this pattern quite closely, hence the genesis of the original theory.

The more usual pattern found experimentally is that shown by B, which is called a sigmoid curve. Here the graph is indicative of a slow initial rate of kill, followed by a faster, approximately linear rate of kill where there is some adherence to first-order reaction kinetics; this is

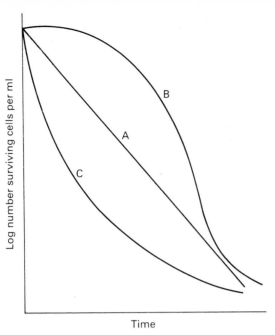

Fig. 12.2 Survivor/time curves for the disinfection process. A, obtained if the disinfection process obeyed the first-order kinetic law. B, sigmoid curve. This shows a slow initial rate of kill, a steady rate and finally a slower rate of kill. This is the form of curve most usually encountered. C, obtained if bacteria are dying more quickly than first-order kinetics would predict. The 'constant', K, diminishes in value continuously during the process.

followed again by a slower rate of kill. This behaviour is compatible with the idea of a population of bacteria which contains a portion of susceptible members which die quite rapidly, an aliquot of average resistance, and a residue of more resistant members which die at a slower rate. When high concentrations of disinfectant are used, i.e. when the rate of death is rapid, a curve of the type shown by C is obtained; here the bacteria are dying more quickly than predicted by first-order kinetics and the rate 'constant' diminishes in value continuously during the disinfection process.

The reason for this varied behaviour is not difficult to find. A population of bacteria does not possess the uniformity of properties inherent in pure chemical substances. This fact, together with the varied manner in which bactericides exert their effect and the complex nature of the bacterial cell, should provide adequate and satisfying reasons why the precise theories of reaction kinetics should have failed to explain the disinfection process.

The application of kinetic data is now being increasingly used in the evaluation of biocidal activity. As pointed out later (section 3.2), for example, data derived from viable counting procedures form the basis of modern suspension test methods.

The effects of temperature, pH and dilution on biocidal activity are of considerable significance and are dealt with in section 2.

2　Factors affecting the disinfection process

Apart from the obvious effect of concentration there are other important factors which affect the action of disinfectants.

2.1　Effect of temperature

In 1880, the bacteriologist Koch had noted that anthrax spores were more rapidly killed by the same concentrations of phenol if the temperature was elevated. A former pharmacopoeial sterilization process 'heating with a bactericide' used an elevated temperature, 80–100°C, maintained for 30 minutes, to ensure that quite low concentrations of bactericides would sterilize parenteral injections and eye-drops.

The idea that disinfection could be treated as a first-order chemical reaction led to ideas equating the effect of heat on the process to the effect of heat on chemical reactions, invoking the Arrhenius equation. For reasons already given, attempts to fit equations derived from chemical reactions to the disinfection process are unrewarding, although as a generalization it is true that as the temperature is increased in arithmetical progression, the rate of disinfection (rate of kill) increases geometrically.

The effect of temperature on bactericidal activity may be expressed quantitatively by means of a temperature coefficient, either the temprature coefficient per degree rise in temperature, denoted by θ, or the coefficient per 10° rise, the Q_{10} value.

θ may be calculated from the equation

$$\theta^{(T_2 - T_1)} = \frac{t_1}{t_2} \tag{12.2}$$

where t_1 is the extinction time at T_1°C, and t_2 the extinction time at T_2°C (i.e. $T_1 + 1$°C).

Q_{10} values may be calculated easily by determining the extinction time at two temperatures differing exactly by 10°C. Then

$$Q_{10} = \frac{\text{Time to kill at } T°}{\text{Time to kill at } (T + 10)°} \tag{12.3}$$

An overall picture of whole process may be obtained by plotting rate of kill against temperature.

The value for Q_{10} of chemical and enzyme-catalysed reactions lies between 2 and 3. The Q_{10} values of disinfectants vary widely; thus, for phenol it is 4, for butanol 28, for ethanol 45, and for ethylene glycol monoethyl ether, nearly 300. These figures alone should suggest that pushing the analogy of disinfection and chemical reaction kinetics too far is unwarranted.

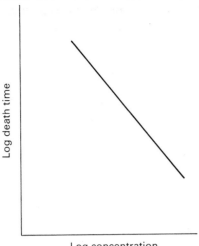

Fig. 12.3 Graphical determination of the concentration exponent, η, of a disinfectant.

Practical meaning of the temperature coefficient

The value of Q_{10} for phenol is 4, which means that over the 10°C range used to determine the Q_{10} (actually 20–30°C) the activity will be increased by a factor of 4.

2.2 **Effect of dilution**

The effects of concentration or dilution of the active ingredient on the activity of a disinfectant are of paramount importance. Failure to be aware of these changes in activity is responsible for many misleading claims concerning the properties of a disinfectant.

It was realized at the end of the nineteenth century that there was an exponential relationship between potency and concentration. Thus, if the $\log_{10}$ of a death time, that is the time to kill a standard inoculum, is plotted against the $\log_{10}$ of the concentration, a straight line is usually obtained, the slope of which is the concentration exponent (η) (Fig. 12.3). Expressed as an equation,

$$\eta = \frac{\text{(Log death time at concentration } C_2) - \text{(log death time at concentration } C_1)}{\text{Log } C_1 - \log C_2} \quad (12.4)$$

Thus, η may be obtained from experimental data either graphically or by substitution in the above equation. Some numerical values of η are given in Table 12.1.

2.2.1 *Practical meaning of the concentration exponent*

Mercuric chloride has a concentration exponent of 1; thus the activity will be reduced by the power of 1 on dilution, and a threefold dilution means

263 *Evaluation of non-antibiotic antimicrobial agents*

Table 12.1 Concentration exponents, η, for some disinfectant substances

Antimicrobial agent	η	Antimicrobial agent	η
Hydrogen peroxide	0.5	Parabens	2.5
Silver nitrate	0.9–1.0	Sorbic acid	2.6–3.2
Mercurials	0.03–3.0	Potassium laurate	2.3
Iodine	0.9	Benzyl alcohol	2.6–4.6
Crystal violet	0.9	Aliphatic alcohols	6.0–12.7
Chlorhexidine	2	Glycolmonophenyl ethers	5.8–6.4
Formaldehyde	1	Glycolmonoalkyl ethers	6.4–15.9
QACs	0.8–2.5	Phenolic agents	4–9.9
Acridines	0.7–1.9		
Formaldehyde donors	0.8–0.9		
Bronopol	0.7		
Polymeric biguanides	1.5–1.6		

the disinfectant activity will be reduced by the value 3^1 or 3, that is to a third. Put another way the disinfection time will be three times as long. In the case of phenol, however, with a concentration exponent of 6, a threefold dilution will mean a decrease in activity of $3^6 = 729$, a figure 243 times the value for mercuric chloride. This explains why phenols may be rapidly inactivated by dilution and should sound a warning bell regarding claims for diluted phenol solutions based on data obtained at high concentrations.

2.3   Effect of pH

During the disinfection process a change of pH can, at one and the same time, affect:
1 the rate of growth of the inoculum;
2 the potency of the antibacterial agent itself;
3 the ability of the drug to combine with sites on the cell surface.

2.3.1   *Rate of growth of the inoculum*

In general, bacterial growth is optimal in the pH range 6–8; on either side of this bracket the rate of growth declines.

2.3.2   *Potency of the antibacterial agent*

If the agent is an acid or a base its degree of ionization will depend on the pH. If its acid dissociation constant, pK_a, is known the degree of ionization at any pH may be calculated or determined by reference to published tables.

It has been shown that in some compounds the active species is the non-ionized molecule while the ion is inactive (benzoic acid, phenols, nitrophenols, salicylic acid, acetic acid). Thus, conditions of pH which

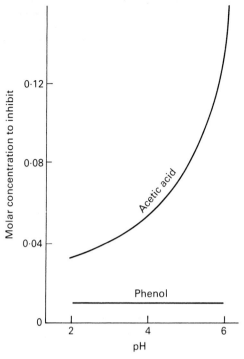

Fig. 12.4 The effect of pH on the concentration of phenol (pK_a 10) and of acetic acid (pK_a 4.7) to inhibit mould growth.

favour the formation of the ions of these compounds will also reduce their activity. The effect of pH on the ability of acetic acid and phenol to inhibit the growth of a mould is shown in Fig. 12.4.

In other cases the activity of the drug is due to the ionized molecule. For example, with the antibacterial acridine dyestuffs it is the cation which is the active agent and factors favouring ionization, all other things being equal, enhance their antibacterial activity (see Chapter 11).

Thus, at pH 7.3, 9-aminoacridine, which exists at this pH entirely as the cation, will inhibit the growth of *Streptococcus pyogenes* at a dilution of 1:160 000; the corresponding figure for 5-aminoacridine-3-carboxylic acid, which does not form cations at pH 7.3, is 1:5000.

Usually the antibacterial activity of cationic detergents such as cetrimide and the acridines increases with increase of pH (section 2.3.3).

2.3.3 *Effect on the cell surface*

Before an antibacterial agent can exert its effect on a cell it must combine with that cell. This process often follows the pattern of an adsorption isotherm. Clearly, factors which affect the state of the cell surface, as the pH of the cell's environment must do, must affect, to some extent, the adsorption process. An increase in the external pH renders the cell

265   *Evaluation of non-antibiotic antimicrobial agents*

surface more negatively charged. Biocidal agents that are cationic in nature thus bind more strongly to the cell surface with a consequent increase in activity.

In most situations of practical disinfection, pH may not be a significant variable but it has long been recognized that phenols are less active in alkaline solution, an effect readily explained by the foregoing account.

2.4 Effect of surface activity

The possession of surface activity *per se* may be an important factor in the antibacterial action of a group of drugs, for example the cationic detergents. The addition of low concentrations of surface-active compounds may potentiate the biological effect of an antibacterial agent. Thus phenols are often more active in the presence of soaps.

2.5 Presence of interfering substances

It has already been stated (section 2.3) that in most instances, before an antibacterial agent can act on a cell, it must first combine with it. It is not difficult to envisage the fact that the presence of other material, often referred to as organic matter, may reduce the effect of such an agent by adsorbing or inactivating it and thus reducing the amount available for combining with the cells it is desired to kill. Extraneous matter may be able to form a protective coat around the cell, thereby preventing the penetration of the active agent to its site of action. The possible influence, therefore, of other matter in the environment should not be overlooked.

2.6 Effect of inoculum size

This variable is often the one least controlled in the performance of tests upon disinfectants. Clearly, if it is postulated that disinfectant substances are first adsorbed on to a cell and thereafter kill it, the number of cells added to a given quantity of disinfectant may well be of significance. This is illustrated in Table 12.2.

Table 12.2 Effect of inoculum size on the minimum inhibitory concentration (MIC) of three antiseptics against *Staphylococcus aureus*

Antiseptic	MIC ($\mu g\,ml^{-1}$) vs. inoculum size		
	1×10^6	4×10^9	Increase (%)
Chlorocresol	225	350	55
Phenylethanol	3250	4750	65
Phenylmercuric acetate	3.5	5	70

In all experiments the inoculum size should be controlled and clearly stated in any account of the experiment.

3   Evaluation of liquid disinfectants

This evaluation may conveniently be classified into suspension tests (section 3.1) and counting methods, although the latter themselves use suspensions of microorganisms and hence are refered to here as quantitative suspension tests (section 3.2).

3.1   Suspension tests

These are essentially tests for sterility (Chapter 23) upon bacterial suspensions performed after treatment with the antibacterial agent for a prescribed time and under controlled conditions. They differ in the manner in which the experimental findings are calculated as well as in the details of experimental procedure.

They may be subdivided into phenol coefficient-type tests, of which there are many, quantitative suspension tests (which measure the rate at which test organisms are killed) and tests carried out at use-dilutions.

3.1.1   *Phenol coefficient tests*

The Rideal–Walker (RW) and Chick–Martin (CM) tests, long quoted, often in inappropriate circumstances, as standards for all disinfectants, were introduced at a time when typhoid fever was endemic. The tests were an attempt to standardize phenolic disinfectant claims or to kill the causal organism (*Salmonella typhi*) of typhoid fever. In the case of the CM test, account was taken of the presence of organic matter. During the first decade of the twentieth century, when the RW and CM tests were first described, it is true to say that these were valid tests. Phenols were the disinfectants almost invariably used, typhoid fever was still a present, although declining, menace to public health and it was utensils, rooms and surfaces at room temperature which were to be disinfected. The greatest single abuse of this type of test has been in the extrapolation of data to other situations and to disinfectants very different from phenol. Thus, it was not uncommon to find a preparation recommended for the treatment of wounds and declared able to kill staphylococci on the skin (at 37°C) in the presence of serum (organic matter), claimed as being six times as effective as pure phenol as judged by the RW test. If it is reiterated that the latter gives information about *Sal. typhi* at 17–18°C in an aqueous environment in the absence of organic matter, the extravagence of the extrapolation is plain. To use a phenol coefficient to evaluate non-phenolic disinfectants also contravened the fundamental concept of a biological assay, that is, that the standard and unknown should be of like mode of action. Not surprisingly, therefore, the RW and CM tests are falling into disuse, but full accounts will be found in the appropriate British Standards (BS 541:1985 and BS 808:1986). Both tests were fully described in the fourth edition (pp. 261–264) of the present book.

Capacity use-dilution test

The Kelsey–Sykes (KS) test.   Having regard to the many disadvantages alleged against the RW and CM tests, attempts were made and published in the early 1960s to find improved test methods. The foundations for the new test were laid by Kelsey *et al.* in 1965, and with the collaboration of the late G. Sykes and of Isobel M. Maurer, the Kelsey–Sykes test was evolved. This test embodied several principles. Firstly, it was a capacity test. Here a bacterial inoculum was added to the disinfectant in three successive lots at 0, 1 and 5 minutes. This is the principle of a capacity test where the capacity or lack of capacity of the disinfectant to destroy successive additions of a bacterial culture is tested.

The total test is performed in separate repeats using four test organisms: *Staph. aureus*, *Escherichia coli*, *Pseudomonas aeruginosa* and *Proteus vulgaris*. These were considered a more realistic choice than *Sal. typhi*, employed as the sole test organism in the RW and CM tests. The organisms, furthermore, are grown on a synthetic medium and survival is tested in a broth containing the non-ionic surface-active agent sorbitan mono-oleate (Tween 80). The disinfectant reaction is at 20°C and recovery of organisms at 32°C. Calibrated and dropping pipettes rather than loops are used for inoculation and other liquid manipulations, and disinfectants diluted at approximately the dilutions recommended for use are made in hard water. The test outlined above is carried out under clean and dirty conditions (compare RW, clean, and CM, dirty), the latter being simulated by dried yeast as in the CM test.

The disinfectant is assessed on its overall performance, namely its ability to kill microorganisms, as judged by subculture recovery or lack of it and not by comparison with phenol, i.e. a disinfectant would pass or fail according to its performance. A use-dilution concentration of a disinfectant must pass the test at three replications.

In summary, therefore, the KS suspension test differs from the RW and CM tests in that it is a capacity test, it reports the data as a pass or fail and not as a numerical cipher, i.e. not as a coefficient, and it uses a range of microorganisms. It combines an individual feature of the RW test in that it can report on disinfectant activity under both clean and dirty conditions.

This account highlights the main features of the test. It is described in detail by Kelsey and Maurer (1974).

Criticisms of the KS test.   An extensive collaborative trial was carried out on the KS test and the conclusion (Cowen, 1978) was that the test was suitable for white and clear, soluble disinfectants providing due care was taken in interpreting the pass concentration. Further modification of the test is necessary before it can be applied to other disinfectants.

3.2   **Quantitative suspension tests**

There is no doubt that the maximum information concerning the fate of a bacterial population is obtained by performing viable counts at selected time intervals. Alternatively, the number of survivors expressed as the percentage remaining viable at the end of a given period of time may be determined by viable counts and this parameter is often used in assessing bactericidal activity.

Viable counting is a technique used in all branches of pure and applied bacteriology. Essentially, the method consists of dispersing the sample in a solid nutrient which is then incubated. Any developing colonies are counted and if the assumption is made that each countable colony arises from a single viable cell in the original sample and that each viable cell is capable of, eventually, producing a colony, the viable bacterial content of that sample is thus determined. Viable numbers are usually expressed as colony-forming units (cfu) per millilitre.

This type of test may be used to investigate bactericidal, sporicidal or fungicidal activity.

It will be recalled (section 1) that research on the time course of the disinfection process was carried out making extensive use of viable counts, and notions concerning the dynamics of the disinfection process were gathered by these means.

A far more useful parameter for practical disinfectant evaluation is to perform a viable count at the end of a chosen period and to determine the concentration of disinfectant to achieve a 99, 99.9, 99.99 or 99.999% kill. The use of a percentage kill calculated to three places of decimals may sound pedantic but these become significant when dealing with large populations. Thus, if 99.999% of a population of bacteria originally containing 1 000 000 cells are killed in a given time there are still 10 survivors. Expressed in another way, 90, 99, 99.9, 99.99 and 99.999% kills represent $\log_{10}$ reductions of 1, 2, 3, 4 and 5 respectively. This aspect provides the basis of a new European Suspension test, currently being designed and still debated. The principle of this method is that a test bacterial suspension is exposed to a test disinfectant; after a specified time the numbers of cells remaining viable are compared with control (untreated) cells. A hypothetical example is provided in Table 12.3 together with an explanation of the calculation involved.

Tests should also be done in the presence of organic matter (e.g. albumin) and in hard water. It is important to remember when performing viable counts that care must be taken to ensure that, at the moment of sampling, the disinfection process is immediately arrested by the use of a suitable inactivator or ensuring inactivation by dilution (Table 12.4). Membrane filtration is an alternative procedure, the principle of which is that treated cells are retained on the filter whilst the disinfectant forms

Table 12.3 Hypothetical example of a quantitative suspension test procedure (disinfectant used for 2 minutes at 20°C)

Subculture dilution	Control series (C)		Disinfectant series (D)	
	Cfu	Cfu ml^{-1}	Cfu	Cfu ml^{-1}
10^{-1}	TNTC*	—	88	8.8×10^2
10^{-2}	TNTC	—	8	8×10^2
10^{-3}	TNTC	—	0	—
10^{-4}	TNTC	—		
10^{-5}	110	1.1×10^7		
10^{-6}	11	1.1×10^7		

* TNTC, too numerous to count.

Microbicidal effect $(M_E) = \log N_C - \log N_D$
$$= \log 1.1 \times 10^7 - \log 8.8 \times 10^2$$
$$= 7.04 - 2.94$$
$$= 4.10 \text{ (after 2 minutes)}$$

where N_C and N_D represent the number of Cfu ml^{-1} in the control and disinfectant series, respectively.

Table 12.4 Neutralizing agents for some antimicrobial agents*

Antimicrobial agent	Neutralizing agent
Benzoic acid and esters of p-hydroxybenzoic acid	Dilution or Tween 80
Bronopol	Cysteine hydrochloride
Chlorhexidine	Lubrol W and egg lecithin or Tween 80 and lecithin
Formaldehyde	Ammonium ions
Glutaraldehyde	Glycine
Halogens	Sodium thiosulphate
Hexachlorophane	Tween 80
Mercurials	Thioglycollic acid
Phenolic disinfectants	Dilution or Tween 80
QACs	Lubrol W and lecithin or Tween 80 and lecithin
Sulphonamides	p-Aminobenzoic acid

* This table should be read in conjunction with Table 23.3 in Chapter 23.

the filtrate. After washing *in situ*, the membrane is transferred to the surface of a solid (agar) recovery medium and the colonies that develop on the membrane are counted.

A major source of error in performing viable counts results from clumping of the organism so that one colony on the final plate may arise, not from one organism, but perhaps from numbers which may be of the order of 100. Unfortunately, many antibacterial agents, by affecting the surface charge on the bacterial cell, actually promote clumping and steps must be taken to overcome this.

Quaternary ammonium compounds (QACs; Chapter 11) such as cetrimide, and also the bisbiguanide, chlorhexidine, are notoriously prone

Fig. 12.6B. The plates are incubated overnight to allow diffusion of the drug and to dry the surface. The test organisms must be streaked in a direction running from the highest to the lowest concentration. Up to six organisms may be tested in this way.

To calculate the result, the length of growth and the total length of the agar surface streaked is measured; then if total length of possible growth is x cm and total length of actual growth is y cm, the inhibitory concentration as determined by this method is:

$$\frac{c \times y}{x} \, \text{mg ml}^{-1}$$

where c is the final concentration, in μg or mg ml^{-1}, of the drug in the *total* volume of the medium. It should always be borne in mind that, in comparing results obtained on solid and in a liquid environment, the factor of drug diffusion may have a bearing on all results using solid environments.

3.7 **Tests for antifungal activity**

Fungi may be potential pathogens or occur as contaminants in pharmaceutical products. In performing tests on potential antifungal preparations the differing culture requirements of the fungi should be borne in mind, otherwise tests similar to those used for antibacterial activity may be employed.

Two typical media for growth of fungi are Sabouraud liquid medium and Czapek–Dox medium. Both these media may be solidified with agar if required.

3.7.1 *Fungicidal activity*

Fungal spores or mycelium may be added to the solution under test. At selected time intervals, samples can be subcultured into suitable media and the presence or absence of growth noted after incubation. A quantitative assessment similar to that described for bactericidal activity (section 3.2, Table 12.3) can also be undertaken.

3.7.2 *Fungistatic activity*

Both the liquid and the solid dilution tests described above for bacteria (sections 3.6.1 and 3.6.4) may be used; suitable media must, of course, be employed.

3.7.3 *Choice of test organism*

For the evaluation of preparations to be used against pathogenic fungi, suitable cultures of these pathogens should be used. To test substances intended to inhibit general contaminants, cultures of common fungi

275 *Evaluation of non-antibiotic antimicrobial agents*

obtained conveniently by exposing Petri dishes of solid media to the atmosphere may be used, or alternatively dust or soil may be used as a source of a mixed inoculum.

3.8   **Virucidal activity**

The testing of disinfectants for virucidal activity is not an easy matter. As pointed out earlier (Chapter 3), viruses are unable to grow in artificial culture media and thus some other system, usually employing living cells, must be considered. One such example is tissue culture, but not all virus types can propagate under such circumstances and so an alternative approach has to be adopted in specific instances. The principles of such methods are given below.

3.8.1   *Tissue culture or egg inoculation*

A standardized viral suspension is exposed, in the presence of yeast suspension, to appropriate dilutions of disinfectant in WHO hard water. At appropriate times, dilutions are made in inactivated horse serum and each dilution is inoculated into tissue cell culture or embryonated eggs (as appropriate for the test virus). The drop in infectivity of the treated virus is compared with that of the control (untreated) virus.

Since disinfectant itself might be toxic to the tissue culture or eggs, a toxicity test must also be carried out. Here, appropriate dilutions of disinfectant are mixed with inactivated horse serum and inoculated into tissue cells or eggs (as appropriate). These are examined daily for damage.

3.8.2   *Plaque assays*

Plaque assays, at present, apply to only a very limited number of viruses, e.g. poliovirus, herpes virus, human rotavirus. The principle of these assays is as follows: test virus is dried on to coverslips which are immersed in various concentrations of test disinfectant for various time intervals and a plaque-counting method used to determine surviving viral particles. The plaques are similar to those described in Chapter 3 (section 6.1 and Fig. 3.4) except that a host cell other than bacteria (Chapter 3) has to be employed.

For assaying herpes virus, monolayers of baby hamster kidney (BHK) cells are used. Virus titre is expressed as the number of plaque-forming units (pfu) per millilitre before and after exposure to a disinfectant, so that the virucidal efficacy of the test agent can be determined. A diagrammatic representation is given in Figure 12.7.

3.8.3   *'Acceptable' animal model*

The hepatitis B virus (HBV) does not grow in tissue culture and an 'acceptable' animal model has been found to be the chimpanzee. This is

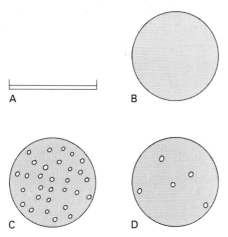

Fig. 12.7 A, diagrammatic representation of plaque assay for evaluating virucidal activity and B, monolayers of baby hamster kidney (BHK) cells; C, virus titre: *untreated* virus (o represents a plaque-forming unit, pfu, in BHK cells); D, virus titre: *disinfectant*-treated virus (before plating onto BHK, the disinfectant must be neutralized in an appropriate manner). Note the greatly reduced number of pfu in D, indicative of fewer uninactivated virus particles than in C.

observed for clinical infection after inoculation with treated and untreated virus, care being taken in the test series that residual disinfectant is removed by adequate means before inoculation into the animal.

This procedure is limited by the number of animals that can be used and by the strictures imposed by a humane approach. HBV has not yet been transmitted to non-primate animals.

3.8.4   *Duck hepatitis B virus: a possible model of infectivity of human hepatitis B virus*

Duck hepatitis B virus (DHBV) has been proposed as a possible model for the inactivation of human HBV by chemical disinfectants. The principle of the test method uses viral DNA polymerase (DNAP) as a target, total inhibition *in vitro* of DNAP by chemical disinfectants being predictive of inactivation of infectivity *in vivo*.

3.8.5   *Immune reaction*

Three types of particles are associated with HBV: small spherical particles, 22 nm in diameter; tubular particles, also having a diameter of 22 nm; and larger spherical particles (42 nm diameter) known as the Dane particles. The Dane particle alone has a typical virus structure and appears to be infectious but is the least common form. It consists of a complex, double-layered sphere with an electron-dense core. It contains partially double-stranded circular DNA and is regarded as the putative virion (for further information on virions see Chapter 3). The Dane particles contain three antigens: hepatitis B surface antigen (HBsAg) which is also present on 22-nm particles, hepatitis B core antigen (HBcAg) found in the inner core and hepatitis B e antigen (HBeAg) found in the core and responsible for infectivity.

The specific immunological detection of the HBV surface antigen (HBsAg) is considered as being presumptive evidence for the presence of viable HBV. The hypothesis, then, on which this method is based is that if the disinfectant can destroy the reactivity of the HBsAg, it can also destroy the infectivity of HBV. A problem with some disinfectants, e.g. formaldehyde and glutaraldehyde, is that their actions are essentially fixative in nature. The HBsAg immunological reaction is thus not destroyed at concentrations known to be high enough to kill the most resistant forms (bacterial spores) of microorganisms. Furthermore, concentrations of disinfectants necessary to inactivate HBsAg within a reasonable period of time are often comparatively high.

This type of procedure may thus suggest that an unnecessarily high disinfectant concentration (so-called overkill) may be employed in practice to achieve a virucidal effect.

3.8.6   *Virus morphology*

The serum from patients with clinical symptoms of hepatitis B commonly contain three distinct structures that possess HBsAg (section 3.8.5 above). The effects of different concentrations of various disinfectants on the structure of Dane particles have been studied, but it is unlikely that morphological changes can be related to virucidal activity.

3.8.7   *Endogenous reverse transcriptase*

The human immunodeficiency virus (HIV; lymphadenopathy-associated virus, LAV; human T-cell lymphotrophic virus type 3, HTLV III) is responsible for acquired immune deficiency syndrome (AIDS; see Chapter 3). Because of the hazard and difficulties of growing the virus outside humans, a different approach has to be examined for determining viral sensitivity to disinfectants.

Studies have demonstrated that one such method is to examine the effects of disinfectants on endogenous RNA-dependent DNA polymerase (i.e. reverse transcriptase) activity. In essence. HIV is an RNA virus; after it enters a cell the RNA is converted to DNA under the influence of reverse transcriptase. The virus induces a cytopathic effect on T lymphocytes, and in the assay reverse transcriptase activity is determined after exposure to different concentrations of various disinfectants. However, it has been suggested that monitoring residual viral reverse transcriptase activity is not a satisfactory alternative to tests whereby infectious HIV can be detected in systems employing fresh human peripheral blood mononuclear cells.

4   Semi-solid antibacterial preparations

The use of the term 'semi-solid' has been coined to embrace a group of pharmaceutical preparations known as pastes, ointments, creams and gels. The chief feature which distinguishes the first three is their viscosity or, to use a more descriptive word, their stiffness, which decreases in the order: paste, ointment, cream. They may consist of an intimate mixture of the active agent with either an oleaginous base or, alternatively, an emulsion with either water or an oleaginous substance as a continuous phase. Gels are preparations in which the base is usually a carbohydrate polymer (starch, pectin, methylcellulose, tragacanth, sterculia gum) and water, or more rarely having a base of protein origin, such as gelatin, with a suitable quantity of water. More recently polyethylene glycols and other organic polymers have been used.

When formulating antibacterial preparations it is imperative to realize that the properties of the base may seriously modify the antibacterial activity of the medicament. It is quite useless to formulate a well-proven antiseptic into an otherwise elegant pharmaceutical preparation without determining if the final formulation is, itself, an effective antibacterial agent.

4.1   Tests for bacteriostatic activity

The first official test was published by the Food, Drug and Insecticide Administration of the US Department of Agriculture, in which portions of the preparation were placed on the surface of nutrient agar inoculated with *Staph. aureus*. After incubation the zones of inhibition, if any, around the preparation were measured. This test was modified later by incorporating 10% of horse serum in the agar 'to simulate conditions in a wound' and a control consisting of unmedicated base was also used in each experiment. This test is known as the cup-plate test (see also section 3.6.3 and Fig. 12.5).

In addition to placing the test preparation onto sectors of seeded agar, it may be placed in a trough cut in uninoculated agar and test organisms streaked in parallel lines up to the edge of the trough. Failure to grow up to the edge is indicative of inhibition.

Thus the cup-plate method is useful to test several preparations or varying formulations of the same preparation against one organism under identical conditions, and the ditch-plate method enables one preparation to be tested against several organisms (Fig. 12.5A, B).

4.2   Tests for bactericidal activity

A number of tests have been described which imitate, at least in part, the principle of the phenol coefficient test for liquid disinfectants. A culture of the test organism is mixed intimately with the semi-solid preparation, and the mixture subcultured by means of a loop into a suitable broth designed to disperse the base and neutralize the antibacterial activity of the medicament.

Thus, the culture may be mixed and transferred to a hypodermic syringe surrounded by a constant-temperature jacket; at desired intervals, the mixture is subcultured by ejecting small volumes from the syringe nozzle into subculture medium.

A technique, devised by one of the authors (W.B.H.), was designed to test the preparation when spread on to an infected surface. The surface of a nutrient agar plate was inoculated evenly with the test organism and incubated to produce an even surface growth. The preparation under test was spread evenly upon this, and at selected time intervals a core of agar, cells, and preparation were removed with a sterile cork-borer and the disc of agar and cell removed by means of a sterile needle and inoculated into recovery medium, which was then incubated. As much of the preparation is removed as is possible and care taken to ensure its dispersal in the medium. The organism should, if still viable, grow through the back of the agar disc to give growth in the subculture tube also.

4.3 Tests on skin

It is possible to also test semi-solid antibacterial preparations on the skin itself, as described for liquid disinfectants (section 3.5.1). A portion of the skin—the backs of the fingers between the joints is a useful spot—is treated with the test organism, the preparation is then applied and after a suitable interval the area is swabbed and the swab incubated in a suitable medium. Alternatively, the method employing pig skin, described in section 3.5.1, may well be adapted to the problem of testing semi-solid skin disinfectants.

4.4 General conclusions

It is suggested that, as a minimum routine for the final test of an alleged antibacterial semi-solid formulation, the following be used.
1 The cup-plate technique for bacteriostatic activity (section 3.6.3).
2 A test for bactericidal activity.
3 A skin test.

For routine assessment of test formulations during development work the cup-plate and ditch-plate methods are adequate.

5 Solid disinfectants

Solid disinfectants (disinfectant powders) usually consist of a disinfectant substance diluted by an inert powder. For example phenolic substances adsorbed onto kieselguhr form the basis of many disinfectant powders, while another widely used powder of respectable antiquity is hypochlorite powder. Disinfectant or antiseptic powders for use in medicine include substances such as acriflavine, or antifungal compounds such as zinc undecenoate or salicylic acid mixed with talc.

Solid disinfectants may be evaluated *in vitro* by applying them to suitable test organisms growing on solid medium. Discs may be cut from the agar and subcultured, observing the usual precautions.

To test their inhibitory power, the powders may be dusted onto the surface of seeded agar plates, using the inert diluent as a control and noting the extent of growth.

Disinfectant and sanitary powders are the subject of a British Standard (BS 1013:1946), now withdrawn, which describes a method of determining the RW coefficient of such powders. A weighed quantity was shaken with distilled water at 18°C for 30 minutes and this suspension was used in the test already described (section 3.1.1).

6   Evaluation of air disinfectants

One of the most potent routes for transmission of bacterial disease is via the air. Cross-infection in hospital wards, infection in operating theatres, the transmission of disease in closed spaces such as cinemas and other places of assembly, in the ward rooms and crew's quarters of ships and in submarines are all well known. Of equal importance is the provision of a bacteria-free environment for aseptic manipulations generally. Clearly, the disinfection of atmospheres is a worthwhile field of study and to this end much research has been done. It is equally clearly important to be able to evaluate preparations claimed to be air disinfectants.

Heretofore the milieu on or in which the disinfectant has been required to act has been either solid or liquid; now antibacterial action in the gas or vapour phase or in the form of aerosol (colloidal) interaction must be considered, and this presents the problem of determining the viable airborne population.

6.1   Determination of viable airborne microorganisms

The simplest way of assessing the viable microbial population of the air is to expose Petri dishes containing a solid nutrient medium to the air, followed by incubation; indeed this method was used in 1881 by Koch. Although this method does depend on the organisms or organism-bearing particles actually falling on the plate by gravity it is a method which is still used to assess the general cleanliness of air in pharmaceutical factories where aseptic operations are taking place, in food processing areas or in hospital wards. More positive data may be obtained, however, if a force other than gravity is used to collect airborne particles.

An early attempt at quantification consisted of placing a Petri dish containing a nutrient agar in a box beneath an inverted funnel, the stem of which passed out of the box into the atmosphere. By applying a partial vacuum to the box, air was drawn in through the stem of the funnel and impinged on the agar. The plate could be incubated directly and developing colonies counted. Provided the air drawn in was metered, a direct quantitative assessment of the viable airborne population could be made. This idea led logically to the development of the slit sampler illustrated in Fig. 12.8. The principle is similar to that described immediately above, but the Petri dish is placed on a turntable which can be revolved at varying speeds and the funnel is replaced by a cylinder in which the end nearest the nutrient medium is furnished with a slit c. 2.5 mm wide. The arrangement is set so that the slit runs parallel to a radius of the dish but leaves a clear space around the circumference and at the centre of the plate. In operation, a vacuum is applied to the chamber containing the turntable, air passes in through the slit and the nutrient medium revolves so that the airborne particles, if any, are trapped on the medium and spread in a sector over the medium.

6.2   **Experimental evaluation**

In brief, the experimental technique is to create a bacterial population in a closed chamber, obtain a quantitative assessment of the viable airborne bacterial population by means of a suitable sampling device, submit the population to the disinfectant action, whether ultraviolet light, chemical vapour or aerosol, and then determine the airborne population at suitable intervals.

7   **Preservatives**

Preservatives may include disinfectant and antiseptic chemicals together with certain compounds used almost exclusively as preservatives. They are added to many industrial, including pharmaceutical, products which may, by their nature, support the growth of bacteria and moulds causing spoilage of the product and possibly infection of the user. In the field of pharmaceutical preservation, addition of an inhibitory substance to a multidose injection (Chapter 21) or the prevention of growth in aqueous suspensions of drugs intended for oral administration (Chapter 19) are prime examples.

Preservatives are widely employed in cosmetic preservation for lotions, creams and shampoos. Preservation is also an important aspect of formulation in emulsion paints and cutting fluids, i.e. fluids used to cool and lubricate lathe and drilling tools.

7.1   **Evaluation of preservatives**

Potential chemical preservatives may be evaluated in the first place by the methods outlined above, especially by determining MIC values (section 3.6) or by viable counts (section 3.2). The RW, CM and KS tests (sections 3.1.1 and 3.1.2) have no relevance in preservative evaluation. It will be recalled (section 2.5) that formula ingredients may reduce the

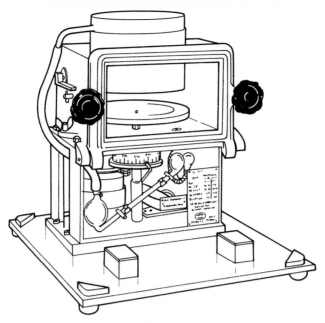

Fig. 12.8 Slit sampler (C.F. Casella & Co., Ltd).

efficiency of a preservative which has shown up well in conventional tests using culture media as the suspending fluid.

Emulsions, especially oil-in-water emulsions which, incidentally, figure widely in cosmetic products, are especially prone to failure because the preservative may partition into the oily phase of the emulsion while contaminants will flourish in the aqueous phase now deprived of preservative by partitioning (see Chapter 18 for further details).

The cardinal requirement, therefore, for preservative efficacy is the evaluation of the finally preserved preparation and this may be performed by means of a challenge test. In essence, the (hopefully) preserved product is deliberately inoculated (challenged) with suitable test organisms and incubated and examined to see if the inoculum has been able to grow or if its growth has been successfully suppressed. There has been extensive debate on challenge testing and the subject has been reviewed by Cowen and Steiger (1976).

The *British Pharmacopoeia* (1988) contains a test for efficacy of preservatives. In essence, the product is deliberately challenged separately by the fungus *Aspergillus niger*, the yeast *Candida albicans* and the bacteria *Ps. aeruginosa* and *Staph. aureus*. These organisms represent potential contaminants in the environment in which products are prepared, stored or used. Other organisms may be used in specified circumstances, e.g. The osmophilic yeast, *Zygosaccharomyces rouxii* for preparations with a high sucrose content, and *E. coli* for oral liquid preparations.

Different performance criteria are laid down for injectable and

283   *Evaluation of non-antibiotic antimicrobial agents*

ophthalmic preparations, topical preparations and oral liquid preparations. Inhibition of the challenge organism is determined by viable counting techniques. The *British Pharmacopoeia* (1988) should be consulted for full details of the experimental procedures to be used.

The *United States Pharmacopeia* (1990, 22nd edn) also gives procedures for evaluating the efficacy of antimicrobial preservatives in pharmaceutical products.

7.2 Preservative combinations

The use of preservative combinations may be used to extend the range and spectrum of preservation. Thus, in the series of alkyl esters of 4-hydroxybenzoic (*p*-hydroxybenzoic) acid (parabens), water solubility decreases in the order: methyl, ethyl, propyl and butyl ester. By combining these products it is possible to achieve a situation where both the aqueous and oil phase of an emulsion are protected.

Combinations may also be used to extend the spectrum of a preservative system. Thus the preservative Germall 115 has an essentially antibacterial activity and very low, if not zero, antifungal activity. By combining Germall 115 with parabens, which possess antifungal activity, a broader spectrum (antibacterial/antifungal) preservative system is obtained.

7.2.1 Synergy in preservative combinations

Very occasionally a combination of antimicrobial agents exhibits synergy. Synergy is measured against a single microorganism and is exhibited when a combination of two compounds exerts a greater inhibitory effect than could be expected from a simple additive effect of the two compounds in the mixture.

7.2.2 Evaluation of synergy

Synergy may be evaluated and displayed by preparing mixtures of the two compounds being investigated and determining their growth inhibitory power by means of an MIC determination (section 3.6.1).

The results may be plotted in the form of a graph (called an isobologram) and an example is given in Fig. 12.9. This graph may be interpreted as follows: 50×10^{-2} mg% and 35 mg% of phenylmercuric acetate and chlorocresol, respectively, used alone, inhibits the growth of *Staph. aureus*. In combination, 20×10^{-2} mg% of phenylmercuric acetate and 5 mg% of chlorocresol inhibit the growth of this organism. Thus, growth inhibition is obtained with a lower total quantity of preservative. If the combinations were merely additive, the isobologram plot would follow the course of the dotted line (A), and if antagonistic the dotted curve (B).

Synergy has been discussed in depth by Denyer *et al.* (1985).

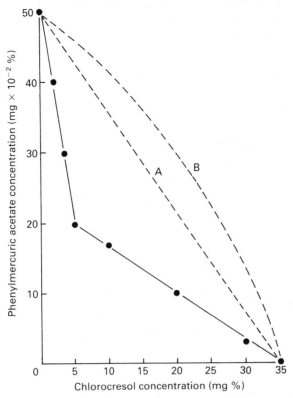

Fig. 12.9 Isobologram (●—●) drawn from minimum growth inhibitory concentrations (MIC values) of chlorocresol and phenylmercuric acetate used alone and in combination against *Staph. aureus*, showing synergy. A, result if combination was merely additive; B, result if combination was antagonistic.

7.2.3   *Rapid methods*

In many cases, especially in the food industry, it would be very useful if the performance of a biocide or the extent of contamination of products, apparatus and working surfaces could be deduced sooner than that provided by a method which depends on visible microbial growth (12–24 hours). Many methods have been devised to secure a more rapid result and have been designated rapid tests. They have recently been reviewed by Denyer (1990).

Two such methods will be mentioned here.

1 *Epifluorescence* depends on the fact that certain dyes, acridine orange being widely used, will stain cellular material. When examined in fluorescent light any living cells present will fluoresce green or greenish yellow, whereas dead cells will appear orange to red. The methods will be found in the literature under direct epifluorescent microscopy (DEM) and direct epifluorescent filter technique (DEFT). In DEM, material suspected of being contaminated or a sample in which living bacteria are sought are examined directly; in DEFT the sample being examined is filtered and the residue on the filter examined as above.

285   *Evaluation of non-antibiotic antimicrobial agents*

2 *Bioluminescence.* In another method, luminous bacteria, or bacteria not normally luminous but which have been manipulated genetically to become luminous, are used. Their death under chemical stress or presence in hygiene studies are assessed in a sensitive light meter. A variant of this method depends on the fact that bacterial adenosine triphosphatase (ATPase), present in bacteria, will catalyse the normal biological light-producing reaction to give detectable light in a sensitive meter.

This is a very brief summary but is included as readers may come across these methods or a reference to rapid methods in their general reading or work experience.

Rapid methods have yet to be validated for official testing.

8 Appendix: British Standards

Method for determination of the Rideal–Walker coefficient of disinfectants. BS 541:1985.

Method for assessing the efficacy of disinfectants by the modified Chick–Martin test. BS 808:1938 (1986).

Method for laboratory evaluation of disinfectant activity of quaternary ammonium compounds by suspension test procedure. BS 3286:1960.

Specification for black and white disinfectant fluids. BS 2462:1961 (1986).

Glossary of terms relating to disinfectants. BS 5283:1976.

Specification: *Aromatic disinfectant fluids.* BS 5197:1976. With amendment 1976 (1985).

Method for estimation of concentration of disinfectants used in 'dirty' conditions in hospitals by the modified Kelsey–Sykes test. BS 6905:1987.

Method for determination of the antimicrobial efficacy of disinfectants for veterinary and agricultural use. BS 6734:1986.

Method for determination of the antimicrobial activity of QAC disinfectant formulations. BS 6471:1984.

Specification for QAC based aromatic disinfectant fluids. BS 6424:1984.

9 Further reading

Akess M.J. & Taylor C.J. (1990) Official methods of preservative evaluation and testing. In *Guide to Microbiological Control in Pharmaceuticals* (Eds S.P. Denyer & R.M. Baird), pp. 292–303. Chichester: Ellis Horwood.

Cowen R.A. (1978) Kelsey–Sykes capacity test: a critical review. *Pharm J.* **220**, 202–204.

Cowen R.A. & Steiger B. (1976) Antimicrobial activity—a critical review of test methods of preservative efficiency. *J Soc Cosmet Chem.* **27**, 467–481.

Croshaw B. (1981) Disinfectant testing—with particular reference to the Rideal–Walker and Kelsey–Sykes tests. In *Disinfectants: Their Use and Evaluation of Effectiveness* (Eds C.H. Collins, M.C. Allwood, S.F. Bloomfield & A. Fox), pp. 1–15. London: Academic Press.

Denyer S.P. (1990) Monitoring microbiological quality: application of rapid microbiological methods to pharmaceuticals. In *Guide to Microbiological Control in Pharmaceuticals* (Eds S.P. Denyer & R.M. Baird), pp. 146–156. Chichester: Ellis Horwood.

Denyer S.P. & Hugo W.B. (Eds) (1991) *Mechanisms of Action of Chemical Biocides.* Society for Applied Bacteriology Technical Series No. 27. Oxford: Blackwell Scientific Publications.

Denyer S.P., Hugo W.B. & Harding V.D. (1985) Synergy in preservative combinations. *Int J Pharm.* **25**, 245–253.

Hodges N.A. & Hanlon G.W. (1991) Detection and measurement of combined biocide action. In *Mechanisms of Action of Chemical Biocides* (Eds S.P. Denyer & W.B. Hugo), Society for Applied Bacteriology Technical Series No. 27, pp. 297–310. Oxford: Blackwell Scientific Publications.

Kelsey J.C. & Maurer I.M. (1974) An improved Kelsey–Sykes test for disinfectants. *Pharm J.* **207**, 528–530.

Leak R.E. & Leech R. (1988) Challenge tests and their predictive stability. In *Microbial Quality Assurance in Pharmaceuticals, Cosmetics and Toiletries* (Eds S.F. Bloomfield, R. Baird, R.E. Leak & R. Leach), pp. 129–146. Chichester: Ellis Horwood.

Orth D.S. (1990) Preservative evaluation and testing: the linear regression method. In *Guide to Microbiological Control in Pharmaceuticals* (Eds S.P. Denyer & R.M. Baird), pp. 304–312. Chichester: Ellis Horwood.

Resnick L., Veren K., Salahuddin S.Z., Tondreau S. & Markham P.D. (1986) Stability and inactivation of HTLV-III/LAV under clinical and laboratory environments. *J Am Med Assoc.* **255**, 1887–1891.

Reybrouck G. (1992) The evaluation of the antimicrobial activity of disinfectants. In *Principles and Practice of Disinfection, Preservation and Sterilization* (Eds A.D. Russell, W.B. Hugo & G.A.J. Ayliffe), 2nd edn, pp. 114–133. Oxford: Blackwell Scientific Publications.

Russell A.D. (1981) Neutralization procedures in the evaluation of bactericidal activity. In *Disinfectants: Their Use and Evaluation of Effectiveness* (Eds C.H. Collins, M.C. Allwood, S.F. Bloomfield & A. Fox), pp. 45–59. London: Academic Press.

Russell A.D. (1982) *The Destruction of Bacterial Spores*. London: Academic Press.

Russell A.D., Hugo W.B. & Ayliffe G.A.J. (Eds) (1992) *Principles and Practice of Disinfection, Preservation and Sterilization*, 2nd edn. Oxford: Blackwell Scientific Publications.

Stannard C.J., Petitt S.B. & Skinner F.A. (Eds) (1989) *Rapid Microbiological Methods for Foods, Beverages and Pharmaceuticals*. Society for Applied Bacteriology Technical Series No. 25. Oxford: Blackwell Scientific Publications.

Tyler R. & Ayliffe G.A.J. (1987) A surface test for virucidal activity of disinfectants: preliminary study with herpes virus. *J Hosp Infect.* **9**, 22–29.

13 Mode of action of non-antibiotic antibacterial agents

1   Cell wall

2   Cytoplasmic membrane
2.1   Action on membrane potentials
2.2   Action on membrane enzymes
2.2.1 Electron transport chain
2.2.2 Adenosine triphosphatase
2.2.3 Enzymes with thiol groups
2.3   Action on general membrane
       permeability

3   Cytoplasm
3.1   General coagulation

3.2   Ribosomes
3.3   Nucleic acids
3.4   Thiol groups
3.5   Amino groups

4   **Highly reactive compounds:**
       multitarget reactors

5   **Conclusions**

6   **Further reading**

This group of drugs has often been classified as non-specific protoplasmic poisons and indeed such views are still expressed today. Such a broad generalization is, however, very far from the true position.

It is convenient to consider the modes of action in terms of the drugs' targets within the bacterial cell and in the following pages various examples will be given. The targets to be considered are the cell wall, the cytoplasmic membrane and the cytoplasm. Much more detailed treatments of the subject will be found in the references at the end of this chapter. Experimental methods for determining the mode of action of an antimicrobial substance have recently been compiled (Denyer & Hugo, 1991).

1   Cell wall

This structure is the traditional target for a group of antibiotics which include the penicillins (Chapter 9), but a little-noticed report which appeared in 1948 showed that low concentrations of disinfectant substances caused cell wall lysis such that a normally turbid suspension of bacteria became clear. It was thought that these low concentrations of disinfectant cause enzymes whose normal role is to synthesize the cell wall to reverse their role in some way and effect its disruption or lysis.

In the original report, the disinfectants (at the following percentages: formalin, 0.12; phenol, 0.32; mercuric chloride, 0.0008; sodium hypochlorite, 0.005 and merthiolate, 0.0004) caused lysis of *Escherichia coli*, streptococci and staphylococci.

Glutaraldehyde also owes part of its mode of action to its ability to react with, and provide irreversible cross-linking in, the cell wall. As a result, other cell functions are impaired. This phenomenon is especially found in Gram-positive cells.

288

2   Cytoplasmic membrane

Actions on the cytoplasmic membrane may be divided into three categories.

1   Action on membrane potentials.
2   Action on membrane enzymes.
3   Action on general membrane permeability.

Action on membrane potentials

Recent work has shown that bacteria, in common with chloroplasts and mitochondria, are able, through the membrane-bound electron transport chain aerobically, or the membrane-bound ATP anerobically, to maintain a gradient of electrical potential and pH such that the interior of the bacterial cell is negative and alkaline. Ths potential gradient and the electrical equivalent of the pH difference (1 pH unit $\equiv$ 58 mV at 37°C) give a potential difference across the membrane of 100–180 mV, with the inside negative. The membrane is impermeable to protons, whose extrusion creates the potential described.

These results may be expressed in the form of an equation, thus:

$$\Delta p = \Delta \psi - Z\Delta pH$$

where Δp is the protonmotive force, $\Delta \psi$ the membrane electrical potential and ΔpH the transmembrane pH gradient, i.e. the pH difference between the inside and outside of the cytoplasmic membrane. Z is a factor converting pH units to millivolts so that all the units of the equation are the same, i.e. millivolts. Z is temperature-dependent and at 37°C has a value of 62.

This potential, or protonmotive force as it is also called, in turn drives a number of energy-requiring functions which include the synthesis of ATP, the coupling of oxidative processes to phosphorylation, a metabolic sequence called oxidative phosphorylation and the transport and concentration in the cell of metabolites such as sugars and amino acids. This, in a few simple words, is the basis of the chemiosmotic theory linking metabolism to energy-requiring processes.

Certain chemical substances have been known for many years to uncouple oxidation from phosphorylation and to inhibit active transport, and for this reason they are named uncoupling agents. They are believed to act by rendering the membrane permeable to protons hence short-circuiting the potential gradient or protonmotive force.

Some examples of antibacterial agents which owe at least a part of their activity to this ability are tetrachlorosalicylanilide (TCS), tricarbanilide, trichlorocarbanilide (TCC), pentachlorophenol, di-(5-chloro-2-hydroxyphenyl) sulphide (fentichlor) and 2-phenoxyethanol.

Action of non-antibiotic antibacterial agents

2.2 Action on membrane enzymes

2.2.1 *Electron transport chain*

Hexachlorophane inhibits the electron transport chain in bacteria and thus will inhibit all metabolic activities in aerobic bacteria.

2.2.2 *Adenosine triphosphatase*

Chlorhexidine has been shown to inhibit the membrane ATPase and could thus inhibit anaerobic processes.

2.2.3 *Enzymes with thiol groups*

Mercuric chloride and other mercury-containing antibacterials will inhibit enzymes in the membrane, and for that matter in the cytoplasm, which contain thiol, —SH, groups. A similar action is shown by 2-bromo-2-nitropropan-1,3-diol (bronopol). Under appropriate conditions the toxic action on cell thiol groups may be reversed by addition of an extrinsic thiol compound, for example cysteine or thioglycollic acid (see also Chapters 12 and 23).

2.3 Action on general membrane permeability

This lesion was recognized early as being one effect of many disinfectant substances. The membrane, as well as providing a dynamic link between metabolism and transport, serves to maintain the pool of metabolites within it.

Treatment of bacterial cells with appropriate concentrations of such substances as cetrimide, chlorhexidine, phenol and hexylresorcinol, causes a leakage of a group of characteristic chemical species. The potassium ion, being a small entity, is the first substance to appear when the cytoplasmic membrane is damaged. Amino acids, purines, pyrimidines and pentoses are examples of other substances which will leak from treated cells.

If the action of the drug is not prolonged or exerted in high concentration the damage may be reversible and leakage may only induce bacteriostasis.

3 Cytoplasm

Within the cytoplasm are a number of important subcellular particles which include the ribosome and oxy- and deoxyribonucleic acids. Enzymes other than those in the membrane are also present in the cytoplasm.

Many early studies measured overall enzyme inhibition in bacterial cultures and a search was made for a peculiarly sensitive enzyme which might be identified as a target, interference with which would cause death. No such enzyme has been found.

3.1 General coagulation

High concentrations of disinfectants, for example chlorhexidine, phenol or mercury salts, will coagulate the cytoplasm and in fact it was this kind of reaction which gave rise to the epithet 'general protoplasmic poison', already referred to, providing an uncritical and dismissive definition of the mode of action of disinfectants. There is little doubt, however, that the disinfectants in use in the 1930s had just this effect when applied at high concentrations.

3.2 Ribosomes

These organelles, the sites of protein synthesis, are well-established targets for antibiotic action.

Both hydrogen peroxide and p-chloromercuribenzoate will dissociate the ribosome into its two constituent parts but whether this is a secondary reaction of the two chemicals is difficult to assess. There is no real evidence that the ribosome is a prime target for disinfectant substances.

3.3 Nucleic acids

Acridine dyes used as antiseptics, i.e. proflavine and acriflavine, will react specifically with nucleic acids, by fitting into the double helical structure of this unique molecule. In so doing they interfere with its function and can thereby cause cell death.

3.4 Thiol groups

Mention has been made of thiol groups in the cytoplasmic membrane as targets for certain antibacterial compounds. Thiol groups also occur in the cytoplasm and these groups will also serve as targets.

Bronopol, chlorine, chlorine-releasing agents, hypochlorites and iodine will oxidize thiol groups.

3.5 Amino groups

Formaldehyde, sulphur dioxide and glutaraldehyde react with amino groups. If these groups are essential for metabolic activity, cell death will follow reactions of this nature.

4 Highly reactive compounds: multitarget reactors

There are one or two chemical sterilants in use whose chemical reactivity is so high that they have a very wide spectrum of cell interactions and it is

Table 13.1 Cellular-targets for non-antibiotic antibacterial drugs

Target or reaction attacked	Acridine dyes	Alcohols	Anilides (TCS, TCC)	Bronopol	Chlorhexidine	Copper II salts	Ethylene oxide	Formaldehyde	Glutaraldehyde	Hexachlorophane	Hydrogen peroxide	Hypochlorites, chlorine releasers	Iodine	Mercury II salts, organic mercurials	Phenols	β-propiolactone	QACs	Silver salts	Sulphur dioxide, sulphites
1 Cell wall								+	+			+		+	+				
2 Cytoplasmic membrane															+				
2.1 Action on membrane potentials			+							+									
2.2 Action on membrane enzymes																			
2.2.1 Electron transport chain						+				+									
2.2.2 Adenosine triphosphatase				+															
2.2.3 Enzymes with thiol groups		+					+												
2.3 Action on general membrane permeability		+	+		+				+		+	+	+	+	++		+		+
3 Cytoplasm																			
3.1 General coagulation					+++	+++			++	+++				+++	+++		+++	+++	
3.2 Ribosomes											+			+					
3.3 Nucleic acids	+																		
3.4 Thiol groups				+		+	+	+	+		+	+	+	+		+		+	
3.5 Amino groups							++	++	+			+				++			
4 High reactive compounds: multitarget reactors							+	+								+			+

Crosses, indicating activity, which appear in several rows for a given compound, demonstrate the multiple actions for the compound concerned. This activity is nearly always concentration dependent, and the number of crosses indicates the order of concentration at which the effect is elicited, i.e. +, elicited at low concentrations; +++, elicited at high concentrations.
When a cross appears in only one target row, this is the only known site of action of the drug.

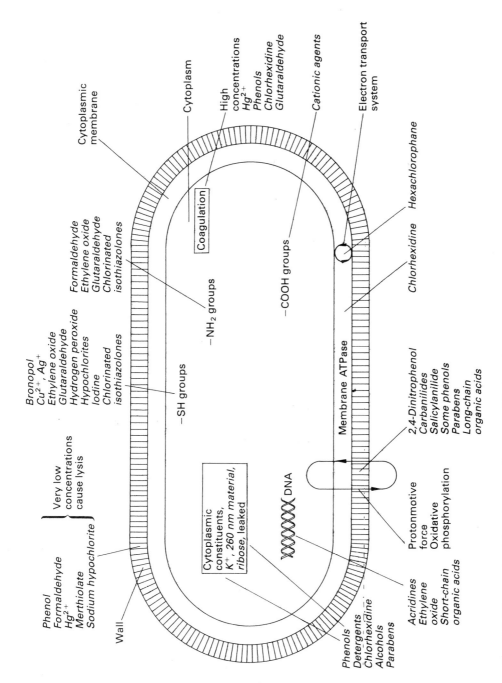

Fig. 13.1 Diagram showing main targets for non-antibiotic antibacterial agents.

difficult to pin-point the fatal reaction. In fact, it is safe to say that there is no single fatal reaction but that death results from the accumulated effects of many reactions; one or two specific reactions of compounds in this category have already been referred to.

β-Propiolactone is one example. It will alkylate amino, imino, hydroxyl and carboxyl groups, all of which occur in proteins, and react also with thiol and disulphide groups responsible for the secondary structure of proteins and the activity of some enzymes.

Another example is ethylene oxide, which has a very similar range of chemical activity.

Sulphur dioxide, sulphites and bisulphites, used as preservatives in fruit juices, ciders and perrys are yet other examples.

5   Conclusions

The above account, Table 13.1 and Fig. 13.1 all indicate the range and complexity of the reactions involved in the action of some non-antibiotic antibacterial agents.

The concentration-dependent dual or even multiple role of many of these substances should be noted. For a more detailed treatment the reader is directed to the references given below.

6   Further reading

Denyer S.P. & Hugo W.B. (Eds) (1991) *Mechanisms of Action of Chemical Biocides: their Study and Exploitation.* Society for Applied Bacteriology Technical Series No. 27. Oxford: Blackwell Scientific Publications.

Hugo W.B. (1967) The mode of action of antiseptics. *J Appl Bacteriol.* **30**, 17–50.

Hugo W.B. (Ed.) (1971) *The Inhibition and Destruction of the Microbial Cell.* London: Academic Press.

Hugo W.B. (1976a) Survival of microbes exposed to chemical stress. In *The Survival of Vegetative Microbes* (Eds T.R.G. Gray & J.R. Postgate), pp. 383–413. 26th Symposium of the Society for General Microbiology. Cambridge: Cambridge University Press.

Hugo W.B. (1976b) The inactivation of vegetative bacteria by chemicals. In *The Inactivation of Vegetative Bacteria* (Eds F.A. Skinner & W.B. Hugo), pp. 1–11. Symposium of the Society of Applied Bacteriology. London: Academic Press.

Hugo W.B. (1980) The mode of action of antiseptics. In *Wirkungmechanisma von Antiseptica* (Eds H. Wigert & W. Weufen), pp. 39–77. Berlin: VEB Verlag.

Hugo W.B. (1992) Disinfection mechanisms. In *Principles and Practice of Disinfection, Preservation and Sterilization*, 2nd edn. (Eds A.D. Russell, W.B. Hugo & G.A.J. Ayliffe), pp. 187–210. Oxford: Blackwell Scientific Publications.

Rusell A.D. & Chopra I. (1990) *Understanding Antibacterial Action and Resistance.* Chichester: Ellis Horwood.

14 Resistance to non-antibiotic antimicrobial agents

1   Introduction

2   Permeability and resistance
2.1 Enterobacteriaceae
2.2 *Pseudomonas aeruginosa*

3   Resistance of other organisms to
     antimicrobial agents
3.1 Gram-negative bacteria
3.2 Gram-positive bacteria
3.3 Bacterial spores

3.4 Mycobacteria
3.5 Fungi

4   Plasmids and resistance to
     antimicrobial agents

5   Intrinsic resistance and disinfection
     policies

6   Further reading

1   Introduction

Table 14.1 illustrates the range of sensitivity of different microorganisms to the non-antibiotic antimicrobial agent chlorhexidine. In general, Gram-negative bacteria are more resistant than Gram-positive bacteria, the difference being due to the more complex envelope structure of Gram-negative organisms (see Figs 1.1 and 1.4 in Chapter 1) which restricts the uptake and penetration of antimicrobial agents into the cells. This 'intrinsic resistance' of Gram-negative bacteria, however, is subject to considerable variation by both genetic and environmental factors. The genetic composition of an organism, which basically determines the structure, composition and properties of the envelope, can be altered by spontaneous mutations, deletions or acquisition of new genes from plasmids, bacteriophages and transposons (see Chapter 10). Thus, it is possible for different strains of a given species to display a wide range of sensitivity to antimicrobials. The minimum inhibitory concentration (MIC) for different isolates of *Ps. aeruginosa* against chlorhexidine can vary from 10 to $500 \mu g \, ml^{-1}$.

This inherent variability in sensitivity has extremely important practical implications for the use of antimicrobials. For example, the population of bacteria making up the normal flora of organisms contaminating the working surfaces, floor, air or water supply in an environment such as a hospital pharmacy will probably contain a very low number of naturally resistant organisms. These might be resistant to the agent used as a disinfectant because they have acquired additional genetic information or lost, by mutation, genes involved in controlling the expression of other genes. In the absence of the antimicrobial agent, the resistant strains would have no competitive advantage over the sensitive strains, and in fact they might grow more slowly and would not predominate in the population. Under the selective pressure introduced by continual use of one kind of disinfectant, resistant strains would predominate as the sensitive strains are eliminated. Eventually the entire population would

295

Table 14.1 Sensitivity of microorganisms to chlorhexidine

Organism	Minimum inhibitory concentration ($\mu g\,ml^{-1}$)
Gram-negative bacteria	
Pseudomonas aeruginosa	10–500
Proteus mirabilis	25–100
Pseudomonas cepacia	5–100
Serratia marcescens	3–50
Salmonella typhimurium	14
Klebsiella aerogenes	1–12
Escherichia coli	1–5
Gram-positive bacteria	
Staphylococcus aureus	1–2
Enterococcus faecalis	1–3
Bacillus subtilis	1–3
Streptococcus mutans	0.1
Mycobacterium tuberculosis	0.7–6
Fungi	
Candida albicans	7–15
Trichophyton mentagrophytes	3
Penicillium notatum	200

be resistant to the disinfectant and a serious contamination hazard would arise. To avoid this situation it is normal practice to adopt a policy in which different antimicrobials are used in rotation.

Environmental factors affect antimicrobial sensitivity by influencing the phenotypic expression of microorganisms. In particular, the composition and properties of microbial envelopes, the main exclusion barriers, are dependent upon the conditions under which the organisms are grown. The nature of the nutrients on which an organism grows is especially important in determining the composition of its envelope. In a natural environment, growth is controlled by the restricted availability of one or more nutrients such as carbon, nitrogen, oxygen, phosphorus, sulphur, or an essential metal ion such as magnesium or potassium. The organism will grow, adapting itself as necessary to the restricted levels of the particular nutrient in short supply, until all of that nutrient is consumed. Growth will then cease. The cells are termed 'nutrient-depleted', e.g. carbon-depleted or nitrogen-depleted, etc. according to which nutrient has become exhausted.

The important consequence of nutrient-depletion is that the cells display properties which are strongly dependent upon the nature of the particular depleted nutrient. Among the properties affected are the envelope composition and hence the sensitivity to antimicrobial agents. It is important to recognize the variation in properties that can result from different nutrient depletions. For example, *Ps. aeruginosa* grown to magnesium depletion is highly resistant to the quaternary ammonium compound (QAC), benzalkonium chloride, whereas the same strain grown under carbon depletion is extremely sensitive. Other parameters such as growth rate and temperature also influence sensitivity to anti-microbials by inducing phenotypic changes in the organism.

One of the major implications of this phenotypic variability in bio-logical properties brought about by the growth conditions is the need for standardization of test conditions for antimicrobial agents. A preservative or disinfectant will appear to perform quite differently depending on the growth conditions used to prepare the test organism. Phenotypic variation in drug resistance is not confined to Gram-negative bacteria; Gram-positive bacteria and fungi show an equal versatility in phenotypic expression.

Another aspect of microbial growth which has an important bearing upon the sensitivity to antimicrobial agents is the formation of biofilms on colonized surfaces. It is now increasingly realized that organisms colonize surfaces by forming films, perhaps up to 100 cells thick which are sur-rounded by polysaccharide material, termed the glycocalyx. Whether in natural environments, i.e. on walls, floors or the surfaces of containers and utensils, on catheters or other devices inserted into the body, or upon host tissues the same features of biofilm formation apply. The most familiar biofilm is the formation of dental plaque, a complex consortium of different organisms and host material adhering to the teeth. Any surface in nature can be colonized by microbes; organisms are continually released from this reservoir to scout for more surfaces to colonize. These free-living or 'planktonic' organisms are generally very susceptible to antimicrobials since they are not protected by the enveloping biofilm. It is the 'sessile' organisms located in the matrix of the biofilm which are extremely difficult to eradicate. The problem is partly a physical one of poor penetration of antimicrobials into the biofilm, and partly a phenotypic problem, since the biofilm organisms are probably very slow-growing. No disinfection evaluation methods take into account the recalcitrant nature of biofilm organisms, yet these are the most important organisms to eradicate since they form a focus for recontamination of surfaces.

2   Permeability and resistance

2.1   Enterobacteriaceae

A great deal of our current understanding of the structure and function of the outer membrane of Gram-negative bacteria has come from studies with *E. coli* and *Sal. typhimurium*. The permeability barrier function of the outer membrane can be demonstrated by treatment of *E. coli* cells with ethylenediamine tetra-acetic acid (EDTA), which greatly enhances their permeability and sensitivity towards antimicrobial agents. By bind-ing metal ions such as magnesium, which is essential for the stability of the outer membrane, EDTA releases 30–50% of the lipopolysaccharide (LPS) from the outer membrane together with some phospholipid and protein. The permeability barrier is effectively removed and the cells, which retain their viability, then become sensitive to large hydrophobic antimicrobials like actinomycin (Chapter 10) against which they are normally resistant. More complete removal of the outer membrane and peptidoglycan with EDTA and lysozyme (a muramidase enzyme which degrades peptidoglycan) produces spheroplasts in Gram-negative bacteria (see also Chapter 1). These osmotically fragile, but viable, cells are equivalent to protoplasts of Gram-positive bacteria, which are cells where the wall has been completely removed with lysozyme. Both spheroplasts and protoplasts are equally sensitive to lysis by membrane-active agents such as QACs, phenols and chlorhexidine. This demonstrates that the difference in sensitivities of whole cells to these agents is not due to a difference in sensitivity of the target cytoplasmic membrane but in the different permeability properties of the overlying wall or envelope structures.

Permeability properties of the outer membrane are influenced by the nature of the LPS (Figs 1.4 and 1.5 in Chapter 1). This has been shown in studies by Nikaido on a group of *Sal. typhimurium* mutants which differ in the length of the polysaccharide chain attached to lipid A. Smooth strains, i.e. those with complete O-polysaccharide chains, are sensitive to hydrophilic antimicrobials but remain resistant to hydrophobic molecules such as fatty acids and phenols. Rough strains, those with no O-antigen but with a short oligosaccharide core region attached to lipid A, show some sensitivity to hydrophobic agents. Deep rough strains which lack the core oligosaccharide are more sensitive to hydrophobic molecules, and heptose-less strains which contain only lipid A and 3-keto-2-deoxyoctonate in their LPS are extremely sensitive to these agents. The entire series of *Sal. typhimurium* mutants are equally sensitive to hydrophilic agents, which are assumed to enter the cells via the aqueous porin channels in the outer membrane. Their passage to the porins through the polysaccharide region of the LPS protruding from the cells is not impeded. By contrast, hydrophobic agents cannot penetrate either the polysaccharide 'coat' of LPS polysaccharide chains or the aqueous porin channels. The hydrated nature of amino acid residues lining the porin channels presents an energetically unfavourable barrier to the passage of hydrophobic molecules. In the rough strains the reduction in the amount of polysaccharide on the cell surface allows hydrophobic molecules to approach the surface of the outer membrane and cross the outer membrane lipid bilayer by passive diffusion. This process is greatly facilitated in the deep rough and heptose-less strains which have phospholipid molecules on the outer face of their outer membranes as well as on the inner face. The exposed areas of phospholipids favour the absorption and penetration of the hydrophobic agents.

This picture of two pathways, one hydrophilic (porin-mediated) and one hydrophobic (diffusion across the membrane) for penetration of antimicrobials across the outer membrane holds for all Gram-negative bacteria. It is especially important for the Enterobacteriaceae which survive the antimicrobial action of hydrophobic bile salts and fatty acids in the gut by the combined effects of the penetration barrier of their smooth LPS and the small size of their porin channels (which restricts passage of hydrophilic molecules to those of molecular weight less than 650). By contrast, an organism like *Neisseria gonorrhoeae*, which does not produce an O-antigen polysaccharide on its LPS and is naturally rough, is very sensitive to hydrophobic molecules. Natural fatty acids help to defend the body against these organisms.

Members of the Enterobacteriaceae are reasonably sensitive to membrane-active antimicrobials with strong surface-active properties, such as the QACs (Chapter 11). These, and other membrane-active agents such as chlorhexidine (Chapter 11), probably mediate their own uptake by damaging the outer membrane, removing segments to allow access to the periplasm and vulnerable cytoplasmic membrane. Their effect can be seen quite dramatically under the electron microscope. Small bulges or blebs appear on the outer face of the outer membrane. The blebs increase in size and are released from the cells as vesicles containing LPS, protein and phospholipid. The outer membrane has a limited capacity to reassemble itself; this it does with phospholipids spontaneously re-forming into a bilayer. If the amount of outer membrane material released is too great to be compensated for by phospholipid, the cells lose their protective barrier, and the agents penetrate to the cytoplasmic membrane and cause irreversible damage.

Studies with porin-deficient mutants of many Gram-negative species have confirmed that detergents do not use the porin channels to gain access to the cytoplasmic membrane. Porin-deficient strains in general show no difference in sensitivity to detergents compared with their parent strains, even though the permeability of their outer membrane to small hydrophilic molecules is reduced up to 100-fold. Other mutations affecting the stability of the outer membrane, such as loss of the lipoprotein which anchors it to the peptidoglycan, are associated with extreme sensitivity to membrane-active agents. This observation is in accord with the concept that these agents damage the outer membrane and thereby gain access to the cytoplasmic membrane. Some mutants of *E. coli* are highly permeable and sensitive to a wide range of antimicrobial agents, but have no major defect in envelope composition. The explanation presumably lies in the way the individual components are organized in the envelope. Since components are not covalently linked together, ionic interactions mediated by divalent metal ions play an important part in maintaining the integrity of the outer membrane. For this reason EDTA is particularly effective in destabilizing the outer membrane and making it permeable to agents. EDTA potentiates the action of many antimicrobials and for this purpose is a valuable additive to preservatives, especially QACs. One disinfectant formulation that has been available commercially has EDTA and the phenolic agent chloroxylenol (Chapter 11) as its active constituents.

Phenotypic modifications in envelope composition and drug sensitivity of Enterobacteriaceae have been demonstrated with nutrient-depleted batch-grown cells. For example, carbon-depleted *E. coli* are significantly more sensitive to chlorophenols and 2-phenoxyethanol than either phosphate- or magnesium-depleted cultures. The presence of the antibiotic resistance plasmid RP1 further enhances the sensitivity of the carbon-depleted cells but has no effect on the other nutrient-depleted cells. These changes in sensitivity are accompanied by some complex alterations in levels of individual phospholipids and polyhydroxybutyrate present in cells with different nutrient depletions. From complex changes such as these it is difficult to determine which factors contribute to the changes in drug resistance. Phenotypic variations in LPS content would also contribute to drug sensitivity, so that the overall effect depends upon a complex interplay of factors, each affecting the permeability of the outer membrane in a different manner.

2.2   *Pseudomonas aeruginosa*

This organism is notorious for its ability to survive in the environment, particularly in moist conditions. It is a dangerous contaminant of medicines, surgical equipment, clothing and dressings, with the ability to cause serious infections in immunocompromised patients. The intrinsic resistance of Gram-negative bacteria is especially apparent with *Ps. aeruginosa*; many disinfectants and preservatives possess insufficient activity against it to be of any use. Added to the problem of natural resistance to antimicrobials is the organism's extensive repertoire of phenotypic variation.

The basis of the greater resistance of *Ps. aeruginosa* compared with other Gram-negative bacteria (Table 14.1) is not at all clear. The answer presumably lies in the properties of the envelope because when this is removed, the resulting spheroplasts are just as sensitive as those of other organisms. The outer membrane is not significantly different from that of other organisms in terms of overall composition. The same components (LPS, proteins, phospholipid, peptidoglycan) are present. One difference is the number of phosphate groups present in the lipid A region of the LPS. This is significantly higher in *Ps. aeruginosa* than in members of the Enterobacteriaceae and might account for the unusual sensitivity of the organism to EDTA. The high phosphate content means that the outer membrane is unusually dependent upon divalent metal ions for stability; their removal by EDTA therefore has a dramatic effect upon cell integrity. Magnesium-depleted cells of *Ps. aeruginosa* are extremely resistant to EDTA. Presumably the lower magnesium content of the cell envelope reflects a decreased phosphorylation of lipid A. Other effects follow from magnesium depletion, including complex changes in lipid composition and increased production of an outer membrane protein known as H1, which is believed to replace magnesium ions in binding together LPS molecules on the cell surface.

3   Resistance of other organisms to antimicrobial agents

Whilst most information on envelope structure and permeability has been gained from studies on organisms such as *E. coli* and *Ps. aeruginosa*, other pharmaceutically relevant organisms in the environment have received relatively little attention.

3.1   Gram-negative bacteria

Alterations in drug sensitivity by changes in growth conditions are not confined to EDTA and *Ps. aeruginosa*. QACs, chlorhexidine, phenols and alcohols show major variations in their activity towards *Ps. aeruginosa*, *Ps. cepacia*, *E. coli*, *Serratia marcescens* and *Pr. mirabilis* depending on how the organisms are grown. Growth rate as well as specific nutrient limitation is important in determining sensitivity. In general, slow-growing cells are more resistant to antimicrobials than fast-growing cells. This inherent variability in sensitivity must be considered when choosing organisms and growth conditions for assessing the performance of antimicrobial compounds as disinfectants or preservatives. Organisms such as *Ps. cepacia* and *Ser. marcescens* have an outstanding capacity to withstand the action of antimicrobials. There are reports of *Ps. cepacia* surviving in concentrated solutions of QACs for 14 years. *Ser. marcescens* has been found to survive in 2% chlorhexidine for over 2 years. These organisms have low virulence but can pose a significant health threat to susceptible patients with low natural defences. Survival of these organisms in hospital environments is recognized as a major problem which must be countered by a carefully planned and enforced policy for use of disinfectants. One important feature of the survival of organisms in the environment is their ability to adhere to surfaces. Adherent biofilms are particularly difficult to remove from the insides of containers. Since the cells are often covered with a protective layer of polysaccharide, disinfectants are inefficient at reaching them and mechanical removal is necessary.

3.2   Gram-positive bacteria

In general Gram-positive bacteria do not pose a major problem of resistance to antimicrobials. However, they do have the potential for phenotypic variation with growth conditions. Increases in resistance can be detected and have been explained in terms of altered lipid composition. For example, the resistance of *Staph. aureus* to phenols and QACs can be varied by manipulation of the cellular lipid content. Lipid-rich cells are more resistant and lipid-depleted cells are more sensitive than 'normal' cells. Lipid is not a normal component of the cell wall of *Staph. aureus* but electrophoretic mobility measurements of whole cells indicate that in the lipid-rich cells the extra lipid is located at or near the surface. This extra lipid is assumed to block the penetration of the agents to the cytoplasmic membrane. These observations have serious practical implications. Multiresistant staphylococci are becoming a major problem as hospital-acquired pathogens. In addition to exhibiting resistance to a wide range of antibiotics, these strains are also resistant to commonly used disinfectants. There is growing concern that if multiresistance to antibiotics and disinfectants is genetically linked continued use of disinfectants will select for multiresistance. These developments should encourage the search for antimicrobials against which similar resistance does not occur, and to the changing of disinfectant types in a hospital environment from time to time instead of using one type, however good initially, all the time.

3.3 Bacterial spores

Spores are several thousand times more resistant to chemical agents than vegetative cells. Chemicals normally used as preservatives and disinfectants, i.e. mercurials, QACs, chlorhexidine, phenols and alcohols, have no useful sporicidal activity, although some inhibit germination (see Chapter 1) and all will kill vegetative cells after germination. Reactive chemicals such as ethylene oxide, β-propiolactone, formaldehyde, glutaraldehyde, hydrogen peroxide and the halogens are required to kill spores (see Chapter 11). Even with these compounds the rate of killing is slow, periods from 30 minutes up to several hours being required to effect sterilization. The spore coats are important in resistance, acting as a barrier to penetration of the agents to the protoplast, and perhaps neutralizing agents such as chlorine and ethylene oxide. Like vegetative cells, spores display considerable genetic and phenotypic variation in resistance. Heat resistance and chemical resistance are not necessarily linked but both properties are clearly dependent on the nutritional conditions of the medium used to grow the cells prior to sporulation.

3.4 Mycobacteria

Mycobacteria are very resistant to disinfectants; if a chemical agent must be used to sterilize items contaminated with mycobacteria, then a phenolic preparation should be chosen. Heat sterilization is recommended wherever possible. The basis of the unusually high resistance to chemical agents lies in the structure and composition of the mycobacterial cell wall which is rich in wax-like lipids (see below) arranged in a series of hydrophobic layers. This property is also responsible for the acid-fast nature of mycobacteria. They fail to take up normal bacteriological stains, requiring several minutes steaming with carbolfuchsin or 18 hours of exposure to the dye at room temperature. Once stained, the cells are difficult to decolorize and resist the action of alcohol and dilute acids; this is the origin of the term 'acid-fast'. The mycobacterial wall contains a layer of peptidoglycan covered by a complex peptidoglycolipid layer. This material contains peptidoglycan linked to a polysaccharide containing a number of lipids in which wax-like mycolic acid residues are a particular feature. On removal of the peptidoglycolipid complex with alkaline ethanol, the cells lose their acid-fast properties and it is likely that this material is also responsible for the resistance to disinfectants.

3.5 Fungi

Surprisingly little is known about the resistance of yeasts, fungi and fungal spores to disinfectants and preservatives. They are a major source of potential contamination in pharmaceutical product preparation and aseptic processing since they abound in the environment. Because vegetative cells and spores of fungi are generally not more difficult to kill than their bacterial counterparts, it is usually assumed that antimicrobials which control bacterial contamination also control fungi. Thus in disinfectant evaluation procedures such as the Kelsey–Sykes capacity dilution test (see Chapter 12) recommended organisms are *Ps. aeruginosa*, *Pr. vulgaris*, *E. coli* and *Staph. aureus*. Fungi are probably a greater potential hazard as contaminants of medicines, therefore both the *British Pharmacopoeia* and *United States Pharmacopeia* challenge tests (Chapter 12) for preservatives specify *Aspergillus niger* and *Candida albicans* as test organisms in addition to bacteria such as *Ps. aeruginosa* and *Staph. aureus*. Like bacteria, fungi and yeasts can display genotypic and phenotypic variations in properties which include resistance to antimicrobials. This variability should therefore be considered in choosing growth conditions for yeasts and fungi to be used in challenge tests so that cells of reproducible sensitivity are used.

4 Plasmids and resistance to antimicrobial agents

Plasmids (Chapter 10) encoding for resistance to mercuric and organomercurial compounds have been described in *Staph. aureus*, *E. coli* and *Ps. aeruginosa*. Two kinds of plasmids are recognized: (i) narrow-spectrum plasmids which promote volatilization of mercury from Hg^{2+}; and (ii) broad-spectrum plasmids which promote volatilization from both Hg^{2+} and organomercurial compounds. Both plasmids encode for a mercuric (mercury II) reductase; this reduces Hg^{2+} to metallic mercury which can volatilize from the growth medium. Broad-spectrum plasmids also encode for one or more hydrolases which break carbon–mercury bonds in organomercurials to form Hg^{2+}. Emergence of these plasmids has important implications for the use of organomercurial compounds as preservatives in pharmaceuticals. A puzzling feature, also of some concern, is the occurrence of mercury resistance on the same plasmids as antibiotic resistance genes.

5 Intrinsic resistance and disinfection policies

Studies on antibiotic resistance patterns of Gram-negative bacteria in urinary tract infections have indicated a correlation between multiple antibiotic resistance and resistance to membrane-active disinfectants. It is not clear how these resistance patterns are linked; possibly a common property relating to intrinsic resistance to membrane-active agents is carried on plasmids encoding multiple antibiotic resistance. Of considerable concern is the implication that a disinfectant policy that relies heavily on membrane-active agents will encourage the emergence of multiple antibiotic resistance. A general policy of ringing the changes on in-house disinfectants has already been alluded to (section 3.2).

6   Further reading

Brown M.R.W. (Ed.) (1975) *Resistance of Pseudomonas aeruginosa*. London: John Wiley.

Brown M.R.W. & Williams P. (1985) Influence of substrate limitation and growth phase on sensitivity to antimicrobial agents. *J Antimicrob Chemother*. **15** (Suppl. A), 7–14.

Brown M.R.W., Collier P.J. & Gilbert P. (1990) Influence of growth rate on susceptibility to antimicrobial agents: modification of the cell envelope and batch and continuous culture studies. *Antimicrob Agents Chemother*. **34**, 1623–1628.

Chopra I. (1982) Plasmids and bacterial resistance. In *Principles and Practice of Disinfection, Preservation and Sterilisation* (Eds A.D. Russell, W.B. Hugo & G.A.J. Ayliffe), pp. 199–206. Oxford: Blackwell Scientific Publications.

Costerton J.W., Cheng K.J., Geesey K.G., Ladd P.I., Nickel J.C., Dasgupta M. & Marrie T.J. (1987) Bacterial biofilms in nature and disease. *Annu Rev Biochem*. **41**, 435–464.

Gilbert P., Collier P.J. & Brown M.R.W. (1990) Influence of growth rate on susceptibility to antimicrobial agents: biofilms, cell cycle, dormancy, and stringent response. *Antimicrob Agents Chemother*. **34**, 1865–1868.

Hancock R.E.W. (1984) Alterations in membrane permeability. *Annu Rev Microbiol*. **38**, 237–264.

Hugo W.B. (1976) Survival of microbes exposed to chemical stress. In *The Survival of Vegetative Microbes* (Eds T.R.G. Gray & J.R. Postgate), pp. 383–413. 26th Symposium of the Society for General Microbiology. Cambridge: Cambridge University Press.

Hugo W.B., Pallent L.J., Grant D.J.W., Denyer S.P. & Davies A. (1986) Factors contributing to the survival of a strain of *Pseudomonas cepacia* in chlorhexidine solutions. *Letters Appl Micobiol*. **2**, 37–42.

Lowbury E.J.L., Ayliffe G.A.J., Geddes A.M. & Williams J.D. (Eds) (1975) *Control of Hospital Infection, A Practical Guide*. London: Chapman & Hall.

Nikaido H. (1989) Outer membrane barrier as a mechanism of antimicrobial resistance. *Antimicrob Agents Chemother*. **33**, 1831–1836.

Nikaido H. & Vaara M. (1985) Molecular basis of bacterial outer membrane permeability. *Microbiol Rev*. **49**, 1–32.

Russell A.D. (1982) *The Destruction of Bacterial Spores*. London: Academic Press.

Russell A.D. (1992) Antifungal activity of biocides. In *Principles and Practice of Disinfection, Preservation and Sterilization* (Eds A.D. Russell, W.B. Hugo & G.A.J. Ayliffe), 2nd edn, pp. 134–149. Oxford: Blackwell Scientific Publications.

Russell A.D. & Chopra I. (1990) *Understanding Antibacterial Action and Resistance*. Chichester: Ellis Horwood.

Stickler D.J. & Thomas B. (1982) Intrinsic resistance to non-antibiotic antibacterial agents. In *Principles and Practice of Disinfection, Preservation and Sterilisation* (Eds A.D. Russell, W.B. Hugo & G.A.J. Ayliffe), pp. 187–198. Oxford: Blackwell Scientific Publications.

Waites W.M. (1982) Resistance of bacterial spores. In *Principles and Practice of Disinfection, Preservation and Sterilisation* (Eds A.D. Russell, W.B. Hugo & G.A.J. Ayliffe), pp. 207–220. Oxford: Blackwell Scientific Publications.

15 Fundamentals of immunology

1	**Introduction**	4.5	Complement
1.1	Historical aspects of immunology	4.6	Cell-mediated immunity (CMI)
1.2	Definitions	4.6.1	Helper T cells (TH cells)
		4.6.2	Suppressor T cells (Ts cells)
2	**Non-specific defence mechanisms**	4.6.3	Cytotoxic T cells (Tc cells)
	(innate immune system)	4.7	Immunoregulation
2.1	Skin and mucous membranes	4.8	Natural killer (NK) cells
2.2	Phagocytosis	4.9	Immunological tolerance
2.2.1	Role of phagocytosis	4.10	Autoimmunity
2.3	The complement system and other		
	soluble factors	**5**	**Hypersensitivity**
2.4	Inflammation	5.1	Type I (anaphylactic) reactions
2.5	Host damage	5.2	Type II (cytolytic or cytotoxic)
2.5.1	Exotoxins		reactions
2.5.2	Endotoxins	5.3	Type III (complex-mediated)
			reactions
3	**Specific defence mechanisms**	5.4	Type IV (delayed
	(adaptive immune system)		hypersensitivity) reactions
3.1	Antigenic structure of the	5.5	Type V (stimulatory
	microbial cell		hypersensitivity) reactions
4	**Cells involved in immunity**	**6**	**Tissue transplantation**
4.1	Humoral immunity	6.1	Immune response to tumours
4.2	Monoclonal antibodies		
4.2.1	Uses of monoclonal antibodies	**7**	**Immunity**
4.3	Immunoglobulin classes	7.1	Natural immunity
4.3.1	Immunoglobulin M (IgM)	7.1.1	Species immunity
4.3.2	Immunoglobulin G (IgG)	7.1.2	Individual immunity
4.3.3	Immunoglobulin A (IgA)	7.2	Acquired immunity
4.3.4	Immunoglobulin D (IgD)	7.2.1	Active acquired immunity
4.3.5	Immunoglobulin E (IgE)	7.2.2	Passive acquired immunity
4.4	Humoral antigen–antibody		
	reactions	**8**	**Further reading**

1 Introduction

The science of immunology is one of the most rapidly expanding sciences and represents a vast area of knowledge and research; thus, in a short chapter it is impossible to deal in depth with its theory and application and a list of further reading is given at the end of the chapter.

1.1 Historical aspects of immunology

From almost the first written observations by man it was recognized that persons who had contracted and recovered from certain diseases were not susceptible (i.e. were immune) to further attacks. Thucydides, over 2500 years ago, described in detail an epidemic in Athens (which could have been typhus or plague) and noted that sufferers were 'touched by the pitying care of those who had recovered because they were themselves free of apprehension, for no-one was ever attacked a second time or with a fatal result'.

Many attempts were made to induce this immune state. In ancient times the process of variolation (the inoculation of live organisms of smallpox obtained from diseased pustules from patients who were re-covering from the disease) was practised extensively in India and China. The success rate was very variable and often depended on the skill of the variolator. The results were sometimes disastrous for the recipient. The father of immunology was Edward Jenner, an English country doctor who lived from 1749 to 1823. He had observed on his rounds the similarity between the pustules of smallpox and those of cowpox, a disease that affected cows' udders. He also observed that milkmaids who had con-tracted cowpox by handling the diseased udders were immune to small-pox. Deliberate inoculation of a young boy with cowpox and a later subsequent challenge, after the boy had recovered, with the contents of a pustule taken from a person who was suffering from smallpox failed to induce the disease and subsequent rechallenges also failed. The process of vaccination (Latin, *vacca*, cow) was adopted as a preventative measure against smallpox, even though the mechanism by which this immunity was induced was not understood.

In 1801, Jenner prophesied the eradication of smallpox by the practice of vaccination. In 1967 the disease infected 10 million people. The World Health Organization (WHO) initiated a programme of confinement and vaccination with the object of eradicating the disease. In Somalia in 1977 the last case of naturally acquired smallpox occurred, and in 1979 the WHO announced the total eradication of smallpox, thus fulfilling Jenner's prophecy.

The science of immunology not only encompasses the body's immune responses to bacteria and viruses but is extensively involved in: tumour recognition and subsequent rejection; the rejection of transplanted organs and tissues; the elimination of parasites from the body; allergies; and autoimmunity (the condition when the body mounts a reaction against its own tissues).

1.2 Definitions

Disease in man and animals may be caused by a variety of micro-organisms, the three most important groups being bacteria, rickettsia and viruses.

An organism which has the ability to cause disease is termed a *pathogen*. The term *virulence* is used to indicate the degree of patho-genicity of a given strain of microorganism. Reduction in the normal virulence of a pathogen is termed *attenuation*; this can eventually result in the organism losing its virulence completely and it is then termed *avirulent*. Conversely, any increase in virulence is termed *exaltation*.

The body possesses an efficient natural defence mechanism which restricts microorganisms to areas where they can be tolerated. A breach of this mechanism, allowing them to reach tissues which are normally inaccessible, results in an infection. Invasion and multiplication of the organism in the infected host may result in a pathological condition, the clinical entity of disease.

2 Non-specific defence mechanisms (innate immune system)

The body possesses a number of non-specific antimicrobial systems which are operative at all times against potentially pathogenic microorganisms. Prior contact with the infectious agent has no intrinsic effect on these systems.

2.1 Skin and mucous membranes

The intact skin is virtually impregnable to microorganisms and only when damage occurs can invasion take place. Furthermore, many microorganisms fail to survive on the skin surface for any length of time due to the inhibitory effects of fatty acids and lactic acid in sweat and sebaceous secretions. Mucus, secreted by the membranes lining the inner surfaces of the body, acts as a protective barrier by trapping microorganisms and other foreign particles and these are subsequently removed by ciliary action linked, in the case of the respiratory tract, with coughing and sneezing.

Many body secretions contain substances that exert a bactericidal action, for example the enzyme lysozyme which is found in tears, nasal secretions and saliva; hydrochloric acid in the stomach which results in a low pH; and basic polypeptides such as spermine which are found in semen.

The body possesses a normal bacterial flora which, by competing for essential nutrients or by the production of inhibitory substances such as monolactams or colicins, suppresses the growth of many potential pathogens.

2.2 Phagocytosis

Metchnikoff (1883) recognized the role of cell types (phagocytes) which were responsible for the engulfment and digestion of microorganisms. They are a major line of defence against microbes that breach the initial barriers described above. Two types of phagocytic cells are found in the blood, both of which are derived from the totipotent bone marrow stem cell.

1 The monocytes, which constitute about 5% of the total blood leucocytes. They migrate into the tissues and mature into macrophages (see below).

2 The neutrophils (also called polymorphonuclear leucocytes, PMNs),

which are the professional phagocytes of the body. They constitute >70% of the total leucocyte population, remaining in the circulatory system for less than 48 hours before migrating into the tissues, in response to a suitable stimulus, where they phagocytose material. They possess receptors for Fc and activated C3 which enhance their phagocytic ability (see later in chapter).

Another group of phagocytic cells are the macrophages. These are large, long-lived cells found in most tissues and lining serous cavities and the lung. Other macrophages recirculate through the secondary lymphoid organs, spleen and lymph nodes where they are advantageously placed to filter out foreign material. The total body pool of macrophages constitutes the so-called reticuloendothelial system (RES). Macrophages are also involved with the presentation of antigen to the appropriate lymphocyte population (see later).

2.2.1   *Role of phagocytosis*

The microorganism initially adheres to the surface of the phagocytic cell, and this is then followed by engulfment of the particle so that it lies within a vacuole (phagosome) within the cell. Lysosomal granules within the phagocyte fuse with the vacuole to form a phagolysosome. These granules contain a variety of bactericidal components which destroy the ingested microorganism by systems that are oxygen dependent or oxygen independent.

When a microorganism breaches the initial barriers and enters the body tissues, the phagocytes form a formidable defence barrier. Phagocytosis is greatly enhanced by a family of proteins called *complement*.

2.3   The complement system and other soluble factors

Complement comprises a group of heat-labile serum proteins which, when activated, are associated with the destruction of bacteria in the body in a variety of ways. It is present in low concentrations in serum but, as its action is linked intimately with a second (specific) set of defence mechanisms, its composition and role will be dealt with later in the chapter.

Proteins produced by virally infected cells have been shown to interfere with viral replication. They also activate leucocytes that can recognize these infected cells and subsequently kill them. These leucocytes are known as natural killer cells and the proteins are termed *interferons* (see also Chapters 3, 5 and 24).

The serum concentration of a number of proteins increases dramatically during infection. Their levels can increase by up to 100-fold compared with normal levels. They are known collectively as acute phase proteins and certain of them have been shown to enhance phagocytosis in conjunction with complement.

Inflammation

One early symptom of injury to tissue due to a microbial infection is inflammation. This begins with the dilatation of local arterioles and capillaries which increases the blood flow to the area and causes characteristic reddening. Fluid accumulates in the area of the injury due to an increase in the permeability of the capillary walls and this leads to localized oedema, which creates a pressure on nerve endings resulting in pain. This early oedema may actually promote bacterial growth. Fibrin is deposited which tends to limit the spread of the microorganisms. Blood phagocytes adhere to the inside of the capillary walls and penetrate through into the surrounding tissue. They are attracted to the focus of the infection by chemotactic substances in the inflammatory exudate originating from complement.

Inflammation is a non-specific reaction which can be induced by a variety of agents apart from microorganisms. Lymphokines and derivatives of arachidonic acid, including prostaglandins, leukotrienes and thromboxanes are probable mediators of the inflammatory response. The release of vasoactive amines such as histamine and serotonin (5-hydroxytryptamine) from activated or damaged cells also contribute to inflammation.

Fever is the most common manifestation. The thermoregulatory centre in the hypothalamus regulates body temperature and this can be affected by endotoxins (heat-stable lipopolysaccharides) of Gram-negative bacteria and also by a monokine secreted by monocytes and macrophages called interleukin-1 (IL-1) which is also termed endogenous pyrogen. Antibody production and T cell proliferation have been shown to be enhanced at elevated body temperatures and thus are beneficial effects of fever.

Host damage

Microorganisms that escape phagocytosis in a local lesion may now be transported to the regional lymph nodes via the lymphatic vessels. If massive invasion occurs with which the resident macrophages are unable to cope, microorganisms may be transported through the thoracic duct into the bloodstream. The appearance of viable microorganisms in the bloodstream is termed bacteraemia and is indicative of an invasive infection and failure of the primary defences.

Pathogenic organisms possess certain properties which enable them to overcome these primary defences. They produce metabolic substances, often enzymic in nature, which facilitate the invasion of the body. The following are examples of these.

1 Hyaluronidase and streptokinase are produced by the haemolytic streptococci and enable the organism to spread rapidly through the tissue. Hyaluronidase dissolves hyaluronic acid (intercellular cement), whereas streptokinase (Chapter 25) dissolves blood clots.

2 Coagulase is produced by many strains of staphylococci and causes the coagulation of plasma surrounding the organism. This can act as a barrier protecting the organism against phagocytosis. The presence of a capsule outside the cell wall serves a similar function. The production of coagulase (Chapter 1) is used as an indication of the pathogenicity of the strain.

3 Lecithinase is produced by *Clostridium perfringens*. This is a calcium-dependent lecithinase whose activity depends on the ability to split lecithin. Since lecithin is present in the membrane of many different kinds of cells, damage can occur throughout the body. Lecithinase causes the hydrolysis of erythrocytes and the necrosis of other tissue cells.

4 Collagenase is also produced by *Cl. perfringens* and this degrades collagen, which is the major protein of fibrous tissue. Its destruction promotes the spread of infection in tissues.

5 Leucocidins kill leucocytes and are produced by many strains of streptococci, most strains of *Staphylococcus aureus* and likewise most strains of pathogenic Gram-negative rods, isolated from sites of infection.

Damage to the host may arise in two ways. First, multiplication of the microorganisms may cause mechanical damage to the tissue cells through interference with the normal cell metabolism, as seen in viral and some bacterial infections. Second, a toxin associated with the microorganism may adversely affect the tissues or organs of the host. Two types of toxins, called exotoxins and endotoxins, are associated with bacteria.

2.5.1   *Exotoxins*

These are produced inside the cell and diffuse out into the surrounding environment. They are produced by both Gram-positive and Gram-negative bacteria. They are extremely toxic and are responsible for the serious effects of certain diseases; for example, the toxin produced by *Cl. tetani* (the causal organism of tetanus) is neurotoxic and causes severe muscular spasms due to impairment of neural control. Examples of other toxins identified are necrotoxins (causing tissue damage), enterotoxins (causing intestinal damage) and haemolysins (causing haemolysis of erythrocytes). Gram-positive bacteria producing exotoxins are certain members of the genera *Clostridium*, *Streptococcus* and *Staphylococcus* whilst an example of a Gram-negative bacterium is *Vibrio cholerae* (the causal organism of cholera). Several exotoxins consist of two moieties: one aids entrance of the exotoxin into the target cell whist the toxic activity is associated with the other fraction.

2.5.2   *Endotoxins*

These are lipopolysaccharide–protein complexes associated mainly with the cell envelope of Gram-negative bacteria (see Chapter 1). They are responsible for the general non-specific toxic and pyrogenic reactions

(Chapters 1 and 18) common to all organisms in this group. The specific toxic reactions for different pathogenic Gram-negative bacteria are due to the production of a toxin *in vivo*. Organisms of interest in this group are those causing cholera, plague, typhoid and paratyphoid fever and whooping cough.

3   Specific defence mechanisms (adaptive immune system)

Microorganisms which successfully overcome the non-specific defence mechanisms then have to contend with a second line of defence, the specific defence mechanisms. These involve the stimulation of a specific immune response by the invading microorganism and are evoked by what are termed *immunogens*. These may cause the appearance in the serum of modified serum globulins called immunoglobulins. The term *antigen* is given to a substance that stimulates immunoglobulins that have the ability to combine with the antigen that stimulated their production. These immunoglobulins are then termed *antibodies*. All antibodies are immunoglobulins but it is not certain that all immunoglobulins have antibody function. Antigens associated with microorganisms consist of proteins, polysaccharides, lipids or mixtures of the three and invariably have a high molecular weight. The antigen–antibody reaction is a highly specific one and this specificity is due to differences in the chemical composition of the outer surfaces of the organism. Bacteria, rickettsia and viruses all have the ability to induce antibody formation. The synthesis and release of free antibody into the blood and other body fluid is termed the *humoral immune response*.

Antigens, however, can induce a second type of response which is known as the *cell-mediated immune response*. The antigenic agent stimulates the appearance of 'sensitized' lymphocytes in the body which confer protection against organisms that have the ability to live and replicate inside the cells of the host. Certain of these lymphocytes are also involved in the rejection of tissue grafts.

3.1   Antigenic structure of the microbial cell

The microbial cell surface constitutes a multiplicity of different antigens. These antigens may be common to different species or types of microorganisms or may be highly specific for that one type only.

Three groups of antigens are found in the intact bacterial cell.

H-antigens.   These are associated with the flagella and are therefore only found on motile bacteria (H, *Hauch*, a film, and refers to the film-like swarming seen originally in cultures of flagellated *Proteus*). The precise chemical composition of flagella can vary between bacteria, resulting in a range of different antibodies being produced and use is made of these differences in the typing of different strains of *Salmonella*.

O-antigens. These are associated with the surface of the bacterial cell wall and are often referred to as the somatic antigens (O, *ohne Hauch*, without film, and refers to non-swarming cultures, i.e. absence of flagella). The specificity of the reaction between these antigens and the corresponding antibodies in Gram-negative bacteria is due to the nature and number of the type-specific polysaccharide side chains attached to the lipid A and core polysaccharide portion of the lipopolysaccharide (LPS) (see Chapter 1). This group of organisms is, however, very liable to mutate during cultivation in artificial media and the resultant mutant may lose the O-specific side chain antigens, resulting in the exposure of the more deep-seated core polysaccharide, the R (rough) antigens, which may share a common structure with other unrelated Gram-negative bacteria and so are no longer type-specific. This change is known as the S → R change and is so called because of an alteration in the appearance of the colonies of the organism from the normal, smooth, glistening colony to a rough-edged, matt colony. This S → R change represents a loss of the type-specific O-antigens with a concomitant loss in the specificity of the antigen–antibody reaction.

The major type-specific antigens of Gram-positive bacteria are the teichoic acid moieties associated with the cell wall (see Chapter 1).

Surface antigens. Many bacteria possess a characteristic polysaccharide capsule external to the cell wall and this too has antigenic properties. Over 80 serological types of the Gram-positive pneumococcus group have been differentiated by immunologically distinct polysaccharides in the capsule. Certain Gram-negative organisms of the enteric bacteria, e.g. salmonellae, may possess a polysaccharide microcapsule which is also antigenic and is thought to be responsible for the virulence of the bacteria. It is termed the Vi antigen and its presence is important in relation to the production of the typhoid vaccines (see Chapter 16).

4   **Cells involved in immunity**

The cells that make up the immune system are distributed throughout the body but are found mainly in the lymphoreticular organs, which may be divided into the primary lymphoid organs, i.e. the thymus and bone marrow, and the secondary or peripheral organs, e.g. lymph nodes, spleen, Peyer's patches (which are collections of lymphoid tissue in the submucosa of the small intestine) and the tonsils.

A large number of cells are involved in the immune response and all are derived from the multipotential stem cells of the bone marrow. The predominant cell is the lymphocyte but monocytes–macrophages, endothelial cells, eosinophils and mast cells are also involved with certain immune responses. The two types of immunity (humoral and cell-mediated) are dependent on two distinct populations of lymphocytes, the B cells and the T cells respectively. Both the humoral and the cell-mediated systems interact to achieve an effective immune response.

Humoral immunity, known as antibody-mediated immunity, is due directly to a reaction between circulating antibody and inducing antigen and may involve complement. The B cells originate in the bone marrow. In chickens a lymphoid organ embryonically derived from gut epithelium and known as the bursa of Fabricius is responsible for the maturation of the B cells into immunocompetent cells, which subsequently can synthesize antibody after stimulation by antigen. The bursal equivalent in humans is the bone marrow itself. An antigen (e.g. a bacterium) may possess multiple determinants (epitopes) and each one of these epitopes will stimulate an antibody which will subsequently react with that and closely related epitopes only. Each B cell is only capable of recognizing one epitope via a specific receptor on its surface. This receptor has been shown to be antibody itself. Activation of the B cell occurs by binding of the antigen to the receptor and the resultant complex is endocytosed. For activation to proceed, additional signals are now required. These are supplied by the secretion of peptide molecules (termed lymphokines) from a subset of the T cell family (the helper T cells, TH cells). These lymphokines stimulate the B cells to proliferate and undergo clonal expansion and subsequently to mature into antibody-synthesizing plasma cells and also into longer living, non-dividing memory cells. The lymphokines responsible are collectively described as B-cell growth factor (BCGF) and B-cell differentiation factor (BCDF) respectively. Antigens requiring the assistance of TH cells are termed T-dependent (TD) antigens.

Subsequent antigenic stimulation results in high antibody titres (secondary or memory response) as there is now an expanded clone of cells with memory of the original antigen available to proliferate into mature plasma cells (Fig. 15.1).

Some antigens, such as type 3 pneumococcal polysaccharide, LPS and other polymeric substances such as dextrans (poly-D-glucose) and levan (poly-D-fructose) can induce antibody synthesis without the assistance of TH cells. These are known as T-independent (TI) antigens. Only one class of immunoglobulin (IgM) is synthesized and there is a weak memory response.

Immunoglobulins are associated with the γ-globulin fraction of plasma proteins but, as stated earlier, not all immunoglobulins exhibit antibody activity.

The immunoglobulin (Ig) molecules can be subdivided into different classes on the basis of their structure, and in humans five major structural types can be distinguished. Each type has been distinguished on the basis of a polypeptide chain structure consisting of one pair of heavy (large) chains and one pair of light (small) chains joined by disulphide bonds. The heavy chains are given the name of the corresponding Greek letter (γ chain in IgG, μ in IgM, α in IgA, δ in IgD and ε in IgE). All classes have similar sets of light chains consisting of one of two types, the kappa (κ) or

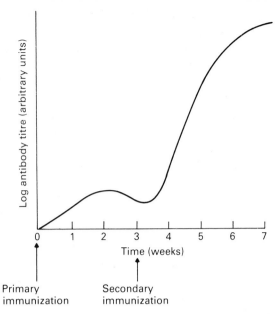

Fig. 15.1 Antibody response to primary and secondary immunization doses.

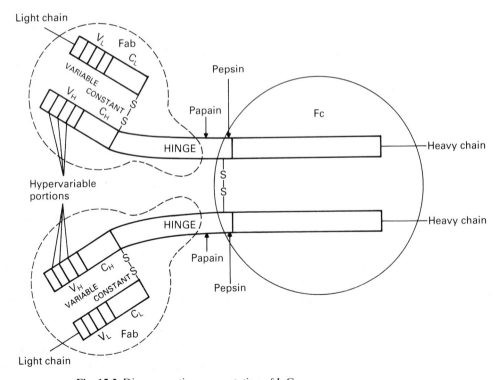

Fig. 15.2 Diagrammatic representation of IgG.

lambda (λ) chains. A suggested ground plan for the most abundant Ig, IgG, is illustrated in Fig. 15.2.

IgG consists of four polypeptide subunits held together by disulphide bonds. Native immunoglobulins are rather resistant to proteolytic digestion but certain enzymes have been useful in elucidating their structure. Papain cleaves the molecule into three fragments of similar size: (i) two Fab fragments each carrying a single antigen-combining site and comprising the variable regions of both chains, the constant region of the light chains and the first constant domain of the heavy chain; and (ii) one Fc fragment composed of the terminal halves of the heavy chains which have no affinity for antigen but can be crystallized.

Cleavage with pepsin yields two fragments only, one consisting of two Fab fragments and the other an Fc fragment which is partially degraded by the enzyme. The variable regions on both the heavy and light chains contribute towards antigen recognition; whilst the constant regions of the heavy chain, particularly the Fc part of the heavy chain backbone, direct the biological activity of the molecule, e.g. complement fixation (see later) and the interaction with a variety of tissue cells, via membrane receptors for the Fc region.

Intrastrand bonding via disulphide links cause the molecule to fold into 'globular domains' and it is these that direct the biological activity of the molecule.

4.2   **Monoclonal antibodies**

After antigenic stimulation, the normal antibody response involves the activation of a large number of clones of antibody-secreting cells (i.e. it is polyclonal). This is due to the fact that antigens possess multiple epitopes. In 1975 Kohler and Milstein successfully developed cell fusion techniques which enabled them to isolate clones of cells which synthesized identical antibody molecules (see Fig. 15.3).

The principles of the technique rely on the fact that an antibody-secreting cell can become cancerous and the unchecked proliferation of such a cell is called a *myeloma*. Progeny of the original transformed cell will continue to secrete a single kind of antibody molecule only. Myeloma cells, like other malignant cells, grow indefinitely in tissue culture. However, the specificity of the antibody is unknown. Mutant myeloma cells have been isolated which have lost the ability to secrete antibody while still retaining their cancerous growth properties.

Mouse myeloma cells are fused with an antibody-secreting cell from the spleen of a mouse immunized with the required antigen. The technique is called somatic cell hybridization and the resultant cell is termed a 'hybridoma'. The rate of successful hybrid formation is low and a technique is necessary which can select these successful fusions. The standard technique is to use a myeloma cell line that has lost the capacity to synthesize hypoxanthine-guanine phosphoribosyl-transferase (HGPRT). This enzyme enables cells to synthesize nucleotides using an extracellular

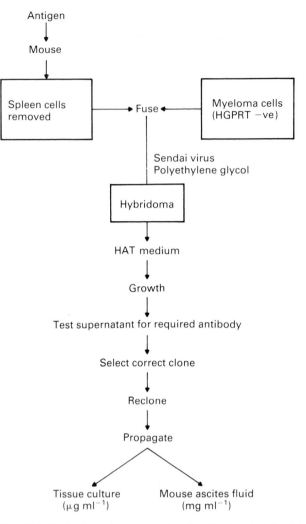

Fig. 15.3 Production of monoclonal antibodies (see text for details).

source of hypoxanthine as a precursor. The absence of HGPRT is normally no problem as cells can use an alternative pathway. When, however, these cells are exposed to aminopterin (a folic acid analogue; see Chapter 9) they are unable to use this other pathway and become fully dependent on HGPRT.

The cell fusion mixture is transferred to a culture medium containing hypoxanthine, aminopterin and thymidine (HAT medium). Unfused myeloma cells are unable to grow as they lack HGPRT. Unfused normal spleen cells can grow but their proliferation is limited and they eventually die out. The hybridoma cell can proliferate in the HAT medium as the normal spleen cell supplies the enzyme which enables the hybridoma to utilize extracellular hypoxanthine.

The hybridoma is now screened for the production of the desired antibody by testing the supernatant from each culture. A single culture, even though positive for antibody production, can contain the progeny of two or more successful fusions. Therefore, it is necessary to dilute positive cultures so that fresh cultures can be started with a single hybridoma cell. When successful, such cultures are truly monoclonal and the antibody is directed against a single epitope on a preselected antigen. Once established, these cell lines are immortal.

The concentration of antibody in tissue cultures of the hybridoma is low $(10–60 \mu g \, ml^{-1})$ but the use of large culture vessels can obviate this. The hybridoma can also be propagated in mice where the antibody concentration in the serum and other body fluids can reach $10 \, mg \, ml^{-1}$.

4.2.1   *Uses of monoclonal antibodies*

Monoclonal antibodies are very sensitive, specific reagents and have applications in many areas of the biological sciences. They revolutionized immunology within a few years of their discovery.

The investigation and characterization of cell surfaces by probing with monoclonal antibodies is one of the most vital areas of application. In this context they have been used in the following ways:

1   To study the ABO and rare blood groups.
2   To detect HLA antigens and consequently to type tissues for transplantation.
3   To classify cell lines, e.g. the T cell subsets, and thence to separate these cell subpopulations.
4   To study cell–cell interactions and differentiation, e.g. embryology.
5   In oncology, to study the relationship between the normal and the tumour cell, to detect tumour-associated antigens (CEA, carcinoembryonic antigen, and AFP, α-fetoprotein) and subsequently to enable cancer therapy to be monitored, to locate tumour metastases, and to deliver cytotoxic drugs, toxins, radionuclides, or liposomes to tumour cells.
6   To identify and characterize bacterial and viral antigens which can then be purified and used to prepare subunit vaccines.

Monoclonal antibodies have further been employed for studying drug and hormone receptors, enzymes and proteins. A whole range of immunoassay techniques using monoclonals have been developed to detect low levels of materials in body fluids, e.g. oxytocin can be detected in human blood using a radioimmunoassay down to $1 \, pmol \, l^{-1}$. Similar assays are used to monitor antibiotic therapy using potentially toxic drugs, e.g. gentamicin. The future of monoclonal antibodies continues to be one of enormous potential and excitement.

4.3   **Immunoglobulin classes**

The synthesis of antibodies belonging to the various classes of immunoglobulin proceeds at different rates after the initial and subsequent antigenic stimuli.

Immunoglobulin M (IgM)

Synthesis of this class occurs after the primary antigenic stimulus. IgMs are polymers of five four-peptide subunits and have a theoretical valency of 10, although against large antigens such as bacteria their effective valency is five. They are extremely effective agglutinating agents and, as they are largely confined to the bloodstream and they appear early in the response to infection, they are of particular importance in bacteraemia.

Serum concentrations lie between 0.5 and 2.5 mg ml^{-1}. IgM can fix complement and a single molecule can initiate the complement cascade. IgM (with IgD) is the major immunoglobulin expressed on the surface of B cells where it acts as an antigen receptor.

4.3.2

Immunoglobulin G (IgG)

This is the major immunoglobulin synthesized during the secondary response and in normal human adults is present at serum concentrations between 10 and 15 mg ml^{-1}. Within this class there are four subclasses, designated IgG$_1$, IgG$_2$, IgG$_3$ and IgG$_4$.

It has the ability to cross the placenta and therefore provides a major line of defence against infection for the newborn. This can be reinforced by transfer of colostral IgG across the gut mucosa of the neonate. It diffuses readily into the extravascular spaces where it can act in the neutralization of bacterial toxins and can bind to microorganisms enhancing the process of phagocytosis (opsonization). This is due to the presence on the phagocytic cell surface of a receptor for Fc.

Complexes of IgG with the bacterial cell activate complement with the resultant advantages to the host.

4.3.3

Immunoglobulin A (IgA)

This occurs in the seromucous secretions such as saliva, tears, nasal secretions, sweat, colostrum and secretions of the lung, urinogenital and gastrointestinal tracts. Its purpose appears to be to protect the external surfaces of the body from microbial attack. It occurs as a dimer in these secretions but as a monomer in human plasma, where its function is not known. The function of IgA appears to be to prevent the adherence of microorganisms to the surface of mucosal cells thus preventing them entering the body tissues. It is protected from proteolysis by combination with another protein—the secretory component.

It is present at serum levels between 0.5 and 3 mg ml^{-1} but higher concentrations are found in secretions. There are two subclasses of this immunoglobulin.

4.3.4

Immunoglobulin D (IgD)

This occurs in normal serum at very low levels (30–50 μg ml^{-1}) but is the predominant surface component of B cells. Immature B cells express surface IgM without IgD but as these cells mature IgD is also expressed. After activation of the B cells, surface IgD can no longer be detected and it would appear that IgD may be involved with the differentiation of B cells.

4.3.5   *Immunoglobulin E (IgE)*

This is a very minor serum component ($0.1-0.3\,\mu g\,ml^{-1}$) but is a major class of immunoglobulins. It binds with very high affinity to mast cells and basophils via a site in the Fc region of the molecule. Cross-linking of the cell-bound IgE antibodies by antigen triggers the degranulation of these cells with the release of histamine, leukotrienes and other vaso-active compounds. This class may play a role in immunity to helminthic parasites but in the western world it is more commonly associated with immediate hypersensitivity reactions such as hay fever and extrinsic asthma.

4.4   Humoral antigen–antibody reactions

Antibody molecules are bivalent whilst antigens can be multivalent. The resultant combination may result in either small, soluble complexes, or large insoluble aggregates, depending on the nature of the two molecules in the system. The following are examples of the reactions that can occur.
1 Neutralization. Small soluble complexes neutralize microbial toxins.
2 Precipitation. The formation of insoluble precipitates which enable the phagocytes to eliminate soluble antigen from the body.
3 Agglutination. The aggregation of bacterial cells into agglutinates enabling phagocytes to eliminate these cells rapidly from the body.
4 Cytotoxic reactions. The antibody and cell react, with resultant lysis of the cell. It was found that the presence of a third component, called complement, was necessary for this reaction to take place.

4.5   Complement

Complement comprises a group of nine serum proteins (C1–C9), which consist of enzyme substrates and precursors acting in sequence, and is responsible for the biological activity of the complement-fixing antibodies. Complement activity was first recognized by Bordet, who showed that the lytic activity of rabbit anti-sheep erythrocyte serum was lost on heating to 56°C but was restored by the addition of fresh serum from an unimmunized rabbit. Thus two factors were necessary, a heat-stable factor, antibody, plus a heat-labile factor, complement, which is present in all sera.

The first component of complement is activated by an immune complex (antibody bound to erythrocyte) which then activates several molecules of the next component in the sequence and so on, thus producing a cascade effect with amplification. The terminal components of the

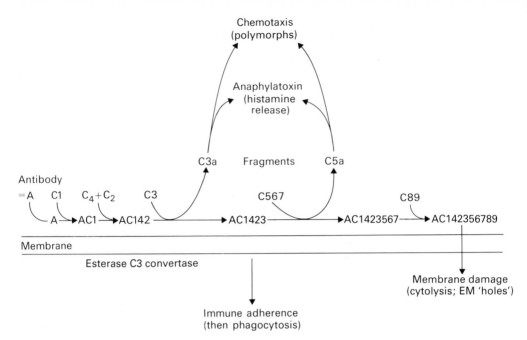

Fig. 15.4 Sequence of classical complement activation by membrane-bound antibody (Roitt, 1984). (Reproduced courtesy of the author.)

complement cascade cause rapid lysis of the cell due to small holes appearing in the membrane. Thus the activation of one C1 molecule can lead to the eventual lysis of a complete cell. Only IgG and IgM can bind C1, and it is the Fc component of the molecule that is responsible for this. The full sequence of classical complement activation by membrane-bound antibody is shown in Fig. 15.4.

The addition of all components is necessary for the lysis of the cell, but during the activation process several small peptide units are produced which exhibit distinct biological properties.

1 C3a and C5a fragments are chemotactic for polymorphs and also have anaphylatoxin activity in that they cause histamine release from mast cells.

2 C3b fragments adhere readily to the target cell surface and there are specific receptors for the membrane-bound C3b on polymorphs and macrophages. This allows immune adherence of the complexes to these cells, thus facilitating subsequent phagocytosis.

Complement can also be activated in the absence of antibody by agents such as bacterial polysaccharides. These convert C3 to C3b by the generation of a C3 convertase. This is known as the alternative pathway and enhances the classical pathway.

Complement plays a significant part in the defence of the body. It can cause lysis of Gram-negative organisms by allowing lysozyme to reach the peptidoglycan layer of the organism. The generation of the C3b complex on the surface of the cell facilitates phagocytosis as the phagocytes possess a receptor for C3b, whilst C3a and C5a cause the release of histamine with the resultant increase in vascular permeability increasing the flow of serum antibody into the infected area. C3a and C5a also attract phagocytic cells to the focus of the infection.

4.6　　　　**Cell-mediated immunity (CMI)**

The term cell-mediated immunity is used to describe the localized re-actions that occur to those microorganisms that have the ability to live and multiply within the cells of the host, for example the tubercle bacil-lus, viruses and protozoal parasites. These reactions are mediated by lymphocytes and phagocytes and antibody plays a subordinate role.

The lymphocytes involved in this type of immunity originate from the multipotential stem cell and are processed by the thymus gland; hence the name 'T' cells. The role of the thymus is to rearrange the genes associated within the T cell receptor (TCR) so that the mature T cells recognize foreign but not self antigens. This receptor has been isolated using mono-clonal antibody probes and has been shown to consist of two disulphide-linked polypeptide chains termed the α and β chain. This receptor is associated with a characteristic cell surface marker, CD3 (CD, clusters of differentiation). Antigen recognition occurs via this membrane structure CD3/TCR. Mature T cells also express other antigenic markers, notably CD4 or CD8. Thymectomized neonate mice do not exhibit the CMI response indicating the importance of the thymus gland.

Infection with a human immunodeficiency virus (HIV-1 and HIV-2; see Chapter 3) can cause the destruction of the TH cell, which is the critical cell of the immune system. This leads to the condition known as AIDS. At present, it is still not known why, in some cases, infection with HIV leaves the immune system intact whereas in others it is irreversibly destroyed, giving rise to AIDS.

The immune system must be able to distinguish between antigens against which an immune response would be beneficial and those where such a response would be harmful to the host, i.e. it must be able to distinguish between 'self' and 'non-self'. This is achieved via molecules of the major histocompatibility complex (MHC). The human MHC is located on chromosome 6 and is known as the HLA (human leucocyte antigen). It is divided into four main regions, designated A, B, C and D. Products of this region are expressed on the surface of cells and these enable cells of the immune system to recognize and signal to each other. Three main groups of these molecules have been identified.

1　Class 1 MHC molecules are integral membrane proteins found on the surface of all nucleated cells and platelets. They are the classical antigens involved in graft rejection.

2　Class 2 MHC molecules are expressed on the surface of B cells, macrophages, monocytes, various antigen-presenting cells and certain cells of the T cell family.

3 Class 3 MHC molecules consist of several complement components.

T cells only respond to protein antigens when the antigen has been processed by the antigen-presenting cells (APCs). The resultant small peptide molecules are then bound to the Class 2 molecules on the surface of the APCs. Monocytes, macrophages, B cells, dendritic cells and some T cells all have the ability to internalize and degrade proteins into peptide fragments and can all therefore act as APCs.

The major T cell classes and their functions are listed below.

4.6.1   *Helper T cells (Tₕ cells)*

These are the central cells of the immune system as they are essential for activation of the other cells associated with an effective immune response by the secretion of peptide mediators termed cytokines. Cytokines produced by macrophages and monocytes are termed monokines whilst those produced by lymphocytes are termed lymphokines. Tₕ cells express CD4 on their surface. Activation of Tₕ cells requires two signals. The first is the binding of CD3/TCR to the Class 2 MHC-antigen complex on the surface of the APC. This stimulates the APC to secrete a monokine (IL-1). This represents the second signal as the now activated Tₕ cell secretes a lymphokine (interleukin-2, IL-2) together with a series of other cytokines associated with cell growth and differentiation. IL-2 induces the growth of cells expressing IL-2 receptors which include the Tₕ cells actually producing it, i.e. an autocatalytic effect. Cytokines secreted by activated Tₕ cells are also associated with the proliferation and differentiation of B cells associated with the humoral response.

T cells responsible for delayed-type hypersensitivity secrete lymphokines which recruit and activate non-specific cells like macrophages into the area of the reaction. Examples of some of these lymphokines are listed below.

1 A macrophage chemotactic factor (MAC) which causes an accumulation of mononuclear phagocytes at the site of the antigen-mediated lymphokine release.

2 A macrophage migration inhibitory factor (MIF) which encourages the macrophages to remain in the area.

3 A macrophage-activating factor (MAF) which enhances the cell's ability to kill ingested intracellular organisms.

These cells were originally classified as Tᴅ cells but as they express CD4 it is now accepted they are a specialized population of Tₕ cells.

4.6.2   *Suppressor T cells (Ts cells)*

These cells suppress ongoing or developing immune responses. They have been found to release specific soluble factors *in vitro* (TsF) which may be directed at either the Tₕ cell or the B cell. The receptors on the Ts cell may recognize antigen which will then act as a bridge between the Ts and its target (the antigen receptor) or, alternatively, the Ts receptor may be a mirror image of the receptor on the target cell and produce direct suppression by binding to it. This recognition is termed 'idiotype recognition'.

These cells express CD8 on their surface.

4.6.3   *Cytotoxic T cells (Tc cells)*

Virally infected host cells, and also tissue grafts from a genetically dissimilar donor, have been shown to stimulate the formation of T cells that are cytotoxic for these cells.

These T cells recognize peptide antigens bound to Class 1 MHC molecules on the surface of the target cell. During viral infections, viral peptides bind to self MHC1 molecules and are subsequently expressed on the cell surface. The MHC1 molecules of transplanted tissues are themselves recognized by the Tc cells.

Like Th cells, two signals are required to activate the Tc cell. The first is an interaction between the TCRs and the Class 1 MHC molecule/foreign epitope complex on the surface of the target cell. The second signal is that of IL-2 produced by the activated Th cell with the resultant release of cytotoxins which destroy the target cell.

Tc cells express CD8 on their cell surface.

4.7   **Immunoregulation**

An ongoing immune response can be regulated by three mechanisms.

Suppressor T cells.   These cells can be specific for the antigen receptors on both B and T cells and thereby can suppress the activity of these two groups of cells.

Antibody feedback.   Antibodies produced in response to an antigen are capable of inhibiting further immune responses to that antigen. This may occur due to diminishing antigen levels as a result of its combination with antibody or through an idiotypic network.

Idiotypic network.   Idiotypic determinants (idiotypes) are unique antigenic epitopes characteristic of the antigen receptors on the surface of T and B cells. They are associated with the variable regions of these receptors. Antibodies produced by B cells as the result of antigenic stimulation can themselves stimulate the production of auto-anti-idiotypic antibodies which have the ability to combine with the B cell receptor (Ig) and thus can dampen down the immune response. Idiotypes may likewise stimulate the production of T cells specific for idiotypic determinants. Jerne (1974) postulated his network hypothesis consisting of a series of complementary anti-idiotypic responses which modulate the immune response.

4.8 Natural killer (NK) cells

NK cells are a subset of lymphocytes found in peripheral blood, bone marrow and the spleen. They are about $15\,\mu m$ in diameter, possess a kidney-shaped nucleus and have two or three large granules in the cytoplasm. They are spontaneously cytolytic for certain tumour lines and also for some virus-infected cells. The membrane receptors responsible for this recognition have not been identified but they have been shown not to possess CD3/TCR or immunoglobulin, which are characteristic of T cells and B cells respectively. They do, however, express the surface markers CD56 and CD16. The most important role of NK cells is to provide a first line of defence against viral infections as they do not require prior exposure to antigen in order to respond. They are therefore effective against virally infected cells prior to the development of antibodies and antigen-specific Tc cells. NK cells operate independently of MHC antigens on the target cell and their activity is markedly enhanced by IL-2, alpha interferon and other agents that activate macrophages, such as BCG vaccine. They may play an important part in controlling the development of neoplastic cells in the body.

NK cells possess a receptor for Fcγ and this enables them to adhere to target cells coated in antibody with the resultant destruction of that cell. This phenomenon is known as antibody-dependent cell-mediated cytotoxicity (ADCC). This was attributed to a separate cell population known as killer (K) cells but these have now been shown to be in effect NK cells.

4.9 Immunological tolerance

The administration of antigenic material does not always evoke an immunological response, a condition termed 'tolerance'. The classic example of this is the exposure of the immature lymphoid system of neonates to antigen, inducing a state of unresponsiveness to later challenge by the same antigen after the animal has reached immunological maturity. This could be the means whereby, during gestation, the body becomes unresponsive to its own constituents enabling the mature lymphoid system to distinguish in later life between 'self' and 'non-self'.

Tolerance can also be induced in adults, but higher doses of the antigen are required where it has been shown that both T and B cells are made unresponsive. As most antibody responses are T-dependent it is likely that it is these cells which are the ones affected. In order to maintain this state of tolerance it is necessary for the antigen to persist in the animal, as in its absence immunocompetent cells which are being produced throughout life are not being rendered tolerant.

Tolerance can occur in several ways.

1 Genetic unresponsiveness. If the animal lacks the necessary genetic ability to recognize antigenic material it will be 'immunologically' silent.

2 T-suppression. Ts cells may be activated more effectively than TH cells, thereby suppressing the immune response.

3 Helplessness. T cells are more readily tolerated than B cells and if they are unable to activate the B cells these cells could be described as 'helpless'.

4 Clonal deletion. Contact with antigen in the neonate results in death or permanent inactivation of the developing lymphocytes.

4.10   **Autoimmunity**

One fundamental property of an animal's immune system is that it does not normally react against its own body constituents, i.e. it exhibits tolerance. However, clinical and experimental evidence shows that certain diseases exist in which the patient apparently destroys his/her own cells. The reactions could involve Tc cells, B cells or NK cells, and the result of the reaction with antigen may result in a pathological condition arising (autoimmune disease). Autoimmunity is the mirror-image of tolerance and reflects the loss of tolerance to 'self'.

Autoimmunity can arise by the following.

1 Evasion of tolerance to self antigens. Hidden or sequestered antigens do exist, for instance spermatozoa and eye-lens tissue. These are confined to anatomical sites which do not have access to lymphoid tissue, and exposure of the above to lymphoid cells as a result of surgery or accident results in the production of the corresponding antibodies.

Drugs frequently bind to blood elements directly (e.g. penicillin to erythrocytes) and the antibodies to the resultant complex react with, and damage, cells coated with the drug. Viruses, especially those that bud, become associated with the host cell surface antigens with the resultant generation of Tc cells.

2 Breakdown of tolerance mechanisms. There are at least two mechanisms for maintaining unresponsiveness to self. The first is by specific deletion of self-reactive clones and the second by suppression. A failure of either of these two may result in an autoimmune disease. In normal, healthy individuals, antigen-binding, self-reactive B cells and the resultant low titres of autoantibodies are not uncommon. The origin of the self-reactive B cells is not clear, but there are four ways in which they may become activated.

(a) Polyclonal activation. High concentrations of polyclonal activators, such as LPS and high molecular weight dextrans, activate B cells irrespective of the immunoglobulin receptor on the B cell surface. Polyclonal activation occurs in parasitic infections and in certain viral infections with the production of a wide spectrum of autoantibodies.

(b) Non-specific helper factors. T cell activation results in the production of a variety of lymphokines which can activate these B cells.

(c) Cross-reactive antigens. These are shared by host and micro-organism and this cross-reaction can activate autoreactive B cells.

(d) Absence of T cell suppression. The sudden depletion or elimination of Ts cells can lead to the spontaneous development of autoantibodies due to the maturation of the autoreactive B cells.

Types of autoimmune diseases vary widely, from 'organ-specific' diseases such as thyroiditis where there may be stimulation (thyrotoxicosis) by antibody against the receptor for pituitary thyroid-stimulating hormone (TSH) or inhibition (myxoedoma) by cell destruction probably mediated by NK cells and autoantibody, through to 'non-organ-specific' diseases such as systemic lupus erythematosus (SLE), where both lesions and autoantibodies are not confined to any one organ. In SLE, antibodies have been detected to DNA, erythrocytes and platelets, and cytotoxic antibodies to T lymphocytes have also been demonstrated. A strong case can be made for rheumatoid arthritis resulting from an autoimmune response to the Fc portion of IgG which gives rise to complexes which are ultimately responsible for the pathological changes characteristic of the rheumatoid joint.

5   Hypersensitivity

Not all antigen–antibody reactions are of benefit to the body, as sometimes the complexes (or their subsequent interaction with body tissues) may result in tissue damage. This must be regarded as a malfunction of the immune system and is known as a hypersensitive reaction. These reactions can be categorized into five main types. The first three involve the interaction between antigen and humoral antibody, and as the onset of the reaction is rapid, the condition is termed *immediate* hypersensitivity. The fourth type (*delayed* hypersensitivity) involves T cells and the symptoms of the reaction appear after 24 hours. The fifth type is where antibody stimulates cell function.

5.1   Type I (anaphylactic) reactions

In these reactions the antigen reacts with antibodies that are bound to the surface of mast cells through the Fc portion. This leads to the degranulation of the mast cells with the resultant release of vasoactive amines which give rise to the characteristic reactions of inflammation. The symptoms of the reaction that appear depend on the distribution of the cell-bound antibody, for example allergic reactions affecting the skin (urticaria), nasal mucosa (rhinitis), eyes (angioneurotic oedema), bronchioles (extrinsic asthma) and the cardiovascular system (anaphylactic shock). Antibodies involved in these reactions are mainly IgE but sometimes IgG. They are called homocytotropic or reaginic antibodies and are responsible for the common allergic reactions that affect nearly 10% of the population.

5.2   Type II (cytolytic or cytotoxic) reactions

These reactions involve damage to particular cells or tissues. The combination of circulating antibody with the antigen on the cell surface results in the destruction of the cell by phagocytosis either by opsonic adherence through Fc or by immune adherence through C3b. Activation of the full complement system results in lysis. ADCC reactions involving NK cells may also occur. Type II reactions include the destruction of erythrocytes by cytolytic antibodies induced by incompatible blood transfusion; Rhesus incompatibility; autoimmune reactions which result in auto-antibodies being produced against the patient's own red cells; and cells whose surface has been altered by sensitizing drugs. Antibodies involved in these reactions are of the IgG and IgM classes.

5.3   Type III (complex-mediated) reactions

These reactions are due to the presence of immune complexes either in the circulation or extravascular space. The complexes may localize in capillary networks (lungs, kidney, joints) where, together with complement and polymorphs, they may produce extensive tissue damage. Two main types of reactions fall into this group.

1   The Arthus reaction. The phenomenon is a local one and occurs if a soluble antigen is introduced into the body when there is a great excess of antibody. The union between the two results in an acute inflammatory reaction which may involve complement, polymorphs, lymphokines or platelet aggregation, all of which enhance the inflammatory response.

2   Serum sickness. This occurs when there is an excess of antigen to antibody, resulting in the formation of soluble complexes. These may circulate and cause systemic reactions or be widely deposited in the kidneys, joints and skin. A rise in temperature, swollen lymph nodes, a generalized urticarial rash and painful swollen joints occur. The repeated administration of foreign serum (e.g. antidiphtheria serum or antitetanus serum prepared in horses) can lead to this condition due to antibodies being produced to the horse protein material.

5.4   Type IV (delayed hypersensitivity) reactions

These reactions are slow to manifest themselves (1–3 days after contact with antigen). Many allergic reactions to bacteria, viruses and fungi, sensitization to simple chemicals and the rejection of transplanted tissues result. The reactions are initiated by reaction between antigen-specific T cells and antigen, with the resultant release of lymphokines that affect a variety of accessory cells, especially macrophages. Antibody and complement are not involved. The classic example of this type of reaction may arise in persons subjected to the tuberculin test. Subjects who have previously been in contact with *Mycobacterium tuberculosis* have T cells sensitized to a protein extract of the tubercle bacillus. Intradermal injections of this protein extract induce an inflammatory response in the skin, at the site of the injection, which appears after 24 hours and may persist for several months. It is taken as an indication of immunity to the disease due to prior exposure to the organism, and a rough indication of the quality of this immunity can be interpreted according to the response to varying concentrations of the protein.

Reaction against virally infected or transplanted cells results in stimulated lymphocytes transforming into Tc cells which can eliminate target cells bearing the sensitizing antigen.

5.5 Type V (stimulatory hypersensitivity) reactions

Cells possess surface receptor sites for the chemical messengers of the body. Should an autoantibody be produced against this site, it can combine with it and cause the same effect as the chemical messenger, e.g. thyrotoxicosis caused by autoantibody to the receptor site to TSH as previously described (section 4.10).

6 Tissue transplantation

The replacement of certain diseased or damaged organs by healthy ones is now a fairly routine occurrence, but the immunological nature of graft rejection was only accepted when it was shown that second grafts from the same donor were more rapidly rejected than first grafts. Tissue transferred from one site to another within the same individual (*autografts*) or between genetically identical individuals, e.g. uniovular twins (*isografts*), are invariably successful. Grafts between genetically different individuals of the same species (*allografts*) or between different species (*xenografts*) evoke an intense immunological response and are rejected.

The specificity of transplantation antigens is under genetic control and these genes can be divided into two categories. The first are those that control the 'strong' transplantation antigens which induce intense allograft reactions where incompatibility between donor and recipient leads to rapid graft rejection. In mice this locus is termed the H-2 complex and in humans the HLA system. These constitute the MHC, which dominates all transplantation reactivity.

As previously described, four principal loci have been identified in the HLA system, namely HLA-A, B, C and D, and their products occur as transmembrane glycoprotein antigens. The second category of histocompatibility genes codes for 'minor' transplantation antigens where differences between donor and recipient lead to relatively slow graft rejection. Successful organ grafting relies on matching donor and recipient antigens as closely as possible but often the clinical urgency of the transplant does not permit this. Graft rejection is mediated by T and/or B cells with their usual associated systems such as complement, NK cells, etc., and the time taken for rejection to occur depends on whether the recipient has previously been sensitized to the antigens of the donor. The major routine measures to prevent graft rejection are the use of anti-inflammatory and immunosuppressive drugs such as steroids, azathioprine and cyclosporin A, where the rationale is to destroy cells responding to antigen. The use of antibodies to host lymphocytes (antilymphocyte serum, ALS) in conjunction with chemotherapy has proved successful in heart grafts.

Tissue-typing studies have revealed that there is a large range of diseases, mostly of presumed immunologic origin, that are associated with the presence of a specific HLA antigen. The most overwhelming relationship occurs in the disease ankylosing spondylitis, where 95% of sufferers possess the HLA-B27 antigen compared to an incidence of only 5% in the controls. The precise reason for the relationship is not known but many divergent theories have been postulated.

6.1 Immune response to tumours

It is suggested that altered cells which could be potentially malignant are recognized by the immune system and eliminated. This must mean that cancer cells possess new antigens on their cell surface. These antigens have been identified and can be categorized into three groups.

1 Virally-induced antigens result from a malignant transformation occurring in the cell, due to an oncogenic virus. These evoke powerful immune responses in experimental animals.

2 Cells transformed by chemical carcinogens possess antigens that evoke a weak response.

3 Naturally occurring tumours evoke little or no immune response in experimental animals. This is disappointing but it must be remembered that these cells have already escaped the normal immune surveillance.

The possible use of immunotherapy for the prevention or treatment of malignant disease relies on the stimulation of the natural immune response and this is an area of exciting research, but has, as yet, proved to have limited success.

7 Immunity

Immunity, the state of relative resistance to an infection can be divided into two main groups, natural and acquired immunity.

7.1 Natural immunity

This is subdivided into the following.

7.1.1 Species immunity

Humans are susceptible to diseases to which other animals are immune and vice versa. This is due to body temperature, biochemical differences, etc.

7.1.2 Individual immunity

Variation in natural immunity between individuals can depend on the state of health, age, hormonal balance, etc.

329   *Fundamentals of immunology*

Acquired immunity

This is subdivided into actively acquired and passively acquired immunity, each of which may be induced naturally or artificially.

7.2.1   *Active acquired immunity*

This is produced as a result of an antigenic stimulus. This stimulus may occur naturally by means of a clinical or subclinical infection, or artificially by the deliberate introduction into the body of the appropriate antigen in the form of a vaccine or toxoid (Chapter 16). This type of immunity is normally long-lasting.

7.2.2   *Passive acquired immunity*

Passively acquired immunity involves no 'work' on the part of the body's defence mechanisms, and produces immediate protection of short duration.

It involves the transfer into the recipient of preformed antibody, i.e. there is no antigenic stimulus. This can occur (i) naturally, by trans-placental passage of antibody from mother to child and also by antibodies being transmitted in breast milk; or (ii) artificially, by means of the administration of antibodies preformed in another human (human γ-globulin) or in animals, e.g. horses, which are used for the production of antitoxic sera (antitoxins such as tetanus, diphtheria, etc.; see Chapter 16). The length of the immunity depends on the rate of degradation of the antibody and is only short-lived.

8   **Further reading**

Abbas A.K., Lichtman A.H. & Pober J.S. (1991) *Cellular and Molecular Immunology*. London: W.B. Saunders.

Baldwin R.W. (1985) Monoclonal antibody targeting of anti-cancer agents. *Eur J Cancer Clin Oncol.* **21**, 1281–1285.

Benjamini E. & Leskowitz S. (1991) *Immunology: A Short Course*, 2nd edn. Chichester: Wiley–Liss.

Clark W.R. (1991) *The Experimental Foundations of Modern Immunology*, 4th edn. Chichester: John Wiley & Sons.

Gallo R. (1991) Recent advances in human medicine. *Pharm J.* **247**, 355.

Jerne N.K. (1974) Towards a network theory of the immune system. *Ann Institut Pasteur Immunol.* **125C**, 373–389.

Kohler G. & Milstein C. (1975) Continuous culture of fused cells secreting antibody of predefined specificity. *Nature*, **256**, 495–497.

Leder P. (1982) The genetics of antibody diversity. *Sci Am.* **246**, 72–83.

Liddell J.E. & Cryer A. (1991) *A Practical Guide to Monoclonal Antibodies*. Chichester: John Wiley & Sons.

Marrack P. & Kappler J. (1987) The T cell receptor. *Science*, **238**, 1073–1079.

Muller G. (Ed.) (1984) Idiotype networks. *Immunol Rev.* **79**.

Playfair J.H.L. (1992) *Immunology at a Glance*, 5th edn. Oxford: Blackwell Scientific Publications.

Roitt I.M. (1991) *Essential Immunology*, 7th edn. Oxford: Blackwell Scientific Publications.

Tomlinson E. & Davis S.S. (Eds) (1986) *Site-specific Drug Delivery*. Chichester: John Wiley. (This deals in part with monoclonal antibodies.)

Trinchieri G. (1989) Biology of natural killer cells. *Adv Immunol.* **47**, 187–376.

16

The manufacture and quality control of immunological products

1	**Introduction**	2.4	Blending
		2.5	Filling and drying
2	**Vaccines**	2.6	Quality control
2.1	The seed lot system	2.6.1	In-process control
2.2	Production of the bacteria and	2.6.2	Final-product control
	bacterial components of bacterial		
	vaccines	**3**	**Immunosera**
2.2.1	Fermentation		
2.2.2	Processing of bacterial harvests	**4**	**Human immunoglobulins**
2.3	Production of the components of		
	viral vaccines	**5**	**Tailpiece**
2.3.1	Growth of viruses		
2.3.2	Processing of viral harvests	**6**	**Further reading**

1 Introduction

Immunological products, especially vaccines, have brought to mankind benefits that are arguably greater than those provided by any other group of pharmaceuticals. Smallpox vaccine, used relentlessly by the World Health Organization (WHO) in a worldwide vaccination campaign, has eradicated for all time a disfiguring and often fatal infection which has been one of mankind's severest scourges. The diphtheria, tetanus, whooping-cough, polio, measles, mumps and rubella vaccines used in public health programmes protect millions of children each year in the industrialized countries and, as the WHO's Expanded Programme of Immunization takes effect, vaccines protect even more millions in the developing countries. Furthermore, in addition to the vaccines that have global application there are a dozen or more which, used selectively, provide immunity for those children and adults who, by reason of their habitats, customs, occupations or impaired health, are at particular risk of one or other of some less common or less threatening infections.

2 Vaccines

The vaccines currently used for the prevention of the infectious diseases of humans all originate, albeit in a variety of ways, from pathogenic microbes. The essence of vaccine manufacture thus consists of procedures which make from dangerous pathogens, their components or their products, the immunogens that are devoid of pathogenic properties but which, none the less, protect from infection the individuals to whom they are given. The techniques that are used by vaccine manufacturers are constrained by costs, by problems of delivery to the vaccinee and, most of all, by the biological properties of the pathogens from which vaccines are derived. Even so the end-products of vaccine manufacture, the final vaccines, fall into seven readily recognizable types.

1 *Live vaccines.* Live vaccines are preparations of live bacteria or viruses which, when administered in an appropriate way, cause symptomless or almost symptomless infections. In the course of such an infection the constituents of the microbes in a vaccine evoke an immune response which provides protection against a serious natural disease. Live vaccines have a long history, for the very first vaccine, smallpox vaccine, was a live vaccine. It was introduced in 1796 by the Gloucestershire doctor, Edward Jenner, who recognized that an attack of the mild condition known as cowpox protected milkmaids from smallpox during epidemics of this dreaded disease. He therefore took some fluid, the lymph, from a cowpox pustule on the hand of a milkmaid and used it to inoculate a small boy. A little later he courageously inoculated the boy with lymph from a case of smallpox—and nothing happened! The boy, protected as he was by the cowpox infection, remained well. Jenner's use of the causative organism of one disease to provide protection against another is paralled by the use of the bacille Calmette–Guérin (BCG) strain of bovine tubercle bacilli to protect against infections with the human strain. However, in this case, there is the difference that the ability of the BCG strain to cause disease, its pathogenicity or virulence, has been reduced by many sequential subcultivations on laboratory media. Live vaccines such as smallpox and BCG vaccines that rely on the phenomenon of 'cross-protection' are exceptions to the generality that most vaccines are derived from the causative organisms of the diseases against which each is intended to provide protection. Thus a virulent typhoid bacillus that was enzymically crippled by the action of nitrosoguanosine on its DNA gave rise to the live typhoid vaccine Ty21a. Likewise polioviruses from human infections were grown in the laboratory in such a way that it was possible to select infectious but innocuous progeny viruses suitable for use in live (oral) poliovaccines. Comparable procedures have been used to obtain the viruses that are currently used in live measles, mumps, rubella and yellow fever vaccines. The microbes with the reduced ability to cause disease that are used in live vaccines are said to have attenuated virulence and are often referred to as attenuated or vaccine strains.

2 *Killed vaccines.* Killed vaccines are suspensions of bacteria or of viruses that have been killed by heat or by disinfectants such as phenol or formaldehyde. Killed microbes do not replicate and cause an infection and so it is necessary that, in each dose of a killed vaccine, there are sufficient microbes to stimulate a vaccinee's immune system. Killed vaccines have therefore to be relatively concentrated suspensions. Even so, such preparations are rather poor antigens and, at the same time, tend to be somewhat toxic. It is thus necessary to divide the total amount of vaccine that is needed to induce protection into two or three doses that are given at intervals of a few days or weeks. Such a course of vaccination takes advantage of the enhanced 'secondary' response that occurs when a vaccine is administered to a vaccinee whose immune system has been sensitized by a previous dose of the same vaccine. The best known killed vaccines are whooping-cough (pertussis), typhoid, cholera, Salk type poliovaccine and rabies vaccine.

3 *Toxoid vaccines.* Toxoid vaccines are preparations derived from the toxins that are secreted by certain species of bacteria. In the manufacture of such vaccines the toxoid is separated from the bacteria and treated in a way that eliminates toxicity without eliminating immunogenicity. Formalin (*ca.* 38% of formaldehyde gas in water) is used for this purpose and consequently the treated toxins are often referred to a formol toxoids. Toxoid vaccines are very effective in the prevention of those diseases such as diphtheria and tetanus in which the harmful effects of the infecting bacteria are due to the deleterious action of bacterial toxins on physiology and biochemistry.

4 *Bacterial cell component vaccines.* Several bacterial vaccines are now available that consist not of whole bacterial cells, as in conventional whooping-cough vaccine, but of selected components of bacterial cells. The intended advantages of such vaccines is that they evoke a more specific immune response and avoid the injection of cellular material which makes no contribution to the immunity and may contribute to undesirable side-effects. Among such vaccines are two consisting of bacterial polysaccharides, the meningococcal polysaccharide vaccine that contains capsular polysaccharides from one to four strains of *Neisseria meningitidis* and the pneumococcal polysaccharide vaccine that contains polysaccharides from as many as 23 strains of *Streptococcus pneumoniae*. Other vaccines made from bacterial components include an *N. meningitidis* Type B vaccine, a *Haemophilus influenzae* Type B vaccine and the acellular pertussis vaccine which contains the two components pertussis toxin and pertussis haemagglutinin and, in some cases, a pertussis agglutinin.

5 *Viral subunit vaccines.* Two viral subunit vaccines are in wide use. The best known is the influenza vaccine, consisting of the two subunits of influenza virus known as the haemagglutinin and neuraminidase. The other is the hepatitis B vaccine that is prepared by extracting from the blood of persons with hepatitis B the virus subunit known as the surface antigen HBsAg (see below also).

6 *Vaccines produced by genetic manipulation.* The techniques of genetic manipulation have been applied to vaccine manufacture in various ways. In the case of bacterial vaccines, precisely engineered modifications of the DNA of bacterial cells have been made to impair an essential pathway of bacterial metabolism. In this way it has been possible to develop strains of *Salmonella typhi,* the Aro$^-$ strains, which are unable to synthesize amino acids of the aromatic series. Such strains are metabolically crippled to the extent that they are avirulent when administered orally. Another method is applicable to vaccines such as diphtheria and tetanus vaccines which consist of toxin converted to non-toxic toxoid by the action of formalin. With these vaccines the procedure is to modify the bacterial DNA responsible for toxin production in a way that results in the production of a molecule that resembles the toxin molecule but is devoid of toxicity. A vaccine made with such a toxin-like material has the advantages that the toxoiding stage of manufacture is eliminated and there is no possibility of reversion to toxicity should the vaccine be improperly stored. In the case of a viral vaccine it has been possible to transfer the gene responsible for the production of the hepatitis B surface antigen, HBsAg, to yeast cells and to achieve expression of the gene in amounts sufficient for vaccine production. The advantage of this method of production is that the need for a human source of the HBsAg is obviated and with it all of the problems associated with the use of material derived from human donors.

7 *Synthetic peptides.* These are potential vaccines now that the amino acid sequences and structures of some protein immunogens have been resolved. Such peptides, being only a small part of the protein immunogen, are likely to be poor antigens and so may need to be linked by physicochemical methods to liposomes, mineral adjuvants or, possibly, tetanus toxoid, to increase their immunogenicity. The advantages of vaccines of this type will be the absence of any ability to replicate in the tissues of the vaccinee and the possibility of quality control by physicochemical methods.

2.1 The seed lot system

The starting point for the production of all microbial vaccines, other than synthetic peptide vaccines, is an isolate of the appropriate microbe. Such isolates have mostly been derived from human infections and in some cases have yielded strains suitable for vaccine production very readily; in other cases a great deal of manipulation and selection in the laboratory has been needed before a suitable strain has been obtained.

Once a suitable strain is available, the practice is to grow, often from a single organism, a sizeable culture which is distributed in small amounts in a large number of ampoules and then stored at $-70°C$ or freeze-dried. This is the seed lot. From this seed lot, one or more ampoules are taken and used as the seed to originate a limited number of batches of vaccine which are first examined exhaustively in the laboratory and then, if found to be satisfactory, tested for safety and efficacy in clinical trials. Satisfactory results in the clinical trials validate the seed lot as the seed from which batches of vaccine for routine use can subsequently be produced.

2.2 Production of the bacteria and bacterial components of bacterial vaccines

The bacteria and bacterial components needed for the manufacture of bacterial vaccines are readily prepared in laboratory media by well-recognized fermentation methods. The end-product of the fermentation, the harvest, is processed to provide a concentrated and purified vaccine component that may be conveniently stored for long periods or even traded as an article of commerce.

Fermentation

The satisfactory growth of a bacterial seed into the bulk material necessary for vaccine production is critically dependent on the provision of optimal cultural conditions. Liquid media, when capable of providing such conditions, are always preferred to solid media for production purposes as they can be used in large fermenter vessels and so obviate the labour-intensive manipulation of the numerous smaller vessels needed for solid media. Media are prepared whenever possible from chemically defined materials, or at least from approximately defined materials such as digested casein, in order to reduce batch variation to a minimum.

The production of a bacterial vaccine batch begins with the breaking-out of the seed and its growth through one or more passages in pre-production media. The organisms grown in the preproduction media are used to inoculate the production medium, which may be either in a large number of relatively small containers or in a single large fermenter. When small containers are used they may be stationary or shaken; when a fermenter is used the contents are continuously mixed and the pH and oxidation–reduction potential of the medium are often monitored and adjusted throughout the growth period. In the case of rapidly growing organisms this period may be as short as a few hours but in the case of slowly growing organisms as long as 2 weeks. At the end of the growth period, cultures in small containers are pooled to provide a large volume of material known as the harvest; cultures in fermenters are often sufficiently large to be used as harvests without pooling.

2.2.2   *Processing of bacterial harvests*

The harvest, i.e. the culture after the incubation period, is a very complex mixture of bacterial cells, metabolic products and exhausted medium. In the case of a live attenuated vaccine it is innocuous and all that is necessary is for the bacteria to be separated and resuspended in an appropriate menstruum, possibly for freeze-drying. With a vaccine made from a pathogen, the harvest may be intensely dangerous and great care is necessary in the following procedures.

1 *Killing.* The process by which the live bacteria in the culture are killed and thus rendered harmless. Heat and disinfectants are employed. Heat and/or formalin are required to kill the cells of *Bordetella pertussis* used to make whooping-cough vaccines, and phenol is used to kill the *Vibro cholerae* in cholera vaccine and the *Sal. typhi* in typhoid vaccine.

2 *Separation.* The process by which the bacterial cells are separated from the culture fluid. Centrifugation using either a batch or continuous flow process is commonly used, but precipitation of the cells by reducing the pH is an alternative. In the case of vaccines prepared from cells, the fluid is discarded and the cells are resuspended in a saline mixture; where vaccines are made from a constituent of the fluid, the cells are discarded.

3 *Fractionation.* The process by which components are extracted from bacterial cells or from the medium in which the bacteria are grown and obtained in a purified form. The polysaccharide antigens of *N. meningitidis* are separated from the bacterial cells by treatment with hexadecyltrimethylammonium bromide and those of *Strep. pneumoniae* with ethanol. The purity of an extracted material may be improved by resolubilization in a suitable solvent and precipitation. After purification a component may be dried to a powder, stored indefinitely and, as required, incorporated into a vaccine in precisely weighed amounts at the blending stage.

4 *Detoxification.* The process by which bacterial toxins are converted to harmless toxoids. Formalin is used to detoxify the toxins of both *Corynebacterium diphtheriae* and *Clostridium tetani*. The detoxification may be performed either on the whole culture in the fermenter or on the purified toxin after fractionation.

2.3   **Production of the components of viral vaccines**

Because viruses can replicate only in living cells the first viral vaccines were made in animals: smallpox vaccine in the dermis of calves and sheep, and rabies vaccine in the spinal cords of rabbits and the brains of mice. Such methods, although still used for rabies vaccine production, are no longer a part of modern vaccine production. Today the only intact host used in advanced production techniques is the developing chick embryo. Almost all virus production for vaccines is preferably, and necessarily, achieved in cell cultures.

2.3.1   *Growth of viruses*

The chick embryo is still the most convenient host for the growth of the virus needed for two vaccines, those against influenza and yellow fever. Influenza viruses accumulate in high titre in the readily harvested allantoic fluid of the infected egg, and yellow fever virus accumulates particularly in the embryo's nervous tissues.

The cell cultures principally used in the past were derived from the kidneys of freshly killed animals such as monkeys and rabbits. The presence in the kidneys of various viruses, including the very dangerous simian herpes B virus, often made such cultures unusable and this disadvantage, together with ethical and environmental considerations, led to an increasing use of human diploid cells for virus production. The use of diploid cells is now paralleled by the use of cultures of cells from continuous cell lines for, although the cells of such lines have characters akin to those of cultured cancer cells, they are considered safe for vaccine production subject to certain rigorous safeguards.

2.3.2   *Processing of viral harvests*

The processing of the virus-containing materials from chick embryos may take one of several different forms. In the case of influenza vaccines

337   *The manufacture and quality control of immunological products*

the allantoic fluid is centrifuged to provide a concentrated and purified suspension of virus. This concentrate may then be killed by the addition of dilute formalin when inactivated influenza vaccine is prepared or, alternatively, it may be treated with ether or other disruptive agents to split the virus into its components when a 'split virion' or surface antigen vaccine is prepared. The chick embryos used in the production of yellow fever vaccine are treated differently in that the embryos are homogenized in water to provide a virus-containing purée. Centrifugation then precipitates the embryonic debris and leaves much of the virus in an aqueous suspension.

Cell cultures provide infected fluids that contain little debris and can generally be satisfactorily clarified by filtration. Because most viral vaccines made from cell cultures consist of live attenuated virus, there is no inactivation stage in their manufacture. There are, however, two important exceptions: inactivated poliomyelitis virus vaccine is inactivated with dilute formalin or β-propiolactone and rabies vaccine is inactivated with β-propiolactone. The preparation of these inactivated vaccines also involves a concentration stage, by adsorption and elution of the virus in the case of poliomyelitis vaccine and by ultrafiltration in the case of rabies vaccine. When processing is complete the bulk materials may be stored until needed for blending into final vaccine. Because of the lability of many viruses, however, it is necessary to store most purified materials at temperatures of $-70°$C.

| 2.4 | **Blending** |

Blending is the process in which the various components of a vaccine are mixed to form a final bulk. It is undertaken in a large, closed vessel fitted with a stirrer and ports for the addition of constituents and withdrawal of the final blend. When bacterial vaccines are blended, the active constituents usually need to be greatly diluted and the vessel is first charged with the diluent, usually containing a preservative such as thiomersal and, if required, an adjuvant such as aluminium hydroxide. A single-component final bulk is then made by adding bacterial suspension, bacterial component or concentrated toxoid in such quantity that it is at the desired concentration in the final product. A multiple-component final bulk of a combined vaccine is made by adding each required component in sequence. When viral vaccines are blended the need to maintain adequate antigenicity or infectivity may preclude dilution, and tissue culture fluids or concentrates made from them are often used undiluted or, in the case of multicomponent vaccines, merely diluted one with another. After thorough mixing a final bulk may be broken down into a number of moderate sized volumes to facilitate handling.

The single-component bacterial vaccines are listed in Table 16.1 with, in the case of each preparation, notes of the basic material from which it is made, the salient production processes, and the specific tests for potency and safety. In addition to these single-component vaccines there

Table 16.1 Bacterial vaccines used for the prevention of infectious disease in humans. Vaccines marked with an asterisk are those used in conventional immunization schedules; those marked with a dagger are widely available for use in special circumstances

Vaccine	Source material	Processing	Potency assay	Safety tests
Anthrax[†]	Medium from culture of *Bacillus anthracis*	Separation of protective antigen from medium	3 + 3 quantal assay in guinea-pigs using challenge with *B. anthracis*	Exclusion of live *B. anthracis* and of anthrax toxin
BCG*	Live BCG cells from cultures in liquid or on solid media	1 Cells centrifuged from medium 2 Resuspension in stabilizer 3 Freeze-drying	Viable count: Induction of sensitivity to tuberculin in guinea-pigs	Exclusion of virulent mycobacteria; absence of excessive dermal reactivity
Diphtheria (plain)*	Diphtheria toxin from liquid cultures of *C. diphtheriae*	1 Separation and concentration of toxin 2 Conversion of toxin to toxoid by action of formaldehyde	Induction of immunity in guinea-pigs tested for by intradermal challenge with toxin	Inoculation of guinea-pigs to exclude residual toxin
Diphtheria (adsorbed)*	Diphtheria toxin from liquid cultures of *C. diphtheriae*	1 Separation and concentration of toxin 2 Conversion of toxin to toxoid 3 Adsorption of toxoid to adjuvant	3 + 3 quantal assay in guinea-pigs using intradermal challenge	Inoculation of guinea-pigs to exclude residual toxin
N. meningitidis Types A, C, W-135 and Y[†]	Polysaccharides from cultures of *N. meningitidis* of one or more serotypes	1 Precipitation with hexadecyl-trimethyammonium bromide 2 Solubilization and purification 3 Freeze-drying	Chemical estimation of polysaccharide	No specific tests
Pneumococcal polysaccharide[†]	*Step. pneumoniae* of up to 23 serotypes	1 Precipitation of polysaccharides with ethanol 2 Blending into polyvalent vaccine	Physicochemical estimation of polysaccharides	No specific tests
Tetanus (plain)*	Tetanus toxin from liquid cultures of *Cl. tetani*	1 Conversion of toxin to toxoid with formalin 2 Separation and purification of toxoid	Induction of immunity in guinea-pigs tested for by measurement of antitoxin	Inoculation of guinea-pigs to exclude presence of untoxoided toxin

Contd over page

Table 16.1 *Continued*

Vaccine	Source material	Processing	Potency assay	Safety tests
Tetanus (adsorbed)*	Tetanus toxin from liquid cultures of *Cl. tetani*	1 Conversion of toxin to toxoid 2 Separation and purification of toxoid 3 Adsorption to adjuvant	3 + 3 quantal assay in mice using subcutaneous challenge with tetanus toxin	Inoculation of guinea-pigs to exclude presence of untoxoided toxin
Typhoid†	Cells of *Sal. typhi* grown on solid or in liquid media	1 Killing with heat or formalin 2 Separation and resuspension in saline	Induction of antibodies in rabbits	No specific tests
Whooping cough (pertussis)*	Cells of *B. pertussis* grown on solid or in liquid media	1 Killing with formalin 2 Separation of bacteria 3 Resuspension in saline	3 + 3 quantal assay in mice using intracerebral challenge with live *B. pertussis*	Estimation of bacteria to limit content to 20×10^9 per human dose

Notes: Diphtheria and whooping cough vaccines are seldom used as monocomponent preparations but as components of diphtheria/tetanus vaccines and diphtheria/tetanus/pertussis vaccines. Vaccines less generally available than those listed in the table include botulism vaccine, *H. influenzae* Type B vaccine combined with *N. meningitidis* Type B vaccine, plague vaccine, *Pseudomonas aeruginosa* vaccine, tularaemia vaccine, typhoid vaccine Ty21A and typhoid vaccine prepared from the typhoid Vi (virulence) antigen.

are several mixed vaccines which are prepared by blending the concentrates used for single-component vaccines one with another. Such mixed preparations induce immunity to two or more diseases simultaneously and are so convenient that some single components, e.g. diphtheria and pertussis, are almost always used in this form. Best known of the mixed bacterial vaccines are the diphtheria/tetanus/pertussis vaccine (DTP/Vac) used to immunize infants, and the diphtheria/tetanus vaccine (DT/Vac) used for children at school entry.

The single-component viral vaccines are listed in Table 16.2 with notes similar to those provided for the bacterial vaccines. The only combined viral vaccine of importance is the MMR (measles/mumps/rubella) vaccine which in many countries is now used routinely for the immunization of infants. In a sense, however, both inactivated (Salk) and live oral (Sabin) poliovaccine are combined vaccines in that they are mixtures of all three of the serotypes of poliovirus. Influenza vaccines, too, are usually combined vaccines in that many contain immunogens from two or three strains of influenza virus.

Vaccines blended without an adjuvant are generally described as 'plain' or, in the case of bacterial toxoids, as 'fluid' vaccines; those blended with an adjuvant are described as 'adsorbed'. Aluminium hydroxide gel is the most common adjuvant and is used in the adsorbed

Table 16.2 Viral vaccines used for the prevention of infectious disease in humans. Vaccines marked with an asterisk are those used in conventional immunization programmes; those marked with a dagger are widely available for use in special circumstances

Vaccine	Source material	Processing	Potency assay	Safety tests
Hepatitis A†	Hepatitis A virus from cultures of human diploid cells	Purification by centrifugation	Estimation of virus by *in vitro* immunoassay	Inoculation of cell cultures to exclude live virus
Hepatitis B†	Hepatitis virus surface antigen expressed in cultures of yeast cells	1 Separation of HBsAg from yeast cell constituents 2 Freeze-drying	Assay of HBsAg	Not necessary
Influenza (inactivated)†	Allantoic fluid from embryonated hens' eggs infected with influenza virus	1 Inactivation with formalin 2 Concentration and purification by centrifugation	Haemagglutination of fowl erythrocytes	Inoculation of embryonated hens' eggs to exclude live virus
Influenza (surface antigen)†	Allantoic fluid from embryonated hens' eggs infected with influenza virus	1 Inactivation and disruption 2 Separation of haemagglutinin and neuraminidase 3 Blending of haemagglutinins of different serotypes	Assay of haemagglutinin	Inoculation of embryonated hens' eggs to exclude live virus
Measles*	Live virus from cultures of chick embryo cells infected with attenuated measles virus	1 Clarification 2 Freeze-drying	Infectivity titration in cell cultures	Seed virus shown to be attenuated
Mumps*	Live virus from cultures of human diploid cells	1 Clarification 2 Freeze-drying	Infectivity titration in cell cultures	Seed virus shown to be attenuated
Poliomyelitis (inactivated)† (Salk type)	Virus from monkey kidney or other cell cultures infected with virus of each serotype	1 Clarification 2 Inactivation with formalin 3 Concentration 4 Blending of virus of each serotype	Induction of antibodies to polioviruses in chicks or guinea-pigs	Inoculation of cell cultures and monkey spinal cords to exclude live virus
Poliomyelitis (live or oral)* (Sabin type)	Live virus from monkey kidney or human diploid cells infected with poliovirus of each of the serotypes	1 Clarification 2 Blending of virus of three serotypes in stabilizing medium	Infectivity titration of each of three virus serotypes	Safety testing by inoculation of spinal cords of monkeys and comparison of lesions with those produced by a reference vaccine

Contd over page

341   *The manufacture and quality control of immunological products*

Table 16.2 *Continued*

Vaccine	Source material	Processing	Potency assay	Safety tests
Rubella* (German measles)	Live virus from cultures of rabbit kidney, chick embryo or human diploid cells	1 Clarification 2 Blending with stabilizer 3 Freeze-drying	Infectivity titration in cell cultures	Seed virus shown to be attenuated
Tick-borne encephalitis†	Fluid from cultures of embryonated hens' eggs infected with TBE virus	1 Clarification 2 Inactivation with formalin 3 Freeze-drying	Infectivity titrations in cell cultures	Inoculation of cell cultures to exclude live virus
Yellow fever†	Aqueous homogenate of chick embryos infected with attenuated yellow fever virus 17D	1 Centrifugation to remove cell debris 2 Freeze-drying	Infectivity titrations in cell cultures	No specific tests

Notes: The use of monocomponent measles, mumps and rubella vaccines is steadily diminishing due to the inclusion of a combined measles/mumps/rubella vaccine in immunization schedules. Viral vaccines less generally available than those listed in the table include Congo Crimean haemorrhagic fever vaccine, dengue fever vaccine, Japanese encephalitis B vaccine, smallpox vaccine, varicella (chicken pox) vaccine, and Venezuelan equine encephalitis vaccine.

diphtheria vaccine, the adsorbed tetanus vaccine, the adsorbed pertussis vaccine and various adsorbed combined vaccines. Aluminium phosphate is a satisfactory alternative, and in some countries calcium phosphate is preferred.

A preservative, almost always the mercury-containing thiomersal (Chapter 11), is usually added to bacterial vaccines and is always added to killed vaccines in multidose containers. The possibility that a contaminant, introduced into the container during the withdrawal of a dose of vaccine, may later infect a vaccinated individual is thereby much reduced (see also Chapter 21).

Live, attenuated viral vaccines lose potency rather quickly in suspension and to alleviate this vaccines are stored frozen or a stabilizer may be added. Live, attenuated poliomyelitis vaccine, for example, is effectively stabilized with $1 \, mol \, l^{-1}$ magnesium chloride or with sucrose.

2.5 Filling and drying

As vaccine is required for distribution, bulk vaccine is dispensed into sterile ampoules or bottles in appropriate volumes. Vaccines dispensed as liquids are sealed or capped in their containers; those dispensed as dry preparations are freeze-dried before sealing.

2.6 Quality control

The quality control of vaccines is intended to provide assurances of both the efficacy and the safety of every batch of every product. It is executed in three ways: (i) in-process control; (ii) final-product control; and (iii) a requirement that for each product the starting materials, intermediates, final product and processing methods are consistent. The results of all quality control tests are always recorded in detail as, in those countries in which the manufacture of vaccines is regulated by law, they are part of the evidence on which control authorities judge the suitability or otherwise of each batch of each preparation.

2.6.1   *In-process control*

In-process quality control is the control exercised over starting materials and intermediates. Its importance stems from the opportunities that it provides for the examination of a product at the stages in its manufacture at which testing is most likely to provide the most meaningful information. The WHO *Requirements* and national authorities stipulate many in-process controls but manufacturers often perform tests in excess of those stipulated, especially sterility tests (Chapter 23) as, by so doing, they obtain assurance that production is proceeding normally and that the final product is likely to be satisfactory. Examples of in-process control abound but three of different types should suffice.

1   The quality control of both diphtheria and tetanus vaccines requires that the products are tested for the presence of free toxin, that is for specific toxicity due to inadequate detoxification with formalin, at the final-product stage. By this stage, however, the toxoid concentrates used in the preparation of the vaccines have been much diluted and, as the volume of vaccine that can be inoculated into the test animals (guinea-pigs) is limited, the tests are relatively insensitive. In-process control, however, provides for tests on the undiluted concentrates and thus increases the sensitivity of the method at least 100-fold.

2   An example from virus vaccine manufacture is the titration, prior to inactivation, of the infectivity of the pools of live poliovirus used to make inactivated poliomyelitis vaccine. Adequate infectivity of the virus from the tissue cultures is an indicator of the adequate virus content of the starting material and, since infectivity is destroyed in the inactivation process, there is no possibility of performing such an estimation after formolization.

3   A more general example from virus vaccine production is the rigorous examination of tissue cultures to exclude contamination with infectious agents from the source animal or, in the cases of human diploid cells or cells from continuous cell lines, to detect cells with abnormal characteristics. Monkey kidney cell cultures are tested for simian herpes B virus, simian virus 40, mycoplasma and tubercle bacilli. Cultures of human diploid cells and continuous line cells are subjected to detailed karyological examination (examination of chromosomes by microscopy) to ensure that the cells have not undergone any changes likely to impair the quality of a vaccine or lead to undesirable side-effects.

343   *The manufacture and quality control of immunological products*

Final-product control is the quality control exercised by the monographs of a pharmacopoeia over products in their final containers. In practice, each product is regulated partly by identity, potency and safety tests unique to itself and closely associated with its biological properties, and partly by tests of general applicability. Combined vaccines are required to pass the tests prescribed for each of the separate components.

Identity tests.   Identity tests ensure that the material in an ampoule corresponds with the legend on the ampoule label. The identities of many vaccines can be checked by rapid *in vitro* agglutination and precipitation methods. Some vaccines are tested by observation of the specific antibody responses that they evoke in vaccinated animals. Live viral vaccines are identified by neutralization of their cytopathic effects by specific antisera.

Potency assay.   The definitive measure of the potency of a vaccine for use in humans is its protective efficacy in a clinical trial. Clearly, it is rarely possible to assay a product in this way and so recourse must almost always be made to laboratory methods. In the cases of a few vaccines in which the active principles are highly purified microbial components it is possible to assess potency by physicochemical or serological techniques. Thus the quality, i.e. the molecular profiles and purity, of the polysaccharides of meningococcal vaccines and of pneumococcal vaccines can be readily assessed by gel filtration techniques and the haemagglutinin of subunit influenza vaccine by single radial diffusion in agarose. Most vaccines, however, still have to be assayed for potency by biological techniques, often by tests in animals. In ideal circumstances a portion of a preparation that has been found to be effective in a clinical trial is adopted as a laboratory standard with which all subsequent batches are compared. Only those batches that exhibit an agreed percentage of the potency of the standard are acceptable.

Because each vaccine contains components unique to itself, each must be tested for potency in a different way. Even so, the potency tests applicable to all vaccines are based on a limited number of basic principles and the tests used for quite different preparations are often similar in design. Where the designs are alike, the ability of the tests to estimate a unique property is dependent on the use of different standard materials, experimental animals, criteria of antibody response and other features.

Most vaccines containing killed microorganisms or their products, such as bacterial toxoids, are tested for potency in assays in which a vaccine's ability to evoke antibody production in a group of laboratory animals is measured. In its older form this type of assay involves vaccination of a group of animals, the abstraction of a blood sample from each animal at a fixed time afterwards, and titration of the samples for antibody. The titres provided by the titrations are combined to a mean which, if greater than an arbitrary level, is considered to indicate a potent product. A rather similar assay is prescribed in the *British Pharmacopoeia* for plain diphtheria vaccine except that, in this case, the antibody response is judged not from antibody titres but from the results of intra-dermal injection of a small quantity of diphtheria toxin into each test animal. Animals with antitoxin neutralize the toxin and show no reaction; those without develop an area of erythema at the injection site. A variant of this test is used for the potency assay of inactivated poliomyelitis vaccine and consists of the determination of the quantity of vaccine that evokes an antibody response in 50% of a group of chicks or guinea-pigs.

Truly quantitative estimates of the potencies of most inactivated vaccines, however, can be obtained only by a formal comparison in animals of the efficacy of a test vaccine with that of a corresponding standard vaccine. The usual format of such a test is the 3 + 3 dose quantal assay as used to estimate the potencies of whooping-cough vaccines (*British Pharmacopoeia*, 1988) and of cholera vaccines (WHO, 1973). Three serial doses of the test vaccine and three of a standard vaccine, carefully chosen so that in both cases the middle doses are sufficient to immunize about 50% of the mice to which they are given, are each used to vaccinate groups of 16 or more mice. Fourteen days later all the mice are artificially infected ('challenged') with the pathogen against which the vaccine is intended to provide protection. The number of protected animals, the survivors, in each of the six groups is counted several days later and used to calculate an estimate of potency of the test vaccine by the statistical technique of probit analysis (Finney, 1971). The estimate of potency is realized either as a percentage of the potency of the standard or, when a standard with a potency assigned in international units (IU) is used, in international units with appropriate confidence limits. Similar tests, involving bacterial toxins rather than live bacteria for the challenge, are prescribed by the *British Pharmacopoeia* (1988) for the potency assay of adsorbed diphtheria and tetanus toxoids.

Vaccines containing live microorganisms are generally tested for potency by counts of their viable particles. In the case of the only live bacterial vaccine in common use, BCG vaccine, dilutions of vaccine are made and dropped in fixed volumes on to solid media capable of supporting the microorganisms' growth. After a fortnight the colonies generated by the drops are counted and the live count of the undiluted vaccine is calculated. The potency of live viral vaccines is estimated in much the same way except that a substrate of living cells is used. Dilutions of vaccine are inoculated into the brains of mice or on to tissue culture monolayers in Petri dishes or in plastic trays, and the live count of the vaccine is calculated from the infectivity of the dilutions and dilution factor involved.

Safety tests. Because many vaccines are derived from basic materials of intense pathogenicity—the lethal dose of a tetanus toxin for a mouse is estimated to be $3 \times 10^{-5}\,\mu g$—safety testing is of paramount importance. Effective testing provides a guarantee of the safety of each batch of every product, and most vaccines in the final container must pass one or more safety tests as prescribed in a pharmacopoeial monograph. This generality does not absolve a manufacturer from the need to perform 'in-process' tests as required, but it is relaxed for those preparations which have a final formulation that makes safety tests on the final product either impractical or meaningless.

Bacterial vaccines are regulated by relatively simple safety tests. Those vaccines composed of killed bacteria or bacterial products must be shown to be completely free from the living microbes used in the production process, and inoculation of appropriate bacteriological media with the final product provides an assurance that all organisms have been killed. Those containing diphtheria and tetanus toxoids require, in addition, a test system capable of revealing inadequately detoxified toxins; inoculation of guinea-pigs, which are exquisitely sensitive to both diphtheria and tetanus toxins, is always used for this purpose. Inoculation of guinea-pigs is also used to exclude the presence of abnormally virulent organisms in BCG vaccine.

Viral vaccines present problems of safety testing far more complex than those experienced with bacterial vaccines. With killed viral vaccines the potential hazards are those due to incomplete virus inactivation and the consequent presence of residual live virus in the preparation. The tests used to detect such live virus consist of the inoculation of susceptible tissue cultures and of susceptible animals. The cultures are examined for cytopathic effects and the animals for symptoms of disease and histological evidence of infection at autopsy. This test is of particular importance in inactivated poliomyelitis vaccine, the vaccine being injected intraspinally into monkeys. At autopsy, sections of brain and spinal cord are examined microscopically for the histological lesions indicative of proliferating poliovirus.

With attenuated viral vaccines the potential hazards are those associated with reversion of the virus during production to a degree of virulence capable of causing disease in vaccinees. To a large extent this possibility is controlled by very careful selection of a stable seed but, especially with live attenuated poliomyelitis vaccine, it is usual to compare the neuro-virulence of the vaccine with that of a vaccine known to be safe in field use. The technique involves the intraspinal inoculation of monkeys with a reference vaccine and with the test vaccine and a comparison of the neurological lesions and symptoms, if any, caused. If the vaccine causes abnormalities in excess of those caused by the reference it fails the test.

Tests of general application. In addition to the tests designed to estimate the potency and to exclude the hazards peculiar to each vaccine there are a number of tests of more general application. These relatively simple tests are as follows.

1 *Sterility*. In general vaccines are required to be sterile. The exceptions to this requirement are smallpox vaccine made from the dermis of animals and bacterial vaccines such as BCG, Ty21A and tularaemia vaccine which consist of living but attenuated microbes. WHO *Requirements* and pharmacopoeial standards stipulate, for vaccine batches of different size, the numbers of containers that must be tested and found to be sterile. The preferred method of sterility testing is membrane filtration as this technique permits the testing of large volumes without dilution of the test media. The test system must be capable of detecting aerobic and anaerobic organisms and fungi (see Chapter 23).

2 *Freedom from abnormal toxicity.* The purpose of this simple test is to exclude the presence in a final container of a highly toxic contaminant. Five mice of 17–22 g and two guinea-pigs of 250–350 g are inoculated with one human dose or 1.0 ml, whichever is less, of the test preparation. All must survive for 7 days without signs of illness.

3 *Presence of aluminium and calcium.* The quantity of aluminium in vaccines containing aluminium hydroxide or aluminium phosphate as an adjuvant is limited to 1.25 mg per dose and it is usually estimated compleximetrically. The quantity of calcium is limited to 1.3 mg per dose and is usually estimated by flame photometry.

4 *Free formalin.* Inactivation of bacterial toxins with formalin may lead to the presence of small amounts of free formalin in the final product. The concentration, as estimated by colour development with acetylacetone, must not exceed 0.02%.

5 *Phenol concentration.* When phenol is used to preserve a vaccine its concentration must not exceed 0.25% w/v or, in the case of some vaccines, 0.5% w/v. Phenol is estimated by the colour reaction with aminophenazone and hexacyanoferrate.

3   **Immunosera**

Immunosera provide immediately available antibody both for the prevention of infection and for the treatment of established disease, and there are monographs for eight such preparations in the *European Pharmacopoeia* (1980–86). However, since two are for use in conditions, diphtheria and tetanus, that have been largely eradicated from industrialized countries by vaccination, and the remainder are for relatively uncommon conditions, the manufacture of immunosera is today conducted much less extensively than in the past.

Horses are chiefly used for the production of immunosera, but cattle, goats and sheep are alternative sources. They are immunized by repeated injections of the appropriate antigen, initially of the toxoid, but when some protective antibody has been raised, often of the toxin, which provides a greater antigenic stimulus. Trial bleeds are taken and when the antibody titre is sufficiently high several litres of blood are withdrawn by venepuncture into citrate solution. After the cells have settled, the plasma is drawn off and the immunoglobulins are precipitated with ammonium sulphate and digested with trypsin to provide a 'refined' antiserum that contains no more than a trace of serum albumin. The final product is filtered, diluted to an appropriate concentration and ampouled.

Table 16.3 Immunosera used in the prevention and treatment of human infections

Immunoserum	Minimal potency (IU ml^{-1})	Potency assay
Botulinum antitoxin	500 Type A 500 Type B 50 Type E	Neutralization of the lethality for mice of botulinum toxins of Types A, B and E
Diphtheria antitoxin	1000 if from horses 500 if from other species	Neutralization of the erythrogenic property of diphtheria toxin in rabbit or guinea-pig skin using the technique of intradermal injection
Gas gangrene (novyi) Gas gangrene (perfringens) Gas gangrene (septicum) Gas gangrene (mixed)	3750 1500 1500 1000 of novyi 1000 of perfringens 500 of septicum	Neutralization of the lethality for mice of the homologous gas gangrene toxins
Rabies antiserum	80	Neutralization of the lethality of rabies virus on injection into the brains of mice
Tetanus antitoxin	1000 for prophylaxis 3000 for treatment	Neutralization of the paralytic effect of tetanus toxin in mice

The quality of immunosera is controlled by potency assays specific for each preparation and by safety tests of general application. In the potency tests an estimate of antitoxin content (in IU ml^{-1}) is obtained by comparing the capacities of an immunoserum and of the corresponding standard preparation calibrated in international units to neutralize a fixed dose of homologous toxin (*British Pharmacopoeia*, 1988). The test methods for each of the immunosera are indicated in Table 16.3, together with the required minimal potencies. The safety tests for immunosera regulate the species of origin, pH, sterility, phenol concentration, abnormal toxicity, total protein content and the presence of contaminating albumin.

4   Human immunoglobulins

Like immunosera, human immunoglobulins provide immediately available antibody but with the advantages that, being homologous, they neither sensitize nor cause reactions in hypersensitive persons to whom they are administered. Additionally, and importantly, they are removed from the blood much more slowly than heterologous immunosera. The source of most immunoglobulins is donated human blood and plasma obtained by plasmapheresis but placentae are occasionally used as an alternative. Pools of plasma from not less than a thousand donors are employed for the preparation of normal immunoglobulin but very much smaller pools are acceptable for the preparation of specific immunoglobulins, these being necessarily made only from the blood of convalescents and recent vaccinees.

Table 16.4 Human immunoglobulins used in the prevention and treatment of human infections

Immunoglobulin	Minimal potency (IU ml^{-1})	Potency assay
Measles	50	Neutralization of the infectivity of measles virus for tissue culture
Normal	Measurable amounts of any one bacterial and any one viral antibody for which there are international standards	Neutralization of a bacterial toxin and of the infectivity of a virus, as required
Tetanus	100	Neutralization of the paralytic effects of tetanus in mice
Vaccinia	500	Neutralization of the property of vaccinia virus to cause pocks on the chorioallantoic membranes of fertile hens' eggs

Several methods have been devised to maximize the yield of the therapeutically useful constituents that can be obtained from human plasma. Fractional precipitation with ethanol in cold conditions, with rigorous control of pH and ionic strength (Cohn *et al.*, 1946), is the basis of the method although numerous modifications using rivanol and ammonium sulphate have been introduced to improve yields or to facilitate production (Watt, 1976). The immunoglobulin fraction obtained may be dispensed freeze-dried or in saline at a concentration 10–20 times that of immunoglobulin in natural plasma. Glycine may be added as a stabilizer and thiomersal as a preservative.

The quality control of immunoglobulins parallels that of immunosera. Potency is determined by neutralization tests and the methods and minimal acceptable potencies for each type of immunoglobulin are indicated in Table 16.4. Safety tests regulate pH, sterility, abnormal toxicity, total protein content, stability and pyrogenicity. In addition, the human origin of immunoglobulins is controlled by precipitin tests and the homogeneity of the immunoglobulin by moving boundary electrophoresis and by determination of sedimentation coefficients in the ultracentrifuge.

In recent years many attempts have been made to replace human immunoglobulins, particularly tetanus immunoglobulin, by the antibodies secreted by monoclones of lymphocytes. The possibilities of contamination of such immunoglobulins with murine viruses and with oncogenic DNA, and the low titres and avidity of monoclonal antibodies produced in tissue culture, have to date precluded the clinical exploitation of such preparations.

5 Tailpiece

The immunological products recommended for the protection of individuals at ordinary risk of infectious disease can be purchased at a cost equivalent to about 20 loaves of bread. That is a small price to pay for virtually life-long protection from diphtheria, tetanus, whooping cough, poliomyelitis, measles, mumps, rubella and tuberculosis. It is a price that belies the extraordinary care for efficacy and safety which is exercised in the manufacture and the quality control of every immunological preparation.

6   Further reading

British National Formulary. London: British Medical Association and Pharmaceutical Press. (This contains a useful section on immunological products.) New editions of the BNF appear at regular intervals.

British Pharmacopoeia (1988). London: HMSO.

Cohn E.J. Strong L.E., Hughes W.L., Hulford D.J., Ashworth J.N., Melin M. & Taylor H.L. (1946) Preparation and properties of serum proteins. IV. *J Am Chem Soc.* **68**, 459–475.

Dudgeon A.J. & Cutting A.M. (Eds) (1991) *Immunization: Principles and Practice.* London: Chapman and Hall Medical.

European Pharmacopoeia (1980–86), 2nd edn. Saint Ruffine, France: Maisonneuve SA.

Finney D.J. (1971) *Probit Analysis.* London: Cambridge University Press.

Germanier R. (1984). *Bacterial Vaccines.* London: Academic Press.

Harms A.J. (1948) The purification of antitoxic plasmas by enzyme treatment and heat denaturation. *Biochem J.* **42**, 340–347.

Hemert P. van (1974) Vaccine production as a unit process. *Prog Ind Microbiol.* **13**, 151–171.

Watt J.G. (1976) Plasma fractionation. *Clin Haematol.* **5**, 95–111.

World Health Organization (1973) Expert committee on biological standardisation *Technical Report Series*, **530**, 18–21.

Part 3 Microbiological aspects of pharmaceutical processing

Many failures in pharmaceutical processing have arisen because of the inability of those responsible for its design to be aware of the distribution and survival potential of microorganisms in the environment and in the raw materials and equipment used in a pharmaceutical factory.

The first chapter in this section provides a unique account of the ecology, i.e. distribution, survival and life-style of microorganisms in the factory environment, and should enable process designers, controllers and quality control personnel to comprehend, trace and eradicate the sources of failure due to extraneous microbial contaminants in the finished product. Much of the information given here is applicable to hospital manufacture also, and this is extended in a contribution (Chapter 19) dealing with contamination in hospital pharmaceutical products and in the home.

The dire consequences of failure to heed the precepts enunciated in Chapter 17 are considered in Chapters 18 and 19 which review the spoilage wreaked upon pharmaceutical products as a consequence of microbial infestation. The wide, and at first sight bizarre, range of substrates used by contaminating microorganisms and the range of biochemical reactions that follow and which, in turn, give rise to the overall picture of spoilage are well documented; it would be no exaggeration to state that microbial spoilage and its prevention, through both good working conditions and the use of preservatives, represents the major problem of pharmaceutical microbiology.

An important group of pharmaceutical products, including those intended for parenteral administration or for instillation into the eye, are required to be free from living microorganisms, and with parenteral products, from those residues of the bacterial cell which may give rise to fever. The principles of their preparation and sterilization are considered in Chapter 21, while the theory of sterilization processes is dealt with in the preceding chapter. These two chapters *in toto* cover the most exacting operation in medicine preparation, and one in which failure has given rise to several disasters ranging from patient death, for instance as a result of sterilization failure in intravenous drips, to blindness in the case of contaminated eyedrops.

Chapter 22 deals with general factory and hospital hygiene and the principles of good manufacturing practice (GMP) which if adhered to go a long way towards compounding the success of the processes described in Chapters 17 and 20.

The subject of quality control and surveillance is discussed in a chapter on sterilization control and sterility testing, which deals with aspects of in-process and post-process control.

Finally, lest it be thought that microorganisms are always harmful, two chapters (24 and 25) describe ways in which they can be harnessed for the benefit of mankind.

17 Ecology of microorganisms as it affects the pharmaceutical industry

1	**Introduction**	**5**	**Raw materials**
2	**Atmosphere**	**6**	**Packaging**
2.1	Microbial content		
2.2	Reduction of microbial count	**7**	**Buildings**
2.3	Compressed air	7.1	Walls and ceilings
		7.2	Floors and drains
3	**Water**	7.3	Doors, windows and fittings
3.1	Raw or mains water		
3.2	Softened water	**8**	**Equipment**
3.3	Deionized or demineralized water	8.1	Pipelines
3.4	Distilled water	8.2	Cleansing
3.5	Water produced by reverse osmosis	8.3	Disinfection and sterilization
3.6	Distribution system	8.4	Microbial checks
3.7	Disinfection of water		
		9	**Cleaning equipment and utensils**
4	**Skin and respiratory-tract flora**		
4.1	Microbial transfer from operators	**10**	**Further reading**
4.2	Hygiene and protective clothing		

1   Introduction

The microbiological quality of pharmaceutical products is influenced by the environment in which they are manufactured and by the materials used in their formulation. With the exception of preparations which are terminally sterilized in their final container, the microflora of the final product may represent the contaminants from the raw materials, from the equipment with which it was made, from the atmosphere, from the person operating the process or from the final container into which it was packed. Some of the contaminants may be pathogenic whilst others may grow even in the presence of preservatives and spoil the product. Any microorganisms which are destroyed by in-process heat treatment may still leave cell residues which may be toxic or pyrogenic (Chapter 1), since the pyrogenic fraction, lipid A, which is present in the cell wall is not destroyed under the same conditions as the organisms.

2   Atmosphere

2.1   Microbial content

Air is not a natural environment for the growth and reproduction of microorganisms since it does not contain the necessary amount of moisture and nutrients in a form which can be utilized. However, almost any sample of untreated air contains suspended bacteria, moulds and yeast, but to survive they must be able to tolerate desiccation and the continuing

353

dry state. Microorganisms commonly isolated from air are the spore-forming bacteria *Bacillus* spp. and *Clostridium* spp., the non-sporing bacteria *Staphylococcus* spp., *Streptococcus* spp. and *Corynebacterium* spp., the moulds *Penicillium* spp., *Cladosporium* spp., *Aspergillus* spp. and *Mucor* spp., as well as the yeast *Rhodotorula* spp.

The number of organisms in the atmosphere depends on the activity in the environment and the amount of dust which is disturbed. An area containing working machinery and an active personnel will have a higher microbial count than one with a still atmosphere, and the air count of a dirty, untidy room will be greater than that of a clean room. The microbial air count is also influenced by humidity. A damp atmosphere usually contains fewer organisms than a dry one as the contaminants are carried down by the droplets of moisture. Thus the air in a cold-store is usually free from microorganisms and air is less contaminated during the wet winter months than in the drier summer months.

Microorganisms are carried into the atmosphere suspended on particles of dust, skin or clothing, or in droplets of moisture or sputum following talking, coughing or sneezing. The size of the particles to which the organisms are attached, together with the humidity of the air, determines the rate at which they will settle out. Bacteria and moulds not attached to suspended matter will settle out slowly in a quiet atmosphere. The rate of settling out will depend upon air currents caused by ventilation, air extraction systems, convection currents above heat sources and the activity in the room.

The microbial content of the air may be increased during the handling of contaminated materials during dispensing, blending and their addition to formulations. In particular, the use of starches and some sugars in the dry state may increase the mould count. Some packaging components, for example card and paperboard, have a microflora of both moulds and bacteria, and this is often reflected in high counts around packaging machines.

Common methods for checking the microbiological quality of air include the following (see also p. 281).

1 The exposure of Petri dishes containing a nutrient agar to the atmosphere for a given length of time. This relies upon microorganisms or dust particles bearing them to settle on the surface.

2 The use of an air-sampling machine which draws a measured volume of air from the environment and impinges it on a nutrient agar surface on either a Petri dish, a plastic strip or a membrane filter which may then be incubated with a nutrient medium. This method provides valuable information in areas of low microbial contamination, particularly if the sample is taken close to the working area.

The type of formulation being prepared determines the microbiological standard of the air supply. In areas where sterile products such as injections and ophthalmic drops, which are not terminally sterilized by autoclaving, are being made the air count should be very low and a count of less than 10 organisms per 1000 litres of air is practicable under good conditions. Although such products are required to pass a test for sterility (Chapter 23), the test itself is destructive, and therefore only relatively few samples are tested. An unsatisfactory air count may lead to the casual contamination of a few containers and be undetected by the test for sterility. The manufacture of liquid or semi-solid preparations for either oral or topical uses requires a clean environment for both the production and filling stages. Whilst many formulations are adequately protected by chemical preservatives or a pH unfavourable to airborne bacteria that may settle in them, preservation against airborne mould spores is more difficult to achieve.

2.2   **Reduction of microbial count**

The microbial count of air may be reduced by filtration, chemical disinfection and to a limited extent by ultraviolet (UV) light. Filtration is the most commonly used method and filters may be made of a variety of materials such as cellulose, glass wool, fibreglass mixtures or polytetrafluoroethylene (PTFE) with resin or acrylic binders. For the most critical aseptic work, it may be necessary to remove all particles in excess of $0.1\,\mu$m in size, but for many operations a standard of less than 100 particles per 3.5 litres $(1.0\,\text{ft}^3)$ of $0.5\,\mu$m or larger (class 100) is adequate. Such fine filtration is usually preceded by a coarse filter stage, or any suspended matter is removed by passing the air through an electrostatic field. To maintain efficiency, all air filters must be kept dry, since microorganisms may be capable of movement along continuous wet films and may be carried through a damp filter.

Filtered air may be used to purge a complete room, or it may be confined to a specific area and incorporate the principle of laminar flow, which permits operations to be carried out in a gentle current of sterile air. The direction of the airflow may be horizontal or vertical, depending upon the type of equipment being used, the type of operation and the material being handled. It is important that there is no obstruction between the air supply and the exposed product, since this may result in the deflection of microorganisms or particulate matter from a non-sterile surface and cause contamination. Airflow gauges are essential to monitor that the correct flow rate is obtained in laminar flow units and in complete suites to ensure that a positive pressure from clean to less clean areas is always maintained.

The integrity of the air-filtration system must be checked regularly. One method is by counting the particulate matter both in the working area and across the surface of the filter. For systems which have complex ducting or where the surfaces of the terminal filters are recessed, smoke tests using a chemical of a known particle size such as dioctylphthalate (DOP), may be introduced just after the main fan and monitored at each outlet. The test has a twofold application as both the terminal filter and any leaks in the ducting can be checked. These methods are useful in conjunction with those for determining the microbiological air count as given earlier.

Chemical disinfectants are limited in their use as air sterilants because of their irritant properties when sprayed. However, some success has been achieved with atomized propylene glycol at a concentration of $0.05-0.5\,\mathrm{mg\,l^{-1}}$ and quaternary ammonium compounds (QACs) at 0.075% may be used. For areas which can be effectively sealed off for fumigation purposes, formaldehyde gas at a concentration of $1-2\,\mathrm{mg\,l^{-1}}$ of air at a relative humidity of 80–90% is effective.

Ultraviolet irradiation at wavelengths between 280 and 240 nm (2800 and 2400 Å) is used to reduce bacterial contamination of air, but is only active at a relatively short distance from source. Bacteria and mould spores, in particular those with heavily pigmented spore coats, are often resistant to such treatment.

2.3   Compressed air

Compressed air has many applications in the manufacture of pharmaceutical products. A few examples of its uses are the conveyance of powders and suspensions, providing aeration for some fermentations and as a power supply for the reduction of particle size by impaction. Unless it is sterilized by filtration or a combination of heat and filtration, microorganisms will be introduced into the product. The microbial content of compressed air may be assessed by bubbling a known volume through a nutrient broth and filtering the broth through a membrane, which is then incubated with a nutrient agar and a total viable count made.

3   Water

The microbial ecology of water is of great importance in the pharmaceutical industry due to its multiple uses as a constituent of many products as well as for various washing and cooling processes. Two main aspects are involved: the quality of the raw water and any processing it receives, and the distribution system.

Microorganisms indigenous to fresh water include *Pseudomonas* spp., *Alcaligenes* spp., *Flavobacterium* spp., *Chromobacter* spp. and *Serratia* spp. Such bacteria are nutritionally undemanding and often have a relatively low optimum growth temperature. Bacteria which are introduced as a result of soil erosion, heavy rainfall and decaying plant matter include *Bacillus subtilis*, *B. megaterium*, *Klebsiella aerogenes* and *Enterobacter cloacae*. Contamination by sewage results in the presence of *Proteus* spp., *Escherichia coli* and other enterobacteria, *Streptococcus faecalis* and *Clostridium* spp. Bacteria which are introduced as a result of animal or plant debris usually die as a result of the unfavourable conditions.

An examination of stored industrial water supplies showed that 98% of the contaminants were Gram-negative bacteria; other organisms isolated were *Micrococcus* spp., *Cytophaga* spp., yeast, yeast-like fungi and actinomycetes.

3.1 Raw or mains water

The quality of the water from the mains supply varies with both the source and the local authority, and whilst it is free from known pathogens and from faecal contaminants such as *E. coli*, it may contain other microorganisms. When the supply is derived from surface water the flora is usually more abundant and faster growing than that of supplies from a deep water source such as a well or spring. This is due to surface waters receiving both microorganisms and nutrients from soil and sewage whilst water from deep sources has its microflora filtered out. On prolonged storage in a reservoir, water-borne organisms tend to settle out, but in industrial storage tanks the intermittent through-put ensures that, unless treated, the contents of the tank serve as a source of infection. The bacterial count may rise rapidly in such tanks during summer months and reach $10^5 - 10^6 \, \text{ml}^{-1}$.

One of the uses of mains water is for washing chemicals used in pharmaceutical preparations to remove impurities or unwanted by-products of a reaction, and although the bacterial count of the water may be low, the volume used is large and the material being washed may be exposed to a considerable number of bacteria.

The microbial count of the mains water will be reflected in both softened and deionized water which may be prepared from it.

3.2 Softened water

This is usually prepared by either a base-exchange method using sodium zeolite, by a lime-soda ash process, or by the addition of sodium hexa-metaphosphate. In addition to the bacteria derived from the mains water, additional flora of *Bacillus* spp. and *Staphylococcus aureus* may be introduced into systems which use brine for regeneration and from the chemical filter beds which, unless treated, can act as a reservoir for bacteria.

Softened water is often used for washing containers before filling with liquid or semi-solid preparations and for cooling systems. Unless precautions are taken, the microbial count in a cooling system or jacketed vessel will rise rapidly and if faults develop in the cooling plates or vessel wall, contamination of the product may occur.

3.3 Deionized or demineralized water

Deionized water is prepared by passing mains water through anion and cation exchange resin beds to remove the ions. Thus, any bacteria present in the mains water will also be present in the deionized water, and beds which are not regenerated frequently with strong acid or alkali are often heavily contaminated and add to the bacterial content of the water. This problem has prompted the development of resins able to resist microbiological contamination. One such resin, a large-pore, strong-base, macroreticular, quaternary ammonium anion exchange resin which permits microorganisms to enter the pore cavity and then electrostatically binds them to the cavity surface, is currently being marketed. The main function is as a final cleaning bed downstream of conventional demineralizing columns.

Deionized water is used in pharmaceutical formulations, for washing containers and plant, and for the preparation of disinfectant solutions.

3.4 Distilled water

As it leaves the still, distilled water is free from microorganisms, and contamination occurs as a result of a fault in the cooling system, the storage vessel or the distribution system. The flora of contaminated distilled water is usually Gram-negative bacteria and since it is introduced after a sterilization process, it is often a pure culture. A level of organisms up to $10^6 \, ml^{-1}$ has been recorded.

Distilled water is often used in the formulation of oral and topical pharmaceutical preparations and a low bacterial count is desirable. It is also used after distillation with a specially designed still, often made of glass, for the manufacture of parenteral preparations and a post-distillation heat sterilization stage is commonly included in the process. Water for such preparations is often stored at 80°C in order to prevent bacterial growth and the production of pyrogenic substances which accompany such growth.

3.5 Water produced by reverse osmosis

Water produced by reverse osmosis (RO) is forced by an osmotic pressure through a semi-permeable membrane which acts as a molecular filter. The diffusion of solubles dissolved in the water is impeded, and those with a molecular weight in excess of 250 do not diffuse at all. The process, which is the reverse of the natural process of osmosis, thus removes microorganisms and their pyrogens. Post-RO contamination may occur if the plant after the membrane, the storage vessel or the distribution system, is not kept free from microorganisms.

3.6 Distribution system

If microorganisms colonize a storage vessel, it then acts as a microbial reservoir and contaminates all water passing through it. It is therefore important that the contents of all storage vessels are tested regularly. Reservoirs of microorganisms may also build up in booster pumps, water meters and unused sections of pipeline. Where a high positive pressure is absent or cannot be continuously maintained, outlets such as cocks and taps may permit bacteria to enter the system.

An optimum system for reducing the growth of microbial flora is one that ensures a constant recirculation of water at a positive pressure through a ring-main without 'dead-legs' (areas which due to their location are not regularly used) and only very short branches to the take-off points. In addition there should be a system to resterilize the water, usually by membrane filtration or UV light treatment, just prior to return to the main storage tank.

Some plumbing materials used for storage vessels, pipework and jointing may support microbial growth. Some plastics, in particular plasticized polyvinylchlorides and resins used in the manufacture of glass-reinforced plastics, have caused serious microbiological problems when used for water storage and distribution systems. Both natural and synthetic rubbers used for washers, O-rings and diaphragms are susceptible to contamination if not sanitized regularly. For jointing, packing and lubricating materials, PTFE and silicone-based compounds are superior to those based on natural products such as vegetable oils or fibres and animal fats, and petroleum-based compounds.

3.7   Disinfection of water

Three methods are used for treating water, namely chemicals, filtration or UV light.

1 Chemical treatment is applicable usually to raw, mains and softened water, but is also used to treat the storage and distribution systems of distilled and deionized water and of water produced by reverse osmosis (section 3.5).

Sodium hypochlorite and chlorine gas are the most common agents for treating the water supply itself, and the concentration employed depends both upon the dwell time and the chlorine demand of the water. For most purposes a free residual chlorine level of 0.5–5 p.p.m. is adequate. For storage vessels, pipelines, pumps and outlets a higher level of 50–100 p.p.m. may be necessary, but it is usually necessary to use a descaling agent before disinfection in areas where the water is hard. Distilled, deionized and RO systems and pipelines may be treated with sodium hypochlorite or 1% formaldehyde solution. With deionized systems it is usual to exhaust the resin beds with brine before sterilization with formaldehyde to prevent its inactivation to paraformaldehyde. If only local contamination occurs, live steam is often effective in eradicating it. During chemical sterilization it is important that no 'dead-legs' remain untreated and that all instruments such as water meters are treated.

2 Membrane filtration is useful where the usage is moderate and a continuous circulation of water can be maintained. Thus, with the exception of that drawn off for use, the water is continually being returned to the storage tank and refiltered. As many water-borne bacteria are small, it is usual to install a 0.22-μm pore-size membrane as the terminal filter and to use coarser prefilters to prolong its life. Membrane filters require regular sterilization to prevent microbial colonization and 'grow through'. They may be treated chemically with the remainder of the storage/distribution system or removed and treated by moist heat. The latter method is usually the most successful for heavily contaminated filters.

3   UV light at a wavelength of 254 nm is useful for the disinfection of water of good optical clarity. Such treatment has an advantage over chemical disinfection as there is no odour or flavour problem and, unlike membrane filters, is not subject to microbial colonization. The siting in the distribution system is important since any insanitary fittings downstream of the unit will recontaminate the water. Industrial in-line units with sanitary type fittings which replace part of the water pipeline are manufactured.

One of the most useful techniques for checking the microbial quality of water is by membrane filtration, since this permits the concentration of a small number of organisms from a large volume of water. When chlorinated water supplies are tested it is necessary to add an inactivating agent such as sodium thiosulphate. Although an incubation temperature of 37°C may be necessary to recover some pathogens or faecal contaminants from water, many indigenous species fail to grow at this temperature, and it is usual to incubate at 20–26°C for their detection.

4   Skin and respiratory-tract flora

4.1   Microbial transfer from operators

Microorganisms may be transferred to pharmaceutical preparations from the process operator. This is undesirable in the case of tablets and powders, and may result in spoilage of solutions or suspensions, but in the case of parenterals it may have serious consequences for the patient. Of the natural skin flora organisms, *Staph. aureus* is perhaps the most undesirable. It is common on the hands and face and, since it resides in the deep layers of the skin, is not eliminated by washing. Other bacteria present are *Sarcina* spp. and diphtheroids, but occasionally Gram-negative rods such as *Mima* spp. (Acinetobacter) and *Alcaligenes* spp. achieve resident status in moist regions. In the fatty and waxy sections of the skin, lipophilic yeast are often present, *Pityrosporum ovale* on the scalp and *P. orbiculare* on glabrous skin. Various dermatophytic fungi such as *Epidermophyton* spp., *Microsporon* spp. and *Trichophyton* spp. may be present. Ear secretions may also contain saprophytic bacteria.

Bacteria other than the natural skin flora may be transferred from the operator as a result of poor personal hygiene, such as faecal organisms from the anal region or bacteria from a wound. Open wounds without clinical manifestation of bacterial growth often support pathogenic bacteria and *Staph. aureus* has been found in 20%; other contaminants include micrococci, enterococci, α-haemolytic and non-haemolytic streptococci, *Clostridium* spp., *Bacillus* spp. and Gram-negative intestinal bacteria. *Clostridium perfringens* in such circumstances is usually present as a saprophyte and dies fairly rapidly. Wounds showing signs of infection may support *Staph. aureus*, *Strep. pyogenes*, enterococci, coliforms, *Proteus* spp. and *Pseudomonas aeruginosa*.

The nasal passages may contain large numbers of *Staph. aureus* and a limited number of *Staph. albus*, whilst the nasopharynx is often colonized by streptococci of the viridans group, *Strep. salivarius* or *Neisseria pharyngis*. Occasionally, pathogens such as *Haemophilus influenzae* and *K. pneumoniae* may be present. The most common organisms secreted during normal respiratory function and speech are saprophytic streptococci of the viridans group.

The hazard of the transfer of microorganisms from humans to pharmaceutical preparations may be reduced by comprehensive training in personal hygiene coupled with regular medical checks to prevent carriers of pathogenic organisms from coming in contact with any product.

4.2   **Hygiene and protective clothing**

Areas designed for the manufacture of products intended for injection and eye or ear preparations usually have washing facilities with foot-operated taps, antiseptic soap and hot-air hand driers at the entrance to the suite, which must be used by all process operators. For the manufacture of such products it is also necessary for the operators to wear sterilized clothing including gowns, trousers, boots, hoods, face masks and gloves. For the production of products for oral and topical use, staff should be made to wash their hands before entering the production area. The requirements for protective clothing are usually less stringent but include clean overalls, hair covering and gloves, and where possible, face masks are an advantage.

5   **Raw materials**

Raw materials account for a high proportion of the microorganisms introduced in the manufacture of pharmaceuticals, and the selection of materials of a good microbiological quality aids in the control of contamination levels in both products and the environment.

Untreated raw materials which are derived from a natural source usually support an extensive and varied microflora. Products from animal sources such as gelatine, desiccated thyroid, pancreas and cochineal may be contaminated with animal-borne pathogens. For this reason some statutory bodies such as the *British Pharmacopoeia* (1988) require freedom of such materials from *E. coli* and *Salmonella* spp. at a stated level before they can be used in the preparation of pharmaceutical products. The microflora of materials of plant origin such as gum acacia and tragacanth, agar, powdered rhubarb and starches may arise from that indigenous to plants and may include bacteria such as *Erwinia* spp., *Pseudomonas* spp., *Lactobacillus* spp., *Bacillus* spp. and streptococci, moulds such as *Cladosporium* spp., *Alternaria* spp. and *Fusarium* spp., and non-mycelated yeasts, or those introduced during cultivation. For example, the use of untreated sewage as a fertilizer may result in animal-borne pathogens such as *Salmonella* spp. being present. Some refining processes modify the microflora of raw materials, for example drying may concentrate the level of spore-forming bacteria and some solubilizing processes may introduce water-borne bacteria such as *E. coli*.

Synthetic raw materials are usually free from all but incidental microbial contamination.

The storage condition of raw materials, particularly hygroscopic substances, is important, and since a minimum water activity (A_w) of 0.70 is required for osmophilic yeasts, 0.80 for most spoilage moulds and 0.91 for most spoilage bacteria, precautions should be taken to ensure that dry materials are held below these levels. Some packaging used for raw materials, such as unlined paper sacks, may absorb moisture and may itself be subject to microbial deterioration and so contaminate the contents. For this reason polythene-lined sacks are preferable. Some liquid or semi-solid raw materials contain preservatives, but others such as syrups depend upon osmotic pressure to prevent the growth of osmophiles which are often present. With this type of material it is important that they are held at a constant temperature since any variation may result in evaporation of some of the water content followed by condensation and dilution of the surface layers to give an A_w value which may permit the growth of osmophiles and spoil the syrup.

The use of natural products with a high non-pathogenic microbial count is possible if a sterilization stage is included either before or during the manufacturing process. Such sterilization procedures (see also Chapter 20) may include heat treatment, filtration, irradiation, recrystallization from a bactericidal solvent such as an alcohol, or for dry products where compatible, ethylene oxide gas. If the raw material is only a minor constituent and the final product is adequately preserved either by lack of A_w, chemically or by virtue of its pH, sugar or alcohol content, an in-process sterilization stage may not be necessary. If, however, the product is intended for parenteral or ophthalmic use a sterilization stage is essential.

The handling of contaminated raw materials as described previously may increase the airborne contamination level, and if there is a central dispensing area precautions may be necessary to prevent airborne cross-contamination, as well as that from infected measuring and weighing equipment. This presents a risk for all materials but in particular those stored in the liquid state where contamination may result in the bulk being spoiled.

6   Packaging

Packaging material has a dual role and acts both to contain the product and to prevent the entry of microorganisms or moisture which may result in spoilage, and it is therefore important that the source of contamination is not the packaging itself. The microflora of packaging materials is dependent upon both its composition and storage conditions. This, and a consideration of the type of pharmaceutical product to be packed, determine whether a sterilization treatment is required.

Glass containers are sterile on leaving the furnace, but are often stored in dusty conditions and packed for transport in cardboard boxes. As a result they may contain mould spores of *Penicillium* spp., *Aspergillus* spp. and bacteria such as *Bacillus* spp. It is commonplace to either airblow or wash glass containers to remove any glass spicules or dust which may be present, and it is often advantageous to include a disinfection stage if the product being filled is a liquid or semi-solid preparation. Plastic bottles which are either blow- or injection-moulded have a very low microbial count and may not require disinfection. They may, however, become contaminated with mould spores if they are transported in a non-sanitary packaging material such as unlined cardboard.

Packaging materials which have a smooth, impervious surface, free from crevices or interstices, such as cellulose acetate, polyethylene, polypropylene, polyvinylchloride, and metal foils and laminates, all have a low surface microbial count. Cardboard and paperboard, unless treated, carry mould spores of *Cladosporium* spp., *Aspergillus* spp. and *Penicillium* spp. and bacteria such as *Bacillus* spp. and *Micrococcus* spp.

Closure liners of pulpboard or cork, unless specially treated with a preservative, foil or wax coating, are often a source of mould contamination for liquid or semi-solid products. A closure with a plastic flowed-in linear is less prone to introduce or support microbial growth than one stuck in with an adhesive, particularly if the latter is based on a natural product such as casein. If required, closures can be sterilized by either formaldehyde or ethylene oxide gas.

The packaging for injectables and ophthalmic preparations which are prepared aseptically, i.e. which do not receive a sterilization treatment in their final container, has to be sterilized. Dry heat at 160–170°C is often used for vials and ampoules. Containers and closures may also be sterilized by moist heat, chemicals, ethylene oxide or formaldehyde gas, or irradiation, but consideration for the destruction or removal of bacterial pyrogens may be necessary.

7 Buildings

7.1 Walls and ceilings

Moulds are the most common flora of walls and ceilings and the species usually found are *Cladosporium* spp., *Aspergillus* spp., in particular *A. niger* and *A. flavus*, *Penicillium* spp. and *Aurebasidium* (*Pullularia*) spp. They are particularly common in poorly ventilated buildings with painted walls. The organisms derive most of their nutrients from the plaster on to which the paint has been applied and a hard gloss finish is more resistant than a softer, matt one. The addition of up to 1% of a fungistat such as pentachlorophenol, 8-hydroxyquinoline or salicylanilide is an advantage. To reduce microbial growth, all walls and ceilings should be smooth, impervious and washable and this requirement may be met by cladding with a laminated plastic. In areas where humidity is high, glazed bricks or tiles are the optimal finish, and where a considerable volume of steam is used, ventilation at ceiling level is essential.

To aid cleaning, all electrical cables and ducting for other services should be installed deep in cavity walls where they are accessible for maintenance but do not collect dust. All pipes which pass through walls should be sealed flush to the surface.

7.2 Floors and drains

To minimize microbial contamination, all floors should be easy to clean, impervious to water and laid on a flat surface. In some areas it may be necesary for the floor to slope towards a drain, in which case the gradient should be such that no pools of water form. Any joints in the floor, necessary for expansion, should be adequately sealed. The floor-to-wall junction should be coved.

The finish of the floor usually relates to the process being carried out and in an area where little moisture or product is liable to be spilt, polyvinylchloride welded sheeting may be satisfactory, but in wet areas or where frequent washing is necessary, brick tiles, sealed concrete or a hard ground and polished surface like terazzo is superior. In areas where acid or alkaline chemicals or cleaning fluids are applied, a resistant sealing and jointing material must be used. If this is neglected the surface becomes pitted and porous and readily harbours microorganisms.

Where floor drainage channels are necessary they should be open if possible, shallow and easy to clean. Connections to drains should be outside areas where sensitive products are being manufactured and, where possible, drains should be avoided in areas where aseptic operations are being carried out. If this cannot be avoided, they must be fitted with effective traps, preferably with electrically operated heat-sterilizing devices.

7.3 Doors, windows and fittings

To prevent dust from collecting, all ledges, doors and windows should fit flush with walls. Doors should be well fitting to reduce the entry of microorganisms, except where a positive air pressure is maintained. Ideally, all windows in manufacturing areas should serve only to permit light entry and not be used for ventilation. In areas where aseptic operations are carried out, an adequate air-control system, other than windows, is essential.

Overhead pipes in all manufacturing areas should be sited away from equipment to prevent condensation and possible contaminants from falling into the product. Unless neglected, stainless steel pipes support little microbial growth, but lagged pipes present a problem and unless they are regularly treated with a disinfectant they will support mould growth.

8   Equipment

Each piece of equipment used to manufacture or pack pharmaceuticals has its own peculiar area where microbial growth may be supported, and knowledge of its weak points may be built up by regular tests for contamination. The type and extent of growth will depend on the source of the contamination, the nutrients available and the environmental conditions, in particular the temperature and pH.

The following points are common to many pieces of plant and serve as a general guide to reduce the risk of microbial colonization.

1   All equipment should be easy to dismantle and clean.

2   All surfaces which are in contact with the product should be smooth, continuous and free from pits, with all sharp corners eliminated and junctions rounded or coved. All internal welding should be polished out and there should be no dead ends. All contact surfaces require routine inspection for damage, particularly those of lagged equipment, and double-walled and lined vessels, since any crack or pinholes in the surface may allow the product to seep into an area where it is protected from cleaning and sterilizing agents, and where microorganisms may grow and contaminate subsequent batches of product.

3   There should be no inside screw threads and all outside threads should be readily accessible for cleaning.

4   Coupling nuts on all pipework and valves should be capable of being taken apart and cleaned.

5   Agitator blades and the shaft should preferably be of one piece and be accessible for cleaning. If the blades are bolted onto the shaft, the product may become entrained between the shaft and blades and support microorganisms. If the shaft is packed into a housing and this fitting is within a manufacturing vessel it also may act as a reservoir of microorganisms.

6   Mechanical seals are preferable to packing boxes since packing material is usually difficult to sterilize and often requires a lubricant which may gain access to the product. The product must also be protected from lubricant used on other moving parts.

7   Valves should be of a sanitary design, and all contact parts must be treated during cleaning and sanitation. If diaphragm valves are used, it is essential to inspect the diaphragm routinely. Worn diaphragms can permit seepage of the product into the seat of the valve, where it is protected from cleaning and sterilizing agents and may act as a growth medium for microorganisms. If diaphragm-type valves are used in very wet areas, a purpose-made cover may be necessary to prevent water seepage and potential microbial contamination from occurring under the diaphragm.

8 All pipelines should slope away from the product source and all process and storage vessels should be self-draining. Run-off valves should be as near to the tank as possible and sampling through them should be avoided, since any nutrient left in the valve may encourage microbial growth which could contaminate the complete batch. A separate sampling cock or hatch is preferable.

9 If a vacuum exhaust system is used to remove air or steam from a preparation vessel, it is necessary to clean and disinfect all fittings regularly. This prevents residues which may be drawn into them from supporting microbial growth, which may later be returned to the vessel in the form of condensate and contaminate subsequent batches of product.

10 If any filters or straining bags made from natural materials such as canvas, muslin or paper are used, care must be taken to ensure that they are cleaned and sterilized regularly to prevent the growth of moulds such as *Cladosporium* spp., *Stachybotrys* spp., and *Aureobasidium* (*Pullularia*) *pullulans*, which utilize cellulose and would impair them.

8.1 **Pipelines**

The most common materials used for pipelines are stainless steel, glass and plastic, and the latter may be rigid or flexible. Continuous sections of pipework are often designed to be cleaned and sterilized in place by the flow of cleansing and sterilizing agents at a velocity of not less than $1.5\,\mathrm{m\,s^{-1}}$ through the pipe of the largest diameter in the system. The speed of flow coupled with a suitable detergent removes microorganisms by a scouring action. To be successful, stainless steel pipes must be welded to form a continuous length and must be polished internally to eliminate any pits or crevices which would provide a harbour for microorganisms. However, as soon as joints and cross-connections are introduced they provide a harbour for microorganisms, particularly behind rubber or teflon O-rings. In the case of plastic pipes, bonded joints can form an area where microorganisms are protected from cleaning and sterilizing agents.

The 'in-place' cleaning system described for pipelines may also be used for both plate and tubular types of heat exchange units, pumps and some homogenizers. However, valves and all T-piece fittings for valves and temperature and pressure gauges may need to be cleaned manually. Tanks and reaction vessels may be cleaned and sterilized automatically by rotary pressure sprays which are sited at a point in the vessel where the maximum area of wall may be treated. If spray balls are incorporated into a system which reuses the cleansing-in-place (CIP) fluids, then it may be necessary to incorporate a filter to remove particles which may block the pores of the spray ball. Fixtures such as agitators, pipe inlets, outlets and vents may have to be cleaned manually. The nature of many products or the plant design often renders cleaning in place impracticable and the plant has to be dismantled for cleaning and sterilizing.

8.2 Cleansing

There are several cleansing agents available to suit the product to be removed, and the agents include acids, alkalis and anionic, cationic and non-ionic detergents. The agent selected must fulfil the following criteria.

1 It must suit the surface to be cleaned and not cause corrosion.
2 It must remove the product without leaving a residue.
3 It must be compatible with the water supply.

Sometimes a combined cleansing and sterilizing solution is desirable, in which case the two agents must be compatible.

8.3 Disinfection and sterilization

Equipment may be sterilized or disinfected by heat, chemical disinfection or a combination of both. Many tanks and reaction vessels are sterilized by steam under pressure, and small pieces of equipment and fittings may be autoclaved, but it is important that the steam has access to all surfaces. Equipment used to manufacture and pack dry powder is often sterilized by dry heat. Chemical disinfectants commonly include sodium hypochlorite and organochlorines at 50–100 p.p.m. free residual chlorine, QACs (0.1–0.2%), 70% (v/v) ethanol in water and 1% (v/v) formaldehyde solution. The method of disinfection may be by total immersion for small objects or by spraying the internal surfaces of larger equipment. When plant is dismantled for cleaning and sterilizing, all fittings such as couplings, valves, gaskets and O-rings also require treatment. The removal of chemical disinfectants is very important in fermentation processes where residues may affect sensitive cultures.

8.4 Microbial checks

The efficiency of CIP systems can be checked by plating out a sample of the final rinse water with a nutrient agar, or by swab tests. Swabs may be made of either sterile cotton wool or calcium alginate. The latter is used in conjunction with a diluent containing 1% sodium hexametaphosphate which dissolves the swab and releases the organisms removed from the equipment; these organisms may then either be plated out with a nutrient agar or an alternative method of enumeration used. Swabs are useful for checking the cleanliness of curved pieces of equipment, pipes, orifices, valves and connections, but unless a measuring guide is used the results cannot be expressed quantitatively. Such measurements can be made by pressing nutrient agar against a flat surface. The agar is usually poured into specially designed Petri dishes or contact plates, or is in the form of a disc sliced from a cylinder of a solid nutrient medium. The nutrient agar plate or section, when incubated, replicates the contamination on the surface tested. Since this technique leaves a nutrient residue on the surface tested, the equipment must be washed and resterilized before use.

9 Cleaning equipment and utensils

The misuse of brooms and mops can substantially increase the microbial count of the atmosphere by raising dust or by splashing with water-borne contaminants. To prevent this, either a correctly designed vacuum cleaner or a broom made of synthetic material, which is washed regularly, may be used. Hospital trials have shown that, when used, a neglected dry mop redistributes microorganisms which it has picked up, but a neglected wet mop redistributes many times the number of organisms it picked up originally, because it provides a suitable environment for their growth. In order to maintain mops and similar non-disposable cleaning equipment in a good hygienic state, it was found to be necessary first to wash and then to boil or autoclave the items, and finally to store them in a dry state. Disinfectant solutions were found to be inadequate.

Many chemical disinfectants (see also Chapter 11), in particular the halogens, some phenolics and QACs, are inactivated in the presence of organic matter and it is essential that all cleaning materials such as buckets and fogging sprays are kept clean. Halogens rapidly deteriorate at their use-dilution levels and QACs are liable to become contaminated with *Ps. aeruginosa* if stored diluted. For such reasons it is preferable to store the bulk of the disinfectant in a concentrated form and to dilute it to the use concentration only as required.

10 Further reading

Anderson J.D. & Cox C.S. (1967) Microbial survival. In *Airborne Microbes* (Eds P.H. Gregory & J.L. Monteith), pp. 203–226. Seventeenth Symposium of the Society for General Microbiology. Cambridge: Cambridge University Press.

Burman N.P. & Colbourne J.S. (1977) Techniques for the assessment of growth of microorganisms on plumbing materials used in contact with potable water supplies. *J Appl Bacteriol*. **43**, 137–144.

Chambers C.W. & Clarke N.A. (1968) Control of bacteria in non-domestic water. *Adv Appl Microbiol*. **8**, 105–143.

Collings V.G. (1964) The freshwater environment and its significance in industry. *J Appl Bacteriol*. **27**, 143–150.

Denyer S.P. & Baird R.M. (Eds) (1990) *Guide to Microbiological Control in Pharmaceuticals*. Chichester: Ellis Horwood.

Favero M.S., McDade J.J., Robertson J.A., Hoffman R.V. & Edward R.W. (1968) Microbiological sampling of surfaces. *J Appl Bacteriol*. **31**, 336–343.

Gregory P.H. (1973) *Microbiology of the Atmosphere*, 2nd edn. London: Leonard Hill.

Maurer I.M. (1985) *Hospital Hygiene*, 3rd edn. London: Edward Arnold.

Nishannon A. & Pokja M.S. (1977) Comparative studies of microbial contamination of surfaces by the contact plate and swab methods. *J Appl Bacteriol*. **42**, 53–63.

Packer M.E. & Litchfield J.H. (1972) *Food Plant Sanitation*. London: Chapman & Hall.

Russell A.D., Hugo W.B. & Ayliffe G.A.J. (Eds) (1992) *Principles and Practice of Disinfection, Preservation and Sterilization*, 2nd edn. Oxford: Blackwell Scientific Publications.

Skinner F.A. & Carr F.G. (Eds) (1974) *The Normal Microbial Flora of Man*. Society for Applied Bacteriology Symposium No. 5. London: Academic Press.

Underwood E. (1992) Good manufacturing practice. In *Principles and Practice of Disinfection, Preservation and Sterilization* (Eds A.D. Russell, W.B. Hugo & G.A.J. Ayliffe), 2nd edn, pp. 274–291. Oxford: Blackwell Scientific Publications.

18 Microbial spoilage and preservation of pharmaceutical products

1   **Microbial spoilage**
1.1   Introduction
1.2   Types of spoilage
1.2.1   Infection induced by contaminated pharmaceutical products
1.2.2   Chemical and physicochemical deterioration of pharmaceutical products
1.2.3   Pharmaceutical ingredients subject to microbial attack
1.2.4   Observable effects of microbial attack in pharmaceutical products
1.3   Factors affecting microbial spoilage of pharmaceutical products
1.3.1   Size and type of contaminant inoculum
1.3.2   Nutritional factors
1.3.3   Moisture content: water activity (A_w)
1.3.4   Redox potential
1.3.5   Storage temperature
1.3.6   pH
1.3.7   Package design
1.3.8   Protection of microorganisms within pharmaceutical products

2   **Preservation of pharmaceutical products using antimicrobial preservatives: basic principals**
2.1   Introduction
2.2   Effects of preservative concentration, temperature and size of inoculum
2.3   Factors affecting the 'availability' of preservatives
2.3.1   Effect of pH
2.3.2   Effect of multiphase systems
2.3.3   Effect of container
2.3.4   Effect of water activity

3   **Quality assurance and the management of microbial spoilage**
3.1   Introduction
3.2   Quality assurance in formulation design and development
3.3   Good manufacturing practice
3.4   Quality control procedures
3.4.1   The assurance of sterility
3.4.2   Estimation of pyrogens and microbial toxins
3.5   Determination of physical or chemical changes in microbial spoilage
3.6   Post-market surveillance

4   **Further reading**

1   Microbial spoilage

1.1   Introduction

The formulation of an elegant, efficacious medicine which is both stable and acceptable to the patient may necessitate the use of a wide variety of ingredients in a complex physical state, aspects of which could create conditions conducive to the survival, and even extensive replication, of contaminant microorganisms that might enter the product during its manufacture or with use by the patient or medical staff. A medicine may be considered microbiologically spoiled if, depending on its intended use: (i) low levels of acutely pathogenic microorganisms, or higher levels of opportunist pathogens, are present; (ii) toxic microbial metabolites persist even after death of the original contaminants; or (iii) microbial growth has initiated significant physical or chemical deterioration of the product. Such spoilage might well result in serious financial problems for the manufacturer, either in the immediate loss of product or in the

369

increasingly expensive cost of litigation should the spoilage cause harm to the user.

The principles underlying the onset of microbial spoilage and methods for its limitation are not unique to pharmacy, and the pharmaceutical formulator could with advantage make use of the extensive experiences of workers in the food, cosmetic, textile and agricultural industries when seeking to solve his/her own problems.

1.2   Types of spoilage

1.2.1   *Infection induced by contaminated pharmaceutical products*

With their fastidious nature, contaminants acutely pathogenic to man are probably unable to replicate in most medicines but could remain viable and infective for an appreciable time. Although infrequently reported as pharmaceutical contaminants, they attract considerable attention when they are. For example, *Salmonella* infections have arisen from tablets and capsules of yeast, carmine, pancreatin, thyroid extract and powdered vegetable drugs. Here the low levels of pathogens encountered in the finished medicines were traced to the raw ingredients of manufacture (see also Chapter 19).

Of greater practical concern are a wider variety of common saprophytic contaminants, usually of limited pathogenicity to healthy individuals, which may replicate readily within some aqueous medicines. To certain groups of unhealthy patients, however, or to particularly damaged tissues some of these opportunist microbes can represent a serious health hazard. For example, whilst the intact cornea is resistant to infection, when scratched, scarred, or surgically incised it offers little resistance to pseudomonads and some other Gram-negative bacteria. Many eyes have been lost following the use of improperly designed ophthalmic solutions containing actively growing *Pseudomonas aeruginosa* as contaminants. The skin of badly burned patients has become infected by pseudomonads contaminating 'antiseptic' solutions, resulting in failure of skin-grafting operations, and even death. Gram-negative microbes in ointments and creams have infected eczematous skin, and *Klebsiella* contaminants in barrier creams have induced fatal respiratory disease in neonates. Oral mixtures and antacid suspensions have become colonized with Gram-negative bacteria, with probably limited consequence to many, but in patients immunocompromised as a result of antineoplastic chemotherapy or as an aid to transplant surgery, life-threatening infections have resulted. *Candida* spp. are a particular hazard to such patients and fatal septicaemias have been associated with these contaminants in oral and intravenous solutions. Extensive growth of bacteria in bladder washout solutions, including those containing quaternary ammonium antiseptics, have been responsible for many extremely painful infections. The high incidence of localized infections at the sites of intravenous catheter insertion may be more related to the patient's skin flora although septicaemias have arisen from contaminated infusion fluids, including life-threatening infections from parenteral nutrition fluids containing *Candida* spp. It is virtually impossible to designate with certainty any contaminant as being harmless!

The major toxic metabolites found in medicines are pyrogens (usually, but not exclusively, lipopolysaccharides; see also Chapters 1 and 20) liberated from Gram-negative bacteria and blue–green algae (Cyanobacteria). These can induce acute febrile shock by the presence of miniscule quantities in infusion fluids entering the bloodstream or from contaminated haemodialysis fluids by diffusing through dialysis membranes. The acute bacterial toxins associated with food poisoning episodes have not been commonly reported in pharmaceutical products, although toxigenic fungi including aflatoxin-producing aspergilli and aflatoxins themselves (see Chapter 2) have been detected. Many of the metabolic products of microbial deterioration are extremely unpleasant, if not toxic, and would deter use of a medicine if present even in small quantities.

1.2.2   *Chemical and physicochemical deterioration of pharmaceutical products*

The terrestrial range of microorganisms shows such metabolic versatility that most organic matter (and much inorganic matter) is subject to the majority of possible chemical modifications under conditions which are very mild compared to those necessary for the reactions to occur by non-biological mechanisms. Under suitable environmental selection pressures novel microbial pathways can emerge, even for newly introduced synthetic chemicals (xenobiotics). Natural microbial communities are often far more effective cooperative biodeteriogens than their individual component species. Nevertheless, adverse physiological conditions, or the absence of particular microorganisms, can result in the localized accumulation of normally biodegradable materials; the build-up of lignins as peat and the persistence of oil slicks at sea exemplify this point. Biological degradative half-lives of chemicals released into the environment can vary from hours (phenol), to months ('hard', i.e. not easily biodegradable surfactants), or even years (halogenated pesticides). The overall rate of deterioration of a chemical will depend on: (i) its chemical structure; (ii) the physicochemical characteristics of a particular environment: (iii) the type and quantity of contaminant microbes present; and (iv) whether the metabolic fragments can yield materials and energy for microbial growth and the synthesis of more degradative enzymes.

Pharmaceutical formulations may be considered as specialized microenvironments and their potential for attack by contaminant microorganisms assessable using conventional ecological criteria. The particular susceptibility of some ingredients, conducive physicochemical characteristics, and the often delicately balanced stability of many medicines, can render them very sensitive to microbial destruction unless the problem is properly managed.

The use of crude animal or vegetable drugs or extracts, with their wide assortment of nutrients and inorganic components, or the reliance upon many modern synthetic ingredients, often designed to be readily biodegradable in the biosphere, can markedly raise a formulation's potential for gross microbial destruction. Also, initial, albeit limited, attack by one group of organisms may modify a product sufficiently to render it more susceptible to a second group able to induce even more devastating damage.

1.2.3   *Pharmaceutical ingredients subject to microbial attack*

Surface active agents.   The widespread introduction of anionic 'detergents' based on tetrapropylenebenzene sulphonates in the 1950s caused an alarming build-up of biologically recalcitrant molecules in the biosphere, with the consequence that subsequent surfactants have been selected for their ready biodegradability by a wide range of microorganisms. This presents a particular nutritional hazard for pharmaceuticals formulated with them.

Of the anionic surfactants used, alkali metal and amine soaps of fatty acids are generally protected by their slightly alkaline formulations, although readily degraded when disposed of in sewage. Alkyl and alkylbenzene sulphonates and sulphate esters are metabolized by ω-oxidation of their terminal methyl groups followed by sequential β-oxidation of the alkyl chains and fission of aromatic rings. The presence of chain branching involves additional α-oxidative processes. Generally, ease of degradation decreases with increasing length and complexity of branching of the alkyl chain. Sulphonate and sulphate ester residues are degraded to sulphate, although sulphonate residues are significantly more recalcitrant than the esters. Reduction of sulphate to hydrogen sulphide by anaerobic sulphate-reducing bacteria has been reported in shampoos where their viscous nature has maintained suitably low oxygen tensions.

Non-ionic surfactants such as alkylpolyoxyethylene alcohol emulsifiers are readily metabolized by a wide variety of microorganisms. Increasing chain length and branching again decreases ease of attack. Alkylphenol polyoxyethylene alcohols are similarly attacked, but are significantly more resistant. Lipolytic release of fatty acids from sorbitan esters, polysorbates and sucrose esters is often followed by degradation of the cyclic nuclei, producing small molecules readily utilizable for microbial growth.

Ampholytic surfactants based on phosphatides, betaines and alkyl-amino-substituted amino acids are an increasingly important group of emulsifying agents and are reported to be readily biodegradable.

The cationic surfactants used as antiseptics and preservatives in pharmacy are usually only slowly metabolized in sewage. At 'use' concentrations they have, however, supported extensive growth of Gram-negative bacteria, both at their own expense and that of other ingredients, in one case the ammonium acetate buffer used as a vehicle.

Organic polymers.   Many of the thickening and suspending agents used in pharmacy are subject to microbial depolymerization by specific classes of extracellular enzymes, yielding nutritive fragments and monomers. Examples of such enzymes, with their substrates in parentheses, are amylases (starches), pectinases (pectins), cellulases (carboxymethyl-celluloses, but not alkylcelluloses), uronidases (tragacanth and acacia), dextranases (dextrans) and proteases (proteins). Agar is used as an inert support for culture media because its enzymic depolymerization is particularly rare, although agar-degrading microorganisms do exist. Polyethylene glycols are readily degraded by sequential oxidation of the hydrocarbon chains, except for the very high molecular weight congeners. Polymers used in plastic packaging are very recalcitrant, with the exception of cellophane which is subject to cellulolytic attack in some circumstances.

Humectants.   Glycerol and sorbitol in pharmaceuticals readily support microbial growth unless present in high concentrations.

Fats and oils.   These hydrophobic materials are usually attacked extensively only when dispersed in aqueous formulations, although fungal growth is reported in condensed moisture films on the surface of oils, or if water droplets contaminate the bulk phase during storage. Lipolytic rupture of triglycerides liberates glycerol and fatty acids, the latter often then undergoing β-oxidation of the alkyl chains with the production of odorous ketones. The microbial metabolism of hydrocarbon oils presents a considerable problem in engineering and fuel technology when water is also present as a contaminant, although there are only limited reports of it taking place within pharmaceutical formulations. *Cladosporium resinae* in particular has induced serious corrosion in aircraft following its metabolism of aviation kerosene contaminated with water, and the accumulation of acidic metabolites.

Sweetening, flavouring and colouring agents.   Many of the sugars and other sweetening agents used in pharmacy are ready substrates for microbial attack. Very concentrated stock solutions of sugars (syrups), with low water activities (see section 1.3.4), are more resistant to attack, although spoilage by growth of osmophilic yeasts in them is sometimes reported and additional preservatives are often added for protection. Aqueous stock solutions of flavouring agents, such as peppermint water and chloroform water, and colouring agents such as amaranth readily support growth of bacteria and yeasts. While the use of simple stock solutions of flavouring agents is no longer recommended, colouring agents are still commonly used in this manner for extemporaneous dispensing.

Potent therapeutic agents.   Laboratory experiments have demonstrated that many drugs are capable of gross degradation by a wide variety of microorganisms with destruction of, or marked change in, therapeutic efficiency. Materials as diverse as alkaloids (morphine, strychnine, atropine), analgesics (aspirin, paracetamol), thalidomide, barbiturates, steroid esters, or mandelic acid can be metabolized and serve as substrates for microbial growth. The formation of irritant salicylic acid from aspirin, or inactive products from penicillin (by β-lactamase; see Chapters 5 and 10) or chloramphenicol (by chloramphenicol acetylase; see Chapter 10) or certain other antibiotics can be demonstrated. The numerous and subtle microbial transformations of steroid molecules form the basis of a valuable biotechnology producing potent drugs from relatively inert steroids. Other drugs such as dicophane (DDT, a chlorinated hydrocarbon) are highly recalcitrant. However, the observation of such attack within actual pharmaceutical formulations is, surprisingly, far less commonly reported. Appreciable loss of potency of some alkaloids (e.g. atropine in eye-drops and in an orally administered medicine) and degradation of aspirin and paracetamol in aqueous formulations has been recorded. β-lactamase-forming bacteria have inactivated penicillin injections. Localized transformation of steroids has been observed around fungal colonies growing on the surface of steroid tablets and in steroidal creams.

Preservatives and disinfectants. Most organic preservatives and disinfectants are metabolized readily by many bacteria and fungi and may even serve as growth substrates, although usually at concentrations well below 'use' levels, as might occur in sewage and effluent after disposal. However, quaternary ammonium antimicrobial agents are only degraded slowly. Halogenation or nitration considerably increases recalcitrance. Of particular environmental concern is the rapid microbial conversion of organomercurial preservatives, used and still discharged in significant quantities in industrial effluent, to the insidiously toxic ethylmercury and methylmercury which can reach humans via an ascending food chain. Microbial degradation at 'use' levels of these antimicrobial agents is far less commonly reported but includes a few accounts of the metabolism of chlorhexidine, cetrimide, phenolics, phenylethylalcohol and benzoic acid. Benzalkonium chloride has been reported occasionally to be utilized at 'use' concentrations. However, the rapid utilization of p-hydroxybenzoate (4-hydroxybenzoate) esters at concentrations formerly recommended for preservation of eye-drops and other aqueous pharmaceutical mixtures is carried out by an assortment of common pseudomonads and related bacteria, including *Ps. aeruginosa*, which often use them as substrates for growth.

1.2.4   *Observable effects of microbial attack in pharmaceutical products*

Appreciable chemical or physicochemical spoilage probably requires significant growth of microbial contaminants, although this may be localized in surface moisture films, or unevenly distributed within the bulk of viscous formulations. Early indicative signs will often be organoleptic, with the release of very unpleasant tasting or smelling metabolites such as 'sour' fatty acids, ketones, 'fishy' amines, rotten eggs (H_2S), ammonia, or bitter, sickly, 'earthy' or alcoholic tastes and smells. Products frequently become unappealingly discoloured by various microbial pigments, and even polythene may exhibit a pink discoloration when colonized by fungi. Depolymerization of thickening agents, such as tragacanth or acacia mucilages or sodium carboxymethylcellulose, results in loss of viscosity and sedimentation of insoluble ingredients. Alternatively, polymerization of sugar or surfactant molecules can produce viscous, slimy, masses in syrups or shampoos, and aggregation in creams can produce a 'gritty' texture. Surfactant attack will result in a rapid loss of stabilizing efficacy. The accumulation of acidic or basic metabolites can produce marked shifts in product pH. For example, yeast attack of acidic ingredients can raise the pH sufficiently to allow subsequent growth of bacterial contaminants previously inhibited by the low pH. Bubbles of gaseous metabolites may accumulate within viscous formulations, or even cause ballooning of flexible plastic packages.

Physicochemically complex formulations such as oil/water (o/w) emulsions may suffer gross and progressive deterioration as metabolism of surfactants and suspending agents accelerates creaming of oil globules. The lipolytic release of fatty acids will lower the pH and encourage coalescence of droplets with 'cracking' of the emulsion and souring of taste, whilst bubbles of metabolite gas may become trapped in the product, and pigment discoloration may be visually unpleasant (Fig. 18.1). When the water is compartmentalized, as in w/o emulsions, ease of growth is often restricted but spoilage does occur (see Gould 1989).

1.3   **Factors affecting microbial spoilage of pharmaceutical products**

Since the ease of establishment of microbial spoilage is influenced considerably by the physicochemical and chemical parameters of a product, their intelligent manipulation during formulation development and process design stages can be used to create conditions as unfavourable as possible for the survival or growth of microbes, within the overall limitations of user acceptability and efficacy. The choice of a particular set of characteristics will determine which contaminant types will survive or grow and they may be considerably different from those expected for optimal growth under artificial laboratory conditions. A contaminant may sufficiently modify a formulation's properties as to allow subsequent attack by other contaminants discouraged by those initial conditions.

1.3.1   *Size and type of contaminant inoculum*

The nature and extent of microbial spoilage which might take place in non-sterile or multidose pharmaceuticals will be determined by the type and quantity of contaminants remaining after manufacturing processes or entering during use of the product, and the degree to which they find the

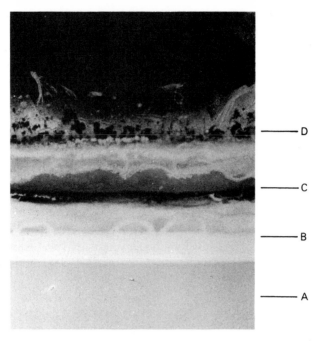

Fig. 18.1 Section (×1.5) through an inadequately preserved olive oil, oil-in-water, emulsion in an advanced state of microbial spoilage showing: A, discoloured, oil-depleted, aqueous phase; B, oil globule-rich creamed layer; C, coalesced oil layer from 'cracked' emulsion; D, fungal mycelial growth on surface. Also present are a foul taste and evil smell!

particular microenvironment conducive to growth or survival. A detailed consideration of the likely contaminant load which could enter a product is necessary to assist in the design of a potentially stable formulation. Should failure subsequently occur, a careful identification of the particular types of spoilage can be used to pinpoint the defective step(s) in the production procedure. Very low levels of contaminants which are unable to replicate in the product would be expected to yield only a minor degree of spoilage. However, a system designed to cope with such anticipated low contaminant loads can become swamped, and spoiled products ensue, if the actual microbial level rises unexpectedly due to (i) particularly heavily contaminated batches of raw materials; (ii) lapse of normal plant cleaning procedures; (iii) a particularly infective or careless operator; (iv) sporadic detachment of large microbial growths from within the plant; (v) the emergence of more aggressive microbial species; or (vi) changes in production procedures which allowed growth during manufacture, for example germination and growth of spores during drying of moist tablet granules at lower than usual temperatures. The product should also be designed to cope with the worst possible scenario for contamination by the user to prevent in-use failure. The types of microorganisms likely to enter pharmaceutical products, their origins, and methods for their limitation are discussed in detail in Chapters 17 and 20.

Upon entry of an aggressive contaminant into a product there is an adaptive lag period before rapid spoilage ensues, which is usually extensive for very low microbial numbers but decreases disproportionately with increasing contaminant loads. Some measure of control over the microbial content of extemporaneous medicines is thus necessary if one is to limit spoilage risk by the standard use of short shelf- and use-lives. The extent of spoilage of many unpreserved foodstuffs is often regulated by such an approach, but this is of limited value for manufactured medicines because of the long delays between manufacture and receipt by the patient.

Finally, the isolation of a particular microbe from a markedly spoiled product does not necessarily mean that it was the initiator of spoilage. It could be due to a secondary opportunist contaminant able to flourish in an already degraded formulation, and which has overgrown the primary spoilage organism.

1.3.2   *Nutritional factors*

The uncomplicated nutritional requirements of most common saprophytic spoilage microorganisms enables them to utilize a wide variety of the simple organic and inorganic trace ingredients present in most medicines as substrates for biosynthesis and cell division. Their specialist degradative abilities to attack the more complex ingredients of the formulations will release a further abundance of substrates. Even good-quality distilled water usually contains sufficient nutrients to allow the ready multiplication of many Gram-negative bacteria including *Ps. aeruginosa*. Demineralized water produced by ion-exchange methods may support even greater amounts of microbial growth. The inclusion of crude vegetable or animal products in a formulation usually provides a particularly nutritious environment. When these contaminants fail to grow in a medicine it is rarely as a consequence of nutrient limitation but due to other, non-supportive, physicochemical or toxic properties.

Most acute pathogens require specific growth factors not often present in pharmaceutical formulations, and are therefore normally unable to multiply in them although they may survive for some time under certain protective conditions.

1.3.3   *Moisture content: water activity (A_w)*

Microorganisms require water in appreciable quantities for growth. Although some medicines such as syrups appear to be 'wet', microbial growth in them is often difficult since the water molecules may not be readily available to the organisms, being complexed via hydrogen bonding with the molecules of the formulation. The concept of water activity (A_w) has been developed to quantify the proportion of uncomplexed water available to equilibrate with any contaminants and be available for growth and metabolism. (It can be estimated from: A_w = vapour pressure

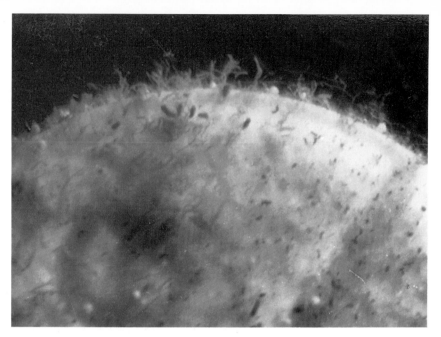

Fig. 18.2 Fungal growth on a tablet which has become damp (raised A_w) during storage. Note the sparseness of mycelia, conidiophores, and cleistothecia which develop in adverse growth conditions.

of solution ÷ vapour pressure of water at same temperature.) The greater the solute concentration the lower the A_w. With the exception of halophilic bacteria, most microorganisms grow best in dilute solutions (high A_w) and as solution concentration increases (lowering A_w) growth rates decline until a minimum, growth-inhibitory A_w is reached. Limiting A_w values are of the order of: Gram-negative rods, 0.95; staphylococci, micrococci and lactobacilli, 0.9; and most yeasts, 0.88. Syrup-fermenting osmotolerant yeasts can occasionally grow at A_w levels as low as 0.73 whilst some filamentous fungi can grow at even lower levels, with *Aspergillus glaucus* as low as 0.61.

Reducing the A_w of formulations below these levels by the addition of high concentrations of sugars or polyethylene glycols is used to protect them from microbial attack. However, even Syrup BP (66% sucrose, $A_w = 0.86$) fails to inhibit the more osmotolerant yeasts occasionally encountered. The *British Pharmacopoeia* intention to remove sucrose from its formulations will cause problems as alternative solutes for lowering A_w such as sodium chloride (for foodstuffs) may have unacceptable taste or toxicity problems. A_w can also be lowered by drying, although the dry, often hygroscopic, products (tablets, capsules, powders) will require suitable packaging to prevent resorption of water and consequent microbial growth (Fig. 18.2). The technique depends on more than a simple reduction in A_w alone as some solutes are more effective at inhibiting growth than others on an equimolar basis.

Condensed moisture films can form on the surface of otherwise 'dry' products such as tablets or bulk oils following exposure to a damp atmosphere and fluctuating storage temperatures, resulting in sufficiently high localized A_w to initiate fungal growth. More, metabolically produced water, may then raise A_w further allowing yet more growth. Dilute aqueous films similarly formed on the surface of viscous syrups, or exuded by synersis from hydrogels, can and do reach sufficiently high A_w to permit yeast and fungal spoilage.

Exactly how reduced A_w causes inhibition of microbial growth is still unclear (see review by Gould 1989).

1.3.4   *Redox potential*

The oxidation–reduction balance of a formulation is determined partly by its oxygen content and partly by its ingredients. Since many aerobic and facultative microbes can multiply, albeit slowly, at relatively low redox potentials, spoilage prevention by sufficient physical removal of oxygen is not a practical proposition for medicines. 'Vacuum-packing' of foods similarly fails in this respect. The inhibitory effect of pressurized carbon dioxide for soft drinks is due to the specific microbicidal action of carbonic acid rather than low oxygen content. The reductive and viscous natures of some foodstuffs which encourage dangerous anaerobic spoilage are uncommon with pharmaceutical products and such attack is less frequently reported, although it has occurred.

1.3.5   *Storage temperature*

Spoilage of pharmaceuticals could occur over a temperature range from about −20 to 60°C, although this occurs much less at the extremes. Storage within specific, narrower temperature ranges will encourage the growth of particular groups of spoilage organisms. Below −20°C growth and spoilage are usually minimal. It should be remembered that, for domestic use, even a room temperature of 20°C (70°F) is 'hot' to a microbe, and a home refrigerator may be as 'warm' as 12°C! It is recommended that distilled water for the preparation of Water for Injections (see also Chapter 21) be maintained at 80°C prior to sterilization, to prevent pyrogen production. The careless storage of an extemporaneously prepared solution near a radiator, rather than in a cool place, could dramatically increase its microbial deterioration. The effect of transportation and storage of products at temperatures ambient in the tropics or subtropics should be considered in this respect.

1.3.6   *pH*

Extremes of pH prevent microbial attack, although feeble growth of mould is commonly observed even in solutions of dilute hydrochloric acid. Around neutrality, bacterial spoilage predominates, while mildly alkaline conditions, for example as found in soap emulsions (pH 8.0–9.0) and mixtures such as those of magnesium hydroxide or aluminium hydroxide gel, discourage attack. Acidic conditions favour fungal and yeast proliferation and inadequately preserved fruit-flavoured pharmaceutical products at pH 3.5 can support good growth. Yeasts can metabolize organic acids raising the pH sufficiently to allow secondary bacterial attack. Protein degradation in acidic formulations can raise the pH by favoured decarboxylase activity, or lower the pH in slightly alkaline formulations by favoured deaminase activity.

1.3.7 *Package design*

The use of unit-dosage packs would eliminate the danger of microbial contamination of residual products inherent in the use of multidose containers, but the high cost involved has limited its application to a greater extent than might be considered desirable. Cap liners of cork or cardboard, formerly widely used, readily absorb moisture and support microbial growth unless impregnated with preservative. Certain packaging plastics aid spoilage by allowing the penetration of moisture vapour and oxygen. Self-sealing rubber wads must be used to prevent microbial entry into multidose injection containers following puncture by a hypodermic needle. Wide-mouthed ointment and cream jars allow the ready entry of contaminants and nutrients via coughs, fingers, etc., increasing the bioburden and reducing the preservative capacity of the formulation.

1.3.8 *Protection of microorganisms within pharmaceutical products*

In addition to reducing directly the efficiency of antimicrobial preservatives, various components of formulations and products may increase the resistance and longevity of contaminant microorganisms. There is some evidence to implicate certain proteins, suspending agents and low concentrations of surfactants such as polysorbates and the cetomacrogols in increasing resistance of certain organisms to phenolic, and other, preservatives. Adsorption of microorganisms onto suspended particles such as kaolin, magnesium trisilicate or aluminium hydroxide gel may increase their longevity. Formulation pH may influence the sensitivity of many organisms towards preservatives (see also Chapter 12).

2 Preservation of pharmaceutical products using antimicrobial preservatives: basic principles

2.1 Introduction

The provision of medicines in multidose containers creates the possibility for contamination during the repeated withdrawal of doses and creates hazards for the remainder of the product. For those with a low infectivity potential, or low growth possibility such as aspirin tablets or talcum powder, the risks may be minimal but for injections or eye-drops they would be high, with an intermediate level of risk for creams, etc. To minimize the risk it is often necessary to incorporate an antimicrobial agent ('preservative') with the specific intention of preventing the replication of contaminants remaining after manufacture or introduced during use or, where necessary, killing them. Ideally such a preservative for universal application should possess: (i) the ability to kill all microorganisms likely to enter the product; (ii) freedom from toxicity or irritancy towards the patient; (iii) sufficient stability to withstand manufacturing procedures, and so remain active throughout the intended life of the product; and (iv) selective reactivity with microbes and minimal affinity for ingredients of the formulation.

Such an ideal preservative is unlikely to exist and one usually has to select the least unsuitable agent for a particular situation. In particular, many potent microbicides are quite irritant or toxic and one may have to settle for agents which kill less efficiently or can only inhibit growth, but which have lower, acceptable toxicities. Thus, numerous antimicrobial agents would adequately kill vegetative contaminants introduced by the patient into a multidose eye-drop formulation before the next dose is due (1–3 hours) but the concentrations required would be markedly irritant to the eye. Quite a few acceptable, efficient preservatives for simple aqueous solutions so react with the varied ingredients of complex systems (such as emulsified products) that often an insufficient quantity remains to inactivate any contaminants present. For example, phenol 0.5% w/v has good antimicrobial activity in water against a wide variety of microorganisms, is stable and, because of rapid dilution, is of reasonable, low toxicity for intravenous injection (but not for instillation into the eye or into cerebrospinal fluid). However, in an o/w emulsion it would interact with most ingredients to such an extent that, to achieve sufficient microbicidal activity, very high phenol concentrations would be needed and a grossly irritant formulation would be produced.

To provide a suitable level of preservation it is necessary to appreciate the nature and complexity of the chemical and physicochemical interactions which may occur between the preservative, the microorganisms and all the ingredients of a formulation, some of the more important of which are considered here.

2.2   **Effects of preservative concentration, temperature and size of inoculum**

Generally there is an exponential relationship between microbicidal activity and preservative concentration, the value of the concentration exponent (η) varying appreciably among antimicrobial agents (see also Chapter 12). For example, halving the concentration of phenol ($\eta = 6$) results in a 64-fold (2^6) reduction in killing rate, while similar dilution of chlorhexidine ($\eta = 2$) reduces killing rate only fourfold (2^2). Increase in temperature almost always increases antimicrobial activity (see also Chapter 12). Temperature coefficients (Q_{10}, rate of change of activity per 10°C) vary with preservative, type of organism, and temperature range. Thus a drop from 30 to 20°C can result in a fivefold and 45-fold reduction in the killing rate of *Escherichia coli* by phenol ($Q_{10} = 5$) and ethanol ($Q_{10} = 45$), respectively. The interaction of concentration exponent and temperature coefficient is complex but it is suggested, for example, that if a 0.1% w/v solution of chlorocresol completely killed an inoculum of *E. coli* at 30°C in 10 minutes ($\eta = 6$, $Q_{10} = 5$) in the laboratory then it would require about 90 minutes if used at 20°C and had lost 10% of its chlorocresol from interaction with the container material. The influences of temperature and concentration changes on microbistatic activities show some similarity with the effects of these on microbicidal activities.

Preservatives are removed from solution and used up during their inactivation of microorganisms. Preservative 'capacity' is a measure of the total quantity of microorganisms a preserved formulation can inactivate before significant deterioration in efficiency becomes evident, and it varies considerably with preservative type. Non-specific interactions of preservative with the general organic component of in-use contamination 'dirt' usually reduces preservative capacity more significantly than interaction with the microbial component does because of its relatively larger mass.

2.3 Factors affecting the 'availability' of preservatives

Most of the commonly used preservatives interact with many of the common ingredients of formulations, via various bonding attractions, as well as with any microorganisms present. An unstable equilibrium is attained in which often only a small proportion of the total preservative is 'available' for equilibration with, and inactivation of, the relatively small microbial mass. The 'unavailable' preservative may still, however, contribute to the general irritancy of the product.

2.3.1 Effect of pH

The activity of ionizable preservatives is usually influenced markedly by the pH of the formulation, since activity resides in either the neutral molecule or the ionized species, but infrequently both. The weakly acidic preservatives are only effective at low pH, where neutral molecules will predominate. Thus, while benzoic acid ($pK_a = 4.2$) is of little value above pH 5.0, the *p*-hydroxybenzoate esters with their non-ionizable, esterified, carboxyl group and poorly ionized hydroxyl group ($pK_a \simeq 8.5$) have modest activity at neutral pH. Sorbic acid ($pK_a = 4.75$) is similarly of limited value above pH 5.0. In the case of quaternary ammonium preservatives and chlorhexidine, activity probably resides mainly with their cations, although this is difficult to confirm due to the low solubility of the neutral molecules.

2.3.2 Effect of multiphase systems

In a complex formulation such as an o/w pharmaceutical emulsion or an elegant cosmetic cream, preservative molecules will distribute themselves to form an unstable equilibrium between the bulk aqueous phase and (i) the oil phase by partition; (ii) the surfactant micelles by solubilization; (iii) non-micellized surfactants, polymeric suspending agents and other solutes by competitive displacement of water of solvation; (iv) solid particles by adsorption onto particulate surfaces; and (v) any micro-organisms present. The preservative molecules immediately available to compete with, and inactivate, the relatively small mass of microbial contaminants is only that proportion, usually small, remaining unbound in the bulk aqueous phase. As this is used up, re-equilibration between the components will, to some extent, slowly replenish levels in the bulk aqueous phase. The extent of o/w distribution is determined by the preservative's partition coefficient for that system and the oil:water ratio. Solubilization is controlled by the relative concentrations of preservative and surfactant and a reaction constant unique for each preservative–surfactant pair. The loss of neutral molecules into the oil or micellar phases is favoured over that of the ionized species. Similarly, reaction constants can be calculated for preservative binding to dissolved solutes and for adsorptive losses onto solid surfaces. The adsorption of certain preservatives, particularly the quaternary ammonium preservatives, onto some surfaces can markedly reduce preservative availability. Although one might expect a corresponding enhancement of lethal activity within the adsorbed layer due to localized preservative concentration, this has not been generally demonstrated.

The selection of a preservative with suitable physicochemical characteristics and the estimation of the overall concentration necessary to give adequate levels of available preservative in such complex formulations present considerable problems. Attempts to improve on the 'add and test' approach have been made by physical pharmaceutists with the development of equations which try to estimate the degree of preservative distribution and binding within multicomponent systems. These make use of physicochemical properties such as partition coefficients and variously derived association constants for preservative–surfactant interaction. However, such equations are of limited value since only some of the relevant parameters are included and very few of the necessary data exist for commonly used ingredients. Additionally, in complex systems the parameter values often differ significantly from those obtained using simplified laboratory model systems with carefully purified ingredients. One method for estimating available preservative which might be of value is to determine the proportion of preservative which is readily dialysable from the complex formulation through a cellophane membrane into water.

2.3.3   *Effect of container*

Volatile preservatives such as chloroform are so readily lost by the routine opening and closing of bulk containers that their use is limited to preservation within a sealed container or of extemporaneous products.

Multidose injection containers must not have closures made of the type of rubber which allows ready permeation of phenolic and other preservatives, with consequent marked decline in preservative concentrations of the solutions within. Appreciable interactions have been observed between certain plastics and preservatives by adsorption and solution within the phenolic preservatives, such as that encountered for nylon and certain rubbers with phenols. Even normal surface adsorption onto glass and plastics can effect significant removal of quaternary ammonium preservatives.

2.3.4   *Effect of water activity*

Preservative efficacy in aqueous formulations is influenced by their relative water activity, and generally high solute concentrations, such as in syrups, impair activity. In 'dry' products (very low A_w) such as tablets or oily solutions preservatives are unlikely to be of any practical value.

3   Quality assurance and the management of microbial spoilage

3.1   Introduction

Quality assurance (QA) relates to a scheme of management which embraces all the procedures necessary to provide a high probability that a medicine will conform consistently to a specified description of quality (fitness for the purpose intended) on every occasion. It includes design and development, good manufacturing practice (GMP), quality control (QC) and post-market surveillance. It is necessary to define and control the impact of contaminant microorganisms upon a medicine at all stages in its life, production, storage and use, and limit the risks to which the product, patient and manufacturer might be subject. Before making such judgements it is necessary to understand how microorganisms might behave, what features might be conducive to their survival and growth, and what might adversely affect them. It is also important to evaluate what the worst-case contamination scenario might be and what consequences could result. It should then be possible to design and manufacture a medicine with maximal resistance to microbes and minimal microbial risk to the patient.

3.2   Quality assurance in formulation design and development

Almost all microbial spoilage hazards arising from contaminants in medicines could be eliminated by the use of sterile unit-dosage forms. There would be no hazard of contaminants left over after manufacture and no risk of contaminating the rest of the medicine when obtaining a dose. It is generally on the grounds of economic expediency that a manufacturer resorts to other, less satisfactory, solutions.

The high risk of infection by microbial contaminants in a medicine for parenteral administration almost always demands a sterile unit-dosage form. With eye-drops it is felt that the risk is lower and multidose packs protected with a 'suitable' preservative are still quite common for domestic use, although less so in hospitals. Oral and topical products generally present low infectivity risks and the emphasis is often more upon preventing microbial attack of the formulation itself.

Provided consultation occurs at an early stage of development it should be possible to create formulations which have some inherent resistance to microbial spoilage by manipulation of physicochemical factors, such as A_w, and by elimination of particularly susceptible ingredients or by the inclusion of suitable preservatives. For 'dry' dosage forms, such as tablets, capsules and powders, it is their very low A_w which is the main protection against microbial attack, and accordingly suitable moisture vapour-resisting packing has to be considered to retain it. A practice of incorporating preservatives into tablets is unlikely to be of much practical use since they would only work effectively in relatively high A_w situations. The suggestion that they would protect 'if the tablets got damp' misses the real solution; keep them dry! Film coating of tablets with a water-vapour impermeant film has been suggested for those intended for humid tropical regions.

Since preservatives are relatively non-specific in their interactions it is very difficult to calculate with any precision what proportion of a preservative added to a complex formulation is 'available' for interaction with, and inactivation of, any contaminant microorganisms (section 2.3).

The only realistic solution for assessment of the preservative efficiency of the complete formulation is a biological test, challenging it with viable microbes. Since it could be disastrous to place a product on the market and wait to see how it performed against natural contaminants, laboratory-based microbial challenge tests have been developed which attempt to mimic the challenges of real life.

Some procedures challenge aliquots of a formulation with individual, fairly large, inocula of selected laboratory cultures and determine their rate of killing by viable counting methods (single challenge technique), whilst others re-inoculate the aliquots repeatedly at set intervals, monitoring the degree of killing until the product fails (multiple challenge technique). Some workers consider that this latter method gives a better measure of the capacity of the formulation to stand up to the repeated challenges that multidose medicines face in real life. The cultures chosen for the tests should bear some relationship to the organisms which might be met in practice, although it is generally recognized that maintenance of microbial inocula with degradative potency similar to that of organisms which have actually caused spoilage in previous products is extremely difficult. Organisms causing marked spoilage of pharmaceuticals commonly become far less aggressive or totally enfeebled when isolated and cultivated on 'normal' laboratory media (nutrient broth, malt agar, etc.). In some cases, reasonable spoilage activity can be maintained by cultivation in media designed to simulate the product originally attacked, such as 'shampoo agar', or by cultivation in fresh aliquots of unpreserved formulation. The use of more than one challenge species at the same time ('mixed cultures') is not generally accepted in pharmacy, although there is considerable evidence of the cooperative effect of microbial species in the wider field of environmental biodegradation.

The *British Pharmacopoeia* preservative challenge test for assessing the probable efficiency of complete formulations to deal with actual in-use contamination is basically a single challenge test using four stock cultures of bacteria, yeast and a mould which have little history of spoilage potential and which are cultivated on 'normal', enfeebling laboratory media. However, extension to multiple challenging is suggested if considered necessary, and the inclusion of known spoilage organisms such as osmotolerant yeasts recommended where this is thought to be a possible in-use problem. The proposed European Pharmacopoeia test is analogous to, but not identical with, the BP test.

Official challenge test procedures such as those in the *British Pharmacopoeia* or *United States Pharmacopeia* rely upon the opinion of a committee of experts rather than absolute scientific criteria to decide what degree of killing should be appropriate for a particular type of formulation to be considered adequately preserved. Nevertheless many manufacturers feel that the tests do provide useful indicators of likely in-use stability.

In an attempt to produce a more quantifiable estimation of adequacy, Orth has applied the concepts of the D-value developed for describing the lethality of sterilization processes (Chapter 20) to the killing efficiency of preserved formulations, the linear regression method. This measures the D-value for the rate of kill of an inoculum in a preserved system and, by extrapolation, estimates the time which would be needed to achieve a prescribed level of inactivation. Problems arise when trying to predict the behaviour of very low levels of survivors, and the method has its detractors as well as followers.

3.3   Good manufacturing practice

Unless considerable care is taken in assessing what microbial contaminants may arise from raw materials or during processing, trouble will be built in from the beginning. If one considers the manufacturing plant and its environs from an ecological and physiological viewpoint it should be possible to identify areas where microbes might accumulate and multiply to create hazards for subsequent batches of medicine, and then manipulate conditions to discourage survival or growth. Thorough drying of plant is a very useful deterrent to growth. If plant is kept wet after cleaning and residues of product remain in nooks and crannies there is a distinct possibility of organisms developing aggressive spoilage capabilities *in situ*.

It may be necessary to include a step in the processing to reduce the bioburden remaining after manufacture and prevent swamping of the preservative during storage. Chapter 17 deals with contaminants and methods for their control and Chapter 22 with details of some microbial GMP procedures.

3.4   **Quality control procedures**

Quality control procedures contribute to the overall assurance of quality by the devising of suitable specifications for product quality and tests to determine whether batches of medicines have met those specifications. There is general agreement on the need to limit the types and overall numbers of microbes in non-sterile medicines to reduce infective or spoilage risks. However, there is good scientific evidence that current methods for counting (or even detecting) viable microorganisms have extremely poor accuracy and precision when used for estimating the true microbial levels in complex formulations. There is also poor agreement as to what constitutes an acceptable microbial content for various medicinal types, or whether the type of contaminant is more important than its quantity.

Acute pathogens usually occur in very low numbers, may be damaged, and prove difficult to isolate or identify. Despite clear microscopic visualization of microbes in markedly spoiled products they frequently yield surprisingly low colony-forming unit (cfu) counts with conventional cultivation techniques. Although present in high numbers a particular microbe may be neither pathogenic nor the primary spoilage agent, but may be inert (e.g. ungerminated spores) or a secondary contaminant which has outgrown the initiating spoiler. Very unevenly distributed growth in viscous formulations, or around particulate ingredients, can provide serious sampling problems. Aggressive spoilage contaminants present in very low numbers at the time of testing can grow to high levels during storage. The composition of culture media for viable counting and incubation temperature greatly influence the number of cfu obtained, counts varying markedly with different media and even with different batches of the same medium.

Recognizing these difficulties, UK food regulatory authorities have generally abandoned the use of microbial counts as legally enforceable standards of food quality. Despite this, the European Pharmacopoeia Commission is considering the introduction of conventional, quantitative microbial limit tests for medicines which may become statutory requirements in some EC member states.

Routine testing in current use in the UK includes the following investigations.

1 Some medicines are tested for the presence of particular pathogens (e.g. *Salmonella* spp. in pancreatin) by incubation in a non-specific culture medium, followed by enrichment culture in a medium favouring the particular pathogen, and cultivation on several 'selective' and diagnostic media. Selectivity is incomplete and additional biochemical tests are usually performed before declaring the presumptive presence of that

387   *Microbial spoilage and preservation of pharmaceutical products*

pathogen. However, a number of the selective media used have been developed for clinical purposes and often display significant toxicity to microbes subjected to the stresses of pharmaceutical processing.

2 Known spoilage organisms such as pseudomonads or 'flat-souring' (acid, but not gas, producing) bacilli are sought for in a similar manner using selective media developed for the food and dairy industries.

3 Most manufacturers perform viable counts to obtain a rough estimate of the actual microbial content of their products, and fluctuations in counts may give a valuable 'in-house' guide to the efficiency of production controls. Commonly, cfu counts are determined using non-specific solid media inoculated with serial dilutions of the test material.

4 Alternative methods under investigation which show promise of greater reliability, sensitivity and speed of operation for estimating microbial content and product quality include: (i) measurement of electrical impedance changes, oxygen consumption, or heat generation (microcalorimetry), as microbes grow within a product; (ii) estimation of total microbial content at low levels (around 10^3 bacteria) by assay of a metabolite present in (relatively) fixed quantities per viable cell. Thus, adenosine triphosphate (ATP) can be measured by light emission with luciferase enzymes, or esterase activity of 'viable' cells by fluorescence assay of fluorescein diacetate hydrolysis; (iii) release of $^{14}CO_2$ from added radiolabelled substrates indicating the level of microbial metabolism; and (iv) direct ultraviolet (UV) epifluorescence microscopic counting of microbes within a product, where 'dead' cells fluoresce orange with ethidium bromide or acridine orange, and 'viable' cells fluoresce green with fluorescein diacetate (q.v.) (Ethidium bromide and acridine orange do not readily penetrate intact ('viable') membranes.) Recent promising developments for the detection of extremely low numbers of microbes have made use of monoclonal antibodies chemically linked to enzymes which yield detectable products such as fluorescent chemicals when reacted with contaminants on a filter membrane.

3.4.1 *The assurance of sterility*

Legal precedent requires that every item in a batch of 'sterile' medicines must be sterile, although it is not possible to guarantee more than a high probability of this being the case. Since 'tests for sterility' (Chapter 23) are badly flawed in concept the main assurance of this probability being as high as possible must depend upon rigid and tightly regulated GMP procedures.

3.4.2 *Estimation of pyrogens and microbial toxins*

Semi-quantitative tests for pyrogens in pharmaceuticals involves the measurement of any febrile response following their injection into rabbits, which appear to have similar sensitivities to pyrogens as do humans. A less expensive, rapid and very sensitive *in vitro* assay relies upon the highly specific interaction of amoebocyte lysate from the horse-shoe crab (*Limulus polyphemus*) with microbial lipopolysacchands (but not other 'pyrogens') yielding a gelling and opacification of the lysate, or a colour with a chromogenic substrate. Some therapeutic agents (e.g. polymyxins) interfere with the test.

Sophisticated, sensitive methods exist for the detection of most microbial toxins in foodstuffs, which use animal inoculation, mass spectroscopic, chromatographic or immunological techniques. They are not commonly applied to pharmaceuticals with the possible exception of the examination of vegetable ingredients for aflatoxins by extraction, concentration with ion-exchange chromatography and detection under UV light by their strong fluorescence.

3.5　　**Determination of physical or chemical changes in microbial spoilage**

Olfactory detection of a variety of common metabolites with extremely intense tastes or smells is generally a sensitive, qualitative indicator of active microbial attack in medicines. Characterization of these metabolites by HPLC (high-performance liquid chromatography) or gas chromatography can be used to distinguish microbial spoilage from other, non-biological deteriorative processes. When attack promotes physicochemical changes, they can be followed with conventional instrumentation such as, for example, surfactant biodeterioration by measurements of reduction in surface tension. Complex emulsion spoilage may be monitored by following changes in viscosity, surface tension, pH and rates of creaming or particle sedimentation. Metabolism of particular ingredients (e.g. alkaloids or steroids) might be assessed with analytical techniques such as thin-layer chromatography.

3.6　　**Post-market surveillance**

Despite extensive development and rigorous testing in the laboratory it is not possible to be totally certain that a medicine will be sufficiently robust to withstand all the rigours of real-life usage. A proper QA system must include a procedure for recording user complaints and following them up with great care to determine what might have happened, why it happened and what needs to be done to prevent it happening again.

4　　**Further reading**

The chapter sections to which each reference is particularly relevant are indicated in parentheses at the end of the reference, although this section is also intended as a general suggestion of routes to material for those who need to develop the topic in more detail.

Andrew M.E.H. & Russell A.D. (Eds) (1984) *The Revival of Injured Microbes.* Society for Applied Bacteriology Symposium Series No. 12. London: Academic Press. (3.5)

Bloomfield S.F., Baird R., Leak R.E. & Leech R. (Eds) (1988) *Microbial Quality Assurance in Pharmaceuticals, Cosmetics and Toiletries*. Chichester: Ellis Horwood Ltd. (2, 3)

Board R.G., Allwood M.C. & Banks J.C. (Eds) (1987) *Preservatives in the Food, Pharmaceutical and Environmental Industries*. Society for Applied Bacteriology Technical Series No. 22. Oxford: Blackwell Scientific Publications. (2)

British Pharmacopoeia (1988) Appendix XVI B: Test for microbial contamination, A195–A200. Appendix XVI C: Efficacy of antimicrobial preservatives in pharmaceutical products, A200–A203. London: HMSO. (3.2, 3.5)

Denyer S. & Baird R. (Eds) (1990) *Guide to Microbiological Control in Pharmaceuticals*. Chichester: Ellis Horwood. (2, 3)

Gould G.W. (Ed) (1989) *Mechanisms of Action of Food Preservation Procedures*. Barking: Elsevier Science Publishers Ltd. (2)

Hopton J.W. & Hill E.C. (Eds) (1987) *Industrial Microbiological Testing*. Society for Applied Bacteriology Technical Series No. 23. Oxford: Blackwell Scientific Publications. (2, 3)

Hugo W.B. (1991) A brief history of heat and chemical preservation and disinfection. *J Appl Bacteriol.* **71**, 9–18. (2)

Jarvis P. & Paulus K. (1982) Food preservation: an example of negative biotechnology. *J Chem Tech Biotech.* **32**, 233–250. (1.3)

Mossel D.A.A. (1983) Essentials and perspectives of the microbial ecology of foods. In *Food Microbiology: Advances and Prospects* (Eds T.A. Roberts & F.A. Skinner), pp. 1–45. Society for Applied Bacteriology Symposium Series No. 11. London: Academic Press. (1.3)

Russell A.D., Hugo W.B. & Ayliffe G.A.G. (Eds) (1992) *Principles and Practice of Disinfection, Sterilization and Preservation*, 2nd edn. Oxford: Blackwell Scientific Publications. (2)

19 Contamination of non-sterile pharmaceuticals in hospital and community environments

1   Introduction

2   The significance of microbial contamination
2.1   Spoilage
2.2   Hazard to health

3   Sources of contamination
3.1   In manufacture
3.1.1   Water
3.1.2   Environment
3.1.3   Packaging
3.2   In use
3.2.1   Human sources
3.2.2   Environmental sources
3.2.3   Equipment sources

4   The extent of microbial contamination
4.1   Contamination in manufacture
4.2   Contamination in use

5   Factors determining the outcome of a medicament-borne infection
5.1   Type and degree of microbial contamination
5.2   The route of administration
5.3   Resistance of the patient

6   Prevention and control of contamination

7   Further reading

1 Introduction

Pharmaceutical products are used in a variety of ways in the prevention, treatment and diagnosis of disease. In recent years, manufacturers of pharmaceuticals have improved the quality of non-sterile products such that today the majority contain only a minimal microbial population. Nevertheless, a few rogue products with an unacceptable level and type of contamination will occasionally escape the quality control net and when used may, ironically, contribute to the spread of disease in patients.

Although the occurrence of product contamination has been well documented in medical literature, the significance for the patient has not always been clear. Evidence accumulated in the past 30 years or so has, however, enabled a better understanding of why and how contamination occurs, its extent and frequency, the factors determining the outcome for the patient and finally what preventive steps may be taken to control the problems.

2 The significance of microbial contamination

2.1 Spoilage

It has been known for many years that microbial contaminants may effect the spoilage of pharmaceutical products through chemical, physical or aesthetic changes in the nature of the product, thereby rendering it unfit for use (see Chapter 18). Active drug constituents may be metabolized to less potent or chemically inactive forms. Physical changes commonly seen

391

are the breakdown of emulsions, visible surface growth on solids and the formation of slimes, pellicles or sediments in liquids, sometimes accompanied by the production of gas, odours or unwanted flavours, thereby rendering the product unacceptable and possibly even dangerous to the patient. It may, indeed, affect patient compliance with the prescribed course of therapy. Finally, spoilage and subsequent wastage of a product have serious economic implications for the manufacturer.

2.2 Hazard to health

Nowadays it is well recognized that a contaminated pharmaceutical product may also present a potential health hazard to the patient. Although isolated outbreaks of medicament-related infections have been reported since the early part of this century, it is only in the past two decades or so that the significance of this contamination to the patient has been more fully understood. Recognition of these infections presents its own problems. It is a fortunate hospital physician who can, at an early stage, recognize contamination manifest as a cluster of infections of rapid onset, such as that following the use of a contaminated intravenous fluid in a hospital ward. The chances of a general practitioner recognizing a medicament-related infection of insidious onset, perhaps spread over several months, in a diverse group of patients in the community, are much more remote. Once recognized, there is of course a moral obligation to withdraw the offending product, and subsequent investigations of the incident therefore become retrospective.

Table 19.1 Contaminants found in pharmaceutical products

Year	Product	Contaminant
1907	Plague vaccine	*Clostridium tetani*
1943	Fluorescein eye-drops	*Pseudomonas aeruginosa*
1946	Talcum powder	*Clostridium tetani*
1948	Serum vaccine	*Staphylococcus aureus*
1955	Chloroxylenol disinfectant	*Pseudomonas aeruginosa*
1966	Thyroid tablets	*Salmonella muenchen*
1966	Antibiotic eye ointment	*Pseudomonas aeruginosa*
1966	Saline solution	*Serratia marcescens*
1967	Carmine powder	*Salmonella cubana*
1967	Hand cream	*Klebsiella pneumoniae*
1969	Peppermint water	*Pseudomonas aeruginosa*
1970	Chlorhexidine—cetrimide antiseptic solution	*Pseudomonas cepacia*
1972	Intravenous fluids	*Pseudomonas*, *Erwinia* and *Enterobacter* spp.
1972	Pancreatin powder	*Salmonella agona*
1977	Contact-lens solution	*Serratia* and *Enterobacter* spp.
1981	Surgical dressings	*Clostridium* spp.
1982	Iodophor solution	*Pseudomonas aeruginosa*
1983	Aqueous soap	*Pseudomona stutzeri*
1984	Thymol mouthwash	*Pseudomonas aeruginosa*
1986	Antiseptic mouthwash	Coliforms

Pharmaceutical products of widely differing forms are susceptible to contamination by a variety of microorganisms, as shown by a few examples given in Table 19.1. Disinfectants, antiseptics, powders, tablets and other products providing an inhospitable environment to invading contaminants are known to be at risk, as well as products with more nutritious components, such as creams and lotions with carbohydrates, amino acids, vitamins and often appreciable quantities of water. Contaminants isolated from products have ranged from true pathogens, such as *Cl. tetani*, to opportunist pathogens, such as *Ps. aeruginosa* and other free-living Gram-negative organisms, which are capable of causing disease under special circumstances. The outcome of using a contaminated product may vary from patient to patient, depending on the type and degree of contamination and how the product is to be used. Undoubtedly the most serious effects have been seen with contaminated injected products where generalized bacteraemic shock and in some cases death of patients have been reported. More likely, a wound or sore in broken skin may become locally infected or colonized by the contaminant. It must be stressed, however, that the majority of cases of medicament-related infections are probably not recognized or reported as such.

3 Sources of contamination

3.1 In manufacture

The same principles of contamination control apply whether manufacture takes place in industry (Chapter 17) or on a smaller scale in the hospital pharmacy. As discussed in Chapter 22, quality must be built into the product at all stages of the process and not simply assessed at the end of manufacture: (i) raw materials, particularly water and those of animal origin, must be of a high microbiological standard; (ii) all processing equipment should be subject to planned preventive maintenance and should be properly cleaned after use to prevent cross-contamination between batches; (iii) manufacture should take place in suitable premises in a clean, tidy work area supplied with filtered air; (iv) staff involved in manufacture should not only have good health but also a sound knowledge of the importance of personal and production hygiene; and (v) the end-product requires suitable packaging which will protect it from contamination during its shelf-life and is itself free from contamination.

Manufacture in hospital premises raises certain additional problems with regard to contamination control.

3.1.1 Water

Mains water in hospitals is frequently stored in large roof tanks, some of which may be relatively inaccessible and poorly maintained. Water for pharmaceutical manufacture requires some further treatment, usually by distillation, reverse osmosis (Chapter 17, section 3.5) or deionization.

Such processes need careful monitoring, as does the microbiological quality of the water after treatment. Storage of water requires particular care, since some Gram-negative opportunist pathogens can survive on traces of organic matter present in treated water and will readily multiply at room temperature; water should therefore be stored at a temperature in excess of 80°C.

3.1.2 Environment

The microbial flora of the pharmacy environment is a reflection of the general hospital environment and the activities undertaken there. Free-living opportunist pathogens, such as *Ps. aeruginosa* can normally be found in all wet sites, such as drains, sinks and taps. Cleaning equipment, such as mops, buckets, cloths and scrubbing machines, may be responsible for distributing these organisms around the pharmacy; if stored wet they provide a convenient niche for microbial growth, resulting in heavy contamination of equipment. Contamination levels in the production environment may, however, be minimized by observing good manufacturing practices, by installing heating traps in sink U-bends, thus destroying one of the main reservoirs of contaminants, and by proper maintenance and storage of equipment, including cleaning equipment. Additionally, cleaning of production units by contractors should be carried out to a pharmaceutical specification.

3.1.3 Packaging

Sacking, cardboard, card liners, corks and paper are unsuitable for packaging pharmaceuticals, as they are heavily contaminated, for example with bacterial or fungal spores. These have now been replaced by non-biodegradable plastic materials. Packaging in hospitals is frequently reused for economic reasons. Large numbers of containers may be returned to the pharmacy, bringing with them microbial contaminants introduced during use in the wards. Particular problems have been encountered in the past with disinfectant solutions where residues of old stock have been 'topped up' with fresh supplies, resulting in the issue of contaminated solutions to wards. Reusable containers must, therefore, be thoroughly washed and dried, and never refilled directly.

Another common practice in hospitals is the repackaging of products purchased in bulk into smaller containers. Increased handling of the product inevitably increases the risk of contamination, as shown by one survey when hospital-repacked items were found to be twice as often contaminated as those in the original pack (Public Health Laboratory Service Report, 1971).

3.2 In use

Pharmaceutical manufacturers may justly argue that their responsibility ends with the supply of a well-preserved product of high microbiological standard in a suitable pack and that the subsequent use, or indeed abuse, of the product is of little concern to them. Although much less is known about how products become contaminated during use, there is reasonable evidence that continued use of such products is undesirable, particularly in hospitals where it may result in the spread of cross-infection.

All multidose products are subject to contamination from a number of sources during use. The sources of contamination are the same whether products are used in hospital or in the community environment, but opportunities for observing it are, of course, greater in the former.

3.2.1   *Human sources*

During normal usage, the patient may contaminate his/her medicine with his/her own microbial flora; subsequent use of the product may or may not result in self-infection (Fig. 19.1). Topical products are considered to be most at risk, since the product will probably be applied by hand thus introducing contaminants from the resident skin flora of staphylococci, *Micrococcus* spp. and diphtheroids. Opportunities for contamination may be reduced by using disposable applicators for topical products or by taking oral products by disposable spoon.

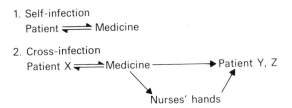

Fig. 19.1 Mechanisms of contamination during use of medicinal products.

In hospitals, multidose products, once contaminated, may serve as a vehicle of cross-contamination or cross-infection between patients. Zinc-based products packed in large stock-pots and used in the treatment and prevention of bed-sores in long-stay and geriatric patients may become contaminated during use with organisms such as *Ps. aeruginosa* and *Staph. aureus*. These unpreserved products will allow multiplication of contaminants, especially if water is present either as part of the for-mulation, for example in oil/water (o/w) emulsions, as a film in w/o emulsions which have undergone local cracking, or as a condensed film from atmospheric water, and appreciable numbers may then be trans-ferred to other patients when reused. Clearly the economics and con-venience of using stock-pots need to be balanced against the risk of spreading cross-infection between patients and the inevitable increase in length of the patients' stay in hospital. Recently, the use of stock-pots in hospitals has noticeably declined.

A further potential source of contamination in hospitals is the nursing staff responsible for medicament administration. During the course of their work nurses' hands become contaminated with opportunist pathogens which are not part of the normal skin flora but are easily removed by thorough handwashing and drying. In busy wards, handwashing between attending to patients may be omitted and any contaminants may subsequently be transferred to medicaments during administration. Hand lotions and creams used to prevent chapping of nurses' hands may similarly become contaminated, especially when packaged in multidose containers and left at the side of the handbasin, frequently without a lid. The importance of thorough handwashing cannot be over-emphasized in the control of hospital cross-infection. Hand lotions and creams should be well preserved and, ideally, packaged in disposable dispensers. Other effective control methods include the supply of products in individual patient's packs and the use of a non-touch technique for medicament administration.

3.2.2   *Environmental sources*

Small numbers of airborne contaminants may settle out in products left open to the atmosphere. Some of these will die during storage, with the rest probably remaining at a static level of about $10^2 - 10^3$ colony forming units (cfu) g^{-1} or ml^{-1}. Larger numbers of water-borne contaminants may be accidentally introduced into topical products by wet hands or by a 'splash-back mechanism', if left at the side of a basin. Such contaminants generally have simple nutritional requirements and, following multiplication, levels of contamination may often exceed $10^6 cfu\, g^{-1}$ or ml^{-1}. This problem is encountered particularly when the product is stored in warm hospital wards or in hot steamy bathroom cupboards at home. Products used in hospitals as soap-substitutes for bathing patients are particularly at risk, and not only soon become contaminated with opportunist pathogens such as *Pseudomonas* spp., but also provide conditions conducive for their multiplication. The problem is compounded by using stocks in multidose pots for use by several patients in the same ward over an extended period of time.

The indigenous microbial population is quite different in the home and in hospitals. Pathogenic organisms are found much more frequently in the latter and consequently are isolated more often from medicines used in hospital. Usually, there are fewer opportunities for contamination in the home, as patients are generally issued with individual supplies in small quantities.

3.2.3   *Equipment sources*

Patients and nursing staff may use a range of applicators (pads, sponges, brushes, spatulas) during medicament administration, particularly for topical products. If reused, these easily become contaminated and may be responsible for perpetuating contamination between fresh stocks of product, as has indeed been shown in studies of cosmetic products. Disposable applicators or swabs should therefore always be used.

In hospitals today a wide variety of complex equipment is used in the course of patient treatment. Humidifiers, incubators, ventilators, resuscitators and other apparatus require proper maintenance and decontamination after use. Chemical disinfectants used for this purpose have in the past through misuse become contaminanted with opportunist pathogens, such as *Ps. aeruginosa*, and ironically have contributed to, rather than reduced, the spread of cross-infection in hospital patients. Disinfectants should only be used for their intended purpose and directions for use must be followed at all times.

4 The extent of microbial contamination

Detailed examination of reports in the literature of medicament-borne contamination reveals that the majority of these are anecdotal in nature, referring to a specific product and isolated incident. Little information is available, however, as to the overall risk of products becoming contaminated and causing patient infections when subsequently used. As with risk analysis (assessment of the hazards of consumption of a contaminated preparation) in food microbiology, this information is considered invaluable not only because it indicates the effectiveness of existing practices and standards, but also because the value of potential improvements in quality from a patient's point of view can be balanced against the inevitable cost of such processes. Thus the old argument that all pharmaceutical products, regardless of their use, should be produced as sterile products, although sound in principle, is kept in perspective by the fact that it cannot be justified on economic grounds alone.

4.1 Contamination in manufacture

Investigations carried out by the Swedish National Board of Health in 1965 revealed some startling findings on the overall microbiological quality immediately after manufacture of non-sterile products made in Sweden. A wide range of products was routinely found to be contaminated with *Bacillus subtilis*, *Staph. albus*, yeasts and moulds, and in addition large numbers of coliforms were found in a variety of tablets. Furthermore, two nationwide outbreaks of infection were subsequently traced to the inadvertent use of contaminated products. Two hundred patients were involved in an outbreak of salmonellosis, caused by thyroid tablets contaminated with *Sal. bareilly* and *Sal. muenchen*; and eight patients had severe eye infections following the use of hydrocortisone eye ointment contaminated with *Ps. aeruginosa*. The results of this investigation have not only been used as a yardstick for comparing the microbiological quality of non-sterile products made in other countries, but also as a baseline upon which international standards could be founded.

In the UK, the microbiological and chemical quality of pharmaceutical products made by industry has since been governed by the Medicines Act 1968. The majority of products have been found to be made to a high standard, although spot checks have occasionally revealed medicines of unacceptable quality and so necessitated product recall. In contrast, the manufacture of pharmaceutical products in hospitals has in the past been much less rigorously controlled, as shown by the results of surveys in the 1970s in which significant numbers of preparations were found to be contaminated with *Ps. aeruginosa*. In 1974 hospital manufacture also came under the terms of the Medicines Act and, as a consequence, considerable improvements have been seen in recent years not only in the conditions and standard of manufacture, but also in the chemical and microbiological quality of finished products. The recent removal of Crown immunity from the NHS authorities has meant that certain manufacturing operations in hospitals are now subject to the full licensing provisions of the Medicines Act.

4.2   Contamination in use

Higher rates of contamination are invariably seen in products after opening and use, and, amongst these, medicines used in hospitals are more likely to be contaminated than those used in the general community. The Public Health Laboratory Service Report of 1971 expressed concern at the overall incidence of contamination in non-sterile products used on hospital wards (327 of 1220 samples) and the proportion of samples found to be heavily contaminated (18% in excess of 10^4 cfu g^{-1} or ml^{-1}). The presence of *Ps. aeruginosa* in 2.7% of samples (mainly oral alkaline mixtures) was considered to be highly undesirable.

In contrast, medicines used in the home are not only less often contaminated but also contain lower levels of contaminants and fewer pathogenic organisms. Generally, there are fewer opportunities for contamination here since smaller quantities are used by individual patients. Medicines in the home may, however, be hoarded and used for extended periods of time. Additionally, storage conditions may be unsuitable and expiry dates ignored and thus problems other than those of microbial contamination may be seen in the home.

5   Factors determining the outcome of a medicament-borne infection

A patient's response to the microbial challenge of a contaminated medicine may be diverse and unpredictable, perhaps with serious consequences. In one patient no clinical reactions may be evident, yet in another these may be indisputable, illustrating one problem in the recognition of medicament-borne infections. Clinical reactions may range from inconvenient local infections of wounds or broken skin, caused possibly from contact with a contaminated cream, to gastrointestinal infections from the ingestion of contaminated oral products, to serious widespread infections, such as a bacteraemia or septicaemia, leading perhaps to death, as have resulted from infusion of contaminated fluids. Undoubtedly, the most serious outbreaks of infection have been seen in the past where contaminated products have been injected directly into the bloodstream of patients whose immunity is already compromised by their underlying disease or therapy. The outcome of any episode is determined by a combination of several factors, amongst which the type and degree of microbial contamination, the route of administration and the patient's resistance are of particular importance.

5.1 Type and degree of microbial contamination

Microorganisms that contaminate medicines and cause disease in patients may be classified as true pathogens or opportunist pathogens. Pathogenic organisms like *Cl. tetani* and *Salmonella* spp. rarely occur in products, but when present cause serious problems. Wound infections from using contaminated dusting powders have been reported, including several cases of neonatal death from talcum powder containing *Cl. tetani*. Outbreaks of salmonellosis have followed the inadvertent ingestion of contaminated thyroid and pancreatic powders. On the other hand, opportunist pathogens like *Ps. aeruginosa*, *Klebsiella*, *Serratia* and other free-living organisms are more frequently isolated from medicinal products and, as their name suggests, may be pathogenic if given the opportunity. The main concern with these organisms is that their simple nutritional requirements enable them to survive in a wide range of pharmaceuticals, and thus they tend to be present in high numbers, perhaps in excess of $10^6 - 10^7 \, \text{cfu} \, \text{g}^{-1}$ or ml^{-1}; nevertheless, the product itself may show no visible sign of contamination. Opportunist pathogens can survive in disinfectants and antiseptic solutions which are normally used in the control of hospital cross-infection but which when contaminated may even perpetuate the spread of infection. Compromised hospital patients, i.e. the elderly, burned, traumatized or immunosuppressed, are considered to be particularly at risk from infection with these organisms, whereas healthy patients in the general community have given small cause for concern.

The critical dose of microorganisms which will initiate an infection is largely unknown and varies not only between species but also within a species. Animal and human volunteer studies have indicated that the infecting dose may be reduced significantly in the presence of trauma or foreign bodies or if accompanied by a drug having a local vasoconstrictive action.

5.2 The route of administration

As stated previously, contaminated products injected directly into the bloodstream or instilled into the eye cause the most serious problems. Intrathecal and epidural injections are potentially hazardous procedures.

In practice, epidural injections are frequently given through a bacterial filter. Injectable and ophthalmic solutions are often simple solutions and provide Gram-negative opportunist pathogens with sufficient nutrients to multiply during storage; if contaminated, numbers in excess of 10^6 cfu and endotoxins should be expected. Total parenteral nutrition fluids, formulated for individual patients' nutritional requirements, can also provide more than adequate nutritional support for invading contaminants. *Ps. aeruginosa*, the notorious contaminant of eye-drops, has caused serious ophthalmic infections, including the loss of sight in some cases. The problem is compounded when the eye is damaged through the improper use of contact lenses or scratched by fingernails or cosmetic applicators.

The fate of contaminants ingested orally in medicines may be determined by several factors, as is seen with contaminated food. The acidity of the stomach may provide a successful barrier, depending on whether the medicine is taken on an empty or full stomach and also on the gastric emptying time. Contaminants in topical products may cause little harm when deposited on intact skin. Not only does the skin itself provide an excellent mechanical barrier but, furthermore, few contaminants normally survive in competition with its resident microbial flora. Skin damaged during surgery or trauma or in patients with burns or pressure sores may, however, be rapidly colonized and subsequently infected by opportunist pathogens. Patients treated with topical steroids are also prone to local infections, particularly if contaminated steroid drugs are inadvertently used.

| 5.3 | **Resistance of the patient** |

A patient's resistance is crucial in determining the outcome of a medicament-borne infection. Hospital patients are more exposed and susceptible to infection than those treated in the general community. Neonates, the elderly, diabetics and patients traumatized by surgery or accident may have impaired defence mechanisms. People suffering from leukaemia and those treated with immunosuppressants are most vulnerable to infection; there is a strong case for providing all medicines in a sterile form for these patients.

| 6 | **Prevention and control of contamination** |

Prevention is undoubtedly better than cure in minimizing the risk of medicament-borne infections. In manufacture the principles of good manufacturing practice must be observed, and control measures must be built in at all stages. Initial stability tests should show that the proposed formulation can withstand an appropriate microbial challenge; raw materials from an authorized supplier should comply with in-house microbial specifications; environmental conditions, appropriate to the production process, require regular microbiological monitoring; finally, end-product analysis should indicate that the product is microbiologically suitable for its intended use and conforms to accepted in-house and international standards.

Based on present knowledge, contaminants, by virtue of their type or number, should not present a potential health hazard to patients when used.

Contamination during use is less easily controlled. Successful measures in the hospital pharmacy have included the packaging of products as individual units, thereby discouraging the use of multidose containers. Unit packaging (one dose per patient) has clear advantages, but economic constraints prevent this desirable procedure from being realized. Ultimately, the most fruitful approach is through the training and education of patients and hospital staff, so that medicines are used only for their intended purpose. The task of implementing this approach inevitably rests with the clinical and community pharmacists of the future.

7 Further reading

Baird R.M. (1981) Drugs and cosmetics. In *Microbial Biodeterioration* (Ed. A.H. Rose), pp. 387–426. London: Academic Press.

Baird R.M. (1985) Microbial contamination of pharmaceutical products made in a hospital pharmacy. *Pharm J*. **234**, 54–55.

Baird R.M. (1985) Microbial contamination of non-sterile pharmaceutical products made in hospitals in the North East Regional Health Authority. *J Clin Hosp Pharm*. **10**, 95–100.

Baird R.M. & Shooter R.A. (1976) *Pseudomonas aeruginosa* infections associated with the use of contaminated medicaments. *Br Med J*. **2**, 349–350.

Baird R.M., Brown W.R.L. & Shooter R.A. (1976) *Pseudomonas aeruginosa* in hospital pharmacies. *Br Med J*. **1**, 511–512.

Baird R.M., Elhag K.M. & Shaw E.J. (1976) *Pseudomonas thomasii* in a hospital distilled water supply. *J Med Microbiol*. **9**, 493–495.

Baird R.M., Parks A. & Awad Z.A. (1977) Control of *Pseudomonas aeruginosa* in pharmacy environments and medicaments. *Pharm J*. **119**, 164–165.

Baird R.M., Crowden C.A., O'Farrell S.M. & Shooter R.A. (1979) Microbial contamination of pharmaceutical products in the home. *J Hyg*. **83**, 277–283.

Bassett D.C.J. (1971) Causes and prevention of sepsis due to Gram-negative bacteria: common sources of outbreaks. *Proc Roy Soc Med*. **64**, 980–986.

Crompton D.O. (1962) Ophthalmic prescribing. *Australas J Pharm*. **43**, 1020–1028.

Denyer S.P. & Baird R.M. (Eds) (1990) *Guide to Microbiological Control in Pharmaceuticals*. Chichester: Ellis Horwood.

Guide to Good Manufacturing Practice (1983). London: HMSO.

Hills S. (1946) The isolation of *Cl. tetani* from infected talc. *N Z Med J*. **45**, 419–423.

Kallings L.O., Ringertz O., Silverstolpe L. & Ernerfeldt F. (1966) Microbiological contamination of medicinal preparations. 1965 Report to the Swedish National Board of Health. *Acta Pharm Suecica*. **3**, 219–228.

Maurer I.M. (1985) *Hospital Hygiene*, 3rd edn. London: Edward Arnold.

Meers P.D., Calder M.W., Mazhar M.M. & Lawrie G.M. (1973) Intravenous infusion of contaminated dextrose solution: the Devonport incident. *Lancet*, **ii**, 1189–1192.

Morse L.J., Williams H.I, Grenn F.P., Eldrige E.F. & Rotta J.R. (1967) Septicaemia due to *Klebsiella pneumoniae* originating from a handcream dispenser. *N Engl J Med*. **277**, 472–473.

Myers G.E. & Pasutto F.M. (1973) Microbial contamination of cosmetics and toiletries. *Can J Pharm Sci*. **8**, 19–23.

Noble W.C. & Savin J.A. (1966) Steroid cream contaminated with *Pseudomonas aeruginosa*. *Lancet*, **i**, 347–349.

Parker M.T. (1972) The clinical significance of the presence of microorganisms in pharmaceutical and cosmetic preparations. *J Soc Cosm Chem*. **23**, 415–426.

Report of the Public Health Laboratory Service Working Party (1971) Microbial contamination of medicines administered to hospital patients. *Pharm J*. **207**, 96–99.

Russell A.D., Hugo W.G. & Ayliffe G.A.J. (Eds) (1992) *Principles and Practice of Disinfection, Preservation and Sterilization*, 2nd edn. Oxford: Blackwell Scientific Publications.

Smart R. & Spooner D.F. (1972) Microbiological spoilage in pharmaceuticals and cosmetics. *J Soc Cosm Chem*. **23**, 721–737.

20 Principles and practice of sterilization

1   **Introduction**

2   **Sensitivity of microorganisms**
2.1   Survivor curves
2.2   Expressions of resistance
2.2.1   *D*-value
2.2.2   *z*-value
2.3   Sterility assurance

3   **Sterilization methods**

4   **Heat sterilization**
4.1   Sterilization process
4.2   Moist heat sterilization
4.2.1   Steam as a sterilizing agent
4.2.2   Sterilizer design and operation
4.3   Dry heat sterilization
4.3.1   Sterilizer design
4.3.2   Sterilizer operation

5   **Gaseous sterilization**
5.1   Ethylene oxide

5.1.1   Sterilizer design and operation
5.2   Formaldehyde
5.2.1   Sterilizer design and operation

6   **Radiation sterilization**
6.1   Sterilizer design and operation
6.1.1   Gamma-ray sterilizers
6.1.2   Electron accelerators
6.1.3   Ultraviolet irradiation

7   **Filtration sterilization**
7.1   Filtration sterilization of liquids
7.2   Filtration sterilization of gases

8   **Conclusions**

9   **Acknowledgements**

10   **Appendix**

11   **Further reading**

1 Introduction

Sterilization is an essential stage in the processing of any product destined for parenteral administration, or for contact with broken skin, mucosal surfaces or internal organs, where the threat of infection exists. In addition, the sterilization of microbiological materials, soiled dressings and other contaminated items is necessary to minimize the health hazard associated with these articles.

Sterilization processes involve the application of a biocidal agent or physical microbial removal process to a product or preparation with the object of killing or removing all microorganisms. These processes may involve elevated temperature, reactive gas, irradiation or filtration through a microorganism-proof filter. The success of the process depends upon a suitable choice of treatment conditions, e.g. temperature and duration of exposure. It must be remembered, however, that with all articles to be sterilized there is a potential risk of product damage, which for a pharmaceutical preparation may result in reduced therapeutic efficacy or patient acceptability. Thus, there is a need to achieve a balance between the maximum acceptable risk of failing to achieve sterility and the maximum level of product damage which is acceptable. This is best determined from a knowledge of the properties of the sterilizing agent, the properties of the product to be sterilized and the nature of the likely contaminants. A

403

suitable sterilization process may then be selected to ensure maximum microbial kill/removal with minimum product deterioration.

2 Sensitivity of microorganisms

The general pattern of resistance of microorganisms to biocidal sterilization processes is independent of the type of agent employed (heat, radiation or gas), with vegetative forms of bacteria and fungi, along with the larger viruses, showing a greater sensitivity to sterilization processes than small viruses and bacterial or fungal spores. The choice of suitable reference organisms for testing the efficiency of sterilization processes (see Chapter 23) is therefore made from the most durable bacterial spores, usually represented by *Bacillus stearothermophilus* for moist heat, certain strains of *B. subtilis* for dry heat and gaseous sterilization, and *B. pumilus* for ionizing radiation.

Ideally, when considering the level of treatment necessary to achieve sterility a knowledge of the type and total number of microorganisms present in a product, together with their likely response to the proposed treatment, is necessary. Without this information, however, it is usually assumed that organisms within the load are no more resistant than the reference spores or than specific resistant product isolates. In the latter case, it must be remembered that resistance may be altered or lost entirely by laboratory subculture and the resistance characteristics of the maintained strain must be regularly checked.

A sterilization process may thus be developed without a full microbiological background to the product, instead being based on the ability to deal with a 'worst case' condition. This is indeed the situation for official sterilization methods which must be capable of general application, and modern pharmacopoeial recommendations are derived from a careful analysis of experimental data on bacterial spore survival following treatments with heat, ionizing radiation or gas.

2.1 Survivor curves

When exposed to a killing process, populations of microorganisms generally lose their viability in an exponential fashion, independent of the initial number of organisms. This can be represented graphically with a 'survivor curve' drawn from a plot of the logarithm of the fraction of survivors against the exposure time or dose (Fig. 20.1). Of the typical curves obtained, all have a linear portion which may be continuous (plot A), or may be modified by an initial shoulder (B) or by a reduced rate of kill at low survivor levels (C). Furthermore, a short activation phase, representing an initial increase in viable count, may be seen during the heat treatment of certain bacterial spores. Survivor curves have been employed principally in the examination of heat sterilization methods, but can equally well be applied to any biocidal process.

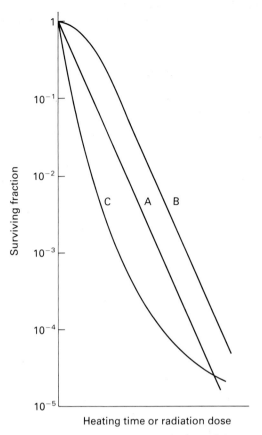

Fig. 20.1 Typical survivor curves for bacterial spores exposed to moist heat or gamma-radiation.

2.2   Expressions of resistance

2.2.1   *D-value*

The resistance of an organism to a sterilizing agent can be described by means of the *D*-value. For heat and radiation treatments, respectively, this is defined as the time taken at a fixed temperature or the radiation dose required to achieve a 90% reduction in viable cells (i.e. a 1 log cycle reduction in survivors; Fig. 20.2A). The calculation of the *D*-value assumes a linear type A survivor curve (Fig. 20.1), and must be corrected to allow for any deviation from linearity with type B or C curves. Some typical *D*-values for resistant bacterial spores are given in Table 23.2 (Chapter 23).

2.2.2   *z-value*

For heat treatment, a *D*-value only refers to the resistance of a micro-organism at a particular temperature. In order to assess the influence of

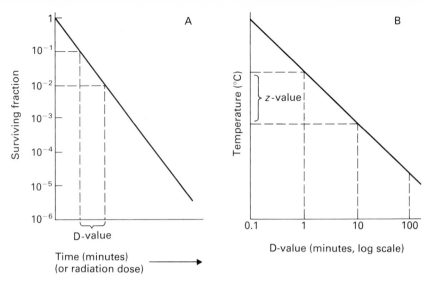

Fig. 20.2 Calculation of: (A) *D*-value; (B) *z*-value.

temperature changes on thermal resistance a relationship between temperature and log *D*-value can be developed leading to the expression of a *z*-value, which represents the increase in temperature needed to reduce the *D*-value of an organism by 90% (i.e. 1 log cycle reduction; Fig. 20.2B). For bacterial spores used as biological indicators for moist heat (*B. stearothermophilus*) and dry heat (*B. subtilis*) sterilization processes, mean *z*-values are given as 10°C and 22°C, respectively. The *z*-value is not truly independent of temperature but may be considered essentially constant over the temperature ranges used in heat sterilization processes.

2.3 Sterility assurance

From the survivor curves presented, it can be seen that the elimination of viable microorganisms from a product is a time-dependent process, and will be influenced by the rate and duration of biocidal action and the initial microbial contamination level. Thus, the likelihood of a product being produced free of microorganisms is best expressed in terms of the probability of an organism surviving the treatment process, a possibility not entertained in the absolute term 'sterile'. From this approach has arisen the concept of sterility assurance or a microbial safety index which gives a numerical value to the probability of a single surviving organism remaining to contaminate a processed product. For pharmaceutical products, the most frequently applied standard is that the probability, post-sterilization, of a non-sterile unit is $\leqslant 1$ in 1 million units processed (i.e. $\leqslant 10^{-6}$). The sterilization protocol necessary to achieve this with any given organism of known *D*-value can be established from the inactivation factor (IF) which may be defined as:

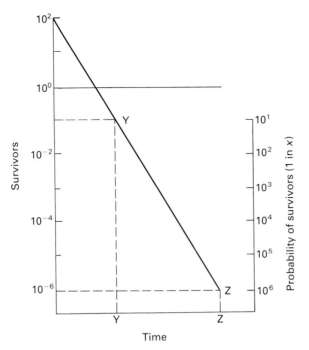

Fig. 20.3 Sterility assurance. At Y, there is (literally) 10^{-1} bacterium in one bottle, i.e. in 10 loads of single containers, there would be one chance in 10 that one load would be positive. Likewise, at Z, there is (literally) 10^{-6} bacterium in one bottle, i.e. in 1 million (10^{6}) loads of single containers, there is one chance in 1 million that one load would be positive.

$$IF = 10^{t/D}$$

where t is the contact time (for a heat or gaseous sterilization process) or dose (for ionizing radiation) and D is the D-value appropriate to the process employed.

Thus, for an initial burden of 10^{2} spores an inactivation factor of 10^{8} will be needed to give the required sterility assurance of 10^{-6} (Fig. 20.3). The sterilization process will therefore need to produce sufficient lethality to achieve an 8 log cycle reduction in viable organisms; this will require exposure of the product to eight times the D-value of the reference organism ($8D$). In practice, it is generally assumed that the contaminant will have the same resistance as the test spores unless full microbiological data are available to indicate otherwise. The inactivation factors associated with certain sterilization protocols and their biological indicator organisms (Chapter 23) are given in Table 20.1.

3   **Sterilization methods**

The *British Pharmacopoeia* recognizes five methods for the sterilization of pharmaceutical products. These are: (i) dry heat; (ii) heating in an

Table 20.1 Inactivation factors (IF) for selected sterilization protocols and their corresponding biological indicator (BI) organisms

Sterilization protocol	BI organism	D-value	IF
Moist heat (121°C for 15 minutes)	B. stearothermophilus	1.5 min	10
Dry heat (160°C for 2 hours)	B. subtilis var. niger	Max. 10 min	Min. 12
Irradiation (25 kGy; 2.5 Mrad)	B. pumilus	3 kGy (0.3 Mrad)	8.3

autoclave (steam sterilization); (iii) filtration; (iv) ethylene oxide gas; and (v) gamma or electron radiation. In addition, other approaches involving steam and formaldehyde and ultraviolet light have evolved for use in certain situations. For each method, the possible permutations of exposure conditions are numerous, but experience and product stability requirements have generally served to limit this choice. Nevertheless, it should be remembered that even the recommended methods and regimens do not necessarily demonstrate equivalent biocidal potential, but simply offer alternative strategies for application to a wide variety of product types. Thus, each should be validated in their application to demonstrate that the minimum required level of sterility assurance can be achieved (section 2.3 and Chapter 23).

In the following sections, factors governing the successful use of these sterilizing methods will be covered and their application to pharmaceutical and medical products considered. Methods for monitoring the efficacy of these processes are discussed in Chapter 23.

4 Heat sterilization

Heat is the most reliable and widely used means for sterilization, affording its antimicrobial activity through destruction of enzymes and other essential cell constituents. These lethal events proceed at their most rapid in a fully hydrated state, thus requiring a lower heat input (temperature

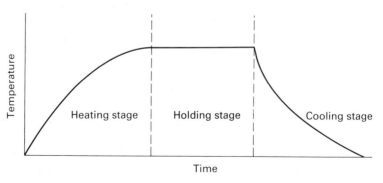

Fig. 20.4 Typical temperature profile of a heat sterilization process.

and time) under conditions of high humidity where denaturation and hydrolysis reactions predominate, rather than in the dry state where oxidative changes take place. This method of sterilization is limited to thermostable products, but can be applied to both moisture-sensitive and moisture-resistant products for which the *British Pharmacopoeia* (1988) recommends dry (160–180°C) and moist (121–134°C) heat sterilization, respectively. Where thermal degradation of a product might possibly occur, it can usually be minimized by selecting the higher temperature range since the shorter exposure times employed generally result in a lower fractional degradation.

4.1

Sterilization process

In any heat sterilization process the articles to be treated must first be raised to sterilization temperature and this involves a heating-up stage. In the traditional approach, timing for the process (the holding time) then begins. It has been recognized, however, that during both the heating-up and cooling-down stages of a sterilization cycle (Fig. 20.4), the product is held at an elevated temperature and these stages may thus contribute to the overall biocidal potential of the process.

A method has been devised to convert all the temperature–time combinations occurring during the heating, sterilizing and cooling stages of a moist heat (steam) sterilization cycle to the equivalent time at 121°C. This involves following the temperature profile of a load, integrating the heat input (as a measure of lethality), and converting it to the equivalent time at the standard temperature of 121°C. Using this approach the overall lethality of any process can be deduced and is defined as the *F*-value, which expresses heat treatment at any temperature as equal to that of a certain number of minutes at 121°C. In other words, if a moist heat sterilization process has an *F*-value of *x*, then it has the same lethal effect on a given organism as heating at 121°C for *x* minutes, irrespective of the actual temperature employed or of any fluctuations in the heating process due to heating and cooling stages. The *F*-value of a process will vary according to the moist heat resistance of the reference organism; when the reference spore is that of *B. stearothermophilus* with a *z*-value of 10°C, then the *F*-value is known as the F_0-value.

A relationship between *F*- and *D*-values, leading to an assessment of the probable number of survivors in a load following heat treatment, can be established from the following equation:

$$F = D(\log N_0 - \log N)$$

in which *D* is the *D*-value at 121°C, and N_0 and *N* represent, respectively, the initial and final number of viable cells per unit volume.

The *F*-concept has evolved from the food industry and principally relates to the sterilization of articles by moist heat. It offers a mechanism by which over-processing of marginally thermolabile products can be reduced without compromising sterility assurance. It has found appli-

Table 20.2 Temperature and pressure relationships in steam sterilization

Temperature (°C)	Steam pressure	
	kPa	psi
115	69	10
121	103	15
126	138	20
134	207	30

cation in the sterilization of medical and pharmaceutical products by moist heat where, for aqueous preparations, the *British Pharmacopoeia* generally requires a minimum F_0-value of 8 from a steam sterilization process. This approach has not been generally applied to dry heat sterilization methods.

4.2　　**Moist heat sterilization**

Moist heat has been recognized as an efficient biocidal agent from the early days of bacteriology, when it was principally developed for the sterilization of culture media. It now finds widespread application in the processing of many thermostable products and devices. In the pharmaceutical and medical sphere it is used in the sterilization of dressings, sheets, surgical and diagnostic equipment, containers and closures, and aqueous injections, ophthalmic preparations and irrigation fluids, in addition to the processing of soiled and contaminated items (Chapter 21).

Sterilization by moist heat usually involves the use of steam at temperatures in the range 121–134°C, and while alternative strategies are available for the processing of products unstable at these high temperatures, they rarely offer the same degree of sterility assurance and should be avoided if at all possible. The elevated temperatures generally associated with moist heat sterilization methods can only be achieved by the generation of steam under pressure. The commonly employed temperatures and their corresponding pure steam pressures are given in Table 20.2.

4.2.1　　*Steam as a sterilizing agent*

To act as an efficient sterilizing agent, steam should be able to provide moisture and heat efficiently to the article to be sterilized. This is most effectively done using saturated steam, which is steam in thermal equilibrium with the water from which it is derived, i.e. steam on the phase boundary (Fig. 20.5). Under these circumstances contact with a cooler surface causes condensation and contraction drawing in fresh steam and leading to the immediate release of the latent heat, which represents a major portion of the heat energy (Table 20.3). In this way

Table 20.3 The contribution of latent and sensible heat energies to the total heat content of steam under pressure

Minimum saturated steam temperature (°C)	Heat content (kJ kg⁻¹) Sensible	Latent	Latent as % of total heat
115	483	2216	82
121	505	2202	81
126	530	2185	80
134	561	2164	79

heat and moisture are imparted rapidly to articles being sterilized and dry porous loads (e.g. surgical dressings) are quickly penetrated by the steam.

Steam for sterilization can either be generated within the sterilizer, as with portable bench or 'instrument and utensil' sterilizers, in which case it is constantly in contact with water and is known as 'wet' steam, or can be supplied under pressure (350–400 kPa) from a separate boiler as 'dry' saturated steam with no entrained water droplets. The killing potential of 'wet' steam is the same as that of 'dry' saturated steam at the same

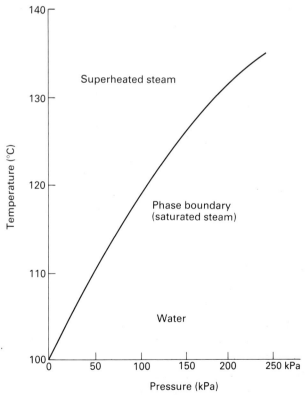

Fig. 20.5 Phase diagram for water vapour.

411   *Principles and practice of sterilization*

temperature, but it is more likely to soak a porous load creating physical difficulties for further steam penetration. Thus major industrial and hospital sterilizers are usually supplied with 'dry' saturated steam and attention is paid to the removal of entrained water droplets within the supply line to prevent introduction of a water 'fog' into the sterilizer.

If the temperature of 'dry' saturated steam is increased, then, in the absence of entrained moisture, the relative humidity or degree of saturation is reduced and the steam becomes superheated (Fig. 20.5). During sterilization this can arise in a number of ways, for example by overheating the steam jacket (see section 4.2.2), by using too dry a steam supply, by excessive pressure reduction during passage of steam from the boiler to the sterilizer chamber, and by evolution of heat of hydration when steaming over-dried cotton fabrics. Superheated steam behaves in the same manner as hot air since condensation and release of latent heat will not occur unless the steam is cooled to the phase boundary temperature. Thus, it proves to be an inefficient sterilizing agent, and although a small degree of transient superheating can be tolerated, a maximum acceptable level of 5° superheat is set, i.e. the temperature of the steam is never greater than 5°C above the phase boundary temperature at that pressure.

The relationship between temperature and pressure holds true only in the presence of pure steam; adulteration with air contributes to a partial pressure but not to the temperature of the steam. Thus, in the presence of air the temperature achieved will reflect the contribution made by the steam and will be lower than that normally attributed to the total pressure recorded. Addition of further steam will raise the temperature but residual air surrounding articles may delay heat penetration or, if a large amount of air is present, it may collect at the bottom of the sterilizer, completely altering the temperature profile of the sterilizer chamber. It is for these reasons that efficient air removal is a major aim in the design and operation of a boiler-fed steam sterilizer.

4.2.2   *Sterilizer design and operation*

Steam sterilizers, or autoclaves as they are sometimes known, are stainless steel vessels designed to withstand the steam pressures employed in sterilization. They can be: (i) 'portable' sterilizers, where they generally have internal electric heaters to produce steam and are used for small pilot or laboratory-scale sterilization and for the treatment of instruments and utensils; or (ii) large-scale sterilizers for routine hospital or industrial use, operating on 'dry' saturated steam from a separate boiler (Fig. 20.6). Because of their widespread use within pharmacy this latter type will be considered in greatest detail.

There are two main types of large sterilizers, those designed for use with porous loads (i.e. dressings) and generally operated at a minimum temperature of 134°C, and those designed as bottled-fluid sterilizers employing a minimum temperature of 121°C. The stages of operation are

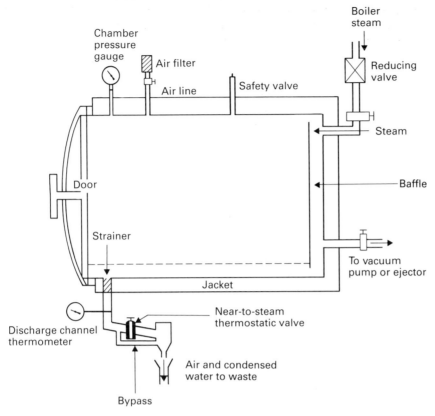

Fig. 20.6 Main constructional features of a large-scale steam sterilizer (autoclave).

common to both and can be summarized as air removal and steam admission, heating-up and exposure, and drying or cooling. Many modifications of design exist and in this section only general features will be considered. Fuller treatments of sterilizer design and operation can be found in Health Technical Memorandum 10 (1980).

General design features. Steam sterilizers are constructed with either cylindrical or oblong chambers, with preferred capacities ranging from 400 to 800 litres. They can be sealed by either a single door or by doors at both ends (to allow through-passage of processed materials; see Chapter 22, section 3.2.3). During sterilization the doors are held closed by a locking mechanism which prevents opening when the chamber is under pressure. In the larger sterilizers the chamber may be surrounded by a steam-jacket which can be used to heat the autoclave chamber. The chamber floor slopes towards a discharge channel through which air and condensate can be removed. Temperature is monitored within the opening of the discharge channel and by thermocouples in dummy packages; jacket and chamber pressures are followed using pressure gauges. In hospitals and industry it is common practice to operate sterilizers on an automatic cycle, each stage of operation being controlled by a timer responding to temperature- or pressure-sensing devices.

Operation

1 *Air removal and steam admission.* Air can be removed from steam sterilizers either by downward displacement with steam, evacuation or a combination of the two. In the downward displacement sterilizer, the heavier cool air is forced out of the discharge channel by incoming hot steam. This has the benefit of warming the load during air removal which aids the heating-up process. It finds widest application in the sterilization of bottled fluids where bottle breakage may occur under the combined stresses of evacuation and high temperature. For more air-retentive loads (i.e. dressings), however, this technique of air removal is unsatisfactory and mechanical evacuation of the air is essential before admission of the steam. This can either be to an extremely high level (e.g. 2.5 kPa) or can involve a period of pulsed evacuation and steam admission, the latter approach improving air extraction from dressings packs. After evacuation, steam penetration into the load is very rapid and heating-up is almost instantaneous. It is axiomatic that packaging and loading of articles within a sterilizer be so organized as to facilitate air removal.

During the sterilization process, small pockets of entrained air may still be released, especially from packages, and this air must be removed. This is achieved with a near-to-steam thermostatic valve incorporated in the discharge channel. The value operates on the principle of an expandable bellows containing a volatile liquid which vaporizes at the temperature of saturated steam thereby closing the valve, and condenses on passage of a cooler air–steam mixture, thus reopening the valve and discharging the air. Condensate generated during the sterilization process can also be removed by this device. Small quantities of air will not, however, lower the temperature sufficiently to operate the valve and so a continual slight flow of steam is maintained through a bypass around the device in order to flush away residual air.

Increasingly, it is becoming common practice to package sterile fluids, especially intravenous fluids, in flexible plastic containers. During sterilization these can develop a considerable internal pressure in the airspace above the fluid and it is therefore necessary to maintain a proportion of air within the sterilizing chamber to produce sufficient overpressure to prevent these containers from bursting. In sterilizers modified or designed to process this type of product, air removal is therefore unnecessary but special attention must be paid to the prevention of air 'layering' within the chamber. This is overcome by the inclusion of a fan or through a continuous spray of hot water within the chamber to mix the air and steam. Air ballasting can also be employed to prevent bottle breakage.

2 *Heating-up and exposure.* When the sterilizer reaches its operating temperature and pressure the sterilization stage begins. The duration of exposure may include a heating-up time in addition to the holding time and this will normally be established using thermocouples in dummy articles. In the absence of this information for bottled fluids, guidance is given in the *Pharmaceutical Handbook* (1980).

3 *Drying or cooling.* Dressings packs and other porous loads may become dampened during the sterilization process and must be dried before removal from the chamber. This is achieved by steam exhaust and application of a vacuum, often assisted by heat from the steam-filled jacket if fitted. After drying, atmospheric pressure within the chamber is restored by admission of sterile filtered air.

For bottled fluids the final stage of the sterilization process is cooling, and this needs to be achieved as rapidly as possible to minimize thermal degradation of the product and to reduce processing time. In modern sterilizers this is achieved by spray-cooling with retained condensate delivered to the surface of the load by nozzles fitted into the roof of the sterilizer chamber. This is often accompanied by the introduction of filtered, compressed air to minimize container breakage due to high internal pressures. Containers must not be removed from the sterilizer until the internal pressure has dropped to a safe level, usually indicated by a temperature of less than 80°C. Occasionally, spray-cooling water may be a source of bacterial contamination and its microbiological quality must be carefully monitored.

4.3   **Dry heat sterilization**

The resistance of bacterial spores to dry heat is generally greater than in the presence of moisture; thus dry heat sterilization usually employs higher temperatures in the range 160–180°C and requires exposure times of up to 2 hours depending upon the temperature employed (section 10). Its application is generally restricted to glassware and metal surgical instruments (where its good penetrability and non-corrosive nature are of benefit), non-aqueous thermostable liquids and thermostable powders (see Chapter 21).

4.3.1   *Sterilizer design*

Dry heat sterilization is usually carried out in a hot air oven which comprises an insulated polished stainless steel chamber, with a usual capacity of up to 250 litres, surrounded by an outer case containing electric heaters located in positions to prevent cool spots developing inside the chamber. A fan is fitted to the rear of the oven to provide circulating air, thus ensuring more rapid equilibration of temperature. Shelves within the chamber are perforated to allow good air flow. Thermocouples can be used to monitor the temperature of both the oven air and articles contained within. A fixed temperature sensor connected to a chart recorder provides a permanent record of the sterilization cycle. Appropriate door-locking controls should be incorporated to prevent interruption of a sterilization cycle once begun.

Recent sterilizer developments have led to the use of dry-heat sterilizing tunnels where heat transfer is achieved by infra-red irradiation or by forced convection in filtered laminar airflow tunnels. Items to be sterilized are placed on a conveyer belt and pass through a high temperature zone (250–300+°C) over a period of several minutes.

4.3.2   *Sterilizer operation*

Articles to be sterilized must be wrapped or enclosed in containers of sufficient strength and integrity to provide good post-sterilization protection against contamination. Suitable materials are paper, cardboard tubes or aluminium containers. Container shape and design must be such that heat penetration is encouraged in order to shorten the heating-up stage; this can be achieved by using narrow containers with dull non-reflecting surfaces. In a hot-air oven, heat is delivered to articles principally by radiation and convection; thus they must be carefully arranged within the chamber to avoid obscuring centrally placed articles from wall radiation or impeding air flow. The temperature variation within the chamber should not exceed ±5°C of the recorded temperature. Heating-up times, which may be as long as 4 hours for articles with poor heat-conducting properties, can be reduced by preheating the oven before loading. Following sterilization, the chamber temperature is usually allowed to fall to around 40°C before removal of sterilized articles; this can be accelerated by the use of forced cooling with filtered air.

5   Gaseous sterilization

The chemically reactive gases ethylene oxide (C_2H_4O) and formaldehyde (methanal, CH_2O) possess broad-spectrum biocidal activity, and have found application in the sterilization of reusable surgical instruments, certain medical, diagnostic and electrical equipment, and the surface sterilization of powders. Ethylene oxide treatment can also be considered as an alternative to radiation sterilization in the commercial production of disposable medical devices (Chapter 21). These techniques do not, however, offer the same degree of sterility assurance as heat methods and are generally reserved for temperature-sensitive items.

The mechanism of antimicrobial action of the two gases is assumed to be through alkylation of sulphydryl, amino, hydroxyl and carboxyl groups on proteins and imino groups of nucleic acids. At the concentrations employed in sterilization protocols, type A survivor curves (section 2.1, Fig. 20.1) are produced, the lethality of these gases increasing in a non-uniform manner with increasing concentration, exposure temperature and humidity. For this reason, sterilization protocols have generally been established by an empirical approach using a standard product load containing suitable biological indicator test strips (Chapter 23). Concentration ranges (given as weight of gas per unit chamber volume) are usually in the order of 800–1200 mg l^{-1} for ethylene oxide and 15–100 mg l^{-1} for formaldehyde, with operating temperatures in the region of

45–63°C and 70–75°C, respectively. Even at the higher concentrations and temperatures the sterilization processes are lengthy and therefore unsuitable for the resterilization of high-turnover articles. A further delay occurs because of the need to remove toxic residues of the gases before release of the items for use.

5.1   Ethylene oxide

Ethylene oxide gas is highly explosive in mixtures of $>3.6\%$ v/v in air; in order to reduce this explosion hazard it is usually supplied for sterilization purposes as a 10 or 12% mix with carbon dioxide or fluorinated hydrocarbons (Freons), respectively. Alternatively, pure ethylene oxide gas can be used at below atmospheric pressure in sterilizer chambers from which all air has been removed. In addition, the gas is a strong vesicant and irritant on contact or inhalation and may also, in common with its chloride reaction product ethylene chlorhydrin, be carcinogenic. It is important, therefore, that ethylene oxide sterilization is carried out in accordance with a strict working protocol and that the environmental gas levels are monitored to ensure the safety of operators.

The efficacy of ethylene oxide treatment depends upon achieving a suitable concentration in each article and this is assisted greatly by the good penetrating powers of the gas, which diffuses readily into many packaging materials including rubber, plastics, fabric and paper. This is not without its drawbacks, however, since the level of ethylene oxide in a sterilizer will decrease due to absorption during the process and the treated articles must undergo a desorption stage to remove toxic residues. Desorption can be allowed to occur naturally on open shelves, in which case complete desorption may take many days, e.g. for materials like PVC, or it may be encouraged by special forced aeration cabinets where flowing, heated air assists gas removal, reducing desorption times to between 2 and 24 hours.

Organisms are more resistant to ethylene oxide treatment in a dried state, as are those protected from the gas by inclusion in crystalline or dried organic deposits. Thus a further condition to be satisfied in ethylene oxide sterilization is attainment of a minimum level of moisture in the immediate product environment. This requires a sterilizer humidity of 30–70% and frequently a preconditioning of the load at relative humidities of greater than 50%.

5.1.1   *Sterilizer design and operation*

An ethylene oxide sterilizer consists of a leak-proof and explosion-proof steel chamber, normally of 100–300 litre capacity, which can be surrounded by a hot-water jacket to provide a uniform chamber temperature. Successful operation of the sterilizer requires removal of air from the chamber by evacuation, humidification and conditioning of the load by passage of subatmospheric pressure steam followed by a further evacuation

417   *Principles and practice of sterilization*

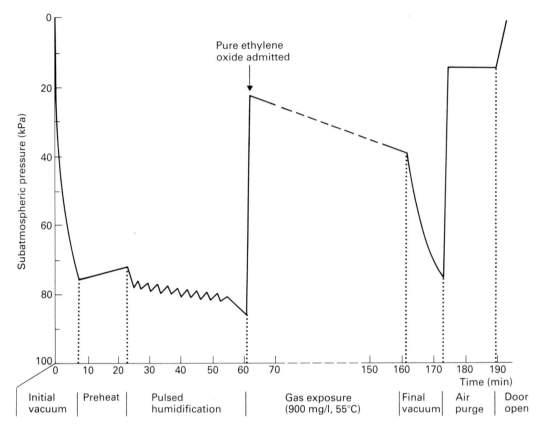

Fig. 20.7 Typical operating cycle for pure ethylene oxide gas.

period and the admission of preheated vaporized ethylene oxide from external pressurized canisters or single-charge cartridges. Forced gas circulation is often employed to minimize variations in conditions throughout the sterilizer chamber. Packaging materials must be air, steam and gas permeable to permit suitable conditions for sterilization to be achieved within individual articles in the load. Absorption of ethylene oxide by the load is compensated for by the introduction of excess gas at the beginning or by the addition of more gas as the pressure drops during the sterilization process. After treatment, the gases are evacuated either directly to the outside atmosphere or through a special exhaust system. Filtered, sterile air is then admitted either for a repeat of the vacuum/air cycle or for air purging until the chamber is opened. In this way, safe removal of the ethylene oxide is achieved reducing the toxic hazard to the operator. Sterilized articles are removed directly from the chamber and arranged for desorption.

The operation of an ethylene oxide sterilizer should be monitored and controlled automatically. A typical operating cycle for pure ethylene oxide gas is given in Fig. 20.7, and general conditions are summarized in section 10.

Formaldehyde

Formaldehyde gas for use in sterilization is produced by heating formalin (37% w/v aqueous solution of formaldehyde) to a temperature of 70–75°C with steam, leading to the process known as low temperature steam and formaldehyde (LTSF). Formaldehyde has a similar toxicity to ethylene oxide and although absorption to materials appears to be lower similar desorption routines are recommended. A major disadvantage of formaldehyde is low penetrating power and this limits the packaging materials that can be employed to principally paper and cotton fabric.

Sterilizer design and operation

An LTSF sterilizer is designed to operate with subatmospheric pressure steam. Air is removed by evacuation and steam admitted to the chamber to allow heating of the load and to assist in air removal. The sterilization period starts with the release of formaldehyde by vaporization from formalin (in a vaporizer with a steam jacket) and continues through either a simple holding stage or through a series of pulsed evacuations and steam and formaldehyde admission cycles. The chamber temperature is maintained by a thermostatically controlled water jacket, and steam and condensate are removed via a drain channel and an evacuated condenser. At the end of the treatment period formaldehyde vapour is expelled by steam flushing and the load dried by alternating stages of evacuation and admission of sterile, filtered air. A typical pulsed cycle of operation is shown in Fig. 20.8 and general conditions are summarized in section 10.

6 Radiation sterilization

Several types of radiation find a sterilizing application in the manufacture of pharmaceutical and medical products, principal among which are accelerated electrons (particulate radiation), gamma-rays and ultraviolet (UV) light (both electromagnetic radiations). The major target for these radiations is believed to be microbial DNA, with damage occurring as a consequence of ionization and free radical production (gamma-rays and electrons) or excitation (UV light). This latter process is less damaging and less lethal than ionization, and so UV irradiation is not as efficient a sterilization method as electron or gamma-irradiation. As mentioned earlier (section 2), vegetative bacteria generally prove to be the most sensitive to irradiation (with notable exceptions, e.g. *Deinococcus (Micrococcus) radiodurans*), followed by moulds and yeasts, with bacterial spores and viruses as the most resistant (except in the case of UV light where mould spores prove to be most resistant). The extent of DNA damage required to produce cell death can vary and this, together with the ability to carry out effective repair, probably decides the resistance of the organism to radiation. With ionizing radiations (gamma-ray and accelerated electrons), microbial resistance decreases with the pre-

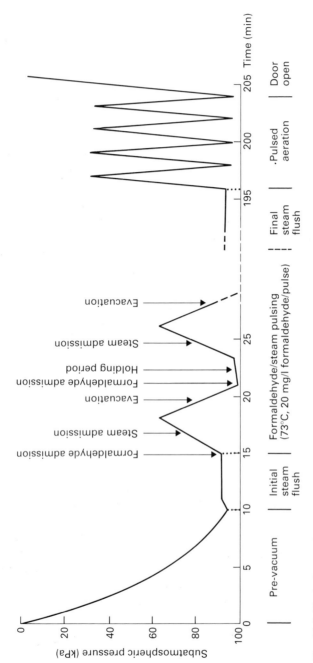

Fig. 20.8 Typical operating cycle for low temperature steam and formaldehyde treatment.

sence of moisture or dissolved oxygen (as a result of increased free radical production) and also with elevated temperatures.

Radiation sterilization with high-energy gamma-rays or accelerated electrons has proved to be a useful method for the industrial sterilization of heat-sensitive products. However, undesirable changes can occur in irradiated preparations, especially those in aqueous solution where radiolysis of water contributes to the damaging processes. In addition, certain glass or plastic (e.g. polypropylene, PTFE) materials used for packaging or for medical devices can also suffer damage. Thus, radiation sterilization is generally applied to articles in the dried state; these include surgical instruments, sutures, prostheses, unit-dose ointments, plastic syringes, and dry pharmaceutical products (Chapter 21). With these radiations, destruction of a microbial population follows the classic survivor curves (see Fig. 20.1) and a *D*-value, given as a radiation dose, can be established for standard bacterial spores (e.g. *B. pumilus*) permitting a suitable sterilizing dose to be calculated. In the UK it is usual to apply a dose of 25 kGy (2.5 Mrad) for pharmaceutical and medical products, although lower doses are employed in the USA and Canada.

UV light, with its much lower energy, causes less damage to microbial DNA. This, coupled with its poor penetrability of normal packaging materials, renders UV light unsuitable for sterilization of pharmaceutical dosage forms. It does find applications, however, in the sterilization of air, for the surface sterilization of aseptic work areas, and for the treatment of manufacturing-grade water.

6.1   Sterilizer design and operation

6.1.1   *Gamma-ray sterilizers*

Gamma-rays for sterilization are usually derived from a cobalt-60 (^{60}Co) source (caesium-137 may also be used), with a half-life of 5.25 years, which on disintegration emits radiation at two energy levels of 1.33 and 1.17 MeV. The isotope is held as pellets packed in metal rods, each rod carefully arranged within the source and containing up to 20 kCi (740 × 10^{12} Bq) of activity; these rods are replaced or rearranged as the activity of the source either drops or becomes unevenly distributed. A typical ^{60}Co installation may contain up to 1 MCi (3.7 × 10^{16} Bq) of activity. For safety reasons, this source is housed within a reinforced concrete building with walls some 2 m thick, and it is only raised from a sunken water-filled tank when required for use. Control devices operate to ensure that the source is raised only when the chamber is locked and that it is immediately lowered if a malfunction occurs. Articles being sterilized are passed through the irradiation chamber on a conveyor belt or monorail system and move around the raised source, the rate of passage regulating the dose absorbed (Fig. 20.9). Radiation monitors are continually employed to detect any radiation leakage during operation or source storage, and to confirm a return to satisfactory background levels within the sterilization

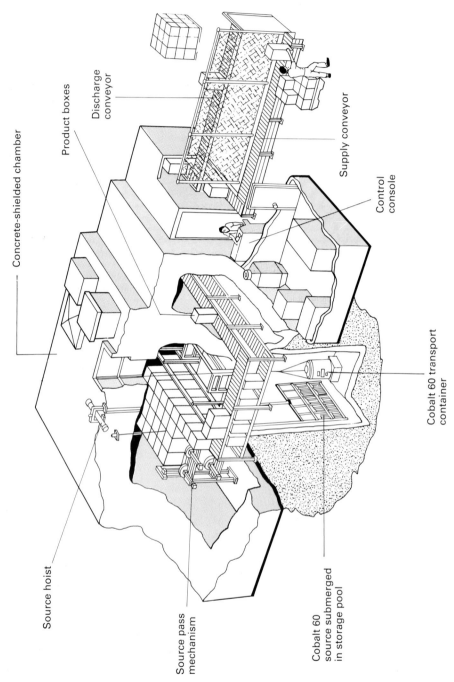

Concrete-shielded chamber

Product boxes

Discharge
conveyor

Supply conveyor

Control
console

Source hoist

Cobalt 60 transport
container

Source pass
mechanism

Cobalt 60
source submerged
in storage pool

Fig. 20.9 Diagram of a typical cobalt-60 irradiation plant.

chamber following operation. The dose delivered is dependent upon source strength and exposure period, with dwell times typically up to 20 hours duration.

<table>
<tr><td>6.1.2</td><td>Electron accelerators</td></tr>
</table>

Two types of electron accelerator machine exist, the electrostatic accelerator and the microwave linear accelerator, producing electrons with maximum energies of 5 MeV and 10 MeV, respectively. In the first, a high-energy electron beam is generated by accelerating electrons from a hot filament down an evacuated tube under high potential difference, while in the second, additional energy is imparted to this beam in a pulsed manner by a synchronized travelling microwave. Articles for treatment are generally limited to small packs and are arranged on a horizontal conveyor belt, usually for irradiation from one side but sometimes from both. The sterilizing dose is delivered more rapidly in an electron accelerator than in a ^{60}Co plant, with exposure times for sterilization usually amounting to only a few seconds or minutes. Varying extents of shielding, depending upon the size of the accelerator, are necessary to protect operators from X-rays generated by the bremsstrahlung effect.

<table>
<tr><td>6.1.3</td><td>Ultraviolet irradiation</td></tr>
</table>

The optimum wavelength for UV sterilization is around 260 nm. A suitable source for UV light in this region is a mercury lamp giving peak emission levels at 254 nm. These sources are generally wall or ceiling mounted for air disinfection, or fixed to vessels for water treatment. Operators present in an irradiated room should wear appropriate protective clothing and eye shields.

<table>
<tr><td>7</td><td>Filtration sterilization</td></tr>
</table>

The process of filtration is unique amongst sterilization techniques in that it removes, rather than destroys, microorganisms. Further, it is capable of preventing the passage of both viable and non-viable particles and can thus be used for both the clarification and sterilization of liquids and gases. The principal applications of sterilizing-grade filters are the treatment of heat-sensitive injections and ophthalmic solutions, biological products and air for supply to aseptic areas (see Chapters 21 and 22). Certain types of filter (membrane filters) also have an important role in sterility testing, where they can be employed to trap and concentrate contaminating organisms from solutions under test. These filters are then placed on the surface of a solid nutrient medium and incubated to encourage colony development (Chapter 23).

The major mechanisms of filtration are sieving, adsorption and trapping within the matrix of the filter material. Of these, only sieving can be regarded as absolute since it ensures the exclusion of all particles above a

Table 20.4 Some characteristics of membrane and depth filters

Characteristic	Membrane	Depth
Absolute retention of microorganisms greater than rated pore size	+	−
Rapid rate of filtration	+	−
High dirt-handling capacity	−	+
Grow-through of microorganisms	Unlikely	+
Shedding of filter components	−	+
Fluid retention	−	+
Solute adsorption	−	+
Good chemical stability	Variable (depends on membrane)	+
Good sterilization characteristics	+	+

+, applicable. −, not applicable.

defined size. It is generally accepted that synthetic membrane filters, derived from cellulose esters or other polymeric materials, approximate most closely to sieve filters, while fibrous pads, sintered glass and sintered ceramic products can be regarded as depth filters relying principally on mechanisms of adsorption and entrapment. Some of the characteristics of filter media are summarized in Table 20.4. The potential hazard of microbial multiplication within a depth filter and subsequent contamination of the filtrate (microbial grow-through) should be recognized.

7.1 Filtration sterilization of liquids

In order to compare favourably with other methods of sterilization the microorganism removal efficiency of filters employed in the processing of liquids must be high. For this reason, membrane filters of 0.2–0.22 μm pore diameter are chiefly used, while sintered filters, of porosity grade 5, are used only in restricted circumstances, i.e. for the processing of corrosive liquids, viscous fluids or organic solvents. In all cases, the filter medium employed must be sterilizable, ideally by steam treatment; in the case of membrane filters this will be for once-only use, while for sintered filters repeat sterilization and reuse will occur. Filtration sterilization is an aseptic process and careful monitoring of filter integrity is necessary as well as final product sterility testing (Chapter 23).

Membrane filters, in the form of discs, can be assembled into pressure-operated filter holders for syringe mounting and in-line use or vacuum filtration tower devices. Filtration under pressure is generally considered most suitable since filling at high flow rates directly into the final containers is possible without problems of foaming, solvent evaporation or air leaks. The filtration capacity of a range of membrane filter discs is given in Table 20.5. To increase the filtration area, and hence process volumes, several discs can be used in parallel in multiple-plate filtration systems or, alternatively, membrane filters can be fabricated into plain or

Table 20.5 Effect of membrane disc filter diameter on filtration volumes

Filter diameter (mm)	Effective filtration area (cm^2)	Typical batch volume (litres)
13	0.8	<0.01
25	3.9	0.05–0.1
47	11.3	0.1–0.3
90	45	0.3–5
142	97	5–20
293	530	>20

pleated cylinders and installed in cartridges. Membrane filters are often used in combination with a coarse-grade fibreglass depth prefilter to improve their dirt-handling capacity.

7.2 Filtration sterilization of gases

The principal application for filtration sterilization of gases is in the provision of sterile air to aseptic manufacturing suites, hospital isolation units and some operating theatres. Filters employed generally consist of pleated sheets in ducts, wall or ceiling panels, overhead canopies, or laminar airflow cabinets (Chapter 22). These high-efficiency particulate air (HEPA) filters can remove up to 99.997% of particles greater than 0.3 μm in diameter and thus are acting as depth filters. In practice their microorganism removal efficiency is rather better since the majority of bacteria are found associated with dust particles and only the larger fungal spores are found in the free state. Air is forced through HEPA filters by blower fans, and prefilters are used to remove larger particles to extend the lifetime of the HEPA filter. The operational efficiency and integrity of a HEPA filter can be monitored by pressure differential and airflow rate measurements, and dioctylphthalate smoke particle penetration tests.

Other applications of filters include the sterilization of venting or displacement air in tissue and microbiological culture (carbon filters and hydrophobic membrane filters), decontamination of air in mechanical ventilators (glass fibre filters), treatment of exhausted air from microbiological safety cabinets (HEPA filters) and the clarification and sterilization of medical gases (glass wool depth filters and hydrophobic membrane filters).

8 Conclusions

A sterilization process should always be considered a compromise between achieving good antimicrobial activity and maintaining product stability. It must, therefore, be validated against a suitable test organism and its efficacy continually monitored during use. Even so, a limit will exist as to the type and size of microbial challenge which can be handled by the process without significant loss of sterility assurance. Thus, sterilization must not be seen as a 'catch-all' or as an alternative to good manufacturing practices but must be considered as only the final stage in a programme of microbiological control.

9 Acknowledgements

The assistance of the following is gratefully acknowledged: F. J. Ley, Isotron plc, Swindon (for discussions and permission to reproduce Fig. 20.9); M. S. Copson, Albert Browne Ltd, Leicester (for discussions and permission to reproduce Fig. 20.6).

10 Appendix

Examples of typical conditions employed in the sterilization of pharmaceutical and medical products

Sterilization method	Conditions
Moist heat (autoclaving)	121°C for 15 min 134°C for 3 min
Dry heat	160°C for 120 min 170°C for 60 min 180°C for 30 min
Ethylene oxide	Gas concentration: 800–1200 mg l^{-1} 45–63°C 30–70% relative humidity 1–4 hours sterilizing time
Low temperature steam and formaldehyde	Gas concentration: 15–100 mg l^{-1} Steam admission to 73°C 40–180 min sterilizing time depending on type of process

Irradiation
  Gamma-rays or accelerated
  electrons
	25 kGy (2.5 Mrad) dose

Filtration	≤0.22 μm pore size, sterile membrane filter

11 Further reading

British Pharmacopoeia (1988) London: HMSO.
British Standards Institution (1961) *Performance of Electrically Heated Sterilising Ovens.* BS 3421. London: BSI.

British Standards Institution (1966) *Sterilisers for Bottled Fluids*. BS 3970, Part 2. London: BSI.

British Standards Institution (1990) *Sterilizing and Disinfecting Equipment for Medical Products*. BS 3970, Parts, 1, 3, 4, 5. London: BSI.

Denyer S.P. & Baird R.M. (Eds) (1990) *Guide to Microbiological Control in Pharmaceuticals*. Chichester: Ellis Horwood. (Chapters 7, 8 and 9 provide additional information.)

Gardner J.F. & Peel M.M. (1991) *Introduction to Sterilisation, Disinfection and Infection Control*. Melbourne: Churchill Livingstone.

Health Technical Memorandum (1980) *Sterilisers*. HTM 10. London: DHSS.

Ley F.J. (1985) *Radiation Sterilisation*. Institute of Sterile Services Management, Official Reference Book. London: Yeoman Publications Ltd.

Pharmaceutical Handbook (1980) 19th edn. London: Pharmaceutical Press.

Pickerill J.K. (1975) Practical system for steam-formaldehyde sterilising. *Lab Pract.* **24**, 401–404.

Russell A.D. (1982) *The Destruction of Bacterial Spores*. London: Academic Press.

Russell A.D., Hugo W.B. & Ayliffe G.A.J. (Eds) (1992) *Principles and Practice of Disinfection, Preservation and Sterilization*, 2nd edn. Oxford: Blackwell Scientific Publications.

Skinner F.A. & Hugo W.B. (Eds) (1976) *The Inactivation of Vegetative Bacteria*. Society for Applied Bacteriology Symposium No. 6. London: Academic Press. (Deals with the effects of heat, radiation and gases on vegetative bacteria.)

Stumbo C.R. (1973) *Thermobacteriology in Food Processing*, 2nd edn. London: Academic Press.

UK Panel on Gamma and Electron Irradiation (1987) Radiation sterilization dose. *Radiat Phys Chem.* **29**, 87–88.

United States Pharmacopeia (1990) 22nd revision. Rockville MD: US Pharmacopeial Convention.

21 Sterile pharmaceutical products

1   **Introduction**

2   **Injections**
2.1   Design philosophy
2.2   Intravenous infusions
2.2.1   Intravenous additives
2.2.2   Total parenteral nutrition (TPN)
2.3   Small-volume aqueous injections
2.3.1   Problems of drug stability
2.4   Small-volume of oily injections

3   **Non-injectable sterile fluids**
3.1   Non-injectable water
3.2   Urological (bladder) irrigation
         solutions
3.3   Peritoneal dialysis and
         haemodialysis solutions
3.4   Inhaler solutions

4   **Ophthalmic preparations**
4.1   Design philosophy
4.2   Eye-drops
4.3   Eye lotions

4.4   Eye ointments
4.5   Contact-lens solutions
4.5.1   Wetting solutions
4.5.2   Cleaning solutions
4.5.3   Soaking solutions

5   **Dressings**

6   **Implants**

7   **Absorbable haemostats**
7.1   Oxidized cellulose
7.2   Absorbable gelatin foam
7.3   Human fibrin foam
7.4   Calcium alginate

8   **Surgical ligatures and sutures**
8.1   Sterilized surgical catgut
8.2   Non-absorbable types

9   **Instruments and equipment**

10   **Further reading**

1   Introduction

Certain forms of drug administration and other pharmaceutical products, such as dressings and sutures, must be sterile in order to avoid the possibility of nosocomial (hospital-induced) infection arising from their usage. This applies particularly to medicines which are administered parenterally but also to any material or instrument likely to contact broken skin or internal organs. While inoculation of human pathogenic bacteria, fungi or viruses poses the most obvious danger to the patient, it should also be realized that microorganisms usually regarded as non-pathogenic which inadvertently gain access to body cavities in sufficiently large numbers can also result in a severe, often fatal, infection. Consequently, injections, ophthalmic preparations, irrigation fluids, dialysis solutions, sutures and ligatures, implants, certain surgical dressings, as well as instruments necessary for their use or administration, must be presented for use in a sterile condition and in such a way that they remain sterile throughout the period of use.

Principles of the methods employed to sterilize pharmaceutical products were described in Chapter 20. The *British Pharmacopoeia* (1988) recommends autoclaving and filtration as suitable methods applicable to aqueous liquids, and dry heat for non-aqueous and dry solid preparations. The choice is determined largely by the ability of the formulation and container to withstand the physical stresses applied by moist heat treatment. The use of ionizing radiation or ethylene oxide is also appropriate in specific instances. The primary considerations relate to the ability of active ingredients to withstand the applied stress and of the container to maintain the product in a sterile condition until use. It should be realized that all products intended to be sterilized must be rendered and kept thoroughly clean and therefore of low microbial content prior to sterilization. Thus the process itself is not overtaxed and is generally well within safety limits to guarantee sterility with minimal stress applied to the product. Because of the clinical consequences (such as granulosae in the lung) of injecting solid particles into the bloodstream, the numbers of particles present in injections and in other solutions used in body cavities must be restricted. The limits imposed on these products depend on the method of operation of the particle-detecting apparatus. The preferred device operates on the electrical zone-sensing principle, and when this device is used, the *British Pharmacopoeia* (1988) sets limits for injections greater than 100 ml of $\leqslant$1000 particles per ml of a size greater than 2.0 μm and $\leqslant$100 particles per ml greater than 5.0 μm.

2 Injections

2.1 Design philosophy

Any injectable product must be designed and produced to the highest possible pharmaceutical standards. Not only must the product have the minimum possible levels of particles and pyrogenic substances, but also the formulation and packaging must maintain product integrity throughout the production processes, the shelf-life and during administration. The formulation must be such as to ensure that the product remains physically and chemically stable over the designated shelf-life. To achieve this, excipients such as buffers and antioxidants may be required to ensure chemical stability, and solubilizers, such as propylene glycol or polysorbates, may be necessary for drugs with poor aqueous solubility to maintain the drug in solution. The packaging must prevent water, excipient or drug loss during sterilization and storage and, in addition, retain microbiological integrity. Axiomatically, ingress of microorganisms must be prevented. The packaging must not contribute any significant amounts of extractable chemicals to the contents, for example vulcanizing agents from rubber or plasticizers from polyvinyl chloride infusion containers.

Most injections are formulated as aqueous solutions, with Water for Injections BP as the vehicle. The formulation of injections depends upon several factors, namely the aqueous solubility of the active ingredient, the dose to be employed, thermal stability of the solution, the route of injection and whether the product is to be prepared as a multidose one (i.e. with a dose or doses removed on different occasions) or in a single-dose form (as the term suggests, only one dose is contained in the injection). Nowadays, most injections are prepared as single-dose forms and this is mandatory for certain routes, e.g. spinal injections such as the intrathecal route and large-volume intravenous infusions (section 2.2). Multidose injections may require the inclusion of a suitable preservative to prevent contamination following the removal of a dose on different occasions. Single-dose injections are usually packed in glass ampoules containing 1, 2 or 5 ml of product; to ensure removal of the correct volume by syringe, it is necessary to add an appropriate overage to an ampoule. Thus, a 1-ml ampoule will actually contain 1.1 ml of product, with 2.15 ml in a 2-ml ampoule. Full details are to be found in the *British Pharmacopoeia* (1988).

Some types of injections must be made iso-osmotic with blood serum. This applies particularly to large-volume intravenous infusions if at all possible; hypotonic solutions cause lysis of red blood corpuscles and thus must not be used for this purpose. Conversely, hypertonic solutions can be employed: these induce shrinkage, but not lysis, of red cells which recover their shape later. Intraspinal injections must also be isotonic, and to reduce pain at the site of injection so should intramuscular and subcutaneous injections. Adjustment to isotonicity can be determined by the following methods.

1 Depression of freezing-point, which depends on the number of dissolved particles present in solution. A useful equation is given by

$$W = \frac{0.52 - a}{b}$$

in which W is the percentage (w/v) of adjusting substance, a the freezing-point of unadjusted solution and b the depression of the freezing-point of water by 1% w/v of adjusting substance.

2 Sodium chloride equivalent, which is produced by dividing the value for the depression of freezing-point produced by a solution of the substance by the corresponding value of a solution of sodium chloride of the same strength.

For details of these and other methods, the *Pharmaceutical Handbook* (1980) should be consulted.

2.2   **Intravenous infusions**

These consist of large-volume injections or drips (500 ml or more) that are infused at various rates (e.g. $50-500 \, \text{ml} \, \text{h}^{-1}$) into the venous system; they are sterilized in an autoclave (see Chapter 20). The most commonly used infusions are isotonic sodium chloride and glucose. These are used to maintain fluid and electrolyte balance, for replacement of extracellular body fluids (e.g. after surgery or during prolonged periods of fluid loss), as a supplementary energy source (1 litre of 5% w/v glucose = 714 kJ) and as a vehicle for drugs. Such solutions are prepared using freshly distilled water as a vehicle under rigidly controlled conditions to minimize pyrogen (see Chapters 1 and 22) and particle content, and filtered to remove remaining particles immediately before distribution to the final clean container.

Other important examples are blood and blood products, which are collected and processed in sterile containers, and plasma substitutes, for example dextrans and degraded gelatin. Dextrans, glucose polymers consisting essentially of $(1 \rightarrow 6)$ α-links, are produced as a result of the biochemical activities of certain bacteria of the genus *Leuconostoc*, e.g. *L. mesenteroides* (see Chapter 25).

A small range of intravenous infusions, for example those containing amino acids or chlormethiazole, are prepared in glass containers. These are sealed with a rubber closure held on by an aluminium screw cap or crimp-on ring. The rubber should be non-fragmenting, not release soluble extractives, and be sufficiently soft and pliable to seal around the giving set needle inserted immediately prior to use. Although bottles are sterilized by autoclaving, it is still possible for the infusion in glass bottles to become contaminated with microorganisms before use. For instance, during the final part of the autoclave process, bottles may be spray-cooled with water to hasten the cooling process and therefore reduce the total autoclaving time. However, due to the poor fit between bottle lip and rubber plug (a skirted insert type is used) it is possible for the spray-cooling water to spread by capillary movement between bottle thread and screw cap and even enter the bottle contents. This process is encouraged if the bottle contains a vacuum as a consequence of rubber seal failure during heating-up. It should also be remembered that autoclaving leads to considerable heat and pressure stresses on the container. Failure may result from any imperfection in the bottle or plug. Microbes may also gain access to the contents of bottles during storage if hair-line cracks (a result of bad handling and rough treatment) are present, through which fluid may seep outwards and microorganisms inwards to contaminate the fluid. Finally, contamination may occur during use due to poor aseptic techniques when setting up the infusion, via an ineffective air inlet (allowing replacement of infused fluid with air) or when changing the giving set or bottle.

Most infusions are now packed in plastic containers. The plastic material should be pliable, thermoresistant, transparent and non-toxic. Plasticized polyvinylchloride (PVC) and polyethylene are commonly used. The former is transparent and very pliable, allowing the pack to collapse as the contents are withdrawn (consequently no air inlet is required). These packs are also amenable to the inclusion of ports into the bag, allowing greater safety during use. Such ports can be protected by sterile overseals. Two problems arise: (i) the possibility of toxic extractives, e.g. diethyl phthalate, from the plastic entering the fluid if poor quality PVC is used; and (ii) moisture permeability leading to loss of water if the packs are not protected by a water-impermeable outer wrap. Bags of high-quality polyethylene are readily moulded (although separate ports cannot be included), translucent and free from potential toxic extractives. As stated, these packs normally collapse readily during infusion. An important advantage of all plastic packs is that the containers are hermetically sealed prior to autoclaving and, therefore, spray-cooling water cannot enter the pack unless there is seal failure, an easily detected occurrence. However, the autoclaving of plastic bags is more complex than that of bottled fluids because a steam–air mixture is necessary to prevent bursting of the bags when heated (air-ballasting); adequate mixing of the steam and air is therefore required to prevent layering of gases inside the chamber.

2.2.1   *Intravenous additives*

A common practice in hospitals is to add drugs to infusions immediately prior to, or during, administration. The most common additives are potassium chloride, lignocaine, heparin, certain vitamins and antibiotics.

Potentially, this can be a hazardous practice. For instance, the drug may precipitate in the infusion fluid because of the pH (e.g. amphotericin) or the presence of calcium salts (e.g. thiopentone). The drug may degrade rapidly (e.g. ampicillin in 5% w/v dextrose). Multiple additions may lead to precipitation of one or both of the drugs or to accelerated degradation. Finally, drug loss may occur because of sorption by the container. For instance, insulin is absorbed by glass or PVC, glyceryl trinitrate and diazepam are absorbed by PVC. Apart from these problems, if the addition is not carried out under strict aseptic conditions the fluid can become contaminated with microorganisms during the procedure. Thus, any addition should be made in a laminar-flow work station or isolator, preferably in the pharmacy, and the fluid administered within 24 hours, unless prepared under strict aseptic conditions.

A novel approach to the problem of providing an intravenous drug additive service is to add the drug to a small-volume (50–100 ml) infusion in a collapsible plastic container and store the preparation at $-20°C$ in a freezer. The infusion can be removed when required and thawed rapidly by microwave. Many antibiotics are stable for several months when stored in minibags at $-20°C$ and are unaffected by the thawing process in a suitable microwave oven. Other antibiotics, e.g. ampicillin, are degraded when frozen.

2.2.2   *Total parenteral nutrition (TPN)*

Total parenteral nutrition is the use of concentrated mixtures of amino acids, vitamins, inorganic salts and an energy source (carbohydrate or fat emulsion, e.g. soyabean oil with lecithin as emulsifying agent) for the long-term feeding of patients who are unconscious or unable to take food. Many hospital pharmacies operate a TPN service. All or most of the ingredients to feed a patient for 1 day are combined in one large (3-litre capacity) collapsible plastic bag. The contents are infused over a 12–24 hour period. Transfer of amino acid, dextrose and electrolyte infusions, and the addition of vitamins and trace elements, must be carried out with great care under aseptic conditions to avoid microbial contamination. These solutions often provide good growth conditions for bacteria and moulds. Fats are administered as oil-in-water emulsions, comprising small droplets of a suitable vegetable oil (e.g. soyabean) emulsified with egg lecithin and sterilized by autoclaving. In many cases, the fat emulsion may be added to the 3-litre bag.

2.3   **Small-volume aqueous injections**

This category comprises single-dose injections, usually of 1–2 ml but as high as 50 ml, dispensed in borosilicate glass ampoules, plastic (poly-ethylene) ampoules or rarely multidose glass vials of 5–15 ml stoppered with a rubber closure through which a hypodermic needle can be inserted, e.g. insulins, vaccines. The closure is designed to reseal after withdrawal of the needle. It is unwise to include too many doses in a multidose container because of the risk of microbial contamination during repeated use. Bactericides must be added to injections in multidose containers to prevent contamination during withdrawal of successive doses, except as detailed below. Bactericides may not be used in injections in which the total volume to be injected at one time exceeds 15 ml. This may occur if the solubility of a drug is such that a therapeutic dose is only soluble in this order of volume (e.g. Bemegride Injection). There is also an absolute prohibition on the inclusion of bactericides in injections of the following categories: intra-arterial, intracardiac, intrathecal or subarachnoid, intra-cisternal and peridural.

Small-volume injections may be sterilized by the following methods.
1 Heating in an autoclave for injections packed in glass ampoules.
2 Filtration followed by aseptic sealing (plastic ampoules, vials). Since the product is not sterilized in its final container, a bactericide may be included to reduce the risks of contamination.

2.3.1   *Problems of drug stability*

1 Thermostability. The choice of sterilization method depends on the thermostability of the active ingredient, autoclaving being applied only to drugs that are heat stable in aqueous solution.
2 Chemical stability. Some medicaments undergo chemical change in aqueous solutions. If the change is due to oxidation, a reducing agent such as sodium metabisulphite is included (e.g. Adrenaline Injection BP).

Aqueous solutions of some drugs are so unstable that chemical stabilization is impossible. In this case the drug itself, not its aqueous solution, is sterilized by dry heat (160°C for 2 hours) in its final container and dissolved immediately before use by the addition of sterile water (Water for Injections BP). For drugs which are both thermolabile and unstable in aqueous solution, a sterile solution of the drug is freeze-dried in the final container and is reconstituted as above just before use (e.g. many anti-biotics, Hyaluronidase BP).

Details of time–temperature regimens as dictated by injection volume and heat transfer to the whole of the product (section 2.2) and of possible interactions between active ingredients and containers must be considered (see also Chapter 20).

2.4 Small-volume oily injections

Certain small-volume injections are available where the drug is dissolved in a viscous oil because it is insoluble in water; non-aqueous solvent must be used. In addition, drugs in non-aqueous solvents provide a depot effect, for example for hormonal compounds. The intramuscular route of injection must be used. The vehicle may be a metabolizable fixed oil such as arachis or sesame oil (but not a mineral oil) or an ester such as ethyl oleate which is also metabolizable. The latter is less viscous and therefore easier to administer but the depot effect is of shorter duration. The drug is normally dissolved in the oil, filtered under pressure and distributed into ampoules. After sealing, the ampoules are sterilized by dry heat at 160°C for 2 hours. A bactericide is probably ineffective in such a medium and therefore offers very little protection against contamination in a multidose oily injection.

3 Non-injectable sterile fluids

There are many other types of solution required in a sterile form for use particularly in hospitals.

3.1 Non-injectable water

This is sterile water, not necessarily of injectable water standards, which is used widely during surgical procedures for wound irrigation, moistening of tissues, washing of surgeons' gloves and instruments during use and, when warmed, as a haemostat. Isotonic saline may also be used. Topical water (as it is often called) is prepared in 500-ml and 1-litre polyethylene or polypropylene containers with a wide neck to allow for ease for pouring, and tear-off cap. Hospitals in the UK probably use larger quantities of topical fluids than of intravenous infusions.

3.2 Urological (bladder) irrigation solutions

These are used for the rinsing of the urinary tract to aid tissue integrity and cleanliness during or after surgery. Either water or glycine solution is used, the latter eliminating the risk of intravascular haemolysis when electrosurgical instruments are used. These are sterile solutions produced in collapsible or semi-rigid plastic containers of up to 3-litre capacity.

Peritoneal dialysis and haemodialysis solutions

Peritoneal dialysis solutions are admitted into the peritoneal cavity as a means of removing accumulated waste or toxic products following renal failure or poisoning. They contain electrolytes and dextrose (1.4–7% w/v) to provide a solution equivalent to potassium-free extracellular fluid. Lactate or acetate is added as a source of bicarbonate ions. Slightly hypertonic solutions are usually employed to avoid increasing the water content of the intravascular compartment. A more hypertonic solution, containing a higher dextrose concentration, is used to achieve a more rapid removal of water. In fact, the peritoneal cavity behaves as if it were separated from the body organs by a semi-permeable membrane. Warm peritoneal solution (up to 5 litres) is perfused into the cavity for 30–90 minutes and then drained out completely. This procedure can then be repeated as often as required. Since the procedure requires large volumes, these fluids are commonly packed in 2.5-litre containers. It is not uncommon to add drugs (for instance potassium chloride or heparin) to the fluid prior to use.

Haemodialysis is the process of circulating the patient's blood through a machine via tubing composed of a semi-permeable material such that waste products permeate into the dialysing fluid and the blood then returns to the patient. Haemodialysis solutions need not be sterile but must be free from heavy bacterial contamination.

3.4 **Inhaler solutions**

In cases of severe acute asthmatic attacks, bronchodilators and steroids for direct delivery to the lungs may be needed in large doses. This is achieved by direct inhalation via a nebulizer device; this converts a liquid into a mist or fine spray. The drug is diluted in small volumes of Water for Injections BP before loading into the reservoir of the machine. This vehicle must be sterile and preservative-free and is therefore prepared as a terminally sterilized unit dose in polyethylene nebules.

4 Ophthalmic preparations

4.1 Design philosophy

Medication intended for instillation on to the surface of the eye is formulated in aqueous solution as eye-drops or lotion or in an oily base as an ointment. Because of the possibility of eye infection occurring, particularly after abrasion or damage to the corneal surface, all ophthalmic preparations must be sterile. Since there is a very poor blood supply to the anterior chamber, defence against microbial invasion is minimal; furthermore, it appears to provide a particularly good environment for growth of bacteria. As well as being sterile, eye products should also be relatively free from particles which might cause damage to the cornea.

However, unlike aqueous injections, the recommended vehicle is purified water since the presence of pyrogens (Chapter 1) is not clinically significant.

Another type of sterile ophthalmic product is the contact lens solution (section 4.5); however, unlike the other types, this is not used for medication purposes but merely as wetting, cleaning and soaking solutions for contact lenses.

4.2   Eye-drops

Eye-drops are presented for use in: (i) sterile single-dose plastic sachets containing 0.3–0.5 ml of liquid; (ii) multidose amber fluted eye-dropper bottles including the rubber teat as part of the closed container or supplied separately (*British Pharmacopoeia* 1988); or (iii) plastic bottles with integral dropper. It should be covered with a breakable seal to indicate that the dropper or cap has not been removed prior to initial use. Although a standard design of bottle is used in hospitals, many proprietary products are manufactured in plastic bottles designed to improve safety and care of use. The maximum volume in each container is limited to 10 ml. Because of the likelihood of microbial contamination of eye-dropper bottles during use (arising from repeated opening or contact of the dropper with infected eye tissue or the hands of the patient), it is essential to protect the product against inopportune contamination. Eye-drops for surgical theatre use should be supplied in single-dose containers (*British Pharmacopoeia* 1988).

The following preservatives can be used: phenylmercuric nitrate or acetate (0.002% w/v), chlorhexidine acetate (0.01% w/v), thiomersal (0.01% w/v) and benzalkonium chloride (0.01% w/v). Chlorocresol is too toxic to the corneal epithelium, but 8-hydroxyquinoline and thiomersal may be used in specific instances. The *United States Pharmacopeia* (1990) includes chlorbutanol (0.5% w/v) and phenethyl alcohol (0.5% v/v). The principal consideration in relation to antimicrobial properties is the activity of the bactericide against *Pseudomonas aeruginosa*, a major source of serious nosocomial eye infections. Although benzalkonium chloride is probably the most active of the recommended preservatives, it cannot always be used because of its incompatibility with many compounds commonly used to treat eye diseases, nor should it be used to preserve eye-drops containing anaesthetics. Since benzalkonium chloride reacts with natural rubber, silicone or butyl rubber teats should be substituted. Since silicone rubber is permeable to water vapour, products should not be stored for more than 3 months after manufacture. As with all rubber components, the rubber teat should be pre-equilibrated with the preservative prior to use. Thermostable eye-drops and lotions are sterilized at 121°C for 15 minutes. For thermolabile drugs, filtration sterilization followed by aseptic filling into sterile containers is necessary. Eye-drops in plastic bottles are prepared aseptically.

In order to lessen the risk of eye-drops becoming heavily contaminated either by repeated inoculation or growth of resistant organisms in the solution, use is restricted, after the container is first opened to 1 month. This is usually reduced to 7 days for hospital ward use on one eye of a single patient. The period is shorter in the hospital environment because of the greater danger of contamination by potential pathogens, particularly pseudomonads.

Finally, eye-drops for use during open-eye surgery must *not* contain a preservative because of their cytotoxicity. Single-dose preparations are, therefore, ideally suited for this purpose.

4.3 Eye lotions

Eye lotions are isotonic solutions used for washing or bathing the eyes. They are sterilized in relatively large-volume containers (100 ml or greater). Eye lotions are sterilized by autoclaving in coloured fluted glass bottles with a rubber closure and screw cap, or plastic container with screw cap or tear-off seal. They may contain a bactericide if intended for intermittent domiciliary use for up to 7 days. If intended for first aid or similar purposes, however, no bactericide is included and any remaining solution discarded after 24 hours.

4.4 Eye ointments

Eye ointments are prepared in a semi-solid base (e.g. Simple Eye Ointment BP, which consists of yellow soft paraffin, eight parts; liquid paraffin, one part; wool fat, one part). The base is filtered when molten to remove particles and sterilized at 160°C for 2 hours. The drug is incorporated prior to sterilization if heat stable, or added aseptically to the sterile base. Finally, the product is aseptically packed in clear sterile aluminium or plastic tubes. Since the product contains virtually no water, the danger of bacteria proliferating in the ointment is negligible. Therefore, there is no recommended maximum period during which they can be used.

4.5 Contact-lens solutions

Most contact lenses are worn for optical reasons as an alternative to spectacles. Contact lenses are of two types, namely hard lenses, which are hydrophobic, and soft lenses, which may be either hydrophilic or hydrophobic. The surfaces of lenses must be wetted before use, and wetting solutions (section 4.5.1) are used for this purpose. Hard and, more especially, soft lenses become heavily contaminated with protein material during use and must therefore be cleaned (section 4.5.2) before disinfection (section 4.5.3). Contact lenses are potential sources of eye infection and consequently microorganisms should be removed before the lens is again inserted into the eye. Lenses must also be clean and easily wettable by the lacrimal secretions. Contact-lens solutions are thus sterile solutions of the various types described below. Apart from achieving their stated functions, either singly or in combination, all solutions must be non-irritating and must protect against microbial contamination during use and storage.

4.5.1 Wetting solutions

These are used to hydrate the surfaces of hard lenses after disinfection. Since they must also cope with chance contamination, they must contain a preservative as well as a wetting agent. They may be isotonic with lacrimal secretions and be formulated to a pH of about 7.2 for compatibility with normal tears.

4.5.2 Cleaning solutions

These are responsible for the removal of ocular debris and protein deposits, and contain a cleaning agent that consists of a surfactant and/or an enzyme product. Since they must also cope with chance contamination, they contain a preservative, are isotonic, and have a pH of about 7.2.

4.5.3 Soaking solutions

These are responsible for disinfection of lenses but also maintain the lenses in a hydrated state. The antimicrobial agents used for disinfecting hard lenses are those used in eye-drops (benzalkonium, chlorhexidine, phenylmercuric acetate or nitrate, thiomersal and chlorbutol). Ethylene diamine tetra-acetic acid (EDTA) is usually present as a synergist (see Chapter 14). Benzalkonium chloride and chlorbutol are strongly bound to hydrophilic soft contact lenses and therefore cannot be used in storage solutions for these. Chlorhexidine and thiomersal are usually employed. It must be added that the concentrations of all preservatives used in contact-lens solutions are lower than those employed in eye-drops, in order to minimize irritancy. Hydrogen peroxide is now becoming commonly used, but must be inactivated prior to lens insertion on to the eye.

Finally, heat may be utilized as an alternative method to disinfect soft contact lenses, especially the hydrophilic type. Lenses are boiled in isotonic saline.

5 Dressings

Dressings and surgical materials are used widely in medicine, both as a means of protecting and providing comfort for wounds and for many associated activities such as cleaning, swabbing, etc. They may or may not be used on areas of broken skin. If there is a potential danger of infection arising from the use of a dressing then it must be sterile. For instance, sterile dressings must be used on all open wounds, both surgical and

Table 21.1 Uses of surgical dressings and methods of sterilization

Dressing	Uses	Method of sterilization
Required to be sterile		
Chlorhexidine gauze dressing	Medicated open-wound dressing, burns, grafts	
Framycetin gauze dressing	Medicated open wound dressing, burns, grafts	
Knitted viscose primary dressing	Ulcerative and granulating wounds	
Paraffin gauze dressing	Burns, scalds, grafts	Any combination of dry heat, gamma-radiation and ethylene oxide
Perforated film absorbent dressing	Postoperative wounds	
Polyurethane foam dressing	Burns, ulcers, grafts, granulating wounds	
Semi-permeable adhesive film	Adhesive dressing for open wounds, i.v. sites, stoma care, etc.	
Sodium fusidate gauze dressing	Medicated open wound dressing, burns, grafts	
May be sterile for use in certain circumstances		
Absorbent cotton wool	Swabbing, cleaning, medication application	Any method
Elastic adhesive dressing	Protective wound dressing	Ethylene oxide or γ-radiation
Plastic wound dressings	Protective dressing (permeable or occlusive)	Ethylene oxide or γ-radiation
Absorbent cotton gauze	Absorbent wound dressing	Any method
Gauze pads	Swabbing, dressing, wound packing	Any method
Absorbent viscose wadding	Wound cleaning, swabbing, skin antiseptic application	Any method

traumatic, on burns and during and after catheterization at a site of injection. It is also important to appreciate that sterile dressings must be packaged in such a way that they can be applied to the wound aseptically.

Dressings are described in the *British Pharmacopoeia* (1988). Methods for their sterilization include autoclaving, dry heat, ethylene oxide and ionizing radiation. Any other effective method may be used. The choice is governed principally by the stability of the dressing constituents to the stress applied and the nature of their components. Most cellulosic and synthetic fibres withstand autoclaving but there are exceptions. For instance, boric acid tenderizes cellulose fibres during autoclaving, and dressings containing waxes cannot be sterilized by moist heat. Certain constituents are also adversely affected on exposure to large doses of gamma-radiation. Those dressings that are required to be sterile are listed in Table 21.1, together with other dressings and materials that may be sterilized when required.

A very important aspect of dressings production is packaging. The packaging material must allow correct sterilization conditions (for example permeation of moisture or ethylene oxide), retain the dressing in a sterile condition and allow for its removal without contamination prior to use. All dressings intended for aseptic handling and application must be double-wrapped. For steam sterilization they may be individually wrapped in fabric, paper or nylon and sterilized in metal drums, cardboard boxes or bleached kraft paper. The choice of method also determines the design of the autoclave cycle. Dressings may be sterilized in downward dis-placement autoclaves, which rely on displacement of air by steam, or in the more modern high prevacuum autoclaves in which virtually all the air is removed before the admission of steam. This method ensures rapid heating-up of the dressings, reduces the time needed to achieve steril-ization (e.g. 134°C for 3 minutes) and shortens the overall sterilization cycle.

A recent development is the use of spray-on dressings. A convenient type is an acrylic polymer dissolved in ethyl acetate and packed as an aerosol. This should be self-sterilizing. The film after application is able to maintain the sterility of a clean wound for up to 2 weeks. However, they can only be used on clean, relatively dry wounds.

6 Implants

Implants are small, sterile cylinders of drug, inserted beneath the skin or into muscle tissue to provide slow absorption and prolonged action therapy. This is principally based on the fact that such drugs, invariably hormones, are almost insoluble in water and yet the implant provides a rate of dissolution sufficient for a therapeutic effect. The *British Pharmacopoeia* (1988) describes one implant, testosterone. The *United States National Formulary* (1990) also includes oestradiol. Implants are made from the pure drug into tablet form by compression or fusion. No other ingredient can be included since this may be insoluble or toxic or, most importantly, may influence the rate of drug release.

Compression of sterile drugs must be conducted under aseptic con-ditions using sterile machine parts and materials. After manufacture, the outer surface of the implant is sterilized by immersion in 0.002% w/v phenylmercuric nitrate at 75°C for 12 hours. After the surface has been dried, each implant is placed aseptically into a sterile glass phial with a cotton-wool plug at both ends. This prevents damage and reduces the risk of glass spicules, formed when the phial is opened, adhering to the implant. This compression process is not ideal. The absence of a lubricant increases the difficulties of making tablets; the hardness of the implant is difficult to regulate, which consequently leads to variations in the rate of drug release. The alternative method, fusion, can be used provided the drug is heat-stable. The pure drug is melted at 5–10°C above its melting temperature and poured into moulds. Note that if the melting tempera-ture is high enough the interior of the implant will automatically be sterilized by this process. It is also possible to sterilize the implant after packaging, by dry heat, provided the melting temperature is above 160°C. Clearly, it is easier to manufacture sterile implants by fusion since the process does not require presterilized ingredients or aseptic processing. The implant hardness is also very consistent.

7 Absorbable haemostats

The reduction of blood loss during or after surgical procedures where suturing or ligature is either impractical or impossible can often be accomplished by the use of sterile, absorbable haemostats. These consist of a soft pad of solid material packed around and over the wound which can be left *in situ*, being absorbed by body tissues over a period of time, usually up to 6 weeks. The principal mechanism of action of these is the ability to encourage platelet fracture because of their fibrous or rough surfaces, and to act as a matrix for complete blood clotting. Four products commonly used are: oxidized cellulose, absorbable gelatin sponge, human fibrin foam and calcium alginate.

7.1 Oxidized cellulose

This consists of cellulosic material which has been partially oxidized. White gauze is the most common form, although lint is also used. It can be absorbed by the body in 2–7 weeks, depending on the size. Its action is based principally on a mechanical effect and it is used in the dry state. Since it inactivates thrombin, its activity cannot be enhanced by thrombin incorporation.

7.2 Absorbable gelatin foam

This is an insoluble gelatin foam produced by whisking warm gelatin solution to a uniform foam, which is then dried. It can be cut into suitable shapes, packed in metal or paper containers and sterilized by dry heat (150°C for 1 hour). Moist heat destroys the physical properties of the material. Immediately before use, it can be moistened with normal saline containing thrombin. It behaves as a mechanical haemostat providing the framework on which blood clotting can occur.

7.3 Human fibrin foam

This is a dry sponge of human fibrin prepared by clotting a foam of human fibrinogen solution with human thrombin. It is then freeze-dried, cut into shapes and sterilized by dry heat at 130°C for 3 hours. Before use, it is saturated with thrombin solution. Blood coagulation occurs in contact with the thrombin in the interstices of the foam.

7.4 **Calcium alginate**

This is composed of the sodium and calcium salts of alginic acid formed into a powder or fibrous material and sterilized by autoclaving. It aids clotting by forming a sodium–calcium alginate complex in contact with tissue fluids, acting principally as a mechanical haemostat. It is relatively slowly absorbed and some residues may occasionally remain in the tissues.

8 Surgical ligatures and sutures

The use of strands of material to tie off blood or other vessels (ligature) and to stitch wounds (suture) is an essential part of surgery. Both absorbable and non-absorbable materials are available for this purpose.

8.1 **Sterilized surgical catgut**

This consists of absorbable strands of collagen derived from mammalian tissue, particularly the intestine of sheep. Because of its source, it is particularly prone to bacterial contamination, and even anaerobic spores may be found in such material. Therefore, sterilization is a particularly difficult process. Since collagen is converted to gelatin when exposed to moist heat, autoclaving cannot be used. The official method is to pack the 'plain' catgut strands (up to 350 cm) on a metal spindle in a glass or other suitable container with a tubing fluid, the purpose of which is to maintain both flexibility and tensile strength after sterilization. Probably the most suitable method is to expose the material to gamma-radiation. There is minimal loss of tensile strength and the container can be overwrapped prior to sterilization to provide a sterile container surface for opening aseptically. The alternative method involves placing the coiled suture immersed in a tubing fluid (commonly 96% ethyl alcohol with or without 0.002% w/v phenylmercuric nitrate) and stored for sufficient time to ensure sterilization. The outer surface of the phial must be sterilized before opening to avoid contamination of the suture when removed. Therefore, the phial is immersed in 1% w/v formaldehyde in ethanol for 24 hours prior to use. It cannot be heated. A non-official method of sterilization is to immerse the catgut in a non-aqueous solvent (naphthalene or toluene) and heat at 160°C for 2 hours. The catgut becomes hard and brittle during this process, and is aseptically transferred to an aqueous tubing fluid to restore its flexibility and tensile strength.

Catgut is packed in single threads, up to 350 cm in length, of various thicknesses related to tensile strength, in single-use glass or plastic containers which cannot be resealed after use. Any remaining material should be discarded. Hardened catgut is prepared by treating strands with certain agents to prolong resistance to digestion. If hardened with chromium compounds, the material is known as 'chromicized' catgut.

Table 21.2 Methods* commonly used to sterilize or disinfect equipment

Equipment	Method of treatment	Disinfection or sterilization	Preferred method	Comments
Syringes (glass)	Dry heat	Sterilization	Dry heat using assembled syringes	Autoclave not recommended: difficulty with steam penetration unless plungers and barrels sterilized separately
Syringes (glass), dismantled	Moist heat	Sterilization		
Syringes (disposable)	γ-radiation Ethylene oxide	Sterilization	γ-radiation	Possibility of 'crazing' of syringes after ethylene oxide
Needles (all metal)	Dry heat	Sterilization	Dry heat	
Needles (disposable)	γ-radiation Ethylene oxide	Sterilization	γ-radiation	
Metal instruments (including scalpels)	Autoclave Dry heat	Sterilization	Dry heat	Cutting edges should be protected from mechanical damage during the process
Disposable instruments	γ-radiation Ethylene oxide	Sterilization	γ-radiation	
Rubber gloves	Autoclave γ-radiation Ethylene oxide	Sterilization	γ-radiation	If autoclave used, care with drying at end of process. Little oxidative degradation when high-vacuum autoclave used
Administration (giving) sets	γ-radiation Ethylene oxide	Sterilization	γ-radiation	
Respirator parts	Moist heat (autoclave)	Sterilization	Sterilization by steam where possible	Chemicals not recommended: may be microbiologically ineffective, may present hazard to patient safety by compromising the safety devices on the machine
	Moist heat (low temperature steam, or hot water at 80°C)	Disinfection		
Dialysis machines	Chemical	Disinfection	Formalin	Ethylene oxide not recommended in NHS for practical reasons
Fragile, heat-sensitive equipment	Ethylene oxide Chemical	Sterilization Disinfection	Ethylene oxide under expert supervision	

* **1** Disposable equipment should not be resterilized or reused.
2 Ethylene oxide is a difficult process to control, and the Department of Health discourages its use in hospitals.
3 Low-temperature steam with formaldehyde is of value in the disinfection/sterilization of some heat-sensitive materials (see also Chapter 20).
4 Chemical agents, e.g. glutaraldehyde, hypochlorite.

Sutures and ligatures are also made from many materials not absorbed by the body tissues. These consist of uniform strands of metal or organic material which will not cause any tissue reactions and are capable of sterilization. Depending on the physical stability of each material, they are preferably sterilized by autoclaving or gamma-radiation. They are packed in single-use glass or plastic containers which may contain a tubing fluid with or without a bactericide. The different materials are described in the *British Pharmacopoeia* (1988). These include linen (adversely affected by gamma-rays), nylon (either monofilament or plaited), silk and stainless steel (monofilament or twisted).

9   Instruments and equipment

The method chosen for sterilization of instruments (see also Table 21.2) depends on the nature of the components and the design of the item. The wide range of instruments that may be required in a sterile condition includes syringes (glass and plastic disposable), needles, giving sets, metal surgical instruments (scalpels, scissors, forceps, etc.), rubber gloves, catheters, etc. Relatively complicated equipment, such as pressure transducers, pacemakers, kidney dialysis equipment, incubators and aerosol machine parts may also be sterilized. Artificial joints could also be included in the vast range of items required in modern medical practice in a sterile condition. The choice of method depends largely on the physical stability of the items and the appropriate technique in particular situations. For instance, incubators necessitate a chemical method of sterilization. On the other hand, even delicate instruments like pressure transducers are now available that can withstand autoclaving. This is a new and developing field of medical technology in which many factors may have to be considered before the choice is made as to the most appropriate method of sterilization in any particular situation.

10   Further reading

Allwood M.C. & Fell J.T. (1980) *Textbook of Hospital Pharmacy*. Oxford: Blackwell Scientific Publications.

British Pharmacopoeia (1988) Addenda (1989, 1990, 1991) London: Pharmaceutical Press.

Denyer S. & Baird R. (Eds) (1990) *Guide to Microbiological Control in Pharmaceuticals*. London: Ellis Horwood.

Pharmaceutical Handbook (1980) 19th edn. London: Pharmaceutical Press.

Phillips I., Meers P.D. & D'Arcy P.F. (1976) *Microbiological Hazards of Infusion Therapy*. Lancaster: MTP Press.

Russell A.D., Hugo W.B. & Ayliffe G.A.J. (1992) *Principles and Practice of Disinfection, Preservation and Sterilization*, 2nd edn. Oxford: Blackwell Scientific Publications.

Turco S. & Young R.E. (1987) *Sterile Dosage Forms*, 3rd edn. Easton, Philadelphia: Lea and Febiger.

22 Factory and hospital hygiene and good manufacturing practice

1 **Introduction**
1.1 Definitions
1.1.1 Manufacture
1.1.2 Quality assurance
1.1.3 Good manufacturing practice (GMP)
1.1.4 Quality control
1.1.5 In-process control

2 **Control of microbial contamination during manufacture: general aspects**
2.1 Environmental cleanliness and hygiene
2.2 Quality of starting materials
2.3 Process design
2.4 Quality control and documentation
2.5 Packaging, storage and transport

3 **Manufacture of sterile products**
3.1 Clean and aseptic areas: general requirements

3.1.1 Design of premises
3.1.2 Internal surfaces, fittings and equipment
3.1.3 Services
3.1.4 Air supply
3.1.5 Clothing
3.1.6 Changing facilities
3.1.7 Cleaning and disinfection
3.1.8 Operation
3.2 Aseptic areas: additional requirements
3.2.1 Clothing
3.2.2 Entry to aseptic areas
3.2.3 Equipment and operation

4 ***Guide to Good Pharmaceutical Manufacturing Practice***

5 **Conclusions**

6 **Further reading**

1 Introduction

The quality of a pharmaceutical product, whether manufactured in industry or in a hospital, cannot be ensured solely by examining in detail a small number of units taken from a completed batch. For instance, a low level or uneven distribution of microbial contamination may not be detected by conventional methods of sampling and sterility testing (Chapter 23). Instead, a product must be manufactured in a suitable environment by a procedure which minimizes the possibility of contamination occurring, at the end of which tests can be performed as an additional safeguard. This chapter describes measures essential for the control, during manufacture, of one important feature of product quality, the level of microbial contamination. It is designed to be read in conjunction with Chapters 17, 18, 20, 21 and 23.

1.1 Definitions

Several terms used in industrial and hospital production must be defined for an understanding of this chapter. These definitions are given in sections 1.1.1–1.1.5.

1.1.1 *Manufacture*

Manufacture is the complete cycle of production of a medical product. This cycle includes the acquisition of all raw materials, their processing into a final product and its subsequent packaging and distribution.

1.1.2 *Quality assurance*

This term refers to the sum total of the arrangements made to ensure that the final product is of the quality required for its intended purpose. It consists of good manufacturing practice plus factors such as original product design and development.

1.1.3 *Good manufacturing practice (GMP)*

Good manufacturing practice comprises that part of quality assurance which is aimed at ensuring that a product is consistently manufactured to a quality appropriate to its intended use. GMP requires that: (i) the manufacturing process is fully defined before it is commenced; and (ii) the necessary facilities are provided. In practice, this means that personnel must be adequately trained, suitable premises and equipment employed, correct materials used, approved procedures adopted, suitable storage and transport facilities available and appropriate records made.

1.1.4 *Quality control*

Quality control refers to that part of GMP which ensures that: (i) at each stage of manufacture the necessary tests are made; and (ii) the product is not released until it has passed these tests.

1.1.5 *In-process control*

This comprises any test on a product, the environment or the equipment that is made during the manufacturing process.

2 Control of microbial contamination during manufacture: general aspects

A pharmaceutical product may become contaminated by various means and at different points during the course of manufacture. There are several important ways in which this risk can be minimized in both industrial and hospital production, and these are considered below.

2.1 Environmental cleanliness and hygiene

Microorganisms may be transferred to a product from working surfaces, fixtures and equipment. In this context, pooled stagnant water is a frequent source of contamination. Thus, all premises, including processing areas, stores and laboratories, should be maintained in a clean, dry and tidy condition. For easy cleaning, walls and ceilings should have an impervious and washable surface, and floors should be made of impervious materials free from cracks and open joints where microorganisms could be harboured. For the same reasons, coving should be used at the junction between walls and floors or ceilings. All services, including pipelines, light fittings and ventilation points, should be sited so that inaccessible recesses are avoided. Procedures for cleaning and disinfection of premises are required and must be enforced. All equipment involved in the manufacturing process should be easy to dismantle and clean. It should be inspected for cleanliness before use.

Fall-out of dust- and droplet-borne microorganisms from the atmosphere is an obvious avenue whereby contamination of products may occur; therefore, 'clean' air is a prerequisite during manufacturing processes. In this context, the spread of dust during manufacturing or packaging must be avoided. Microorganisms may thrive in certain liquid preparations and in creams and ointments (Chapter 18; see also *Pharmaceutical Codex* 1979). The manufacture of such products should thus, as far as possible, be in a closed system; this serves a dual purpose in that it protects the product not only against airborne microbial contamination but also against evaporative loss.

Potentially harmful organisms could be transferred to a product by its direct contact with personnel. High standards of personal hygiene are therefore very important, especially where sterile products (section 3) are being manufactured. Consequently, operatives should be free from communicable diseases and should have no open lesions on the exposed body surfaces. To ensure high standards of personal cleanliness, adequate handwashing facilities and protective garments, including headgear, must be provided. Direct contact between the materials and the operative's hands must be avoided; where necessary gloves should be worn.

Staff should be trained in the principles of GMP and in the practice (and theory) of the tasks assigned to them. Personnel employed in the manufacture of sterile products (section 3) should also receive basic training in microbiology.

2.2   **Quality of starting materials**

Raw materials, including water supplies, are an important source of microorganisms in the manufacturing area (Chapter 17) and can lead to the contamination of both the environment and the final product. Materials of natural origin are usually associated with an extensive microbial flora and require careful storage to prevent growth of the organisms and spoilage of the material. If stable, natural products with a high microbial count may undergo sterilization before use. Staff handling raw materials must receive adequate training to prevent the transfer of contaminants from one raw material to another or to the final product (cross-contamination).

Water for manufacturing may be potable mains water, water purified by ion exchange or reverse osmosis or distillation, or water suitable for injection purposes (Water for Injections BP). When required for parenteral products it must be pyrogen-free (apyrogenic) and is usually prepared in a specially designed still. Although pyrogens are not volatile, they are not removed by ordinary distillation since some will be carried over mechanically into the distillate with the entrainment (spray). Thus a spray trap, consisting of a series of baffles, is fitted to the distilling flask to prevent spray and pyrogens from entering the condenser tubes. Water prepared in this manner can be used immediately for the preparation of injections, provided that these are sterilized within 4 hours of water collection. Alternatively, the water can be kept for longer periods at a temperature above 65°C (usually 80°C) to prevent bacterial growth with consequent pyrogen production. Ultraviolet irradiation (Chapter 20) may be useful in reducing the bacterial content but it is not to be regarded as being a sterilizing agent.

2.3 Process design

The manufacturing process must be fully defined and capable of yielding, with the facilities available, a product that is microbiologically acceptable and conforms to its specifications. This demands that a process be sufficiently evaluated before commencement to ensure that it is suitable for routine production operations. Processes and procedures should be subject to frequent reappraisal and should be re-evaluated when any significant changes are made in the equipment or materials used.

2.4 Quality control and documentation

Selection of starting materials with a low microbial content aids in the control of contamination levels in the environment and the final product. One aspect of quality control is to set acceptable microbiological standards for all raw materials, together with microbial limits for in-process samples and the final product. Further microbiological quality control covers the validation of cleaning and disinfecting procedures and the monitoring of the production environment by microbial counts. Such monitoring should be carried out whilst normal production operations are in progress. In addition, sterile product manufacture will require extra safeguards in the form of tests on the operator's aseptic technique and the monitoring of both air filter and sterilizer efficiency (Chapter 23). Sterility testing (Chapter 23) on the finished product constitutes the final check on the sterilization process. Injectable products may also be tested for pyrogens.

A system of documentation should exist such that the history of each batch of the product, including details of starting materials, packaging materials, and intermediate, bulk and finished products, may be determined. Distribution records must be kept. This information is of paramount importance should a defective batch need to be recalled.

2.5   **Packaging, storage and transport**

Even when a product has been prepared under stringent conditions such as those outlined above, contamination could still arise during storage and transport. For this reason, the packaging used and the conditions employed during storage and transportation should be such as to minimize or, preferably, prevent deterioration or contamination.

3   Manufacture of sterile products

Sterilization methods have been discussed in Chapter 20 and the various types of sterile products have been described in Chapter 21. For manufacturing purposes an important distinction exists between a sterile product which is terminally sterilized and one which is not. Terminally sterilized means that, after preparation, the product is transferred to containers which are sealed and then immediately sterilized by heat (or radiation or ethylene oxide, as appropriate). In general, such a product must be prepared in a clean area (sections 3.1.1–3.1.8). A product which is not to be terminally sterilized is prepared under aseptic conditions either from previously sterilized materials or by filtration sterilization. In either case, filling into sterilized final containers is a post-sterilization manipulation. Strict aseptic conditions are needed throughout (sections 3.2.1–3.2.3).

Vaccines consisting of dead microorganisms, microbial extracts or inactivated viruses (see Chapter 16) may be filled in the same premises as other sterile medicinal products. The completeness of inactivation (or killing or removal of live organisms) must be proven before processing. Separate premises are needed for the filling of live or attenuated vaccines (Chapter 16). Non-sterile products should not be processed in the same areas as sterile products.

3.1   **Clean and aseptic areas: general requirements**

3.1.1   *Design of premises*

Sterile production should be carried out in a purpose-built unit separated from other manufacturing areas and thoroughfares. The unit should be designed to encourage the segregation of each stage of production but should ensure a safe and organized workflow (Fig. 22.1). Sterilized products held in quarantine pending sterility test results (Chapter 23) must be kept separate from those awaiting sterilization.

3.1.2   *Internal surfaces, fittings and equipment*

Particulate, as well as microbial, contamination must be guarded against when sterile products are being manufactured. Thus walls, ceilings and floors should possess smooth, impervious surfaces which will: (i) prevent

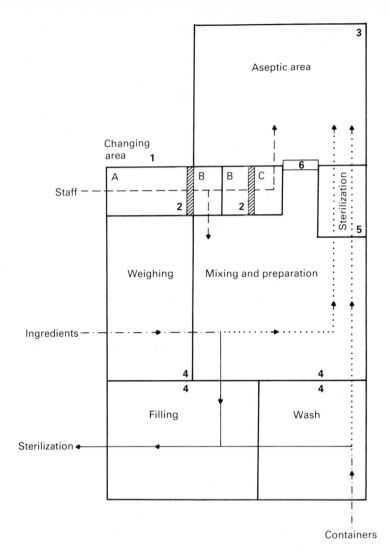

Fig. 22.1 Example of a diagrammatic representation of the layout and workflow of a sterile products manufacturing unit: **1**, The changing area in this example is built on the black (A)–grey (B)–white (C) principle; passage into the clean area is through A and B (see section 3.1.6) whereas entry to the aseptic area is first through A and B followed by C (see section 3.2.2). **2**, Dividing step-over sill. **3**, For details of aseptic area requirements, see text. **4**, These areas are clean areas. In filling rooms for terminally sterilized products, care should be exercised to protect containers from airborne contamination. The final rinse point (i.e. where the containers are finally washed) should be sited as near as possible to the filling point. **5**, Articles which are to be transferred directly to the aseptic area from elsewhere must be sterilized by passage through a double-ended sterilizer. Solutions manufactured in the clean area may be brought into the aseptic area through a sterilizing-grade membrane filter. **6** Double-doored hatchway through which presterilized articles may be passed into the aseptic area (see section 3.2.3).

Note: Inspection, holding and final packaging areas have been omitted. Direction of workflow: ⟶, for terminally sterilized products; · · · ➤ · · · , for aseptically prepared products; –·–➤·–, shared stages of preparation.

the accumulation of dust or other particulate matter; and (ii) allow for easy and repeated cleaning and disinfection. For the same reasons, where walls and floors or ceilings meet, covings should be used.

Suitable materials for floors include welded sheets of PVC and terrazzo, but cracks and open joints, which may harbour dirt and micro-organisms, must be eliminated. While an acceptable finish for walls and ceilings can be achieved by using hardboard which, after the joints have been sealed, is plaster skimmed and painted with a hard gloss paint, the preferred surface materials are plastic, epoxy-coated plaster, plastic fibreglass or glass-reinforced polyester. Frequently, the final finish for floor, wall and ceiling is achieved using continuous welded PVC sheeting. False ceilings must be adequately sealed to prevent contamination from the space above them. Use should be made of well-sealed glass panels, especially in dividing walls, to ensure good visibility and satisfactory supervision. Doors and windows should fit flush with the walls. Windows should not be openable.

Internal fittings such as cupboards, drawers and shelves must be kept to a minimum. Laminated plastic, which may be easily cleaned or dis-infected, is a better surface than wood for work benches; bare wood is to be avoided, although painted or otherwise sealed woodwork is satis-factory. Stainless steel trolleys can be used to transport equipment and materials within the clean and aseptic areas but these must remain confined to their respective units. Equipment should be so designed as to be easily cleaned and sterilized (or disinfected).

3.1.3 *Services*

Clean and aseptic areas must be adequately illuminated; lights are best housed above transparent panels set in a false ceiling. Electrical switches and sockets should fit flush to the wall. When required, gases should be piped into the area from outside the unit. Pipes and ducts, if they have to be brought into the clean area, must be effectively sealed through the walls. Additionally they must either be boxed in (which prevents dust accumulation) or readily cleanable. Alternatively, pipes and ducts may be sited above false ceilings.

Sinks supplied to clean areas should be made of stainless steel and have no overflow, and the water should be of at least potable quality. Wherever possible, drains in clean areas should be avoided. If installed, however, they should be fitted with effective, easily cleanable traps and with air breaks to prevent backflow. Any floor channels in a clean area should be open, shallow and cleanable and should be connected to drains outside the area. They should be monitored microbiologically. Sinks and drains should be excluded from aseptic areas except where radiopharma-ceutical products are being processed when sinks are a requirement.

Table 22.1 Basic environmental standards for the manufacture of sterile products

BS 5295 (1989) classification	Numerical limits (m⁻³) of particles greater than or equal to sizes shown (μm)				Type of environment
	0.5	5.0	10.0	25.0	
F	3 500	0	—	—	Aseptic room
J	350 000	2 000	450	0	Clean room

3.1.4

Air supply

Filtered air (Chapter 17) should be maintained at positive pressure throughout a clean or aseptic area, with the highest pressure in the most critical rooms (aseptic or clean filling rooms) and a progressive reduction through the preparation and changing rooms (Fig. 22.1); a minimum 10-kPa pressure differential is required between each class of room. A minimum of 20 air changes per hour is usual in clean and aseptic rooms and the air quality is defined by particle size limits (Table 22.1). The air inlet points should be situated in or near the ceiling, with the final filters placed as close as possible to the point of input to the room.

The greatest risk of contamination of a pharmaceutical product comes from its immediate environment. Additional protection from particulate and microbial contamination is therefore essential in both the filling area of the clean room and in the aseptic unit. This can be provided by a protective work unit supplied with a unidirectional flow of filtered sterile air. Such a cabinet is known as a laminar airflow (LAF) unit in which the displacement of air is either horizontal (i.e. from back to front) or vertical (i.e. from top to bottom) with minimum airflow rates of $0.45\,\mathrm{m\,s^{-1}}$ and $0.3\,\mathrm{m\,s^{-1}}$, respectively. Thus airborne contamination is not added to the work space and any generated by manipulations within that area is swept away by the laminar air currents.

The efficacy of the filters through which the air is passed should be monitored at predetermined intervals (Chapter 17).

3.1.5

Clothing

Cotton material is comfortable to wear but because of the possibility of the shedding of fibres it is regarded as being unsuitable in the present context. Terylene, which sheds virtually no fibres, is suitable. Airborne particulate and microbial contamination is reduced when trouser suits, close-fitting at the neck, wrists and ankles, are worn. Clean suits for clean areas should be provided at least once daily, but fresh headwear, over-shoes and powder-free gloves are necessary for each working session. Special laundering facilities for this clothing is desirable. Additional requirements for clothing worn in aseptic areas are considered in section 3.2.1

Changing facilities

Entry to clean or aseptic areas should be through a changing room fitted with interlocking doors; this acts as an airlock to prevent the influx of air from outside. This access route is intended for personnel only and does not constitute a means for regularly transferring materials and equipment into these areas. Staff entering the changing rooms should already be clad in the standard factory or hospital protective garments.

For a clean area, passage through the changing room should be from a 'black' area to a 'grey' area, via a dividing step-over sill (Fig. 22.1). Movement through these areas and finally into the clean room is permitted only on observance of a strict protocol. In this, outer garments are removed in the 'black' area and clean-room trouser suits donned in the 'grey' area. After handwashing in a sink fitted with hand or foot-operated taps the operator may enter the clean room.

The changing procedure for personnel entering an aseptic area is dealt with in section 3.2.2.

Cleaning and disinfection

A strict cleaning and disinfection policy is essential if microbial contamination is to be kept to a minimum. Cleaning agents include alkaline detergents and ionic and non-ionic surfactants. A wide range of disinfectants is available commercially (Chapter 11) and a selection of those suitable for use in the sterile product manufacturing environment is given in Table 22.2. Different types of disinfectants should be employed in rotation to help prevent the development of resistant strains of microorganisms. In-use dilutions should not be stored unless sterilized.

As already mentioned, smooth, polished surfaces are cleaned most easily. Floors and horizontal surfaces should be cleaned and disinfected daily, walls and ceilings as often as required, but the interval should not exceed 1 month. Regular microbiological monitoring should be carried out to determine the efficacy of disinfection procedures. Records should be kept and immediate remedial action taken should normal levels for that area be exceeded.

Table 22.2 Disinfectants used during the manufacture of sterile products

Disinfectant	Application
Clear soluble phenols	Interior surfaces and fittings
Halogens, e.g. sodium hypochlorite	Working surfaces (limited use)
Alcohols: ethanol or isopropanol (usually as 70% solutions)	Working surfaces, equipment, gloved hands (rapid action)
Cationic agents (usually in 70% alcohol), e.g. cetrimide, chlorhexidine	Skin, gloved hands (rapid action with residual activity)

3.1.8   *Operation*

The number of persons involved in sterile manufacturing should be as small as possible so as to avoid the inevitable turbulence and shedding of particles and organisms associated with operatives. All operations should be undertaken in a controlled and methodical manner as excessive activity may also increase turbulence and shedding of particles and organisms.

Containers made from fibrous materials such as paper, cardboard and sacking, are generally heavily contaminated (especially with moulds and bacterial spores) and should not be taken into clean or aseptic areas where fibres or microorganisms shed from them could contaminate the product. Ingredients which must be brought into clean areas must first be transferred to suitable metal or plastic containers.

Containers and closures for terminally sterilized products must be thoroughly cleaned before use and should undergo a final washing and rinsing process in apyrogenic distilled water (which has been passed through a bacteria-proof membrane filter) immediately prior to filling. Those containers and closures destined for use in aseptic manufacture must, in addition, be sterilized after washing and rinsing in preparation for aseptic filling.

3.2   **Aseptic areas: additional requirements**

Additional requirements for aseptic areas, over and above those discussed in sections 3.1.1–3.1.8, are considered below.

3.2.1   *Clothing*

Section 3.1.5 considered the general requirements for clothing. Additional requirements are demanded for aseptic areas. Since the operative is a potential source of contamination, it is axiomatic that steps must be taken to minimize this. Accordingly, the operative must wear sterile protective clothing including headwear (which should totally enclose hair and beard), powder-free rubber or plastic gloves, a non-fibre-shedding face-mask (to prevent the release of droplets) and footwear. A suitable garment is a one or two-piece trouser suit. Fresh sterile clothing should normally be provided each time a person enters an aseptic area.

3.2.2   *Entry to aseptic areas*

Entry to an aseptic suite is usually by a 'black–grey–white' changing procedure. In this scheme, progress through from 'black' to 'white' represents passage into areas of increasing cleanliness, with the 'grey' area acting as an intermediate stage before entry to the 'white' (aseptic) changing area. Movement from 'black' to 'white' is generally through two changing rooms, the 'grey' area also serving as an entry to the clean room

(Fig. 22.1 and section 3.1.6). In the 'black' area, the operative removes outer shoes and clothing, swings the legs over a dividing sill and dons slippers. He or she then enters the 'grey' area where, after washing, hands and forearms are dried (with sterile, non-fibre-shedding cloth towels or a hot air dryer), a sterile hood and mask donned, and the hands and forearms rewashed and redried. The operative next enters the 'white' area where a sterile-area suit, overboots and gloves are put on; the gloved hands are rinsed in a disinfectant solution. The aseptic area may then be entered and work commenced.

<table>
<tr><td>3.2.3</td><td></td></tr>
</table>

3.2.3   *Equipment and operation*

Articles which are to be discharged from the clean room (or elsewhere) to the aseptic area must be sterilized. To achieve this they should be transferred via a double-ended sterilizer (i.e. with a door at each end). If it is not possible, or required, that they be discharged directly to the aseptic area, they should be (i) double-wrapped before sterilization; (ii) transferred immediately after sterilization to a clean environment until required; and (iii) transferred from this clean environment via a double-doored hatch (where the outer wrapping is removed) to the aseptic area (where the inner wrapper is removed at the workbench). Hatchways and sterilizers should be arranged so that only one side of the entry into an aseptic area may be opened at any one time. Solutions manufactured in the clean room may be brought into the aseptic area through a sterile 0.22-μm bacteria-proof membrane filter.

Workbenches, including laminar airflow cabinets, and equipment, should be disinfected immediately before and after each work period. Equipment used should be of the simplest design possible commensurate with the operation being undertaken.

Aseptic manipulations should be performed in the sterile air of a laminar airflow unit. Speed, accuracy and simplicity of movement, in accordance with a complete understanding of what is required, are essential features of a good aseptic technique.

Under no circumstances should living cultures of microorganisms, whether they be for vaccine preparation (Chapter 16) or for use in monitoring sterilization processes (Chapter 23), be taken into aseptic areas. As already pointed out, separate premises are needed for the aseptic filling of live or of attenuated vaccines.

4   *Guide to Good Pharmaceutical Manufacturing Practice*

For a number of years the essential features of GMP have been covered in the UK by the *Guide to Good Pharmaceutical Manufacturing Practice* (1981, 1983). These Guides were prepared by the Medicines Inspectorate of the, then, Department of Health and Social Security (now Department of Health) in consultation with industrial, hospital, professional and other interested parties. The principles of these national Guides have now been assimilated into the *EC Guide to Good Manufacturing Practice for Medicinal Products* (1989).

Compliance with the principles of GMP is one of the major factors considered by the Licensing Authority when examining an application for a licence to manufacture under the Medicines Act (1968). Similar codes, with or without the force of law, exist in the USA and other countries.

5   Conclusions

The manufacture of non-sterile pharmaceutical products requires that certain criteria of cleanliness, personal hygiene, production methods, storage, etc. must be met. Many such products are for oral and topical use and the question may fairly be posed as to the point of what are now quite stringent conditions. Nevertheless, some carefully controlled hospital studies have indeed shown that both types of medicine may be associated with nosocomial (hospital-acquired) infections and that the bacterium *Pseudomonas aeruginosa* may be a particular problem in this context.

A greater degree of stringency is required for the production of terminally sterilized products. Again, as the final product is subjected to a sterilization process (usually thermal), it may be asked why so much emphasis is placed upon controlling cleanliness, etc. One answer is obviously to reduce particulate matter, and a second important reason is that the lower the initial microbial load, the easier it is to achieve sterilization (Chapter 20). It must also be realized (as reiterated in Chapter 23) that it is far better to control a process from beginning to end, i.e. with frequent checks all along the line, than to rely solely on tests which can only determine whether a small proportion of the final products in a batch are satisfactory.

Even further criteria must be satisfied when products are being prepared aseptically. It is essential that operatives have a sound working knowledge of the properties of microorganisms, and that they appreciate the importance of personal hygiene, of the techniques that will be adopted, and of the possible sources of contamination and error.

6   Further reading

British Pharmacopoeia (1988) London: HMSO. (Note the section dealing with tests for microbial contamination.)

British Standards Institution (1989) *Environmental Cleanliness in Enclosed Spaces.* BS5295. Parts 0, 1, 2, 3 and 4. London: BSI.

Denyer S.P. (1988) Clinical consequences of microbial action on medicines. In *Biodeterioration* (Eds D.R. Houghton, R.N. Smith & H.O.W. Eggins), vol. 7, pp. 146–151. London: Elsevier Applied Science.

Denyer S.P. (1992) Filtration sterilization. In *Principles and Practice of Disinfection, Preservation and Sterilization*, 2nd edn (Eds A.D. Russell, W.B. Hugo & G.A.J. Ayliffe), pp. 573–604. Oxford: Blackwell Scientific Publications.

Denyer S.P. & Baird R.M. (Eds) (1990) *Guide to Microbiological Control in Phar-*

maceuticals. Chichester: Ellis Horwood. (Chapters 4 and 5 provide additional information.)

EC Guide (1989) *The Rules Governing Medicinal Products in the European Community vol. IV. Guide to Good Manufacturing Practice for Medicinal Products*. Document III/2244/87-EN, Rev. 3. Commission of the European Communities. London: HMSO.

Guide to Good Pharmaceutical Manufacturing Practice (1983) 3rd edn. London: HMSO.

Guide to Good Pharmaceutical Manufacturing Practice for Sterile Single-Use Medical Devices and Surgical Products (1981) London: HMSO.

Hambleton R. & Myers J.A. (1980) Special aspects of sterile fluid production. In *Textbook of Hospital Pharmacy* (Eds M.C. Allwood & J.T. Fell), pp. 144–169. Oxford: Blackwell Scientific Publications.

Kay J.B. (1980) Manufacture of pharmaceutical preparations. In *Textbook of Hospital Pharmacy* (Eds M.C. Allwood & J.T. Fell), pp. 71–143. Oxford: Blackwell Scientific Publications.

Pharmaceutical Codex (1979) 11th edn. London: Pharmaceutical Press. (Note the microbial contamination aspects included in the formulary section.)

Ringertz O. & Ringertz S.H. (1982) The clinical significance of microbial contamination in pharmaceutical and allied products. *Adv Pharm Sci*. **5**, 201–226.

Underwood E. (1992) Good manufacturing practice. In *Principles and Practice of Disinfection, Preservation and Sterilization*, 2nd edn (Eds A.D. Russell, W.B. Hugo & G.A.J. Ayliffe), pp. 274–291. Oxford: Blackwell Scientific Publications.

United States Pharmacopeia (1990) 22nd revision. Rockville, MD: US Pharmacopeial Convention. (Note the section dealing with microbial limit tests.)

23   Sterilization control and sterility testing

1	**Introduction**
2	**Sterilization monitors**
2.1	Physical indicators
2.1.1	Heat sterilization
2.1.2	Gaseous sterilization
2.1.3	Radiation sterilization
2.1.4	Filtration sterilization
2.2	Chemical indicators
2.3	Biological indicators
3	**Sterility testing**
3.1	Methods
3.1.1	Direct inoculation of culture media
3.1.2	Membrane filtration

3.1.3	Introduction of concentrated culture medium
3.2	Antimicrobial agents
3.2.1	Specific inactivation
3.2.2	Dilution
3.2.3	Membrane filtration
3.3	Positive controls
3.4	Specific cases
3.5	Sampling
4	**Conclusions**
5	**Acknowledgements**
6	**Further reading**

1   Introduction

A product to be labelled 'sterile' must be free of viable microorganisms. To achieve this, the product, or its ingredients, must undergo a sterilization process of sufficient microbiocidal capacity to ensure a minimum level of sterility assurance (Chapter 20). It is essential that the required conditions for sterilization be achieved and maintained through every operation of the sterilizer. The necessary confidence in the sterilization process, and ultimately in the microbiological quality of the product, can be gained by regular and routine use of appropriate indicators of sterilizer performance, combined with a final microbiological assessment of the product itself, namely the sterility test (section 3).

This chapter will discuss briefly the principles and applications of the various methods of monitoring and validating sterilization processes.

2   Sterilization monitors

Monitoring of the sterilization process can be achieved by the use of physical, chemical or biological indicators of sterilizer performance. Such indicators are frequently employed in combination.

2.1   Physical indicators

In well-understood and well-characterized sterilization processes (e.g. heat and irradiation), where physical measurements may be accurately made, sterility can be assured by monitoring only the physical conditions of the process. When such a process conforms exactly to its physical specification it is said to have satisfied the required parameters, thereby permitting the *parametric release* of processed products. For other sterilization methods, physical indicators form only part of the overall routine monitoring programme.

2.1.1 Heat sterilization

A temperature record chart is made of each sterilization cycle with both dry and moist heat (i.e. autoclave) sterilizers; this chart forms part of the batch documentation and is compared against a master temperature record (MTR). It is recommended that the temperature be taken at the coolest part of the loaded sterilizer. Further information on heat distribution and penetration within a sterilizer can be gained by the use of thermocouples placed at selected sites in the chamber or inserted directly into test packs or bottles. Since autoclaving depends also upon steam under pressure as well as temperature, pressure measurements form an essential part of the physical monitoring of this process. In addition, periodic leak tests are performed on prevacuum steam sterilizers to assess the efficiency of air removal prior to the introduction of steam.

2.1.2 Gaseous sterilization

Elevated temperatures are monitored for each sterilization cycle by temperature probes, and routine leak tests are performed to ensure gas-tight seals. Pressure and humidity measurements are recorded. Gas concentration is measured independently of pressure rise, often by reference to weight of gas used.

2.1.3 Radiation sterilization

A plastic (often perspex) dosimeter which gradually darkens in proportion to the radiation absorbed gives an accurate measure of the radiation dose and is considered to be the best technique currently available for following the radiosterilization process.

2.1.4 Filtration sterilization

A bubble point pressure test, which is a technique employed for determining the pore size of filters, may be used to check the integrity of certain types of filter device (candle, membrane and sintered glass; see Chapter 20) immediately after use. The principle of the test is that the wetted filter, in its assembled unit, is subjected to an increasing air or nitrogen gas pressure differential. The pressure difference recorded when the first bubble of gas breaks away from the filter is related to the *maximum* pore size. When the gas pressure is further increased slowly, there is a general eruption of bubbles over the entire surface. The pressure difference here is related to the *mean* pore size. A pressure differential below the expected value would signify a damaged or faulty filter. A modification to this test for membrane filters involves measuring the diffusion of gas through a wetted filter at pressures below the bubble point pressure (diffusion rate test); a faster diffusion rate than expected would again indicate a loss of filter integrity. In addition, a filter is considered ineffective when an unusually rapid rate of filtration occurs.

Efficiency testing of high-efficiency particulate air (HEPA) filters used for the supply of sterile air to aseptic workplaces (Chapter 22) is normally achieved by the generation upstream of dioctylphthalate (DOP) or sodium chloride particles of known dimension, followed by detection in downstream filtered air. Retention efficiency is recorded as the percentage of particles removed under defined test conditions. Microbiological tests are not normally performed.

2.2   Chemical indicators

Chemical monitoring of a sterilization process is based on the ability of heat, steam, sterilant gases and ionizing radiation to alter the chemical and/or physical characteristics of a variety of chemical substances. Ideally, this change should take place only when satisfactory conditions for sterilization prevail, thus confirming that a sterilization cycle has been successfully completed. In practice, however, the ideal indicator response is not always achieved and so a necessary distinction is made between (i) those chemical indicators which integrate several sterilization parameters (i.e. temperature, time and saturated steam) and closely approach the ideal; and (ii) those which measure only one parameter and consequently can only be used to distinguish processed from unprocessed articles. Thus indicators which rely on the melting of a chemical substance show that the temperature has been attained but not necessarily maintained.

Chemical indicators generally undergo melting or colour changes (Fig. 23.1), the relationship of this change to the sterilization process being influenced by the design of the test device (Table 23.1). It must be remembered, however, that the changes recorded do not necessarily correspond to microbiological sterility and consequently the devices should never be employed as sole indicators in a sterilization process. Nevertheless, when included in strategically placed containers or packages, chemical indicators are valuable monitors of the conditions prevailing at the coolest or most inaccessible parts of a sterilizer.

2.3   Biological indicators

Biological indicators (BIs) for use in thermal, chemical or radiation sterilization processes consist of standardized bacterial spore preparations which are usually in the form of inoculated paper or foil strips. As with chemical indicators, they are usually placed in dummy packs located at strategic sites in the sterilizer. Alternatively, for gaseous sterilization these may also be placed within a tubular helix (Line–Pickerill) device. After the sterilization process these strips are aseptically transferred to an

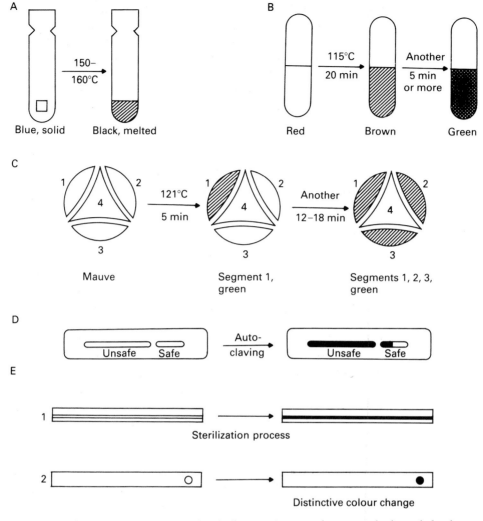

A

150–
160°C

Blue, solid   Black, melted

B

115°C
20 min   Another
5 min
or more

Red   Brown   Green

C

1   2
4
3

Mauve

121°C
5 min

1   2
4
3

Segment 1,
green

Another
12–18 min

1   2
4
3

Segments 1, 2, 3,
green

D

Unsafe   Safe

Auto-
claving

Unsafe   Safe

E

1

Sterilization process

2   O   ●

Distinctive colour change

Fig. 23.1 Examples of chemical indicators: A, temptubes: example shown is for dry heat sterilization, other versions are available for autoclave processes (Propper Mfg. Co., Long Island, USA). B, Browne's tubes: example shown is the type 1 (black spot) tube; other versions are available for alternative autoclaving and dry heat processes (A. Browne Ltd, Leicester, UK). C, timecards: the colour change occurs one segment at a time; segment 2 changes colour after 10–13 min total exposure and segment 4 after 25 min total exposure (Propper Mfg. Co.). D, Thermalog S: a satisfactory autoclaving process is indicated when the blue band enters the 'safe' region. It is claimed the device can be used over a wide range of autoclaving conditions (Info-Chem Inc., New Jersey, USA). E, examples of reactive chemical 'ink' devices used to monitor steam or gaseous sterilization processes, e.g. Strate-line (1) and Gas-chex (2) for steam and ethylene oxide, respectively (Propper Mfg. Co.). (For further details refer to Table 23.1.)

appropriate nutrient medium which is then incubated and periodically examined for signs of growth. To avoid the need for strict asepsis, self-contained units have been developed where the spore strip and nutrient medium are present in the same device ready for mixing after use.

461   *Sterilization control and sterility testing*

Table 23.1 Chemical indicators for monitoring sterilization processes

Sterilization method	Principle	Device	Parameter(s) monitored
Heat			
Autoclaving or dry heat	Fusible pellet in sealed glass tube	Chemical pellet having a specific melting-point and offering a characteristic colour change on melting, e.g. Temptubes	Temperature
	Temperature-sensitive coloured solution	Sealed tubes partly filled with a solution which changes colour at elevated temperatures; rate of colour change is proportional to temperature, e.g. Browne's tubes	Temperature, time
Dry heat only	Temperature-sensitive chemical	Usually a temperature-sensitive white wax concealing a black marked or printed (paper) surface; at a predetermined temperature the wax rapidly melts exposing the background mark(s)	Temperature
Heating in an autoclave only	Steam-sensitive chemical	Usually an organic chemical in a printing ink base impregnated into a carrier material. A combination of moisture and heat produces a darkening of the ink, e.g. autoclave tape. Devices of this sort can be used within dressings packs to confirm adequate removal of air and penetration of saturated steam (Bowie Dick test)	Saturated steam
	Thermochromic chemical	On exposure to saturated steam at predetermined temperatures the ink undergoes a colour change; with certain devices, e.g. Tempcards, TST device, a progressive colour change with time is recorded	Temperature, saturated steam, time (selected devices)
	Capillary principle (Thermalog S)	Consists of a blue dye in a waxy pellet, the melting-point of which is depressed in the presence of saturated steam. At autoclaving temperatures, and in the continued presence of steam, the pellet melts and travels along a paper wick forming a blue band the length of which is dependent upon both exposure time and temperature	Temperature, saturated steam, time
Gaseous sterilization			
Ethylene oxide (EO)	Reactive chemical	Indicator paper impregnated with a reactive chemical which undergoes a distinct colour change on reaction with EO in the presence of heat and moisture. With some devices rate of colour development varies with temperature and EO concentration	Gas concentration, temperature, time (selected devices); NB a minimum relative humidity (rh) is required for device to function

Table 23.1 *Continued*

Sterilization method	Principle	Device	Parameter(s) monitored
Ethylene oxide (EO) (*contd*)	Capillary principle (Thermalog G)	Based on the same 'migration along wick' principle as Thermalog S. Optimum response in a cycle of $600\,mg\,l^{-1}$ EO, temperature 54°C, rh 40–80%. Lower EO levels and/or temperature will slow response time, blue colour of band is fugitive at rh <30%	Gas concentration, temperature, time (selected cycles)
	Royce's sachet	Polythene sachet containing acidified magnesium chloride and a pH indicator. Reaction of EO with sachet contents causes production of ethylene chlorhydrin and magnesium hydroxide, changing the pH from acid to alkaline with a corresponding pH indicator colour change. EO penetration is controlled by the size and thickness of the sachet, temperature, concentration and exposure time	Gas concentration, temperature, time
Low temperature steam and formaldehyde	Reactive chemical	Indicator paper impregnated with a formaldehyde-, steam- and temperature-sensitive reactive chemical which changes colour during the sterilization process	Gas concentration, temperature, time (selected cycles)
Radiation sterilization	Radiochromic chemical	Plastic devices impregnated with radiosensitive chemicals which undergo colour changes at relatively low radiation doses	Only indicate exposure to radiation
	Dosimeter device	Acidified ferric ammonium sulphate or ceric sulphate solutions respond to irradiation by dose-related changes in their optical density (see also section 2.1.3)	Accurately measure radiation doses

The bacterial species to be used in a BI must be selected carefully, since it must be non-pathogenic and should possess above average resistance to the particular sterilization process. Resistance is adjudged from the spore destruction curve obtained upon exposure to the sterilization process; recommended BI spores and their decimal reduction times (*D*-values; Chapter 20) are shown in Table 23.2. Great care must be taken in the preparation of BIs to ensure a standardized response to sterilization processes. Indeed, while certainly offering the most direct method of monitoring sterilization processes, it should be realized that BIs may be less reliable monitors than physical methods and are not recommended for routine use, except in the case of gaseous sterilization.

463   *Sterilization control and sterility testing*

Table 23.2 Biological indicators for monitoring sterilization processes[*]

Sterilization process	Species	Inoculum size	D-value
Heating in an autoclave (121°C)	*Bacillus stearothermophilus*	$>10^5$	1.5 min
	Clostridium sporogenes	$>10^5$	0.8 min
Dry heat (160°C)	*Bacillus subtilis* var. *niger*	$>10^5$	5–10 min
Ethylene oxide (EO)[†] (EO 475 mg l^{-1}, temperature 29°C, no humidification)	*Bacillus subtilis* var. *niger*	10^6–10^7	9 min
Low temperature steam (73°C) and formaldehyde (12 mg l^{-1})[‡]	*Bacillus stearothermophilus*	—	5 min
Ionizing radiation	*Bacillus pumilus*	10^7–10^8	3 kGy (0.3 Mrad)

[*] *British Pharmacopoeia* (1988).
[†] DHSS specification TSS/S/330.012.
[‡] Soper & Davies (1990).

Filtration sterilization requires a different approach from biological monitoring, the test effectively measuring the ability of a filter to produce a sterile filtrate from a culture of a suitable organism. For this purpose, *Serratia marcescens*, a small Gram-negative rod-shaped bacterium (minimum dimension 0.5 μm), has been recommended in the *Pharmaceutical Codex* (1979). The bacterial challenge test is the most severe to which a filter of any construction can be subjected. In the membrane-filter industry, the test using *Ser. marcescens* is usually reserved for filters of 0.45-μm pore size, and a more rigorous test involving *Pseudomonas diminuta* (minimum dimension 0.3 μm) is applied to filters of 0.22-μm pore size. The latter filters are defined as those capable of completely removing *Ps. diminuta* from suspension. In this test, using this organism, a realistic inoculum level must be adopted, since the probability of bacteria appearing in the filtrate rises as the concentration of *Ps. diminuta* cells in the test challenge increases. The extent of the passage of this organism through membrane filters is enhanced by increasing the filtration pressure. Thus sterile filtration depends markedly on the challenge conditions.

3 Sterility testing

A sterility test is basically a test which assesses whether a sterilized pharmaceutical or medical product is free from contaminating micro-organisms, either by incubation of the whole or a part of that product with a nutrient medium. It thus becomes a destructive test and raises the question as to its suitability for testing large, expensive or delicate products or equipment. Furthermore, by its very nature such a test is a statistical process in which part of a batch is randomly* sampled and the chance of the batch being passed for use then depends on the sample passing the sterility test. Nevertheless, the test does have an important application in monitoring the microbiological quality of filter-sterilized, aseptically filled products and does offer a final check on terminally sterilized articles. In the UK, test procedures laid down by the *British Pharmacopoeia* must be followed; this provides details of the sample sizes to be adopted in particular cases. The principles of these tests are discussed in brief below.

3.1 **Methods**

The methods currently employed when carrying out sterility tests are as follows (sections 3.1.1–3.1.3).

3.1.1 *Direct inoculation of culture media*

This involves introducing test samples directly into nutrient media. The *British Pharmacopoeia* recommends two media: (i) fluid mercaptoacetate medium, which contains glucose and sodium mercaptoacetate (sodium thioglycollate) and is particularly suitable for the cultivation of anaerobic organisms (incubation temperature 30–35°C); and (ii) soyabean casein digest medium, which will support the growth of both aerobic bacteria (incubation temperature 30–35°C) and fungi (incubation temperature 20–25°C). Other media may be used provided they can be shown to be suitable alternatives.

3.1.2 *Membrane filtration*

This is the technique recommended by most pharmacopoeias and involves filtration of fluids through a sterile membrane filter (pore size ≤0.45 µm), any microorganism present being retained on the surface of the filter. After washing *in situ*, the filter is divided aseptically and portions transferred to suitable culture media which are then incubated at the appropriate temperature for the required period of time. Water-soluble solids can be dissolved in a suitable diluent and processed in this way.

3.1.3 *Introduction of concentrated culture medium*

A sensitive method for detecting low levels of contamination in intravenous infusion fluids involves the addition of a concentrated culture

---

* It has been proposed that random sampling be applied to products which have been processed and filled aseptically. With products sterilized in their final containers, samples should be taken from the potentially coolest or least sterilant-accessible part of the load.

medium to the fluid in its original container, such that the resultant mixture is equivalent to single strength culture medium. In this way, sampling of the entire volume is achieved.

With the techniques discussed above (sections 3.1.1–3.1.3) the media employed should previously have been assessed for nutritive (growth-supporting) properties and a lack of toxicity using specified organisms. It must be remembered that any survivors of a sterilization process may be damaged and thus must be given the best possible conditions for growth.

As a precaution against accidental contamination, product testing must be carried out under conditions of strict asepsis using, for example, a laminar airflow cabinet to provide a suitable environment (Chapter 22).

3.2 Antimicrobial agents

Where an antimicrobial agent comprises the product or forms part of the product, for example as a preservative, its activity must be nullified in some way during sterility testing so that an inhibitory action in preventing the growth of any contaminating microorganisms is overcome. This is achieved by the following methods (sections 3.2.1–3.2.3).

3.2.1 Specific inactivation

An appropriate inactivating (neutralizing) agent (Table 23.3) is incorporated into the culture media. The inactivating agent must be non-toxic to microorganisms as must any product resulting from an interaction of the inactivator and the antimicrobial agent.

Although Table 23.3 lists only benzylpenicillin and ampicillin as being inactivated by β-lactamase (from *B. cereus*), other β-lactams may also be hydrolysed by their appropriate β-lactamase. Other antibiotic-inactivating enzymes are also known (Chapter 10) and have been considered as possible

Table 23.3 Inactivating agents*

Inhibitory agents	Inactivating agents
Phenols, cresols	None (dilution)
Alcohols	None (dilution)
Parabens	Dilution and Tween
Mercury compounds	-SH compounds
Quaternary ammonium compounds	{ Lecithin + Lubrol W; { Lecithin + Tween (Letheen)
Benzylpenicillin[†] } Ampicillin }	β-Lactamase from *B. cereus*
Other antibiotics[†]	None (membrane filtration)
Sulphonamides	*p*-Aminobenzoic acid

* See also Table 12.1 (Chapter 12).
[†] See text.

inactivating agents, e.g. chloramphenicol acetyltransferase (inactivates chloramphenicol) and enzymes that modify aminoglycoside antibiotics. In addition, encouraging results have been obtained by the use of antibiotic-absorbing resins.

<table>
<tr><td>3.2.2</td><td>

Dilution

The antimicrobial agent is diluted in the culture medium to a level at which it ceases to have any activity, for example phenols, cresols and alcohols (see Chapter 12). This method applies to substances with a high dilution coefficient, η.

</td></tr>
</table>

3.2.2   *Dilution*

The antimicrobial agent is diluted in the culture medium to a level at which it ceases to have any activity, for example phenols, cresols and alcohols (see Chapter 12). This method applies to substances with a high dilution coefficient, η.

3.2.3   *Membrane filtration*

This method has traditionally been used to overcome the activity of antibiotics for which there are no inactivating agents, although it could be extended to cover other products if necessary. Basically, a solution of the product is filtered through a hydrophobic-edged membrane filter which will retain any contaminating microorganisms. The membrane is washed *in situ* to remove any traces of antibiotic adhering to the membrane and is then transferred to appropriate culture media.

3.3   **Positive controls**

It is essential to show that microorganisms will actually grow under the conditions of the test. For this reason positive controls have to be carried out; in these, the ability of small numbers of suitable microorganisms to grow in media in the presence of the sample is assessed. The microorganism used for positive control tests with a product containing or comprising an antimicrobial agent must, if at all possible, be sensitive to that agent, so that growth of the organism indicates a satisfactory inactivation, dilution or removal of the agent. The *British Pharmacopoeia* suggests that use of appropriate strains of *Staphylococcus aureus*, *B. subtilis*, *Cl. sporogenes* and *Candida albicans* for vegetative and spore-forming aerobic, anaerobic and fungal positive controls, respectively.

3.4   **Specific cases**

Specific details of the sterility testing of parenteral products, ophthalmic and other non-injectable preparations, catgut, surgical dressings and dusting powders will be found in the British and European pharmacopoeias.

3.5   **Sampling**

A sterility test attempts to infer the state (sterile or non-sterile) of a batch from the results of an examination of part of a batch, and is thus a statistical operation.

Table 23.4 Sampling in sterility testing

	Infected items in batch (%)					
	0.1	1	5	10	20	50
p	0.001	0.01	0.05	0.1	0.2	0.5
q	0.999	0.99	0.95	0.9	0.8	0.5
Probability, P, of drawing 20 consecutive sterile items:						
First sterility test*	0.98	0.82	0.36	0.12	0.012	<0.00001
First retest[†]	0.99	0.99	0.84	0.58	0.11	0.002

* Calculated from $P = (1 - p)^{20} = q^{20}$.
[†] Calculated from $P = (1 - p)^{20} [2 - (1 - p)^{20}]$.

Suppose that p represents the proportion of infected containers in a batch and q the proportion of non-infected containers. Then, $p + q = 1$ or $q = 1 - p$.

Suppose also that a sample of two items is taken from a large batch containing 10% infected containers. The probability of a single item taken at random being infected is $p = 0.1$ (10% = 0.1), whereas the probability of such an item being non-infected is given by $q = 1 - p = 0.9$.

The probability of both items being infected is $p^2 = 0.01$, and of both items being non-infected, $q^2 = (1 - p)^2 = 0.81$. The probability of obtaining one infected item and one non-infected item is $1 - (0.01 + 0.81) = 0.18 = 2pq$.

In a sterility test involving a sample size of n containers, the probability p of obtaining n consecutive 'steriles' is given by $q^n = (1 - p)^n$. Values for various levels of p (i.e. proportion of infected containers in a batch) with a constant sample size are given in Table 23.4 which shows that the test cannot detect low levels of contamination. Similarly, if different sample sizes are employed (based also upon $(1 - p)^n$) it can be shown that as the sample size increases, the probability of the batch being passed as sterile decreases.

The *British Pharmacopoeia* makes an allowance for accidental contamination which may arise during the execution of a sterility test by allowing the test to be repeated. Under these circumstances the following rules apply.

1 If no growth occurs with fresh samples, the batch passes the test.

2 If growth occurs, but the organism differs from that found previously, the test is repeated on a third sample from the batch using double the number of containers of product.

3 If no growth occurs with the third sample, the batch passes the sterility test; if, however, any microorganism is found, the batch is treated as non-sterile, unless or until the material has been resterilized and has passed the above tests.

In actual fact, however, these additional tests increase the chances of passing a batch containing a proportion of infected items (Table 23.4, first retest). This may be deduced by using the mathematical formula

$$(1 - p)^n[2 - (1 - p)^n]$$

which gives the chance in the first retest of passing a batch containing a proportion p of infected containers.

It can be seen from the above that a sterility test can only show that a proportion of the products in a batch is sterile. Thus, the correct conclusion to be drawn from a satisfactory test result is that the batch has passed the sterility test *not* that the batch is sterile.

4   Conclusions

The techniques discussed in this chapter comprise an attempt to achieve, as far as possible, the continuous monitoring of a particular sterilization process. The sterility test *on its own* provides no guarantee as to the sterility of a batch; however, it is an additional check, and continued compliance with the test does give confidence as to the efficacy of a sterilization or aseptic process. Failure to carry out a sterility test, despite the major criticism of its inability to detect other than gross contamination, may have important legal and moral consequences.

5   Acknowledgements

The assistance of the following is gratefully acknowledged: Dr B. Kirk, Principal Pharmacist, South Western Regional Health Authority; Albert Browne Ltd, Leicester; Arnold R. Horwell Ltd, London; Bennett & Company (1986) Ltd, Andover.

6   Further reading

Anderson J. & Breeze A.S. (1985) The differential binding of antibiotics to resins and its use in sterility testing. *J Pharm Pharmacol.* **37**, 62.

Breeze A.S. & Simpson A.M. (1983) Actual and potential methods of testing antibiotics for sterility. In *Antibiotics: Assessment of Antimicrobial Activity and Resistance* (Eds A.D. Russell & L.B. Quesnel), pp. 339–348. Society for Applied Bacteriology Technical Series No. 18. London: Academic Press.

British Pharmacopoeia (1988) London: HMSO.

Brown M.R.W. & Gilbert P. (1977) Increasing the probability of sterility of medicinal products. *J Pharm Pharmacol.* **29**, 517–523.

Bunn J.L. & Sykes I.K. (1980) A chemical indicator for the rapid measurement of F_0 values. *J Appl Bacteriol.* **51**, 143–147.

Department of Health and Social Security (1985) *Specification for Biological Monitors for the Control of Ethylene Oxide Sterilisation.* Specification No. TSS/SS/330.012. London: DHSS.

Denyer S.P (1982) In-use contamination in intravenous therapy—the scale of the problem. In *Infusions and Infection. The Hazards of In-use Contamination in Intravenous Therapy* (Ed. P.F. D'Arcy), pp. 1–16. Oxford: Medicine Publishing Foundation.

Denyer S.P. (1992) Filtration sterilization. In *Principles and Practice of Disinfection,*

Preservation and Sterilization, 2nd edn (Eds A.D. Russell, W.B. Hugo & G.A.J. Ayliffe), pp. 573–604. Oxford: Blackwell Scientific Publications.

Denyer S.P. & Baird R.M. (Eds) (1990) *Guide to Microbiological Control in Pharmaceuticals*. Chichester: Ellis Horwood. (Chapters 7, 8 and 9 provide additional information.)

EC Guide (1989) *The Rules Governing Medicinal Products in the European Community Vol. IV. Guide to Good Manufacturing Practice for Medicinal Products*. Document III/2244/87-EN, Rev. 3. Commission of the European Communities. London: HMSO.

European Pharmacopoeia (1980) *Part 1: Test for Sterility*, 2nd edn. Saint Ruffine, France: Maisonneuve SA.

Gardner J.F. & Peel M.M. (1991) *Introduction to Sterilisation, Disinfection and Infection Control*, 2nd edn. Melbourne: Churchill Livingstone.

Greene V.N. (1992) Control of sterilization processes. In *Principles and Practice of Disinfection, Preservation and Sterilization*, 2nd edn (Eds. A.D. Russell, W.B. Hugo & G.A.J. Ayliffe), pp. 605–624. Oxford: Blackwell Scientific Publications.

Hambleton R. (1983) T.S.T.—a new range of indicator devices for autoclaving processes. *Sterile World*, **5**, 7–9.

Hibbert H.R. & Spencer R. (1970) An investigation of the inhibitory properties of sodium thioglycollate in media for the recovery of clostridial spores. *J Hgy.* **68**, 131–135.

Health Technical Memorandum (1980) *Sterilisers*. HTM 10. London: DHSS.

Hoxey E.V., Soper C.J. & Davies D.J.G. (1984) The effect of temperature and formaldehyde concentration on the inactivation of *Bacillus stearothermophilus* spores by LTSF. *J Pharm Pharmacol.* **36**, 60.

Kelsey J.C. (1972) The myth of surgical sterility. *Lancet*, **ii**, 1301–1303.

Line S.J. & Pickerell J.K. (1973) Testing a steam-formaldehyde sterilizer for gas penetration efficiency. *J Clin Path.* **26**, 716–720.

Pharmaceutical Codex (1979) London: Pharmaceutical Press.

Soper C.J. & Davies D.J.G. (1990) Principles of sterilization. In *Guide to Microbiological Control in Pharmaceuticals* (Eds S.P. Denyer & R.M. Baird), pp. 157–181. Chichester: Ellis Horwood.

United States Pharmacopeia (1990) 22nd revision. Rockville, MD: US Pharmacopeial Convention.

24 Production of therapeutically useful substances by recombinant DNA technology

1   Introduction

2   The basic principles of
     recombinant DNA technology
2.1 Introduction to cloning
2.2 Expression of cloned genes
2.2.1 Transcription
2.2.2 Translation
2.2.3 Post-translational modification
2.3 Maximizing gene expression
2.4 Choice of cloning host

3   Production of medically important
     polypeptides and proteins

4   Authenticity and efficacy of drugs
     produced by recombinant DNA
     technology

5   Future trends

6   Glossary

7   Further reading

1 Introduction

Natural products of pharmaceutical interest are synthesized by a wide variety of organisms, ranging from prokaryotes such as bacteria to eukaryotes such as yeast, other fungi, flowering plants, animals and man. The commercial production of compounds from microbes is relatively simple since the organism in question can be grown on a large scale and high-yielding variants can be isolated following many successive rounds of mutation and selection. A good example is penicillin production by *Penicillium chrysogenum* (Chapter 7), where the wild strain yield of a few milligrams per litre has been raised to over $20 \, g \, l^{-1}$. Commercial production of compounds from plants is less easy since synthesis may be tissue or organ specific and may only occur at a certain developmental stage. If the genetics of the producing organism have not been studied then selection of high-yielding variants is extremely difficult. Molecules of pharmacological interest from higher animals are by definition extremely potent, for example hormones, and so are synthesized in relatively minute quantities. This is a serious limitation if the producing organisms are animals such as cattle or pigs, as in the case of insulin, but production is well nigh impossible if the only source is man himself, as with human growth hormone.

In order that demand should meet supply, or to reduce production costs, it would be of great benefit if microorganisms could be induced to synthesize pharmacologically active molecules whose production is normally limited to higher plants and animals. With the advent of recombinant DNA technology, often called genetic engineering, this is now possible and synthesis no longer is restricted to polypeptides.

471

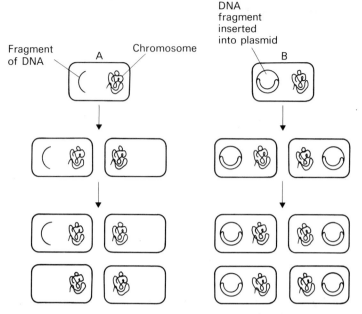

Fragment of DNA A Chromosome

DNA fragment inserted into plasmid B

Fig. 24.1 The requirement for a cloning vector: A, fragments of DNA introduced into the bacterium by transformation do not undergo replication and gradually are diluted out of the population; B, DNA fragments introduced into plasmids are inherited by both daughter progeny at cell division.

The advantages of recombinant DNA technology are enormous, as the following example shows. Somatostatin is a hormone that inhibits the secretion of pituitary growth hormone. The researchers who first isolated somatostatin required nearly half a million sheep brains to produce 5 mg of the substance. Using a chemically synthesized gene, 9 litres of bacterial culture, costing just a few pounds or dollars, produced the same amount. Development work has already led to the production of numerous biologically active human agents in clinically significant amounts, and a number of them are commercially available (Table 24.2).

2   The basic principles of recombinant DNA technology

2.1   Introduction to cloning

Let us suppose that we wish to construct a bacterium which produces human insulin. Naively, it might be thought that all that is required is to introduce the human insulin gene into its new host. The fallacy with this idea is that foreign genes are not maintained in cells, since they are not replicated. With recombinant DNA technology this problem is solved by inserting the insulin gene into a cloning vector. The latter is simply a DNA molecule that can replicate *in vivo*. Cloning vectors are usually

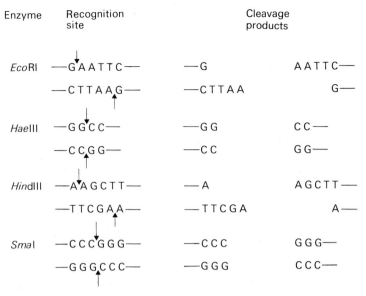

Enzyme	Recognition site	Cleavage products	
*Eco*RI	—G↓AATTC— —CTTAA↑G—	—G —CTTAA	AATTC— G—
*Hae*III	—GG↓CC— —CC↑GG—	—GG —CC	CC— GG—
*Hind*III	—A↓AGCTT— —TTCGA↑A—	—A —TTCGA	AGCTT— A—
*Sma*I	—CCC↓GGG— —GGG↑CCC—	—CCC —GGG	GGG— CCC—

Fig. 24.2 The recognition sites of some common restriction endonucleases. The arrows indicate the cleavage points.

plasmids (Chapter 10), which are extrachromosomal, autonomously replicating DNA molecules (Fig. 24.1).

In order to insert foreign DNA into a plasmid, use is made of special enzymes known as restriction endonucleases. These enzymes cut large DNA molecules into shorter fragments by cleavage at specific recognition sites (Fig. 24.2), i.e. they are highly specific deoxyribonucleases (DNases). Some of these enzymes generate fragments with single-strand protrusions called 'sticky-ends' because their bases are complementary. Fragments of the foreign DNA are inserted into plasmid vectors cut open with the same enzyme, which therefore have matching ends (Fig. 24.3). The resulting recombinants or *chimeras* are transformed into the new host microbe. Since each transformant may contain a different fragment of the foreign genome it is necessary to select those with the desired gene. In practice this can be the most difficult step, but the screening methods used are outside the scope of this chapter. Suffice it to say that a necessary prerequisite usually is a sensitive test for the desired protein product.

Theoretically it is possible to clone any desired gene by 'shotgunning'. This is done by inserting into plasmids a random mixture of fragments from total human DNA, in the case of the insulin gene, and then selecting the appropriate clone. However, if introduced into a bacterium this clone would not make human insulin. The reason for this is that many genes of eukaryotes, including the human insulin gene, are a mixture of coding regions (called *exons*) and non-coding regions (called *introns*). In eukaryotes, genes containing introns are transcribed into messenger RNA (mRNA) in the usual manner but then the corresponding intron

473   *Recombinant DNA technology*

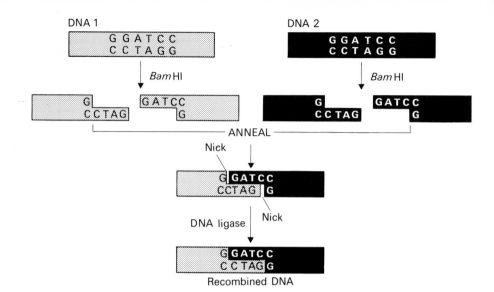

Fig. 24.3 The construction of a chimeric (or recombinant) DNA molecule by joining together two DNA fragments produced by cleavage of different parental DNA molecules with the same restriction endonuclease.

sequences are spliced out (Fig. 24.4). Unfortunately not all bacteria can splice out introns.

A solution to the problem of introns is to isolate mRNA extracted from the human pancreas cells that make insulin. These cells are rich in insulin mRNA from which introns have already been spliced out. Using the enzyme *reverse transcriptase* it is possible to convert this spliced messenger RNA into a DNA copy. This copy DNA (cDNA), which carries the uninterrupted genetic information for insulin, can be cloned. Although yeast cells (*Saccharomyces*) can splice out introns it is normal practice to eliminate them anyway by cDNA cloning.

An alternative approach is to synthesize an artificial gene in the test-tube starting with the appropriate deoxyribonucleotides. This approach, which demands that the entire amino acid sequence be known, has been used to clone genes encoding proteins 200 amino acids long.

2.2   **Expression of cloned genes**

Once a gene is cloned it is necessary to convert the information contained in it into a functional protein. There are a number of steps in gene expression: (i) transcription of DNA into mRNA; (ii) translation of the mRNA into a protein sequence; and (iii) in some instances, post-translational modification of the protein. In discussing these steps in more detail, expression of a cloned insulin gene will be used as an example.

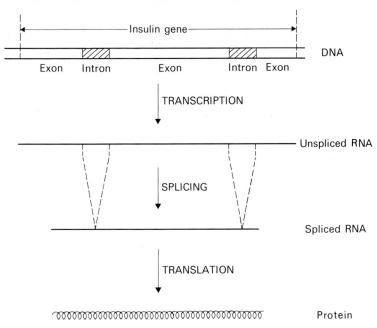

Fig. 24.4 Splicing of a messenger RNA molecule transcribed from a hypothetical insulin gene containing two introns.

2.2.1   *Transcription*

Transcription of DNA into mRNA is mediated by the enzyme RNA polymerase. The first stage is binding of the RNA polymerase to recognition sites on the DNA which are called *promoters*. After binding, the RNA polymerase proceeds along the DNA molecule until a termination signal is encountered. It follows that a gene which does not lie between a promoter and a termination signal will not be transcribed. This would be the case with a cloned insulin gene, since neither a cDNA gene nor an artificially synthesized gene will carry a promoter. The solution is to clone the gene into a vector close to a bacterial promotor. An example is shown in Fig. 24.5.

2.2.2   *Translation*

Translation of mRNA into protein is a complex process which involves interaction of the messenger with ribosomes. One prerequisite for this is that the mRNA must carry a ribosome binding site (RBS) in front of the gene to be translated. After binding, the ribosome moves along the mRNA and initiates protein synthesis at the first AUG codon it encounters. A synthetic insulin gene will lack an RBS and if a cDNA is used as the starting material the RBS may be lost in the process of cloning. The solution is to utilize a vector in which the insulin gene can be inserted downstream from a promoter and RBS (Fig. 24.6).

475   *Recombinant DNA technology*

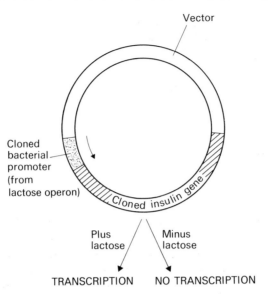

Vector

Cloned
bacterial
promoter
(from
lactose operon)

Cloned insulin gene

| Plus | Minus |
| lactose | lactose |

TRANSCRIPTION   NO TRANSCRIPTION

Fig. 24.5 Insertion of a cloned insulin gene into a vector carrying a bacterial promoter. The arrow indicates the direction of transcription. If we suppose the bacterial promoter is derived from the lactose operon then transcription will be initiated only in the presence of lactose.

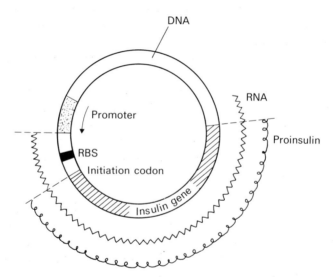

DNA

RNA

Promoter

Proinsulin

RBS

Initiation codon

Insulin gene

Fig. 24.6 The use of a vector carrying a promoter and adjacent ribosome binding site (RBS) and initiation codon to obtain synthesis of proinsulin from a synthetic gene. The arrow indicates the direction of transcription.

2.2.3   *Post-translational modification*

A number of proteins undergo post-translational modifications and insulin is one of these. Proteins which are to be secreted are synthesized with an extra 15–30 amino acids at the N-terminus. These extra amino

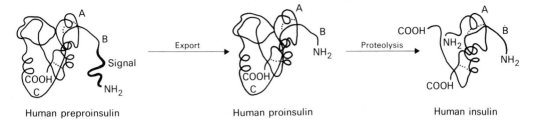

Fig. 24.7 The conversion of preproinsulin to insulin by sequential removal of the signal peptide and the C fragment.

acids are referred to as a *signal sequence* and a common feature of these sequences is that they have a central core of hydrophobic amino acids flanked by polar or hydrophilic residues. During passage through the membrane the signal sequence is cleaved off (Fig. 24.7). If the insulin gene were cloned by the cDNA method then the signal sequence would be present and, in *Escherichia coli* at least, the insulin would be transported through the cytoplasmic membrane (*exported*). Using the synthetic gene approach a signal sequence would be present on the protein only if the corresponding coding sequence had been incorporated at the time of construction. Sometimes it is desirable for the bacterium to export the protein, in which case a signal sequence is incorporated; with other proteins it may be desirable that they are retained within the cell.

In *E. coli* cells the presence of a signal sequence usually results in export of a protein into the periplasmic space rather than into the growth medium. Unfortunately, many recombinant proteins are rapidly and extensively degraded in the periplasmic space because of the presence there of numerous proteases. In Gram-positive bacteria and eukaryotic microorganisms, signal sequences direct proteins into the growth medium. Filamentous organisms such as fungi or actinomycetes might be particularly favourable for export because of their high surface area to volume ratio.

A small number of proteins, and again insulin is an example, are synthesized as pro-proteins with an additional amino acid sequence which dictates the final three-dimensional structure. In the case of proinsulin, proteolytic attack cleaves out a stretch of 35 amino acids in the middle of the molecule to generate insulin. The peptide that is removed is known as the C chain. The other chains, A and B, remain cross-linked and thus locked in a stable tertiary structure by the disulphide bridges formed when the molecule originally folded as proinsulin. Bacteria have no mechanism for specifically cutting out the folding sequences from pro-hormones and the way of solving this problem is described in a later section.

Another modification which can be made *in vivo* is glycosylation, for example that of β and γ interferons, although the biological role of the sugar residues is not known. Bacteria cannot glycosylate the products of cloned mammalian genes. These non-glycosylated proteins retain their pharmacological activity but their pharmacokinetics and *in vitro* stability may be different. Yeast cells can glycosylate proteins but the pattern of glycosylation may well be different from that seen in the normal host of the gene. Non-glycosylated or wrongly glycosylated proteins may provoke the formation of antibodies following administration.

2.3   Maximizing gene expression

From a commercial point of view it is desirable to maximize the yield of protein in a fermentation. This means maximizing gene expression and important factors are:

1   the number of copies of the plasmid vector per unit cell (*copy number*);
2   the strength of the promoter;
3   the sequences of the RBS and flanking DNA;
4   proteolysis.

The limiting factor in expression is the initiation of protein synthesis. Increasing the copy number of the plasmid increases the number of mRNA molecules transcribed from the cloned gene and this results in increased protein synthesis. Similarly, the stronger the promoter (see Fig. 24.5), the more mRNA molecules are synthesized. The base sequence of the RBS (see Fig. 24.6) and the length and sequence of the DNA between the RBS and the initiating AUG codon are so important that a single base change, addition or deletion can affect the level of translation up to 1000-fold.

Proteolysis does not affect transcription and translation but by degrad-

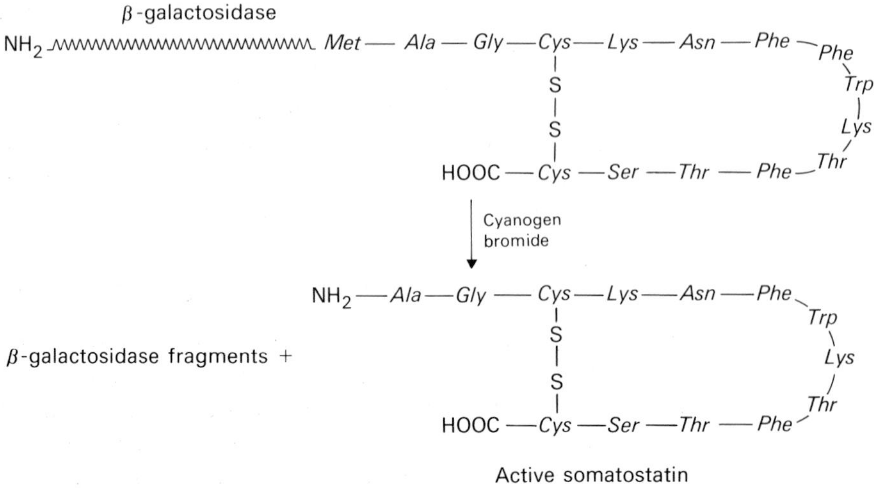

Fig. 24.8 Release of somatostatin from a hybrid protein by cyanogen bromide cleavage. Somatostatin can be purified free of cyanogen bromide and fragments of β-galactosidase.

478   *Chapter 24*

ing the desired product it influences the apparent rate of gene expression. Although proteolysis can be reduced it is difficult to eliminate it completely. One approach is to use protease-deficient mutants and another is to protect the desired protein by fusion to an *E. coli* protein (see below).

Somatostatin was the first human peptide to be synthesized in a bacterial cell. It is only 14 amino acids long and genes for polypeptides of this size are very amenable to direct chemical synthesis. However, small peptides are rapidly degraded in *E. coli* and for this reason the synthetic gene was fused to the 5′ end of the *β*-galactosidase gene. This results in the synthesis of a fusion protein which is relatively stable in *E. coli*. Somatostatin does not contain any methionine residues, so the synthetic gene was constructed in such a way that a methionine was incorporated at the junction of the fusion peptide. By treatment with cyanogen bromide, which breaks proteins into polypeptide fragments at methionine residues, authentic somatostatin could be recovered (Fig. 24.8). Although in this particular instance, and in the case of insulin and *β*-endorphin, the fusion protein contained a remnant of the *E. coli* *β*-galactosidase gene at the N-terminus, other bacterial proteins have been used, for example tryptophan synthetase, *β*-lactamase, etc.

2.4   Choice of cloning host

A number of cloning hosts are in widespread use. *E. coli* is still the most popular organism for initial genetic manipulations and is used for the commercial production of a number of therapeutic proteins. *Bacillus subtilis* has not lived up to its initial promise of high-level protein secretion and interest in it is declining. *Saccharomyces cerevisiae* is widely used but faces competition from recombinant animal cells; progress with the latter has been impressive and high-level expression and secretion systems are available. Good progress has also been made in developing cloning systems for filamentous fungi and actinomycetes, two groups of organisms which have been long used in the production of low molecular weight pharmaceuticals. More recently, there has been growing interest in the development of cloning systems for the more unusual organisms used in the pharmaceutical industry. The advantages and disadvantages of the main cloning systems are shown in Table 24.1.

3   Production of medically important polypeptides and proteins

The overproduction of a wide variety of proteins has now been achieved in *E. coli* and other cloning hosts. Many of these proteins are in clinical trials and, as indicated earlier, a few are already on the market. The current status of many of these proteins is summarized in Table 24.2. The efficacy of many of the proteins listed remains to be determined because until the advent of recombinant DNA technology sufficient quantities were not available to enable clinical trials to be undertaken. It should be noted that clinical efficacy alone is not sufficient. Market size is just as

Table 24.1 Comparison of different organisms as cloning hosts

Organism	Advantages	Disadvantages
E. coli	Ease of manipulation Promoters and gene regulation well understood Easy to culture on large scale Already used in manufacture of insulin, interferon and human somatotrophin	Do not usually get export of proteins into growth medium Over-expressed foreign proteins often form aggregates ('inclusions') of denatured protein Many foreign proteins rapidly degraded Many post-translational modifications do not occur
B. subtilis	Many proteins naturally exported into growth medium Non-pathogenic Easy to culture Some *Bacillus* enzymes excreted at high level ($>5\,g\,l^{-1}$)	Still not much known about gene regulation Good, high-level expression vectors lacking High-level export of heterologous proteins not achieved
S. cerevisiae	Widely used industrial organism which is easy to culture Glycosylates proteins Can get export into growth medium of heterologous proteins High-level expression systems developed Heterologous proteins inside cell do *not* form inclusions	Much still to be learned about control of gene expression Post-translational modifications of proteins not necessarily the same as those in the animal cell
Filamentous fungi	Large surface area to volume ratio should favour protein export Have been used in industrial microbiology for over 40 years	Promoters/gene regulation poorly understood but may be similar to yeast Good expression systems lacking Rheology of fermentations important
Actinomycetes	Large surface area to volume ratio should favour protein export Widely used in industrial microbiology Good expression systems being developed	Promoters/gene regulation still poorly understood Rheology of fermentations important
Mammalian cells	Get export of proteins Get desired post-translational modifications and products not likely to be immunogenic to humans Good expression systems available	Large-scale growth of animal cells costly Great care needed to avoid contamination of cultures

important since it can cost up to £50 million to bring a new drug to the market place and company shareholders expect a good return on their investment.

One of the advantages of recombinant DNA technology is that it enables *analogues* of human proteins to be produced. Thus numerous groups have produced $\alpha-\alpha$ and $\alpha-\beta$ hybrid interferons. Some of these hybrids have altered properties *in vitro* but whether this will translate into a clinical benefit remains to be determined. In some instances the analogues have only a single amino acid change. Thus changing cysteine residue 17 in interferon β to a serine residue yields a protein with improved half-life and *in vitro* stability. Changing methionine residue 358 in α_1-antitrypsin to valine yields a more oxidation-resistant enzyme.

4 Authenticity and efficacy of drugs produced by recombinant DNA technology

To demonstrate the safety and efficacy of any polypeptide drug, regardless of whether it is made by recombinant DNA technology, organic synthesis or extraction from a natural source, a number of quality criteria need to be met. Not only must the protein be produced in accordance with good manufacturing practice but it must also meet specification. Although the absolute specification will vary depending on the identity of the protein, the therapeutic target and the route and period of administration, certain quality guidelines have been adopted by most countries. This core specification is shown in Table 24.3.

Although many of the quality control tests used are designed to assess *purity* they often give data which confirms the *identity* of the protein, e.g. chromatographic behaviour (HPLC), electrophoretic mobility and amino acid composition. However, most of the analytical techniques in current use give no indication of the three-dimensional structure of the protein and hence no indication of biological activity. Thus the absolute specific activity of the protein needs to be determined in a biological test. Determination of the specific activity is particularly important with proteins overproduced in *E. coli*, for such proteins exist in aggregates with nucleic acid often called 'inclusions'. The protein in these aggregates has to be extracted with denaturing agents such as urea, sodium dodecyl sulphate or guanidinium hydrochloride and then renatured, a process akin to recreating native egg white from a meringue.

As indicated earlier, recombinant DNA technology can be used to deliberately produce desired analogues of natural proteins. However, undesirable analogues may also be produced inadvertently during the production process. For example, when the human somatotrophin gene is expressed in *E. coli*, the resultant protein has an additional methionine residue at the N-terminus. Other foreign gene products may or may not carry this additional methionine residue. Recently, methods have been developed for enzymatically removing this N-terminal methionine and for mediating another post-translational modification, C-terminal amidation.

Table 24.2 Current status of selected recombinant proteins

Protein	Size/structure	Expression system	Clinical indications	Comments
Human insulin	Two peptide chains: A, 21 amino acids long, and B, 30 amino acids long	*E. coli*	Juvenile onset diabetes	Approved for sale A and B chains made separately as fusion proteins and joined *in vitro* Compared with animal insulins some undesirable side-effects have been noted
Human somatotrophin	191 amino acids	*E. coli*	Pituitary dwarfism	Approved for sale If useful in treatment of osteoporosis then market size will be much larger Has additional methionine residue at N-terminus, but technology for removing this now available
Interferon-α_{2a} Interferon-α_{2b}	166 amino acids	*E. coli*	Various cancers and viral diseases	Approved for sale Over 80% success in treatment of hairy cell leukemia; success with other cancers lower and more variable Market size may be limited Unpleasant ('flu-like') side-effects
Interferon-γ	143 amino acids, glycosylated	*E. coli*	Chronic granulomatous disease	Approved for sale In clinical trials for treatment of cancer and viral diseases
Tissue plasminogen activator	530 amino acids, glycosylated	*E. coli* Yeast Animal cells	Acute myocardial infarct Pulmonary embolism	Approved for sale Animal cell culture most effective way of producing active enzyme
Relaxin	53 amino acids; insulin-like (two protein chains)	*E. coli*	Facilitates childbirth	Prepares endometrium for parturition and reduces fetal distress Pig relaxin shown to be clinically effective
α-Antitrypsin	394 amino acids, glycosylated	*E. coli* Yeast	Treatment of emphysema	Prevents cumulative damage to lung tissue caused by leucocyte elastase In clinical trials

Product	Structure	Expression system	Application	Status/Notes
Interleukin-2	133 amino acids	*E. coli* / Animal cells	Treatment of cancer	Approved for sale. Very toxic and side-effects severe
Tumour necrosis factor	157 amino acids	*E. coli* / Animal cells	Treatment of cancer	
Human serum albumin	582 amino acids; 17 disulphide bridges	Yeast	Plasma replacement therapy	Normally obtained from plasma but now concern over potential contamination with AIDS virus
Factor VIII	2332 amino acids	Mammalian cells	Treatment of haemophilia	Normally obtained from plasma but now concern over potential contamination with AIDS virus
Factor IX	415 amino acids glycosylated; modified residues	Mammalian cells	Treatment of Christmas disease	Approved for sale. Must be made in mammalian cells since glycosylation and conversion of first 12 glutamate residues to pyroglutamate essential for activity
Erythropoietin	166 amino acids glycosylated	Mammalian cells	Treatment of anaemia associated with dialysis and AZT/AIDS	Approved for sale. Without glycosylation protein is cleared very quickly from plasma
Hepatitis B surface antigen	Monomer has 226 amino acids	Yeast / Mammalian cells	Vaccination	Approved for sale. Monomer self-assembles into structure resembling virus particles
Granulocyte colony stimulating factor	127 amino acids	*E. coli*	Adjunct to cancer chemotherapy	Approved for sale. By stimulating white blood cell formation, aids recovery
Granulocyte-macrophage colony stimulating factor	127 amino acids	*E. coli*	Improved bone marrow transplant	Approved for sale

Table 24.3 Specification of therapeutic proteins to be administered parenterally

Criterion	Appropriate analytical methods
Greater than 95% pure	Gel electrophoresis; PHLC*
Microheterogeneity below specified level	Polyacrylamide gel electrophoresis C- and N-terminal analysis. Amino acid composition
Endotoxin below specified level	*Limulus* amoebocyte lysate method (see also Chapter 18)
Contaminating DNA below specified level (<10 pg dose^{-1})	Hybridization
Toxic chemicals used in purification below specified level	Appropriate methods
IgG below specified limit (if monoclonal antibodies used in purification)	ELISA[†] or RIA[‡]
Absence of microorganisms	Sterility test (see also Chapter 23)

* HPLC, high-performance liquid chromatography.
[†] ELISA, enzyme-linked immunosorbent assay method.
[‡] RIA, radioimmunoassay method.

Another undesirable modification is removal of some amino acids residues from the C-terminus and/or the N-terminus by microbial exoproteases. Care needs to be taken during the fermentation and extraction stages to minimize proteolytic damage and any 'nibbled' molecules should be removed during purification.

5 Future trends

Already twelve recombinant-derived therapeutically useful proteins are being marketed and at least a dozen more are in clinical trials. So, what of the future? There are two disadvantages with developing proteins as therapeutic entities. First, most of them are not active when given orally and parenteral administration is almost *de rigeur*. Some proteins can be administered in other ways; e.g. insulin can be given *per rectum* but this is not a route which is favoured in many countries outside of France and Japan. Other proteins may be active if given sublingually or if administered by aerosol. Clearly, much work needs to be done in developing new dosage forms particularly suited to the administration of proteins. Second, most of the proteins being marketed or currently in clinical trials were obvious candidates for development, e.g. insulin and interferons. Identifying the next generation of therapeutically useful proteins will be much more difficult. There are hundreds of human proteins about which relatively little is known but only a few are likely to be worth developing. For example, a factor which promotes bone growth would have many clinical benefits but the candidate proteins have yet to be identified.

One trend which has become obvious is that many pharmaceutical companies are turning their attention to the application of recombinant DNA technology in the production of small molecules. Microorganisms are widely used in the production of drugs, e.g. antibiotics and steroid transformations. Where the rate-limiting step in production has been identified, cloning the relevant gene could well facilitate synthesis and give increased yields and/or decreased production times. In addition, novel metabolic pathways can be introduced into microorganisms and this could eliminate more conventional but complex production processes involving plants.

6   Glossary

Clone   (Noun) A group of cells all descended from a common ancestor; in genetic engineering, usually refers to a cell carrying a foreign gene. (Verb) To use the techniques of gene manipulation *in vitro* to transfer a gene from one organism to another.

Cloning vector   Plasmid (*vide infra*) into which a foreign gene is placed to ensure its replication in a new host cell.

Exon   Portion of DNA that codes for the final mRNA.

Fusion protein   Covalent linkage of two distinct protein entities, e.g. β-galactosidase and somatostatin. A fusion protein need not retain the different biological properties of its two components.

Genome   Haploid sets of chromosomes with their associated genes.

Intron   An intervening sequence in DNA.

Operon   Two or more genes subject to coordinate regulation by an operator and a repressor.

Plasmid   An extrachromosomal circular DNA molecule in bacteria. Often used as cloning vector.

Promoter   Region of a DNA molecule at which RNA polymerase binds and initiates transcription.

Restriction endonuclease   A deoxyribonuclease which cuts DNA at specific sequences which exhibit twofold symmetry about a point. Name derives from the fact that their presence in a bacterial cell prevents (restricts) the growth of many infecting bacteriophages.

Reverse transcriptase   An enzyme coded by certain RNA viruses which is able to make complementary single-stranded DNA chains from RNA templates and then to convert these DNA chains to double-helical form.

'Sticky-ends'   Complementary single-stranded tails projecting from otherwise double-helical nucleic acid molecules.

'Shotgunning'   Cloning of a complete set of DNA fragments from a particular genome.

Signal sequence   Amino acid sequence in protein, whose function is to direct its final intracellular or extracellular location.

Splicing   (1) Gene splicing: manipulations, the object of which is to attach one DNA molecule to another;

(2) RNA splicing: removal of introns from mRNA precursors.

Vector *See* Cloning vector.

7   Further reading

Bollon A.P. (1984) *Recombinant DNA Products: Insulin, Interferon and Growth Hormone.* Boca Raton, Florida: CRC Press.

Bonnen E.M. & Spiegel R.J. (1984) Interferon-alpha: current status and future promise. *J Biol Resp Mod.* **3**, 580–598.

Courtney M., Jallat S., Tessier L-H., Benavente A., Crystal R.G. & Lecocq J-P. (1985) Synthesis in *E. coli* of alpha$_1$-antitrypsin variants of therapeutic potential for emphysema and thrombosis. *Nature*, **313**, 149–151.

Goeddel D.V., Heyneker H.L., Hoxumi T., Arentzen R., Itakura K., Yansura D.G., Ross M.J., Miozzari G., Crea R. & Seeburg P.H. (1979) Expression in *Escherichia coli* of a DNA sequence coding from human growth hormone. *Nature*, **281**, 544–548.

Goeddel D.V., Kleid D.G., Bolivar F., Heyneker H.L., Yansura D.G., Crea R., Hirose T., Kraszewski A., Itakura K. & Riggs A.D. (1979) Expression in *Escherichia coli* of chemically synthesised genes for human insulin. *Proc Nat Acad Sci USA.* **76**, 106–110.

Goeddel D.V., Yelverton E., Ullrich A., Heyneker H.L., Miozzari G., Holmes W., Seeburg P.J., Dull T., May L., Stebbing N., Crea R., Maeda S., McCandliss R., Sloma A., Tabor J.M., Gross M., Familletti P.C. & Pestka S. (1980) Human leukocyte interferon produced by *E. coli* is biologically active. *Nature*, **287**, 411–416.

Guerigan J.L. (Ed.) (1981) *Insulins, Growth Hormone and Recombinant DNA Technology.* New York: Raven Press.

Guerigan J.L., Bransome E.D. & Outschoorn A.S. (1982) *Hormone Drugs.* Rockville: US Pharmacopeial Convention, Inc.

Itakura K., Hirose, T., Crea R., Riggs A.D., Heyneker H.L., Bolivar F. & Boyer H.W. (1977) Expressions in *Escherichia coli* of a chemically synthesised gene for the hormone somatostatin. *Science*, **198**, 1056–1063.

Old R.W. & Primrose S.B. (1989) *Principles of Gene Manipulation: An Introduction to Genetic Engineering*, 4th edn. Oxford: Blackwell Scientific Publications.

Primrose S.B. (1991) *Molecular Biotechnology*, 2nd edn. Oxford: Blackwell Scientific Publications.

Shine J., Fettes, I., Lan N.C.Y., Roberts J.L. & Baxter J.D. (1980) Expression of cloned β-endorphin gene sequences by *Escherichia coli*. *Nature*, **285**, 456–461.

Tomlinson E. (1991) Impact of the new biologies on the medical and pharmaceutical sciences. *Pharm J.* **247**, 335–344.

25 Additional applications of microorganisms in the pharmaceutical sciences

1   **Introduction**

2   **Pharmaceuticals produced by microorganisms**
2.1   Dextrans
2.2   Vitamins, amino acids and organic acids
2.3   Iron-chelating agents
2.4   Enzymes
2.4.1 Streptokinase and streptodornase
2.4.2 L-Asparaginase
2.4.3 Neuraminidase
2.4.4 β-Lactamases

3   **Use of microorganisms and their products in assays**
3.1   Vitamin and amino acid bioassays

3.2   Phenylketonuria
3.3   Carcinogen and mutagen testing
3.4   Use of microbial enzymes in sterility testing
3.5   Immobilized enzyme technology

4   **Use of microorganisms as models of mammalian drug metabolism**

5   **Applications of microorganisms in the partial synthesis of pharmaceuticals**

6   **Insecticides**

7   **Further reading**

1   Introduction

The exploitation of microorganisms and their products has assumed an increasingly prominent role in the diagnosis, treatment and prevention of human diseases. This chapter will present to the reader areas and horizons, some novel, some well established, in which they find applications in medicine and in the pharmaceutical sciences.

The earliest uses of microorganisms to treat human disease can be traced to the belief that formation of pus in some way drained off noxious humours responsible for systemic conditions. Although the spontaneous appearance of pus in their patients' wounds satisfied most physicians, deliberate contamination of wounds was also practised. Bizarre concoctions of bacteria such as 'ointment of pigs' dung and herb sclerata' were favoured during the Middle Ages. Both early central European and South American civilizations cultivated various fungi for application to wounds. In the nineteenth century, sophisticated concepts of microbial antagonism were developed following Pasteur's experiments demonstrating inhibition of anthrax bacteria by 'common bacteria' simultaneously introduced into the same culture medium. Patients suffering with diseases such as diphtheria, tuberculosis and syphilis were treated by deliberate infection with what were then thought to be harmless bacteria such as staphylococci, *Escherichia coli* and lactobacilli.

Whilst these efforts produced little in the way of clinical reward, they were undoubtedly a necessary prologue to the discovery of antibiotics.

Currently, perhaps the only clinically useful procedure employing microbial antagonism involves the preferential colonization of the urogenital tract by lactobacilli which prevents adherence and subsequent infection with pathogens such as *Candida albicans*.

Some of the most important and widespread uses of microorganisms in the pharmaceutical sciences are the production of antibiotics, vaccines and the use of microorganisms in the recombinant DNA industry. These are described in Chapters 7, 16 and 24. However, there are a variety of other medicinal agents derived from microorganisms including vitamins, amino acids, dextrans, iron-chelating agents and enzymes. Microorganisms as whole or subcellular fractions, in suspension or immobilized in an inert matrix, are employed in a variety of assays. One important example is the use of *Salmonella* mutants in the Ames test for evaluating the potential mutagenicity or carcinogenicity of a chemical. Microorganisms have also been used in the pharmaceutical industry to achieve specific modifications of complex drug molecules such as steroids, in situations where synthetic routes are difficult and expensive to carry out. Recently, there has been much interest in the exploitation of insecticidal agents produced by microorganisms which offer promise as commercial biocontrol agents.

2   Pharmaceuticals produced by microorganisms

2.1   Dextrans

Dextrans are polysaccharides produced by lactic acid bacteria, in particular members of the genus *Leuconostoc* (e.g. *L. dextranicus* and *L. mesenteroides*) following growth on sucrose. These polymers of glucose first came to the attention of industrial microbiologists because of their nuisance in sugar refineries where large gummy masses of dextran clogged pipelines. Dextran is essentially a glucose polymer consisting of $(1 \rightarrow 6)$-α-links of high but variable molecular weight ($15\,000 – 20\,000\,000$; Fig. 25.1). Growth of the dextran producer strain is carried out in large fermenters in media with a low nitrogen but high carbohydrate content. The average molecular weight of the dextrans produced will vary with the strain used. This is important because dextrans for clinical use must have defined molecular weights which will depend on their use. Two main methods are employed for obtaining dextrans of a suitable molecular weight. The first involves acid hydrolysis of very high molecular weight polymers, whilst the second utilizes preformed dextrans of small size which are added to the culture fluid. These appear to act as 'templates' for the polymerization, so that the dextrans are produced with much shorter chain lengths. Once formed, dextrans of the required molecular weight are obtained by precipitation with organic solvents prior to formulation.

Dextrans are produced commercially for use as plasma substitutes (plasma expanders) which can be administered by intravenous injection

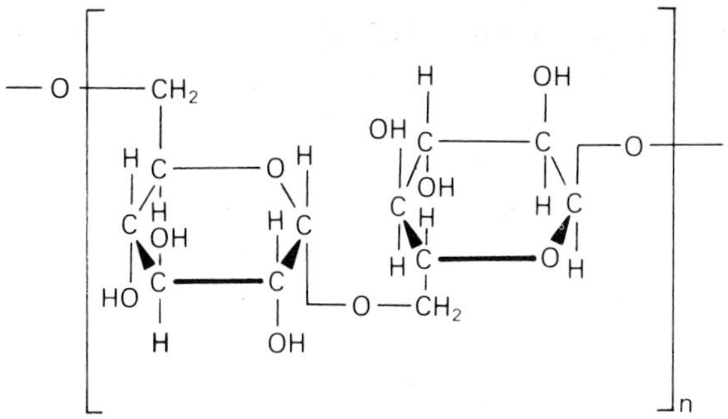

Fig. 25.1 Structure of dextran showing $(1 \rightarrow 6)$-α-linkage.

to maintain or restore the blood volume. They can be used in applications to ulcers or burn wounds where they form a hydrophilic layer which absorbs fluid exudates. Iron dextran injection contains a complex of iron hydroxide with dextrans of average molecular weight between 5000 and 7000, and is used for the treatment of iron-deficiency anaemia in situations where oral therapy is ineffective or impractical. The sodium salt of sulphuric acid esters of dextran, i.e. dextran sodium sulphate, has anticoagulant properties comparable with heparin and is formulated as an injection for intravenous use.

Dextrans for clinical use as plasma expanders must have molecular weights between 40000 (= 220 glucose units) and 300000. Polymers below the minimum are excreted too rapidly from the kidneys, whilst those above the maximum are potentially dangerous because of retention in the body. In practice, infusions containing dextrans of average molecular weights of 40000, 70000 and 110000 are commonly encountered.

Once isolated by the methods described above, dextrans for use as plasma substitutes are dissolved in water to the extent of 6–10% containing 0.9% sodium chloride or 5% glucose. The whole solution is sterilized by heating in an autoclave and samples tested for pyrogenicity (see Chapter 18), toxicity and sterility (see Chapter 23). Since the dextrans are polydisperse in molecular weight, an assessment of the average molecular weight distribution of dextrans for transfusion is usually made. Dextrans which have been chemically cross-linked to render them water-insoluble are widely employed in many chromatographic techniques for fractionation and purification of a wide variety of chemicals, biologicals and pharmaceuticals. Such dextrans are marketed under the name Sephadex.

2.2   **Vitamins, amino acids and organic acids**

Chapter 1 describes how solvents such as butanol and acetone may be produced by bacterial metabolism; these reactions were widely used for industrial production before the advent of the petrochemical industry. A number of chemicals used in medicinal products continue to be produced

Table 25.1 Examples of vitamins, amino acids and organic acids produced by microorganisms

Pharmaceutical	Producer organism	Use
Riboflavin (vitamin B_2)	*Eremothecium ashbyii* *Ashbya gossypii*	Treatment of vitamin B_2 deficiency disease
Cyanocobalamin (vitamin B_{12})	*Propionibacterium freudenreichii* *Propionibacterium shermanii* *Pseudomonas denitrificans*	Treatment of pernicious anaemia
Amino acids, e.g. glutamate, lysine	*Corynebacterium glutamicum* *Brevibacterium flavum*	Supplementation of feeds/food; i.v. infusion fluid constituents
Organic acids Citric acid	*Aspergillus niger*	Effervescent products; sodium citrate used as an anticoagulant; potassium citrate used to treat cystitis
Lactic acid	*Lactobacillus delbrueckii* *Rhizopus oryzae*	Calcium lactate is a convenient source of Ca^{2+} for oral administration; constituent of intraperitoneal dialysis solutions
Gluconic acid	*Gluconobacter suboxydans* *Aspergillus niger*	Calcium gluconate is a source of Ca^{2+} for oral administration; gluconates are used to render bases more soluble, e.g. chlorhexidine gluconate

by fermentation (Table 25.1). These include organic acids such as citric and lactic acids, which also have widespread uses in the food and drink and plastics industries, respectively. Gluconic acid is also used as a metal chelating agent in, for example, detergent products.

Vitamin B_2 (riboflavin) is a constituent of yeast extract and incorporated into many vitamin preparations. Vitamin B_2 deficiency is characterized by symptoms which include an inflamed tongue, dermatitis and a sensation of burning in the feet. In genuine cases of malnutrition, these symptoms will accompany those induced by other vitamin deficiencies. Riboflavin is produced commercially in good yields by the moulds *Eremothecium ashbyii* and *Ashbya gossypii* grown on a protein-digest medium.

Pernicious anaemia was a fatal disease first reported in 1880. It was not until 1926 that it was discovered that eating raw liver effected a remission. The active principle was later isolated and called vitamin B_{12} or cyanocobalamin. It was initially obtained from liver but during the 1960s it was noted that it could be obtained as a by-product of microbial metabolism (Table 25.1). Hydroxycobalamin is the form of choice for therapeutic use and can be derived either by chemical transformation of cyanocobalamin or directly as a fermentation product.

Amino acids find applications as ingredients of infusion solutions for parenteral nutrition and individually for treatment of specific conditions. They are obtained either by fermentation processes similar to those used for antibiotics or in cell-free extracts employing enzymes isolated from bacteria (Table 25.1). Details of the many and varied processes reported

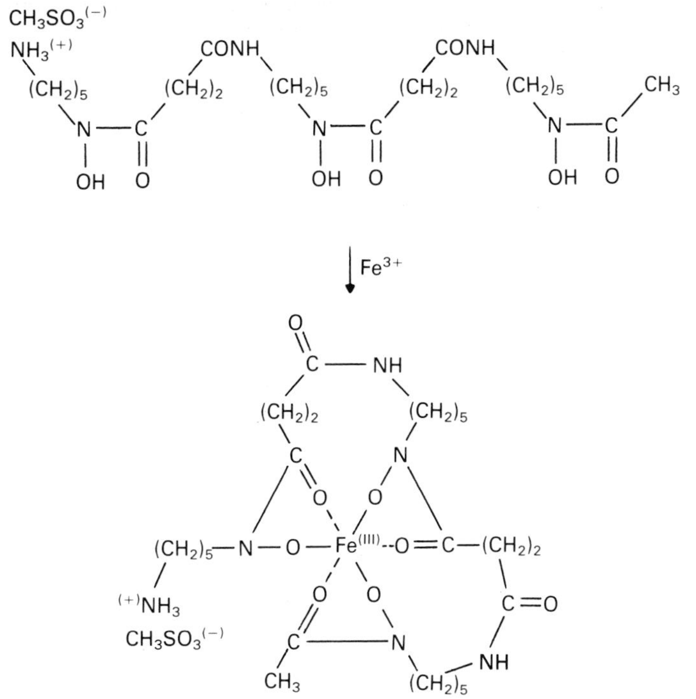

Fig. 25.2 Structure of desferrioxamine B (Desferal) and its corresponding iron chelate.

in the literature will be found in the appropriate references at the end of the chapter.

2.3   **Iron-chelating agents**

Growth of many microorganisms in iron-deficient growth media results in the secretion of low molecular weight iron-chelating agents called siderophores, which are usually phenolate or hydroxamate compounds. The therapeutic potential of these compounds has generated considerable interest in recent years. Uncomplicated iron deficiency can be treated with oral preparations of ferrous (iron II) sulphate but such treatment is not without hazard and iron salts are common causes of poisoning in children. The accidental consumption of around 3 g of ferrous sulphate by a small child leads to acidosis, coma and heart failure amongst a variety of other symptoms which, if untreated, are fatal. Desferrioxamine B (Fig. 25.2), the deferrated form of a siderophore produced by *Streptomyces pilosus* is a highly effective antidote for the treatment of acute iron poisoning. Desferrioxamine is available commercially as Desferal and owes its effectiveness both to its high affinity for ferric iron (its binding constant is in excess of 10^{30}) and because the iron–desferrioxamine complex is highly water-soluble and is readily excreted through the kidneys. In haemolytic anaemias such as thalassaemia, Desferal is used together

491   *Additional pharmaceutical applications of microorganisms*

with blood transfusions to maintain normal blood levels of free iron and haemoglobin. Desferal is prepared as a sterile powder for use as an injection, but it is also administered orally in acute iron poisoning to remove unabsorbed iron from the gut.

The important role played by iron availability during infections in vertebrate hosts has only been recognized relatively recently. The ability of the host to withhold growth-essential iron from microbial and, indeed, neoplastic invaders whilst retaining its own access to this metal, has lead to suggestions that microbial iron chelators or their semisynthetic derivatives may be of use in antimicrobial and anticancer chemotherapy. Preliminary work has shown some encouraging results. The bacterial siderophores parabactin and compound II secreted by *Paracoccus denitrificans* have been shown to inhibit the growth of leukaemia cells in culture and in experimental animals. They also appear capable of inhibiting the replication of RNA viruses.

Siderophores like Desferal may, therefore, find increasing applications not only in the treatment of iron poisoning and iron-overloaded disease states but also as chemotherapeutic agents.

2.4	**Enzymes**

Enzymology is an important branch of biochemistry and as a result of much research a lot is now known about the nature, catalytic activity and sequence of degradative and synthetic reactions in which enzymes act as catalysts in the living cell. This section is concerned with those enzymes that are used therapeutically. The applications of microbially derived enzymes in diagnostic assays and for inactivation of antibiotics in sterility testing will be dealt with in sections 3.3 and 3.4.

2.4.1	*Streptokinase and streptodornase*

Mammalian blood will clot spontaneously if allowed to stand; however, on further standing, this clot may dissolve as a result of the action of a proteolytic enzyme called plasmin. Plasmin is normally present as its inactive precursor, plasminogen. Certain strains of streptococci were found to produce a substance which was capable of activating plasminogen, a phenomenon which suggested a potential use in liquefying blood clots. This substance was isolated, found to be an enzyme and called streptokinase. This enzyme is an effective agent in the treatment of deep vein thromboses, acute arterial occlusions and acute pulmonary emboli. It has recently been tested for its ability to promote thrombolysis in acute myocardial infarction following intracoronary administration. The results of the clinical trials suggested that the improved survival rate over that of conventional therapy was related to a reduction in infarct size and preservation of left ventricular function. Reduction in infarct size, however, appears to be highly dependent upon speed of administration, the most effective results being achieved within 1 hour of symptom onset.

A second enzyme, streptodornase, present in streptococcal culture filtrates was observed to liquefy pus. Streptodornase is a deoxyribonuclease which breaks down deoxyribonucleoprotein and DNA, both constituents of pus, with a consequent reduction in viscosity. Streptokinase and streptodornase together have been used to facilitate drainage by liquefying blood clots and/or pus in the chest cavity. The combination can also be applied topically to wounds which have excessive suppuration.

Streptokinase and streptodornase are isolated following growth of non-pathogenic streptococcal producer strains in media containing excess glucose. They are obtained as a crude mixture from the culture filtrate and can be prepared relatively free of each other. They are commercially available as either streptokinase injection or as a combination of streptokinase and streptodornase.

2.4.2   L-Asparaginase

L-Asparaginase, an enzyme derived from *E. coli* or *Erwinia carotovora*, has been employed in cancer chemotherapy where its selectivity depends upon the essential requirement of some tumours for the amino acid L-asparagine. Normal tissues do not require this amino acid and thus the enzyme is administered with the intention of depleting tumour cells of asparagine by converting it to aspartic acid and ammonia. Whilst L-asparaginase showed promise in a variety of experimentally induced tumours, it is only useful in humans for the treatment of acute lymphoblastic leukaemia, although it is sometimes used for myeloid leukaemia.

2.4.3   Neuraminidase

Neuraminidase derived from *Vibrio cholerae* has been used experimentally to increase the immunogenicity of tumour cells. It appears capable of removing *N*-acetylneuraminic (sialic) acid residues from the outer surface of certain tumour cells, thereby exposing new antigens which may be tumour specific together with a concomitant increase in their immunogenicity. In laboratory animals administration of neuraminidase-treated tumour cells was found to be effective against a variety of mouse leukaemias. Preliminary investigations in acute myelocytic leukaemia patients has suggested that treatment of the tumour cells with neuraminidase in combination with conventional chemotherapy may increase remission rates.

2.4.4   β-Lactamases

β-Lactamase enzymes, whilst being a considerable nuisance because of their ability to confer bacterial resistance by inactivating penicillins and cephalosporins (see Chapter 10), are useful in the sterility testing of certain antibiotics (see section 3.4) and, prior to culture, in inactivating various β-lactams in blood or urine samples in patients undergoing

493   *Additional pharmaceutical applications of microorganisms*

therapy with these drugs. One other important therapeutic application is in the rescue of patients presenting symptoms of a severe allergic reaction following administration of a β-lactamase-sensitive penicillin. In such cases, a highly purified penicillinase obtained from *Bacillus cereus* is administered either intramuscularly or intravenously and in combination with other supportive measures such as adrenaline or antihistamines.

3   Use of microorganisms and their products in assays

Microorganisms have found widespread uses in the performance of bioassays to determine the concentration of certain compounds in complex chemical mixtures, in the diagnosis of certain diseases and in the testing of chemicals to determine potential mutagenicity or carcinogenicity. A comparatively recent development is the use of immobilized enzymes for *in vivo* and *in vitro* monitoring purposes as part of enzyme electrodes.

The use of microorganisms in the bioassay of antibiotics was discussed in Chapter 8 and will thus not be considered here.

3.1   Vitamin and amino acid bioassays

The principle of microbiological bioassays for growth factors such as vitamins and amino acids is quite simple. Unlike antibiotic assays (see Chapter 8) which are based on studies of growth inhibition, these assays are based on growth exhibition. All that is required is a culture medium which is nutritionally adequate for the test microorganism in all essential growth factors except the one being assayed. If a range of limiting concentrations of the test substance is added, the growth of the test microorganism will be proportional to the amount added. A calibration curve of concentration of substance being assayed against some parameter of microbial growth (such as cell dry weight) can be plotted, and from this the concentration of growth factor in the unknown solution can be determined. One example of this is the assay for pyridoxine (vitamin B_6) which can be assayed using a pyridoxine-requiring mutant of the mould *Neurospora*. Lactic acid bacteria have extensive growth requirements and are often used in bioassays. It is possible to assay a variety of different growth factors with a single test organism simply by preparing a basal media with different growth-limiting nutrients. Table 25.2 summarizes some of the vitamin and amino acid bioassays currently available. In practice, only vitamins are assayed by bioassay procedures, because most amino acids are currently determined chemically.

3.2   Phenylketonuria

Phenylketonuria (PKU) is an inborn error of metabolism by which the body is unable to convert surplus phenylalanine (PA) to tyrosine for use in the biosynthesis of, for example, thyroxine, adrenaline and noradrenaline. This condition is readily diagnosed by the detection of

Table 25.2 Some examples of microorganisms used as bioassays for vitamins and amino acids

Assay microorganism	Vitamin or amino acid
Lactobacillus casei	Biotin
L. arabinosus	Calcium pantothenate
L. leichmannii	Cyanocobalamin
L. casei	Folic acid
Saccharomyces uvarum	Inositol
L. arabinosus	Nicotinic acid
Acetobacter suboxydans	Pantothenol
L. casei	Pyridoxal
Neurospora crassa or	Pyridoxine
S. carlsbergiensis	
L. casei	Riboflavine
L. viridans	Thiamine

phenylpyruvic acid (PPVA), a phenylketone, in urine and elevated levels of both PA and PPVA in blood. Control of PKU can be achieved simply by resorting to a low PA-containing diet. Failure to diagnose PKU, however, will result in mental deficiency.

In 1968, the UK Medical Research Council Working Party on PKU recommended the adoption of the Guthrie test as a convenient method for screening newborn infants. This assay employs *B. subtilis* as the test organism. In minimal culture media, growth of this bacterium is inhibited by β-2-thienylalanine. This growth inhibition is reversed in the presence of phenylalanine or phenylpyruvic acid. The use of filter-paper discs impregnated with blood or urine permits the detection of elevated levels of PA and PPVA. The test can be quantitated by the measurement of the diameter of the growth zone around the filter-paper disc and comparing it with a calibration curve constructed from known concentrations of PA or PPVA.

3.3   Carcinogen and mutagen testing

The Ames test is used to screen a wide variety of chemicals for potential carcinogenicity or as potential cancer chemotherapeutic agents. The test enables a large number of compounds to be screened rapidly by examining their ability to induce mutagenesis in several specially constructed bacterial mutants derived from *Salmonella typhimurium*. The test strains contain mutations in the histidine operon so that they cannot synthesize the amino acid histidine. Two additional mutations increase further the sensitivity of the system. The first is a defect in their lipopolysaccharide structure (Chapter 1) such that they are in fact deep rough mutants possessing only 2-keto-3-deoxyoctonate (KDO) linked to lipid A. This mutation increases the permeability of the mutants to large hydrophobic molecules. The second mutation concerns a DNA excision repair system which prevents the organism repairing its damaged DNA following exposure to a mutagen.

The assay method involves treatment of a large population of these mutant tester strains with the test compound. This can be carried out by incorporating both the test strain and test compound in molten agar (at 45°C), which is then poured onto a minimal glucose agar plate. Alternatively, the mutagens can be applied to the surface of the top agar as a liquid or as a few crystals. The medium used for the top agar contains a trace of histidine which permits all the bacteria on the plate to undergo several divisions, since for many mutagens some growth is a necessary prerequisite for mutagenesis to occur. After incubation for 2 days at 37°C the number of revertant colonies can be counted and compared with control plates from which the test compound has been omitted. Each revertant colony is assumed to be derived from a cell which has mutated back to the wild type and thus can now synthesize its own histidine.

A further refinement to the Ames test permits screening of agents which require metabolic activation before their mutagenicity or carcinogenicity is apparent. This is achieved by incorporating into the top agar layer, along with the bacteria, homogenates of rat (or human) liver whose activating enzyme systems have been induced by exposure to polychlorinated biphenyl mixtures. This test is sometimes referred to as the *Salmonella*/microsome assay since the fraction of liver homogenate used, called the S9 fraction, contains predominantly liver microsomes.

It is important to realize that this test is flexible and is still undergoing modification and development. Almost all the known human carcinogens have been tested and shown to be positive. These include agents such as β-naphthylamine, cigarette smoke condensates, aflatoxin B and vinylchloride, as well as drugs used in cancer treatment such as adriamycin, daunomycin and mitomycin C. Whilst the test is not perfect for the prediction of mammalian carcinogenicity or mutagenicity and for making definitive conclusions about potential toxicity or lack of toxicity in humans, it nevertheless represents a significant advance providing useful information rapidly and cheaply.

3.4   **Use of microbial enzymes in sterility testing**

Whole microorganisms, and in particular bacterial spores, are commonly used in the validation of sterilization cycles. Bacterially derived antibiotic-inactivating enzymes have also found use in sterility testing. Sterile pharmaceutical preparations must be tested for the presence of fungal and bacterial contamination before use (see Chapters 18 and 23). If the preparation contains an antibiotic, it must be removed or inactivated. Membrane filtration is the usual recommended method. However, this technique has certain disadvantages. Accidental contamination is a problem, as is the retention of the antibiotic on the filter and its subsequent liberation into the nutrient medium.

Enzymic inactivation of the antibiotic (see also Chapter 10) prior to testing would provide an elegant solution to this problem. Currently, the only pharmacopoeial method permitted is that of using an appropriate β-lactamase to inactivate penicillins and cephalosporins. Other antibiotics which are susceptible to inactivating enzymes are chloramphenicol (by chloramphenicol acetyltransferase) and the aminoglycosides, e.g. gentamicin, which can be inactivated by phosphorylation, acetylation or adenylylation. A method for acetylating and consequently inactivating aminoglycosides prior to testing and using 3-N-acetyltransferase (an enzyme with wide substrate specificity) in combination with acetyl coenzyme A has been described (see references). This method has not yet been adopted by the pharmacopoeias.

3.5 Immobilized enzyme technology

The therapeutic uses of microbially derived enzymes have already been examined (section 2.4). However, enzymes also form the basis of many diagnostic tests used in clinical medicine. For example, glucose oxidase, an enzyme used in blood glucose analysis, is obtained commercially from *Aspergillus niger*. Future development and improvement of such diagnostic tests is likely to involve the immobilization of enzymes in enzyme electrodes. Several types of glucose oxidase electrodes have been developed, although none is yet in clinical use. One basic system employs glucose oxidase layered over a platinum electrode. As the reaction proceeds and oxygen is consumed, i.e. glucose + oxygen → gluconic acid + hydrogen peroxide, the reduction in oxygen levels is detected by the underlying electrode. However, problems of enzyme inactivation *in vivo*, competition between glucose and oxygen in body fluids and calibration have prevented the adoption of this system as an implantable glucose monitor in diabetic patients. However, there are currently a number of major research efforts in this area and it is likely that biosensors employing immobilized enzymes which are potentially useful for monitoring many substances of clinical importance will become readily available in the not too distant future.

4 Use of microorganisms as models of mammalian drug metabolism

The safety and efficacy of a drug must be exhaustively evaluated prior to its approval for use in the treatment of human diseases. Investigations of the manner in which a drug is metabolized are extremely valuable since they provide information on its mode of action, why it exhibits toxicity and how it is distributed, excreted and stored in the body. Traditionally, drug metabolism studies have relied on the use of animal models and to a lesser extent, liver microsomal preparations, tissue culture and perfused organ systems. Each of these models has certain advantages and disadvantages. Animals in particular are expensive to purchase and maintain

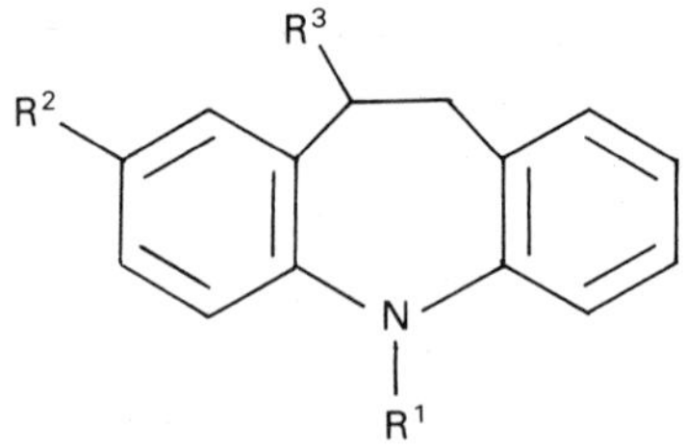

Imipramine	$R^1=(CH_2)_3N(CH_3)_2; R^2=R^3=H$
Desipramine	$R^1=(CH_2)_3NHCH_3; R^2=R^3=H$
2-hydroxyimipramine	$R^1=(CH_2)_3N(CH_3)_2; R^2=OH; R^3=H$
10-hydroxyimipramine	$R^1=(CH_2)_3N(CH_3)_2; R^2=OH; R^3=H$
Iminodibenzyl	$R^1=R^2=R^3=H$
Imipramine-N-oxide	$R^1=(CH_2)_3N(CH_3)_2; R^2=R^3=H$

Fig. 25.3 Structure of imipramine and its metabolites.

and there is considerable pressure from animal welfare groups to curb the use of animals in scientific research.

The use of microbial systems as *in vitro* models for drug metabolism in humans has been proposed since there are many similarities between certain microbial enzyme systems and mammalian liver enzyme systems. The major advantages of using microorganisms is their ability to produce significant quantities of metabolites that would otherwise be difficult to obtain from animal systems or by chemical synthesis, and the considerable reduction in operating costs compared with animal studies.

Microbial drug metabolism studies are usually carried out by firstly screening a large number of microorganisms for their ability to metabolize a drug substrate. The organism is usually grown in a medium such as peptone glucose in flasks which are shaken to ensure good aeration. Drugs as substrates are generally added after 24 hours of growth and are then sampled for the presence of metabolites at intervals up to 14 days after substrate addition. Once it has been determined that a microorganism can metabolize a drug, the whole process can be scaled up for the production of large quantities of metabolites for the determination of their structure and biological properties.

As an example of this the metabolism of the antidepressant drug imipramine can be considered. In mammalian systems, this is metabolized to five major metabolites: 2-hydroxyimipramine, 10-hydroxyimipramine, iminodibenzyl, imipramine-*N*-oxide and desipramine (Fig. 25.3). For microbial metabolism studies, a large number of fungi are screened, from which several are chosen for the preparative scale production of imipramine metabolites. *Cunninghamella blakesleeana* produces the hydroxylated metabolites 2-hydroxyimipramine and 10-hydroximipramine;

Aspergillus flavipes and *Fusarium oxysporum* f. sp. *cepae* yield the N-oxide derivative and iminodibenzyl, respectively; whilst the pharmacologically active metabolite desipramine is produced by *Mucor griseocyanus* together with the 10-hydroxy and N-oxide metabolites. By scaling up this procedure, significant quantities of the metabolites that are formed during mammalian metabolism can be obtained.

Microorganisms thus have considerable potential as tools in the study of drug metabolism. Whilst they cannot completely replace animals they are extremely useful as predictive models for initial studies.

5   Applications of microorganisms in the partial synthesis of pharmaceuticals

Whole microbial cells as well as microbially derived enzymes have played a significant role in the production of novel antibiotics. The potential of microorganisms as chemical catalysts, however, was first fully realized in the synthesis of industrially important steroids. These reactions have assumed increasing importance following the discovery that certain steroids such as hydrocortisone have anti-inflammatory activity, whilst derivatives of the steroidal sex hormones are useful as oral contraceptive agents.

Since steroid hormones can only be obtained in small quantities directly from mammals, attempts were made to synthesize them from plant sterols which can be obtained cheaply and economically in large quantities. However, all adrenocortical steroids are characterized by the presence of an oxygen at position 11 in the steroid nucleus. Thus, although it is easy to hydroxylate a steroidal compound it is extremely difficult to obtain site-specific hydroxylation, so that many of the routes used for synthesizing the desired steroid are lengthy, complex and consequently expensive. This problem was overcome when it was realized that many microorganisms are capable of performing limited oxidations with both stereo- and regio-specificity. Thus, by simply adding a steroid to growing cultures of the appropriate microorganism, specific site-directed chemical changes can be introduced into the molecule. In 1952, the first commercially employed process involving the conversion of progesterone to 11α-hydroxyprogesterone by the fungus *Rhizopus nigricans* was introduced (Fig. 25.4). This reaction is an important stage in the manufacture of cortisone and hydrocortisone from more readily available steroids. Table 25.3 gives several other examples of microbially directed oxidations employed in the manufacture of steroidal drugs.

More recent advances involving the employment of microorganisms in biotransformation reactions utilize immobilized cells (both living and dead). Immobilization of microbial cells, usually by entrapment in a polymer gel matrix, has several important advantages. Whole microbial cells contain complex multistep enzyme systems and there is therefore no longer a need to extract enzymes or enzyme systems which may be inactivated during purification procedures. It also increases the stability of

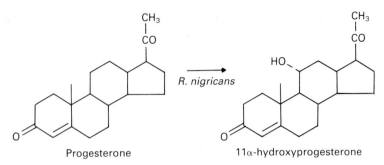

Fig. 25.4 Conversion of progesterone to 11α-hydroxyprogesterone by *Rhizopus nigricans*.

Table 25.3 Examples of biological transformations of steroids

Starting material	Product	Type of reaction
Progesterone	11α-Hydroxyprogesterone	Hydroxylation
Compound S*	Hydrocortisone	Hydroxylation
11α-Hydroxyprogesterone	Δ'-11α-Hydroxyprogesterone	Dehydrogenation
Hydrocortisone	Prednisolone	Dehydrogenation
Cortisone	Prednisone	Dehydrogenation

* Derived from diosgenin by chemical transformation.

membrane-associated enzymes which are unstable in the solubilized state, as well as permitting the conversion of water-insoluble compounds like steroids in two-phase water–organic solvent systems. Immobilized micro-organisms have now been used with considerable success in the partial synthesis of steroids and antibiotics and in the production of the antiviral compound adenine arabinoside. This last manufacturing process involves a transglycosylation reaction to convert uracil arabinoside and adenine into adenine arabinoside by *Enterobacter aerogenes* entrapped in a urethane prepolymer, a reaction which takes place over 30 days at 60°C and in a buffer containing 40% dimethylsulphoxide.

In the antibiotics industry, the hydrolysis of benzylpenicillin to give 6-aminopenicillanic acid by the enzyme penicillin acylase is an important stage in the synthesis of many clinically useful penicillins (see Chapters 5 and 7). The combination of genetic engineering techniques to produce hybrid microorganisms with significantly higher acylase levels, together with their entrapment in gel matrices (which appears to improve the stability of the hybrids), has resulted in considerable increases in 6-aminopenicillanic acid yields.

6   Insecticides

Like animals, insects are susceptible to infections which may be caused by viruses, fungi, bacteria or protozoa. The use of microorganisms to spread diseases to particular insect pests offers an attractive method of bio-control, particularly in view of the ever-increasing incidence of resistance to chemical insecticides. However, any microorganism used in this way must be highly virulent, specific for the target pest but non-pathogenic to animals, man or plants. It must be economical to produce, stable on storage and preferably rapidly acting. Bacterial and viral pathogens have so far shown the most promise.

Perhaps the best studied, commercially available insecticidal agent is *B. thuringiensis*. This insect pathogen contains two toxins of major importance. The δ-endotoxin is a protein present inside the bacterial cell as a crystalline inclusion within the spore case. This toxin is primarily active against the larvae of lepidopteran insects (moths and butterflies). To be effective, this toxin must be ingested, whereupon it dissolves in the alkaline conditions of the larval gut and is partially hydrolysed. This modified protein attacks the insect gut wall so that the gut contents diffuse into the bloodstream inducing a general paralysis, death ultimately resulting from tissue invasion by commensal microorganisms. The commercially available preparations of *B. thuringiensis* are spore-crystal mixtures prepared as dusting powders. They are used primarily to protect commercial crops from destruction by caterpillars and are surprisingly non-toxic to man and animals. Although the currently available preparation has a rather narrow spectrum of activity, a variant *B. thuringiensis* strain has recently been isolated and found to produce a different δ-endotoxin with activity against coleopteran insects (beetles) rather than lepidopteran or dipteran (flies and mosquitoes) insects.

The second *B. thuringiensis* toxin, the β-exotoxin has a much broader spectrum encompassing the Lepidoptera, Coleoptera and Diptera. It is an adenine nucleotide, probably an ATP analogue which acts by competitively inhibiting enzymes which catalyse the hydrolysis of ATP and pyrophosphate. This compound, however, is toxic when administered to mammals so that commercial preparations of the *B. thuringiensis* δ-endotoxin are obtained from strains which do not produce the β-exotoxin.

Other insect pathogens are currently being evaluated for activity against insects which are vectors for diseases such as malaria and sleeping sickness, as well as those which cause damage to crops. Viruses may well have the greatest potential for insect control since they are host-specific and highly virulent, and one infected insect can release vast numbers of virus particles into the environment. They have already been used with considerable success against the spruce sawfly and pine moth.

7   Further reading

Ames B.N., McCann J. & Yamasaki E. (1975) Methods for detecting carcinogens and mutagens with the *Salmonella*/mammalian microsome mutagenicity test. *Mut Res.* **31**, 347–364.

Bergeron R.J., Cavanaugh P.F., Kline S.J., Hughes R.C., Elliott, G.T. & Porter

C.W. (1984) Antineoplastic and antiherpetic activity of spermidine catecholamide iron chelators. *Biochem Biophys Res Commun.* **121**, 848–854.

Breeze A.S. & Simpson A.M. (1982) An improved method using acetyl-coenzyme A regeneration for the enzymic inactivation of aminoglycosides prior to sterility testing. *J Appl Bacteriol.* **53**, 277–284.

Burgess H.D. (1981) *Microbial Control of Pests and Plant Diseases 1970–1980.* New York: Academic Press.

Clark A.M., McChesney J.D. & Hufford C.D. (1985) The use of micro-organisms for the study of drug metabolism. *Med Res Rev.* **5**, 231–253.

Conference (1973) Streptokinase in clinical practice. *Postgrad Med J.* **49**, 3–142.

Data J.L. & Nies A.S. (1974) Dextran 40. *Ann Int Med.* **81**, 500–504.

Davis G., Green, M.J. & Hill H.A.O. (1986) Detection of ATP and creatinine kinase using an enzyme electrode. *Enzyme Microb Tech.* **8**, 349–352.

Demain A.L., Somkuti G.A., Hunter-Cevera J.C. & Rossmore H.W. (1989) *Novel Microbial Products for Medicine and Agriculture.* Amsterdam: Elsevier.

Doenicke A., Grote B. & Lorenz W. (1977) Blood and blood substitutes. *Br J Anaesth.* **49**, 681–688.

Fukui S. & Tanaka A. (1982) Immobilized microbial cells. *Annu Rev Microbiol.* **36**, 145–172.

Gruppo Italiano per Studio della Streptokinasi Nell'infarto Miocardio (GISSI) (1986) Effectiveness of intravenous thrombolytic treatment in acute myocardial infarction. *Lancet*, **i**, 397–401.

Halliday J.W. & Bassett M.L. (1980) Treatment of iron storage disorders. *Drugs*, **20**, 207–215.

Higgins I.J., Best D.J. & Jones J. (1985) *Biotechnology, Principles and Applications.* Oxford: Blackwell Scientific Publications.

Jones R.L. & Grady R.W. (1983) Siderophores as antimicrobial agents. *Eur J Clin Microbiol.* **2**, 411–413.

Kier D.K. (1985) Use of the Ames test in toxicology. *Reg Toxicol Pharmacol.* **5**, 59–64.

Mackowiack P.A. (1979) Clinical uses of micro-organisms and their products. *Am J Med.* **67**, 293–306.

Price S.A. (1967) Assay of vitamins and amino acids. In *Progress in Microbiological Techniques* (Ed. C.H. Collins), pp. 55–84. London: Butterworth & Co.

Ricketts C.R. (1966) Molecular composition of infusion dextran. *Br Med J.* **2**, 1423–1426.

Scientific American (1981) Issue on industrial microbiology, vol. 245, No. 3. (An excellent series of papers describing the manufacture by microorganisms of products useful to mankind.)

Smith R.V. & Rosazza J.P. (1975) Microbial models of mammalian metabolism. *J Pharm Sci.* **64**, 1737–1759.

Turner A.P.F. & Pickup J.C. (1985) Diabetes mellitus: biosensors for research and management. *Biosensors*, **1**, 85–115.

Verall M.S. (1985) *Discovery and Isolation of Microbial Products.* Chichester: Ellis Horwood.

Weinberg E.D. (1984) Iron witholding: a defence against infection and neoplasia. *Physiol Rev.* **64**, 65–107.

White R.J. (1982) Microbiological models as screening tools for anticancer agents: potentials and limitations. *Annu Rev Microbiol.* **36**, 415–433.

Index

Abdominal sepsis 138
Abortion 36
  caused by infection 37
  due to toxic moulds 54
Abscesses 90
  brain 149–50
  lung 143
Acetone–butanol fermentation *21*
Acetylmethylcarbinol production *21*, 30
Acid-fast organisms 40, 302
Acinetobacter spp. 38
Acremonium chrysogenum 161, 163
Acridine dyes 255
  mode of action 291, *292*
Acriflavine 255
  mode of action 291
Acrosoxacin 125, *126*
Actinomycetes, as cloning hosts 479, *480*
Acute cardiac beriberi 56
Acycloguanosine (acyclovir) 78, *130*
Acyclovir (acycloguanosine) 78, *130*
Adenine arabinoside 500
Adenosine arabinoside 129
Adenosine triphosphatase 290, *292*
Adenoviruses *62*, 64, *71*, 76
  oncogenic 79
Administration sets, sterilization *443*
Adriamycin 496
Aflatoxins 56–7, 371
Agar 18
  attacked by microbes 373
Agar diffusion assays 168–73
Agglutinins 86
Aggregates, bacterial 4, 16
AIDS *73*, 80, 321
  see also HIV virus
Air
  compressed 356
  contamination 362, 368
  disinfection 281–2, 356
  filtration 355–6, 425
  microbial content 353–5, 362
  sampling microorganisms in 281–2, 354
  supply, clean areas 452
  UV radiation 356
D-Alanyl-D-alanine, similar to β-lactams 194, *195*, 196–7
Alcaligenes spp. 360
Alcohols
  beer and wine production 49–50
  as disinfectants 240, 242–3, *292*
  as preservatives 243
  virucidal effect 64
Aldehydes, as disinfectants 64, 237, *240*, 243–5
Alexidine 245–6
Algae, blue-green 3, 371
Alimentary toxic aleukia (ATA) 54, 55–6
Alimentary tract

entry for disease 85
exit route for pathogens 95
gastrointestinal infections 146–7
toxic effects of pathogens 89
Allergy, penicillin 110–11, 139
Aluminium
  antimicrobial activity 249
  in vaccines 347
Amanita spp., poisoning from 53
Amantadine hydrochloride 129, *130*
Ames TDA assays 186, 187
Ames test 488, 495–6
6-β-Amidinopenicillanic acid *101*, *102*, 104
Amikacin *114*–15, *136*, 224
  assay for 174–5
Aminacrine 255
Amino acid decarboxylases 27
Amino acid oxidases 24
Amino acids
  bioassays 494, *495*
  from microorganisms 489–91
  metabolism 21–7
7-Aminocephalosporanic acid (7-ACA) 104, *106*
Aminoglycosides 113–15
  for cystic fibrosis 143
  immunoassays for 183–7
  inactivation 175, 223–4, 497
  lipid solubility 137
  mode of action *190*, 197–9
  radioenzymatic assays for 174–6
  resistance to 223–*4*
  toxicity 139
  uptake by bacteria 199
3-Aminomonobactamic acid (3-AMA) 109–*10*
6-Aminopenicillanic acid 500
  production 100–1
Aminopeptidases 23
p-Aminosalicylic acid (PAS) 113, 123, *124*
Ammonium compounds, virucidal effects 64
Amoebiasis, intestinal, treatment 113
Amoxycillin *101*, *102*, *136*, 140, 142
  /clavulanic acid combination 105, 132
Amphotericin *120*
  B 47, 119
    mode of action 205
    toxicity risk 138
Amphotericin methyl ester *120*
Ampicillin *101*, *102*, 103, 104, *136*, 150
  candidiasis development 140
  Klebsiella aerogenes resistance 151
  neutralizing agent for *466*
  for urinary infections 145
Ampicillin esters *101*, *102*, 103
Ampicillins, substituted *101*, *102*
Amylases 90
Anaemia
  haemolytic 491
  pernicious 490

Anilides, sites of antibacterial action *292*
Animals
  disease carriers 84, 95
  inoculation with viruses 75
  model for HBV testing 276–7
  reservoir for resistant bacteria 212
Ankylosing spondylitis 329
Antacids, contaminated 370
Antagonism, drug–drug 132
Anthrax 35, 60, 65, *339*
Antibiotics
  aminoglycoside aminocyclitol 113–15
  antifungal 119–*20*, 128–9
  assays 166–8
    HPLC 176–83
    immunoassays 183–7
    microbiological 168–74
    radioenzymatic 174–6
  clinical use 141–50
  combinations 131–2, 138–9
  definition 99
  glycopeptide 116–17
  inactivation before sterility testing 496–7
  β-lactam 100–11, 154, *190*, 194–7
  macrolide 115–16
  manufacture 154–64
    benzylpenicillin 154–63
    griseofulvin 155
    streptomycin 155
  miscellaneous antibacterial 117–19
  mode of action 189
    on cell walls 189–97
    on chromosome function *190*, 205–6
    on cytoplasmic membranes *190*, 204–6
    on folate metabolism *190*, 202–4
    on ribosome function *190*, 197–200
  peri-operative 140
  permeability, reduction of 226–8
  policies for use 150–3
  polypeptide 116
  potency 167
  prescription 134–5
    policies for 150–3
  principles of use 135–41
  resistance to *see* Resistance, drug
  rifamycin 112–13
  sources 100
    microorganisms 35, 50, 99
    semisynthetic 100, 154, 499
    synthetic 100, 120–9
  tetracycline *111*
  transport processes 227–8
  see also Drugs; Pharmaceutical products; *and*
    individual antibiotics
Antibodies 82, 311
  monoclonal
    culture 315–17
    uses 317
  response to immunization *314*
Antifungal antibiotics 119–*20*, 128–9
Antigen-presenting cells (APCs) 322
Antigens 311
  in bacterial cells 311–12
  epitopes of 313, 315
  HLA 328–9

Antimicrobial agents
  antibiotic *see* Antibiotics
  natural body-produced 307
  non-antibiotic 231–2
    antibacterial 233–5
    antifungal *235*
    antiviral 235–6
    compounds used 238–56
    definitions 258–9
    disinfection policies 256–7, 303–4
    dynamics of disinfection 259–62
    evaluation 267–82
    factors affecting choice 232–8
    mode of action 288–94
    neutralizing agents *270*, 466–7
    physical factors affecting 262–7
    resistance to *see* Resistance
    toxicity 237–8
    see also Antiseptics; Disinfectants; Disinfection;
      Preservatives
  synthetic 120–9
Antiseptics 232, 259
  see also Antimicrobial agents, non-antibiotic
α-Antitrypsin 59
  recombinant *482*
Antitubercular compounds 123, *124*
  see also Tuberculosis
APCs *see* Antigen-presenting cells
API system 30
Archaebacteria 4
Arrhenius equation 262
Arthritis, rheumatoid 326
Arthrospores 48
Arthus reaction 327
Aseptic areas 454–5
Ashbya gossypia 490
L-Asparaginase 493
Aspergillus niger, antimicrobial success *235*
Aspergillus spp.
  mycotoxicoses from 56–7
  spores *48*
  toxin 54
Aspirin, degradation by microbes 374
Assays
  antibiotic *see* Antibiotics
  microbiological 168–74
  use of microorganisms 494–7
Asthma, extrinsic 319
Athlete's foot 51
Atmosphere *see* Air
Attenuation, definition 306
Autoclaving 412–15, 426
  monitoring 459, *461–2*
Autoimmunity 325–6
Avirulent, definition 306
A_w (water activity) of pharmaceuticals 377–9, 384,
  385
Azidothymidine (AZT) 129
Azlocillin *101, 102*, 104
Aztreonam 109, *110*
  mode of action 194

B cells 313
  self-reactive 325
Bacampicillin *101, 102*

Bacillus anthracis   11, 35, *339*
Bacillus cereus, food poisoning   93
Bacillus pumilus spores   404, 421, *464*
Bacillus spp.   360
  antibiotics from   100
  properties   35
  spore formation   11
  tryptophan breakdown   24
Bacillus stearothermophilus spores   404, 406, *408*, *464*
Bacillus subtilis   *296*, *464*
  as a cloning host   479, *480*
  spores   404, 406
Bacillus thuringiensis   501
Bacitracin   35, 100, 116
Bacteraemia   309, 318
Bacteria
  adhesive substances   86
  antibiotics from   35, 99–100
  bacteriophage infection *see* Bacteriophages
  capsules and slime   11
  facultative   17
  Gram-negative *see* Gram-negative bacteria
  Gram-positive *see* Gram-positive bacteria
  growth
    curves   32–3, *260–1*
    energy provision   18–27
    measurement   31–2
    requirements for   16–18
  identification   27–31
    by phage infection   70
  intracellular   199
  microaerophilic   17
  mortality curves   259–60
  obligate aerobes and anaerobes   17
  pigments   11
  properties of selected species   33–41
  replica plating of colonies   *210*
  reproduction   15–16
  spores   11–*14*
    antimicrobial resistance   302
    effect of antimicrobials   *234–5*
    for sterilization monitoring   460
    sterilization resistance   404, 406
  structure and form   4–*14*
  targets of non-antibiotic antimicrobials   288–94
  toxins   14, 90, 91–4, 310–11
  vaccines, production   335–7
  see also Escherichia coli; Resistance; Viable count
Bacterial cell component vaccines   334
Bactericidal activity, tests   279–80
Bactericidal environment, death curve for   *260*
Bactericide, definition   258
Bacteriophages   65–70
  staphylococcal   *62*
  temperate   66, 68
  transduction by   16, 69, 216–17
  virulent   66–7
Bacteriostasis   *260*
  estimation   272–5
Bacteriostats
  definition   258
  selection of   135
  tests for semi-solid   279
Bacteriuria   144–5
Bacteroides fragilis

aztreonam resistance   109, *110*
  drug sensitivity   *136*
Bacteroides spp.   38
  drug sensitivity   *136*
Basidiomycetes   44
BCG vaccine   333, *339*, 345
Bed-sores, contamination of   395
Benethamine   103
Benzalkonium chloride   *234*, *234–5*, 374
  eye-drop preservation   436
  properties   *238*
Benzathine penicillin   103
Benzoic acid
  antimicrobial activity   239
  neutralizing agents for   *270*
  pH and activity   382
Benzyl alcohol   243
Benzylpenicillin (penicillin G)   100, *101*, *102*, 103, *136*
  bacteria protected against   8
  hydrolysis   500
  manufacture   154–63
  neutralizing agent for   *466*
Biguanides, as antimicrobials   *240*, 245–6
  see also Chlorhexidine
Biliary disease, drug concentration   137
Binary fission
  in bacteria   15
  in yeasts   48
Biocide   259
Biofilms   84
  adherent   83, 297
Biological indicators   460–4
Bioluminescence   286
Bisbiguanides   245–6
  mode of action   205
  see also Chlorhexidine
Bismuth sulphite agar   28–9
Bisphenols   *251*, 253
  antibacterial activity   *234*
Bisulphites   294
Bites, as entry for disease   84
Black Death   36
Black fluids (phenolic)   *251*, 253
  properties   *238*
Bladder infections   85, 144
Bladder irrigation solutions   434
  contaminated   370
Blastomycosis   52
Blind therapy   139
Blood, sterilized   431
Blood clots, liquefaction   492, 493
Blood poisoning after childbirth   34
Blood transfusion, incompatible   327
Boils   33, 148
Bordetella pertussis   37, 86, 88
  vaccine from   336, *340*
  see also Whooping cough
Borrelia spp.   41
Botrytis cinerea   52
Botulinum antitoxin   *348*
Botulism   35, 93
Bowie Dick test   *462*
Brain abscesses   149–50
Branhamella catarrhalis   34
Brilliant green   255

British Pharmacopoeia 27, 29, 283–4
  sterilization methods 407–8
  sterility test contamination 468
  sterility testing procedures 465
β-Bromopenicillanic acid *110*
Bromovinyldeoxyuridine 131
Bronchiectasis 142
Bronchitis 37, 142
Bronchopneumonia 38
Bronopol *242, 243*
  mode of action 290, 291
  neutralizing agent *270*
  sites of antibacterial action *292*
Brooms, and contaminants 368
Browne's tubes *461–2*
Brucella abortus 86
  survival after phagocytosis 88
Brucella spp. 37
Bubble point pressure test 459
Bubonic plague 36
Budding, in yeasts 48
Buildings, microbes associated with 363–5
Burkitt's lymphoma *71*, 79
Burns
  dextran application 489
  infected 36, 149
  treated with infected preparations 370
  treatment 124
Butylphenols *251*

C₁₈ packings 180
Calcium, in vaccines 347
Calcium alginate 442
Campylobacter spp. 38–9
  gut infection 147
Candida albicans
  antibiotic treatment 57, 119, 120, 128
  chlorhexidine sensitivity *296*
  disinfectant success *235*
  infection 51
  pathogenicity 46
Candida spp., contaminants in pharmaceuticals
    370–1
Cap liners 380
Capacity use-dilution test 268–9
Capreomycin 116, 123
Capsids, assembly 61, 77
Capsomeres 61–3
Capsules, bacterial 11, 87
  antigenic properties 312
1-Carbapenems 107–9
Carbenicillin *101, 102, 103, 136*
  with gentamycin 113
Carbenicillin esters *101*, 103
  see also Carfecillin; Carindacillin
Carbohydrate metabolism 19–*23*
Carbolic acid *see* Phenol
Carbon, for benzylpenicillin production 159–60
Carbon dioxide, pressurized 379
Carboxypeptidases 23
Carbuncles 148
Carcinogens
  assay for 495–6
  detection 70
  fungal 54

  viral 79
Carcinoma
  caused by toxins 56–7
  caused by viruses 79–80
Cardboard, microbes on 363
Carfecillin *101, 102*
Carindacillin *101, 102*
Catgut, sterilized surgical 442
Catheterization, infection from 85, 144
Cefaclor *106*
Cefamandole *106, 136*
Cafazolin *106, 136*
Cefoperazone *107*
Cefotaxime *107, 136*, 146, 150, *220*–1
  assay 177, *178*
Cefotetan *107*
Cefoxitin *106, 136*
  mode of action 194
Cefsulodin *106*
Ceftazidime *107, 136*, 150
Ceftizoxime *107*
Ceftriaxone *107*
Cefuroxime 105, *106, 136*, 146
  atexil 105
Ceilings, microbes 363–4
Cell membrane *see* Cytoplasmic membrane
Cell wall
  action of disinfectants on 281, 292
  bacterial 4–9
  and drug resistance 137
  fungal 45–6
  phagocytosis resistance 87
Cellulitis 148
Cellulose
  oxidized 441
  structure 46
Central nervous system, infections 93, 149–50
Cephacetrile 105, *106*
Cephalexin *106*, 163
Cephaloridine *106*
Cephalosporins 57, 104–8
  C 104, *107*, 163–4
  inactivation 218–19
  mode of action 194
  N 104
  P 104
  and penicillin-allergic patients 138–9
  problems with testing 152
  sources 100
  see also individual cephalosporins
Cephalosporium spp., antibiotics from 104
Cephalothin 105, *106*
Cephamycins 162, 164
Cephapirin 105, *106*
Cephradine *106*
Cerebrospinal fluid (CSF)
  aminoglycoside penetration 137
  antibiotic levels 149
Cetrimide 29, *254*–5
  antibacterial activity *234*
  antifungal activity *235*
  antimicrobial uses 241
  mode of action 290
  properties *238*
Cetylpyridinium chloride *254*

Cetyltrimethylammonium bromide *see* Cetrimide
Chaetomium, toxins from   54
Challenge tests, for preservation evaluation   303, 385–6
Cheese, fungi used in manufacture   50
Chemical indicators   460, *461–3*
Chemoprophylaxis   140–1
Chemosterilant   232
Chick embryo, for viral culture   74–5, 337–8
Chick–Martin (CM) coefficient   267
Chick–Martin (CM) test   267–8
  and preservative evaluation   282
Chickenpox   *71*
  outbreaks   96
  vaccine   *342*
Chimaeras   473–4
Chitin, structure   45, *46*
Chlamydia spp.   39
  drug sensitivity   *136*
Chlamydospores   48
Chloramines, antibacterial activity   *234*, 248
Chloramphenicol   100, 117–*18*, *136*, 141, 150
  bacterial sensitivity to   135
  bone marrow suppression   139
  inactivation   175, 221–3
  mode of action   137, *190*, 199–200
  radioenzymatic assays for   174–6
Chloramphenicol acetyltransferases (CAT)   117, 175, 221–3, 467
Chlorbutol   *242*, 243
Chlorhexidine   245–6
  antibacterial activity   *234*
  antimicrobial uses   *240*
  eye-drops   436
  microorganism sensitivity to   *296*
  modes of action   290, 291, 299
  neutralizing agents for   *270*
  properties   *238*
  sites of antibacterial action   *292, 293*
Chlorhexidine gluconate, antifungal activity   *235*
Chlorine
  antimicrobial uses   *240*
  disinfectant action   247, 291
  organic compounds of   248
  virucidal effects   64
  water disinfection   359
Chlorine-releasing agents, as disinfectants   291
Chlorocresol   *251*, 253
  /phenylmercuric acetate synergy   284, *285*
Chloroform
  antimicrobial activity   248
  containers for   383
p-Chloromercuribenzoate, mode of action   291
Chloroxylenol   *251*, 253
  antibacterial activity   *234*
  antimicrobial uses   241
  /EDTA combination   299
  properties   *238*
Chlortetracycline   *111*
  production   163
Cholera   36, 65, 147, 310
  vaccines   326, 334
    testing   345
Chromatography, liquid (LC/HPLC)   176–8, 389
  calibration   181–2

controlling precision   182–3
  equipment   *177*, 178–9
  mobile phase formation   180
  paired-ion   180
  reverse phase (RPLC)   179–80
  sample preparation   181
  stationary phases   179
Chromosomes
  bacterial
    replication   201
    transcription   201
  target for antibiotics   *190*, 201
Chytridiomycetes   44
Cilofungin   129
Cinoxacin   *126*
Ciprofloxacin   125, *126*, *136*, 145, 226
  mode of action   201
Cirrhosis, childhood liver   57
Citric acid, from *Aspergillus*   *490*
Cladosporium resinae   373
Clarithromycin   116
Clavams   105
Claviceps spp.   54
Clavulanic acid   105, *109*, 220–1
  drug combinations   132
Cleaning
  to control microbial contamination   453
  pipelines   366
Cleaning equipment
  source of contamination   368, 394
  sterilization of   368
Cleansing-in-place (CIP) fluids   366–7
Clear soluble fluids (phenolic)   *251*, 252
  antibacterial activity   *234*
  antimicrobial uses   *241*
  properties   *238*
Clindamycin   117, *118*–19, *136*
  colitis development   140
  mode of action   *190*, 200
Clomocycline   *111*
Cloning   472–4, 485
  hosts   479, *480*
  in yeasts   58–9
Cloning vectors *see* Plasmids
Clostridium botulinum   14, 35
  ingested with food   83
  toxin   93
Clostridium difficile   35, 116
  overgrowth   140
Clostridium perfringens   361
  drug sensitivity   *136*
  toxins   90, 310
Clostridium sphenoides   35
Clostridium sporogenes   26, 35, *464*
Clostridium spp.   35, 360
  spore formation   11
Clostridium tetani   35
  damage   92, 310
  in dusting powder   399
  vaccine from   337, *339–40*
Closures   363
Clothing
  for pharmaceutical manufacture   447, 452, 454
  protective   361, 447
Clotrimazole   *127*, 128

Cloxacillin   *101*, *102*, *136*
Club root   53
Coagulase   310
Cobalt-60 irradiation plant   421, *422*
Cocci   33–4
   aggregates   4
Coccidioidomycosis   52
Colectomy, postoperative infection   140
Colistin   116
Colitis
   from antibiotic treatment   140
   pseudomembranous   35, 117
Collagenases   90, 310
Colony-forming units (cfu) counts   31
Colouring agents, attack by microbes   373
Columns, liquid chromatography   179
Combinations of drugs   131–2, 138–9
Complement   308, 319–21
Concentration exponent   *263–4*
Conjugation   15, 211–16
Conjunctiva, entry for disease   85–6
Conjunctivitis   34, 37, 39
Contact-lens solutions   437–8
   cleaning   438
   soaking   438
   wetting   438
Contamination
   from pharmaceuticals   392–3, 397–8
   in antibiotic fermentation   160
   in hospitals   393–7
   of pharmaceuticals see Pharmaceutical products
   of plumbing materials   364, 393–4
   of water   40, 359–60, 393–4
Copper, antimicrobial activity   249
Copper II salts, sites of antibacterial action   *292*
Corn steep liquor (CSL), for antibiotic
   fermentation   159–60
Corynebacterium diphtheriae   35, 86
   conversion to toxic strain   69–70
   damage   92
   invasion by   88–9
   vaccine from   337, *339*
   see also Diphtheria
Cosmetics, preservatives   283
Co-trimoxazole   122, 141, 145, 147
Coughing   91
Coumarin compounds   57
Cowpox   306, 333
Coxsackie virus   72
Creams, antimicrobial   279
Cresols   *251*, *252*, *466*
Cryptococcus neoformans   52
   treatment   120, 138
Cryptococcus spp., antibiotics against   128
Crystal violet   255
Culture media, and bacterial growth   18
Cup-plate technique, measuring bacteriostasis   273,
   274, 279
Cyanocobalamin   *490*
Cycloserine   123, *190*, 194, 196–7
Cysteine   290
Cystic fibrosis   143
Cystitis   37
Cytarabine   129, *130*
Cytomegalovirus   *71*
Cytopathic effect (CPE)   74

Cytoplasm
   action of disinfectants on   290–1, *292*
   bacterial   10
   fungal   47
Cytoplasmic membrane   204–5
   action of disinfectants on   289–90, *292*
   bacterial   9–10, 204
   fungal   46–7, 204
   potential difference across   289
   target for antibiotics   *190*, 204–6
Cytosine arabinoside   129, *130*
Cytotoxic T cells (Tc cells)   323
Czapek–Dox medium   275

D-value   14, 386, 405, *406*, 409
Dakin's solution   247
Dane particle   277–8
Dapsone   *121*, 122
Daptomycin   *190*, 196
Daunomycin   496
Decimal reduction time (DRT; *D*-value)   14, 386,
   405, *406*, 409
Defoaming agents, during fermentation   158
Delayed-type hypersensitivity (DTH) reactions   322
Demethylchlortetracycline   *111*, 112
Deoxyribonucleases   90
Dequalinium chloride   256
Dermatophytoses   51
Desferal   *491–2*
Desferrioxamine   *491–2*
Detectors, liquid chromatography   179
Deuteromycetes   50
Dextrans   11, 431, 488–9
DHFR see Dihydrofolate reductase
Dialysis machines, sterilization   *443*
Diamidines, as antimicrobials   255
Diaminobenzylpyrimidines, unsubstituted   122
Diaminopyrimidine derivatives   *121*, 122
Diarrhoea   88, 93
   travellers'   147
   treatment   124
Dichloroamine-T   248
Dicloxacillin   *136*
Dicophane   374
Diethylphenols   *251*
Dihydrofolate reductase (DHFR)   122, 225
Dihydropteroate synthetase   122
Dihydropteroic acid synthesis   203
Dihydrostreptomycin   113
Dilution coefficient   *263–4*
Dipeptidases   23
Dipicolinic acid (DPA)   *13*
Diphtheria   35, 60, 65, 92
   immunoserum   347–8
   vaccine   *339*, 340, 343, 346
      testing   345
   see also Corynebacterium diphtheriae
Diplococci   4, 34
Diplococcus pneumoniae   4
   diseases from   34
   and transformation   15
Disease
   epidemiology   95–6
   pathogenicity see Pathogenicity, microbial
Disinfectants   231–2

contaminated in hospitals 394, 397
definition 259
effects on bacteria 8
effects on viruses 64
efficiency measured 259–61
evaluation
    liquid 267–78
    semi-solid 279–80
    solid 280–1
factors affecting choice 232–8
food for *Ps. aeruginosa* 36
microbial growth substrates 374
powders 280–1
for sterile product manufacture *453*
'topped up' 394
types of 238–56
see also Antimicrobial agents, non-antibiotic;
        Disinfection
Disinfection
of air 281–2, 356
dynamics of 259–62
effect of dilution 263–4
effect of inoculum size 266–7
effect of interfering substances 236–7, 266
effect of pH *238*, 264–6
effect of surface activity 266
effect of temperature 262–3
hands 272
pharmaceutical manufacturing equipment 367
physical factors affecting 262–7
policies 256–7, 303–4
of water 359–60
see also Disinfectants
Distilled water, bacteria in 379
Ditch-plate technique, measuring bacteriostasis 273,
        274, 279
5,5'-Dithiobis(2-nitrobenzoic acid) (DTNB) 176
DNA
effect of acridine dyes on 291, *292*
viral 61
DNA gyrase 201–2
and quinoloes *190*, 201–2
DNA viruses *71–2*
inhibition 129
oncogenic activity 79
replication 76
Dosimeters 459
Doxycycline *111*–12
Drains 364, 451
Dressings
spray-on 440
sterile 438–40
Drips, sterile 430–3
Drugs
absorption 137
antimicrobial 134–5
    adverse reactions 139
    antibiotic policies 150–3
    bactericidal 135
    bacteriostatic 135
    clinical use 141–50
    combinations 138–9
    principles of use 135–41
    prophylactic 140–1
    superinfection from 140
    see also Antibiotics

antiviral 129–31
combinations 131–2, 138–9
in the home 398
induction of immunogenic 184
metabolism evaluated 497–9
resistance *see* Resistance, drug
toxicity 139
    in excretory organs 137
    reduced by combinations 138
    selective 189
    testing 273, 347
see also Pharmaceutical products
DRT *see* Decimal reduction time
Dry heat sterilization 415–6, *426*
DTNB *see* 5,5'-Dithiobis(2-nitrobenzoic acid)
Dusting powders, contaminated 399
Dyes, antimicrobial 255, 291, *292*
    see also Acridine dyes
Dysentery
bacillary 38, 89
    treatment 113
Dysuria 144

Ear infections, treatment 141
Echoviruses *72*
Econazole *127*, 128
Eczema, infected preparations treating 370
EDTA *see* Ethylenediamine tetra-acetic acid
Effluent treatment by fungi 50
Eggs
    for testing disinfectants 272
    for testing virucides 276
    for viral culture 74–5, 337
Eijkman (E) test 30
Electron accelerators for sterilization 419, 423
Electron transport chain, bacterial 9, 290, *292*
Elongation factors 200
Embden–Meyerhof pathway 19, *20–1*
EMIT 186, 187
Emphysema 59
Emulsions
    deterioration 375, *376*
    oil/water 383
    preservatives in 283
Encephalitis 149
    vaccine *342*
Endocarditis
    enterococcal, treatment 138
    prevention 140
Endogenous reverse transcriptase 278
Endopeptidases 23
Endospores *see* Bacteria, spores
Endotoxins 93–4, 310–11
Energy provision 18–27
Enoxacin 125
Entamoeba histolytica, antibiotic for 113, 128
Enteric fever 91
Enteritis 37, 39
Enterobacteriaceae 228
    acid and gas production *21* ·
    furazolidone against 124
    identification 30
    isolation 28
    nalidixic acid against 125

Enterobacteriaceae (*continued*)
  permeability and resistance 297–300
  properties of 37
  see also Escherichia coli; and other individual
    organisms
Enterococcus, drug sensitivity *136*
Enterococcus faecalis 117, 118, 119, 121
Enterotube 30
Enteroviruses 70
Enzymes
  antibiotic-immobilizing 103–4, 175, 497
  microbial, in sterility tests 496–7
  penicillin-sensitive (PSEs) 104, 191
  prepared for microbiological assays 175
  proteolytic 21–4
Epidemics 95–6
Epidemiology, microbial 95–6
Epidermophyton 51, 360
Epifluorescence 285
Episomes 10
Epitopes, antigen 313, 315
Epstein–Barr (EB) virus *71*
  oncogenic activity 79
Equipment, surgical, sterile *443*, 444
Eremothecium ashbyii 490
Ergometrine 57
Ergosterol 204
  antibiotic affinity for 205
Ergotism 54–5
Erysipelas 148
Erythromycin *115*
  mode of action *190*, 200
  resistance to *136*, 225
Erythromycin estolate 115
Erythropoietin, recombinant *483*
Escherichia coli 296
  bacteriophages *66*, 68
  causing meningitis 149–50
  causing travellers' diarrhoea 147
  chromosome replication 201
  as a cloning host *479*, 480
  drug sensitivity *136*
  effect of bismuth sulphate 28
  effect on the gut 89, 93
  erosive lesions 89
  ethanol production *21*
  eukaryotic gene expression 57
  identification *21*, 28, 29–30
  interferon production 78
  penicillin-binding proteins (PBPs) 195–6
  peptidoglycan *6*, 191–4
  permeability 297
  phagocytosis avoidance 87
  phospholipid structure *8*
  pili *214*
  properties of 37
  proteins from 57, *478*, 479
  sulphonamides against 121
Esters 238–42
  p-hydroxybenzoic acid *239*
Ethambutol 123
Ethanol
  antimicrobial uses *242–3*
  as a bactericidal agent *234*
  production by yeast *21*

properties *238*
Ether, effect on viruses 64
Ethionamide 123, *124*
Ethylene oxide
  mode of action 294
  sites of antibacterial action *292*
  for sterilization 417–18, *426*
    monitoring *461*, *462*
  virucidal effect 64
Ethylenediamine tetra-acetic acid (EDTA) 9, 10
  in contact-lens solutions 438
  effect on Enterobacteriaceae 297–8, 299
  effect on *Ps. aeruginosa* *300*
Ethylphenol *251*, 252
Eukaryotes 4, *5*
European Pharmacopoeia 35
European suspension test 269
Exaltation, definition 306
Exons 473, 485
Exopeptidases 23
Exotoxins 93, 310
Eye infections 34, 36
  conjunctivitis 34, 37, 39
Eye-drops *see* Ophthalmic preparations

F (fertility) factor, in bacteria 15
F-pili 10, 15
F-value 14, 409
Factor VIII, recombinant *483*
Factor IX, recombinant *483*
Farmer's lung disease 51
Fats, microbial attack 373
Febrile shock 371
Fentichlor 289
Fermentation
  for antibiotic production 155, 156–62
  for vaccine production 336
Fermenters 156–7
Fibrin clots 90
Fibrin deposition 90
Fibrin foam 441
Fibrinolysins 90
Filters 366
  high-efficiency particulate air (HEPA) 425
    testing of 460
  membrane 424, *425*, 465
  sintered 424
Filtration
  air 355–6, 425, *450*
  membrane, for sterility testing 465, 467
  monitoring 459–60
  sterilization 423–5, *426*, 464
  water 359–60
Fimbriae (pili) 10, 86, 213–*14*
Fish, food poisoning from raw 36
Flagella
  bacterial 10
    antigens associated with 311
  fungal 47–8
Flavobacterium spp. 38
Flavouring agents, attack by microbes 373
Floors
  microbial contamination 364
  in sterile areas 449–51
Flucloxacillin *101*, *102*, 138

Fluconazole   *127*, 128
Flucytosine   *127*–8, 138
Flumequine   125, *126*
Fluorimeters   187
Fluoroimmunoassays   185–6
Fluoroquinolone derivatives   125, 126
Folate metabolism   202–4
   target for antibiotics   190
Food
   fungal contamination   54–7
   microbes ingested with   83, 88, 93
     see also Food poisoning organisms
Food poisoning organisms   34, 35, 36, 37, 39
   Bacillus cereus   93
   Clostridium botulinum   83, 93
   Salmonella   89, 147
   Staphylococcus aureus   93
Formaldehyde   237, 244
   antibacterial activity   *234*
   for fumigation   356
   mode of disinfectant action   291
   neutralizing agent for   *270*
   sites of antibacterial action   *292*
   for sterilization   419, *420*
   water disinfection   359
Formaldehyde-releasing agents   245
Formalin   244, 334
   in vaccines   347
N-Formimidoylthienamycin (imipenem)   109
Framycetin   114
Francisella tularensis   37
Fumigation   356
Fungi   43–*4*
   antibiotics against   119–*20*, 128–9
   antifungal disinfectants   235
   brain infections   149
     see also Meningitis
   as cloning hosts   479, *480*
   dermatophytic   360
   habitat   45
   industrial importance   49–50
   medical importance   50–2
   molecular biology of   57–9
   pharmaceutical contaminants   303, 360
   plant pathogens   52–3
   polyenes active against   205
   reproduction   47–9
   resistance to antimicrobials   *296*, 303
   selective media for   29
   source of antibiotics   50
   spores   47–8
   steroid transformation   499–500
   structure   44–7
   tests for antifungal preparations   275–6
   toxic   53–7, 371
Fungi Imperfecti   50
Fungicides   235, 259
   tests for   275
Fungistats   259
   tests for   275
Furaltadone   124, *125*
Furan   *125*
Furazolidone   124, *125*
Fusarium aquaeductum   50
Fusarium spp., toxigenic species   54, 55

Fusidic acid   117–*18*, 138
   mode of action   *190*, 200

Galls   53
Gamma-rays, for sterilization   419, 421–3, *426*
Ganciclovir   131
Gangrene
   from ergotism   55
   gas   53
     immunosera   *348*
Gardnerella vaginalis   35, 128
Gas-chex   *461*
Gastroenteritis
   infantile   *72*
   infective, treatment   146–7
Gastrointestinal infections   146–7
Gelatin
   foam, absorbable   441
   liquefaction   23
Gels, antimicrobial   279
Gene cloning *see* Cloning
Gene vectors *see* Plasmids
Genes
   'jumping'   215
   transfer of bacterial   69
Genetic engineering *see* Recombinant DNA
     technology
Genetic information transfer
   by conjugation   15, 211–16
   by transduction   16, 69, 216–17
   by transformation   15, 217
Genetic manipulation, vaccines   334
Genome   485
Gentamicin   113, *114*, *136*
   for CNS infections   150
   inactivation   224
   /penicillin combination   138
   radioimmunoassay   174–5, *181*
   source   100
Gentian violet   255
Geotrichum candidum   50, 52
Germall   115, combined with parabens   284
Germination, spore   13, *14*
Giardia lamblia   128
Glandular fever   *71*, 79
Glass bottles, contamination after sterilization   363,
   431
Glucan   47
   structure   45, *46*
Gluconic acid, from microorganisms   *490*
Gluconobacter spp., antibiotics from   100
Glucose oxidase   497
Glutaraldehyde   237, 244
   antibacterial activity   *234*, 288, 291, *292*
   antimicrobial uses   *240*
   antiviral activities   *236*
   neutralizing agent for   *270*
   properties   *238*
Glycerol, microbial attack   373
Glycocalyx   11
Glycopeptide antibiotics   116–17
Glycosylation   477
Gonococci, sulphonamide resistance   121
Gonorrhoea   34

Good manufacturing practice (GMP)   386–7, 446,
    455–6
Gradient-plate technique, measuring
    bacteriostasis   274–5
Graft rejection   328
Gram-negative bacteria   5
    antibiotic resistance   208, 213–15
    antimicrobial resistance   301
    cell envelope   6–8, 208
    cocci   34
    drugs active against   *101*, 104, 109, 208, *209*
    β-lactamase inhibition   109
    β-lactamases   219–21
    penetration of antimicrobials   299
    rods   36–40
    sensitivity to chlorhexidine   295, *296*
Gram-positive bacteria   5, 6, 8, 9, 216
    antimicrobial resistance   301–2
    cocci   33–4
    drugs active against   *101*, 104, 109, 112–13, *209*
    β-lactamases   219–21
    rods   34–6
    sensitivity to chlorhexidine   295, *296*
    transformation in   217
Granulocyte colony stimulating factor   *483*
Granulocyte-monocyte colony stimulating
    factor   *483*
Griseofulvin   57, 119, *120*
    manufacture   155
    sources   100
Growth curves, bacterial   *32–3*
Guide to Good Pharmaceutical Manufacturing
    Practice   455–6
Guthrie test   495

H-antigens   311
Haemodialysis   435
    contaminated fluids   371
Haemolysins   90
Haemophilus influenzae   37, 334
    ampicillin resistance   150
    antibiotics against   115, 142, 221
    causing meningitis   149
    in cystic fibrosis   143
    drug sensitivity   *136*
    sulphonamide resistance   121
Haemostats, absorbable, sterile   441–2
Halazone   *248*
Half-lives, of chemicals   371
Halogens
    as antimicrobials   247–9
    neutralizing agent for   *270*
Handwashing, in hospitals   272, 396
Hay fever   319
Heat, effect on viruses   64
Heat resistance, bacterial   13–14, *405–6*
Heat sterilization   408–16
    dry   415–16, *426*
    moist   410–15, *426*
    monitoring   459
Heavy metals, as antimicrobials   249–50
HeLa cells   74
Helicobacter spp.   39
Helper T cells (Tʜ cells)   322
Hepatitis B surface antigen (HBsAg)

inactivation   276–8
recombinant   *483*
Hepatitis viruses   71, *73*
    A   *73*
    B   57, *71*, 79
        morphology   276–8
        success of antimicrobials   236
        testing virucides for   276–8
        vaccine   334, *341*
Herd immunity   96
Herpes keratitis, treatment   129
Herpes simplex   62, *71*, 131, 148, 149
Herpesviruses   *71*, 276
    drugs against   129
    eye infection treatment   129
    replication inhibition   78
Hexachlorophane   *251*, 253
    antimicrobial uses   241
    mode of action   290
    neutralizing agent for   *270*
    sites of antibacterial action   *292*
Hexylresorcinol, mode of action   290
Histamine   309, 319
Histoplasmosis   52
HIV (HTLV-III, LAV)   *73*, 80, 129, 278, 321
    success of antimicrobials   *236*
HLA system   321, 328–9
Hospitals
    antimicrobial multiresistant organisms in   302
    disinfection policies   256–7
    incidence of contamination   398
    microbial population   396
    pharmaceutical contamination in   392
    sources of contamination   395–7
HPLC assays   176–83
HTLV-III *see* HIV
Human immunodeficiency virus *see* HIV
Hyaluronidase   90, 309
Hybridoma   315–17
Hydrochloric acid   307
Hydrogen peroxide
    antimicrobial activity   *234, 240*, 250
    mode of action   291
    sites of antibacterial action   *292*
Hydrogen sulphide production   26–7
p-Hydroxybenzoates, pH and activity   *382*
p-Hydroxybenzoic acid esters   *239, 242*
    combinations for preservatives   284
    neutralizing agents for   *270*
Hydroxycobalamin   490
11α-Hydroxyprogesterone   499, *500*
Hygiene
    hospital   392, 393–7
    in pharmaceutical manufacture   361, 445–56
Hypersensitivity   139, 326–8
    delayed   322, 327–8
    fungal   50–1
    penicillin   110–11, 139
Hypochlorite   247–8
    antibacterial activity   *234*
    antifungal activity   *235*
    antimicrobial uses   *240*
    antiviral activity   64, *236*
    powder   280
    properties   238
    sites of antibacterial action   291, *292*

Hypochlorous acid   247

Idiotypes   323
Idoxuridine   129, *130*
Imidazole derivatives   125, *127*, 128
Imidazoles, mode of action   *190*, 205–6
Imipenem (*N*-formimidoylthienamycin)   109, 221
   mode of action   194
Imipramine, metabolism of   *498–9*
Immunization, antibody response to   *314*
Immunity
   acquired   330
   adaptive immune system   311–12
   autoimmunity   325–6
   cell-mediated (CMI)   311, 321–3
   cells involved in   312–26
   herd   96
   humoral   311, 313–15, 319
   hypersensitivity   139, 326–8
      fungal   50–1
      penicillin   110–11, 139
   innate immune system   307–11, 329
   natural   329
   tissue transplantation   328–9
Immunoassay   183–7
   antibody preparation   184
   calibration   186–7
   controlling precision   187
   drug labels   183–6
Immunogens   311
   induction of   184
Immunoglobulins   311
   classes   317–19
   human   *349*
      extraction   348–9
   IgG   313–15
      autoimmune response to   326
   structure   313–14
Immunology
   definitions   306–7
   historical aspects   305–6
   see also Immunity
Immunoregulation   323
Immunosera   347–*8*
   production   347–*8*
Impetigo   34, 148
Implants, sterile   440–1
IMVEC and IMViC tests   30
In-process control   446
Inactivation factors (IF)   406–*8*
Incubators, sterilization   444
Indanyl carbenicillin   *101, 102*
Indole   24–5, 29
Infections
   central nervous system   93, 149–50
   chemoprophylaxis   140–1
   from contaminated pharmaceuticals   370–1,
      397–400
   gastrointestinal   146–7
   microbial   82–96
   nosocomial   233, 256
   respiratory tract   141–3
   skin   84
      and soft tissue   147–9
      see also Wounds
   urinary tract   125, 144–6

Inflammation   309
   caused by pathogens   94
   and drug penetration   137
   modulation by pathogens   87
Influenza   37
   vaccines   334, 337, *341*
   see also Influenza viruses
Influenza viruses   62, 64, 70, *72*
   A virus   129
   combat herd immunity   96
   cultivation   74–5
   production and release   76, 77
   see also Influenza
Infusion fluids, contaminated   371
Inhaler solutions, sterile   435
Injections
   contaminated   399–400
   design philosophy   429–30
   intravenous additives   432
   intravenous infusions   430–3
   packaging for   363
   particles in   429
   production, microbial air count   354
   small-volume aqueous   433–4
   small-volume oily   434
   sterile   429–33
   total parenteral nutrition (TPN)   432–3
Injectors, liquid chromatography   178–9
Insect bites   91
   as entry for disease   84
Insecticides
   fungi as   50
   microorganisms as   500–1
Instruments, surgical, sterile   *443*, 444
Insulin   59
   recombinant   472, *473–6*, *482*
Interference
   between viruses   78
   drug–drug   132
Interferons   59, 78–9, 131, 308
   recombinant   481, *482*
Interleukin-1   309
Interleukins   59, 322
   recombinant   *483*
Intravenous catheter insertion, infection from   371
Intravenous infusion, sterile   430–2
   additives   432
Introns   473, 485
Iodine   247
   antimicrobial activity   *234–5, 241*, 248–9, 291, *292*
   antimicrobial uses   *241*
   /povidone combination   249
   virucidal effect   64
Iodophors (iodophores)
   antimicrobial activity   249
   properties   *238*
Ionizing radiation *see* Radiation sterilization
Iron
   availability in infections   492
   deficiency   491
   poisoning   491
Iron-chelating agents, from microorganisms   *491–2*
Iron dextran injections   489
Isobologram   284, *285*
Isoniazid   113, 123, *124*, 138
   neuropathy from   139

Isopropanol 243
  properties *238*
  see also Isopropyl alcohol
Isopropyl alcohol
  antibacterial activity *234*
  antimicrobial uses *240*, 243
  see also Isopropanol

Jenner, Edward 306, 333

Kanamycin 113, *114*, 123
  inactivation *224*
Kelsey–Sykes (KS) test 268–9, 303
  and preservation evaluation 282
Ketoconazole *127*, 128
  mode of action 205–6
Kidney
  infections 85, 144, 145
  inflammation 34
Kieselguhr 280
Killed vaccines 333–4
Killer (K) cells 324
Klebsiella aerogenes 38, *296*
  antibiotic resistance 151
  identification *21*, 27, 28, 29–30
Klebsiella spp.
  contaminants in barrier cream 370
  contaminating pharmaceuticals 399
  drug sensitivity *136*
  sulphonamide resistance *121*
  urinary tract infection 145
Krebs citric acid cycle 19, 21, *22*

Labels, drug 184–6
β-Lactam antibiotics 100–11, 154
  mode of action *190*, 194–7
β-Lactamases 105, 109, 110
  drug degradation by 103, 104, 219–20
  inhibitors 221
  synthesis 219–21
  uses 493–4
Lactic acid, from microorganisms *490*
Lactobacilli, vaginal 85, 488
Laminar airflow (LAF) unit 452
Latamoxef (moxalactam) 105, *109*
  mode of action 194
LAV *see* HIV
Lecithinase 310
Legionella pneumophila 40, 143
  antibiotics against 115
Legionella spp. 40
Legionnaires' disease 40
  cause of pneumonia 143
  treatment 137, 143
Leprosy 40, 122
Leptospira spp. 41
Leucocidins 90, 310
Leucocytes 307
Leuconostoc spp., dextrans from 11, 488
Leukaemia
  cell inhibition 493
  treatment 493
Ligatures, sterile 442–4

Lincomycin 117, *118*–19
  mode of action *190*, 200
Lipid solubility of drugs 137
Lipopolysaccharide (LPS), bacterial 6–8, 298, 299, 300
Listeria monocytogenes 36
  survival after phagocytosis 88
Listeriosis 36
Live vaccines 333
Lividomycin, inactivation *224*
Lockjaw 35, 92
Low temperature steam with formaldehyde (LTSF)
  treatment 419, *420*, *426*
  monitoring *463*
LPS *see* Lipopolysaccharide
Luciferase assay 174
Lung abscess 143
Lyme disease 41
Lymphocytes *see* B cells; T cells
Lymphokines 322
Lysogeny 68–70, 216
Lysol 252
Lysozyme 307
  bacteria protected against 8
  used to burst bacteria 10
Lytic growth cycle 66–7

MacConkey's medium 28
Macrolide antibiotics 115–16
Macrophages 308
Madura foot (maduromycosis) 52
Major histocompatibility complex (MHC) 321–2, 328
Malachite green 255
Malta fever 37
Mammalian cells as cloning hosts *480*
Mannan 45, *46*, 47
Matrix, drug 166, 167
Mean generation time 32
Measles
  immunoglobulin *349*
  outbreaks 96
  vaccine 340, *341*
  virus *72*
Mechanism of action *see* Mode of action
Mecillinams 100–4
Media, selective and diagnostic 28–9
Medicines Act 1968 397
Membrane, cell *see* Cytoplasmic membrane
Membrane filtration 359, 465
Meningitis 34
  bacterial 149
    drug dosage 137
  fungal 52, 138
  infantile 37
  meningococcal 34, 122
    prevention 141
  tuberculous 113, 123
  viral 149–50
Mercurials
  antimicrobial activity of mercury *241*, 249–50
  modes of action 290, 291
  neutralizing agent for *270*, *466*
  resistance-encoding plasmids 303
  sites of antibacterial action 292

see also Organomercurials
Mercury lamps   423
Mesophiles   17
Metabisulphites, antimicrobial activity   239
Methacycline   *111*–12
Methicillin   *101*, *102*, 103
    resistance to   *136*, 209, 225
Methisazone   129, *130*
Methotrexate   122
Methyl red (MR) test   30
Methylindanols   *251*
Methylresorcinols   *251*
Metronidazole   125, *127*, 128, 150
    assay   177
    drug combination   138
    mode of action   *190*, 202
Mezlocillin   *101*, *102*, 104
Miconazole   *127*, 128
    mode of action   205
Micromonospora purpurea, antibiotic from   100
Microorganisms
    antibiotics from   100
    and the pharmaceutical industry *see* Pharmaceutical
        industry
Microsporon   360
Microsporum   51
Mildew, powdery   52
Milk, digestion of clotted   23
Mima spp.   360
Minimum inhibitory concentration (MIC)   272, 273
    of preservatives   282
    of synergistic compounds   284
Minocycline   *111*, 112, 228
Mitomycin C   496
Monobactams   109–10
    source   100
Mode of action
    antibiotics   189–206
        on cell walls   189–97
        on chromosome function   *190*, 205–6
        on cytoplasmic membranes   *190*, 204–6
        on folate metabolism   *190*, 202–4
        on ribosome function   *190*, 197–200
    non-antibiotics   288–94
        on cell walls   288
        on cytoplasm   290–1
        on cytoplasmic membranes   289–90
Moist heat sterilization   410–15, *426*
    see also Autoclaving
Monoclonal antibodies   315–17
Monocytes   307
Monokines   322
Mops, and contaminants   368
Moulds
    associated with buildings   363–4
    infected plants for pharmaceuticals   362
    in packaging materials   363
    structure   44–5
    toxins from   53–7
Moxalactam (latamoxef)   105, *136*, 194
Mucous membranes   84, 307
Mucus   307
Mumps
    vaccine   340, *341*
    virus   62, 64, *72*

Mupirocin   *118*, 119, *190*, 200
Muscarine   54
Mushroom poisonings (mycetismus)   53–4
Mutagens, assay for   495–6
Mutations, affecting resistance   137, 210–11, 227
Mycetismus   53–4
Mycobactericidal activity, testing   271
Mycobacterium spp.   40
    antimicrobial resistance   302–3
    staining   302
    survival after phagocytosis   88
Mycobacterium tuberculosis   40, *296*
    antibiotics against   112, 113, 116
    antimicrobials against   *234*
    bactericide resistance   234
    isoniazid resistance   138
    survival after phagocytosis   88
    see also Tuberculosis
Mycoplasma pneumoniae   143
    causing autoimmunity   94
    drug sensitivity   *136*
Mycotoxicoses   56–7
Mycotoxins   53, 54–7
Myeloma cells   315
Myxoedema   326
Myxoviruses   *72*, 76

Nafcillin   *136*
Naftidine   *190*, 206
Nalidixic acid   29, 124–5, *126*, 145, 201, 226
Naphthols   *251*
Natural killer (NK) cells   308, 324
Nebulizers   435
Needles, sterilization   *443*
Neisseria gonorrhoea   34
    drug sensitivity   *136*
    obligate pathogen   82
    transmission between hosts   95
Neisseria meningitidis   34, 149
    drug sensitivity   *136*
    vaccine from   334, 337, *339*
Neisseria spp.   34, 115
Neomycin   113
    inactivation   *224*
    resistance to   209
Nerve poison, from *Clostridium botulinum*   35
Netilmicin   115, *136*
    assay for   174–5
Neuraminidase   493
Neurotoxins   14
Neutrophils   307
Nitrofuran compounds   123–5
Nitrofurantoin   124, *125*, 145
    mode of action   *190*, 202
Nitrofurazone   124, *125*
5-Nitrofurfural   *125*
Nitrogen source, for benzylpenicillin
        production   159–60
Nocardicins   109, *110*
Norfloxacin   125, *126*, 226
    mode of action   201
Noxythiolin   245
Nucleus, bacterial   10
Nutrient depletion   296–7
Nystatin   119, *120*

Nystatin (*continued*)
  mode of action   205

O-antigens   312
ODS packings   180
Ofloxacin   125, *126*
Oidia   48
Oils, microbial attack   373
Ointments
  antimicrobial   279
  eye   437
Oleandomycin   115, 116
Olivanic acids   *109*
Operon   485
Ophthalmic preparations
  contact-lens solution   437–8
    cleaning   438
    soaking   438
    wetting   438
  contaminated   370, 399–400
    with *Ps. aeruginosa*   397
  design philosophy   435–6
  eye-drops   436–7
  eye lotions   437
  eye ointments   437
  packaging   363
  preservatives   381
  production, microbial air count   354
  sterile   435–8
Opsonization, avoidance   87
Organic acids, from microorganisms   489–91
Organic matter, effect on disinfectants   237, *238*
Organic polymers, attacked by microbes   373
Organomercurials   *250*, 303
  microbial degradation   374
  see also Mercurials
Ornithosis   39
Osteomyelitis, staphylococcal   33
Otitis media, treatment   141
Outgrowth, spore   13, *14*
1-Oxacephems   105
Oxacillin   *101*, *102*, *136*
Oxidative deamination   24
Oxolinic acid   125, *126*, 226
Oxygen
  in antibiotic fermentation   157
  for bacteria   17
Oxytetracycline   *111*
Oxytoxin, detection   317

Packaging
  for dressings   439–40
  for pharmaceuticals   362–3, 380
  plastic, and microbial attack   373
  raw materials for pharmaceuticals   362
Palatin   57
Pandemics   96
Paperboard, microbes on   363
Papilloma virus   *72*, 79
Papovaviruses   72
Papulosporia, toxins from   54
Parabactin   492
Paracoccus denitrificans   492
Paramyxoviruses   72
Paratyphoid fever   28, 37, 147, 311

Paromomycin   113, *224*
Pastes, antimicrobial   279
Pasteurella pestis   36
Pasteurella tularensis   37
Pathogen, definition   306
Pathogenicity, microbial   82–3
  manifestation of disease   88–91
  portals of entry   *83*–6
  recovery from infection   94–5
  survival within the host   86–8
  tissue damage   91–4
Pathogens   306
  obligate   82, 95
  opportunistic   82, 95
PBPs *see* penicillin-binding proteins
Penicillanic acid sulphone   *110*
Penicillin amidases   101
Penicillin-binding proteins (PBPs)   104, 191, 195–6
Penicillin-sensitive enzymes (PSEs)   104, 191
Penicillinic acid   57
Penicillin(s)   *100–4*
  allergy to   110–11, 139, 494
  G *see* benzylpenicillin
  /gentamycin combination   138
  inactivation   103, 218–19, 374
  manufacture   154–63
  mode of action   194–6
  sources   100
  V   100, *136*, 163
  see also individual penicillins
Penicillium chrysogenum   100
  in benzylpenicillin manufacture   155, 163, 471
Penicillium citreoviride, mycotoxicosis from   56
Penicillium griseofulvum   119
Penicillium notatum   100, 155
  chlorhexidine sensitivity   *296*
Penicillium spp.
  antibiotics from   100
  in cheese manufacture   50
  spores   *48*
Penicilloic acid   *103*, *218*
Pentachlorophenol   289
Peptic ulcer   39
Peptidases   90
Peptidoglycans   5–6, 189–91
  biosynthesis   191–4
  synthesis inhibition   *190*, 194–7
Peracetic acid, antimicrobial activity   250
Peracid compounds, as antimicrobials   250
Peritoneal dialysis solutions, sterile   435
Peritonitis   34
Pertussis, vaccine   340
pH
  for bacterial growth   17
  effect on disinfectants   264–6
  effect on preservatives   382
  optimum disinfectant   *238*
  for pharmaceuticals   379–80
Phage therapy   65
Phages *see* Bacteriophages
Phagocytes   307–8
  involved in inflammation   309
  killing   88
  penetration by drugs   135
Pharmaceutical manufacture

affected by microorganisms from
   the atmosphere 353–6, 452
   buildings 363–5, 449–51
   equipment 365–7, 394, 396–7
   operators 360–1, 454
   packaging 362–3, 394
   raw materials 361–2, 447–8
   water 356–60, 393–4
  cleaning equipment 368, 394
  control of contamination 446–9
  controlled in hospitals 398
  definitions 446
  partial synthesis by microorganisms 499–500
  of sterile products 449–55
  see also Pharmaceutical products
Pharmaceutical products
  contamination 38, *392–3*
   effect on patients 392
   extent 397–8
   prevention and control 400–1
   sources 353–67, 393–97
  deterioration 371–2
  formulation design and development 384–6
  microbial spoilage 369–70, 391–2
   assessment 389
   factors affecting 375–80
   management 384–9
   observable effects 374–5
   types of 370–5
   see also Pharmaceutical manufacture
  preservation 380–4
  produced by microbes
   amino acids 489–91
   antibiotics 99, 155, 500
   dextrans 488–9
   enzymes 492–4
   iron-chelating agents *491–2*
   organic acids 489–91
   recombinant 479–81
   vitamins 489–91
   see also Vaccines
  quality assurance 384–9
  sampling *see* Sterility testing
  sterile 384–5, 428–9
   dressings 438–40
   haemostats 441–2
   implants 440–1
   injections 429–33
   instruments and equipment *443*
   ligatures and sutures 442–4
   manufacture 449–55
   non-injectible fluids 434–5
   ophthalmic preparations 435–8
   sterilization methods 407–*26*
   sterility assurance 406–7, 458–69
   types 428–44
   see also Drugs; Pharmaceutical manufacture
Pharyngitis 141
Phenol
  preservative concentrations 381
  sites of antibacterial action *292, 293*
  in vaccines 347
Phenol (carbolic acid) *251, 252*
  MIC determination *273*
  modes of action 290, 291

Phenol coefficient 252, 267
  tests 267–8
Phenols
  antifungal activity *235*
  antiviral activity 64
  as disinfectants 250–3
  neutralizing agent for *270, 466*
  properties *238, 241*
  sites of antibacterial action *292*
Phenoxyethanol *242, 243*, 289
Phenoxymethylpenicillin (penicillin V) 100, *101, 102, 136*, 163
Phenylactic acid (PAA), in antibiotic
  manufacture 161
Phenylethanol *242, 243*
Phenylketonuria (PKU), assay for 494–5
Phenylmercuric acetate, chlorocresol synergy 284, *285*
Phenylmercuric nitrate, antimicrobial uses *241*
Phenylmercuric salts, antibacterial activity *250*
PHMB *see* Polyhexamethylene biguanides
Phospholipases 90
Phospholipids (PL), bacterial 7, *8, 9,*
Phosphonoacetic acid *130*, 131
Phosphonoformate *130*, 131
Phthalylsulphathiazole 122
Picornaviruses 72
Pili 10, 86, *213–14*
Pipelines 366
Pipemidic acid *126*
Piperacillin *101, 102*, 104
Piromidic acid *126*
Pithomyces chastarum 54
Pityrosporum spp. 360
Pivampicillin *101, 102*
Pivmecillinam *101, 102*, 104
Plague 36, 311
Plants
  as contaminated raw materials 362
  fungal pathogens 52–3
Plaques, viral 68
  assays 276, *277*
Plasma substitutes 431
  dextrans 488–9
Plasmids 10, 58, *213–16*
  as cloning vectors *472*, 485
  resistance 138, 303
Plasminogen
  activator, recombinant *482*
  inactivation 492
Plastics
  autoclaving containers 431
  bottles 363
  infusion containers 431
  interaction with preservatives 384
  used in packaging pharmaceuticals 394
Plumbing materials, contamination 359
Pneumococci
  erythromycin against 115
  phagocytosis avoidance 87
  transformation in 217
Pneumocystis carinii 122
Pneumonia
  organisms responsible for 34, 36, *142–3*
  treatment 122, *142–3*

Pneumonic plague   36
Poaefusarin   55
Poliomyelitis vaccines   333, 337, 340, *341*, 342, 343
  testing   345, 346
Poliovirus   *62*, 64, 70, *72*
  replication   76
Polyenes   119–20
  mode of action   *190*, 205
Polyethylene glycols, to protect pharmaceuticals   373
Polyhexamethylene biguanides (PHMB)   246
Polyhydroxybutyric acid   10
Polymyxins   35, 100, 116
  mode of action   205
Polynoxylin   245
Polypeptide antibiotics   116
Polyphosphate granules, bacterial   10
Porins   8, 227, 298, 299
Post-market surveillance   389
Potatoes, warts on   53
Potency, drug   167
Povidone-iodine
  antifungal activity   *235*
  antimicrobial uses   *241*, 249
Powders
  disinfectant and sanitary   280–1
  surface sterilization   416
Poxviruses   *62*, *71*
  drugs against   129
Pradimicins   129
Preservation   380–4
  contact-lens solutions   438
  eye-drops   437
  injections   430
  ophthalmic preparations   381
  see also Preservatives
Preservatives   232, 282–6
  alcohols as   243
  for bacterial vaccines   338
  combinations of   284–6
  evaluation   282–4
  microbial growth substrates   374
  pharmaceutical   380–4
  for vaccines   347
Pressure sore infection   148
Primaquine, causing haemolysis   139
Prions   81
Procaine   103
Proflavine   255
  mode of action   291
Progesterone, conversion   499–500
Prokaryotes   4, 5
Promotors, DNA   475, 485
Prontosil rubrum   120–*1*
Propamidine   255
Prophages   66
β-Propiolactone
  mode of action   294
  sites of antibacterial action   *292*
Propylene glycol, air disinfection   356
Propylphenols   *251*
Prostaglandins   309
Prostate, infections   146
Prostheses, sterile   421
Protein A   87
Proteins

acute phase   308
from genetic engineering   58–9, 479–*83*
fusion   485
metabolism   21–7
synthesis   197
  target for antibiotics   *190*, 197–200
Proteolysis   478–9
Proteus mirabilis   296, 301
  for microbiological assays   174
  R-factor increase   213
  sulphonamide resistance   121
Proteus spp.   38, 116, 144, 361
  drug sensitivity   *136*
  sulphonamide resistance   121
Prothionamide   123, *124*
Proton motive force   289
Provirus   79
PSEs *see* Penicillin-sensitive enzymes
Pseudomembranous colitis   35, 117
Pseudomonas acidophila, antibiotics from   100
Pseudomonas aeruginosa   36, 351
  antibiotic resistance   121, 208
  antibiotics against   113, 116, 125
  antimicrobial supersensitive strain   296
  contaminating pharmaceuticals   399
  in cystic fibrosis   143
  disinfectant resistance   233, 300, 301
  eye infection with   436
  growth on preservatives   374
  in hospital environment   393, 394, 395, 456
  nutrient depletions   296–7
  opportunistic pathogen   82
  sensitivity to chlorhexidine   295, *296*
  skin infection   148–9
  urinary tract infection   145
Pseudomonas cepacia, antimicrobial resistance 296,
  301
Pseudomonas diminuta, for sterilization
  monitoring   464
Pseudomonas spp.
  drug sensitivity   *136*
  selective media for   29
  trytophan breakdown   24
Pseudomonic acid   *190*, 200
  A   *118*, 119
  see also Mupirocin
Psittacosis   39
  cause of pneumonia   143
Psychrophiles   17
Public Health Laboratory Service Report (1971)   398
Puerperal sepsis   34
Pumps, liquid chromatography   178
Pus   34, 487
  liquefaction of   493
Pustules, skin   148
Putrefaction   21
PVC
  drugs adsorbed by   432
  for infusion containers   431–2
Pyelitis   37
Pyelonephritis   37
Pyrazinamide   123, *124*
Pyridine 2,6-dicarboxylic acid (dipicolinic acid)
  (DPA)   *13*
Pyridoxine, assay for   494

Pyrimethamine   122
Pyrimidine   122
Pyrogens   94, 371
  estimation in pharmaceuticals   388–9
Pyruvic acid, metabolism   19–21
Pythium sp.   52

Q-fever   39, 143
QACs *see* Quaternary ammonium compounds
Quality assurance   384–6, 446
Quality control   446, 448
  procedures   387–9
Quaternary ammonium compounds (QACs)   64,
      254–5
  air disinfection   356
  antimicrobial uses   *241*
  degradation by microbes   374
  mode of action   205, 299
  neutralizing agents   *270*–1, *466*
  sites of antibacterial action   *292*
  stored diluted   368
  surface adsorption   236
  see also Cetrimide; Cetylpyridinium chloride
Quinolone drugs   124–5, *126*
  mode of action   *190*, 201–2
  resistance to   213, 226, 228

R-factor, in drug resistance   213, 215–16
Rabies   60, 70
  immunoserum   *348*
  vaccine   337
  virus   *62, 72*
Radiation resistance, bacterial   13
Radiation sterilization   419–23
  accelerated electrons   419, 423
  gamma-rays   419, 421–3, *426*
  monitoring   459, *463*
  UV   419, 423
Radioenzymatic assays   174–6
Radioimmunoassay   184
Random patterns, in microbiological assays   170, 171
Recombinant DNA technology   471–85
  antibiotic production   161–2
  authenticity of drugs   481, 484
  future trends   484–5
  interferon   79, 131
  medical proteins and polypeptides   479, 481, *482–3*
  proteins from fungi   57–9
  terms used in   485–6
  vaccines produced by   334–5
Recovery, from disease   94–5
Relaxin, recombinant   *482*
Replica plating, of bacterial colonies   210
Replication, bacterial chromosome   201
Reproduction
  bacterial   15–16
  fungal   47–9
Resistance
  drug   136, 137–8, 151
    acquired   137, 208–17
    aminoglycosides counteracting   114–15
    to antitubercular compounds   123
    biochemical mechanisms of   217–28
    efforts to overcome   228–9
    inherent   208–9

  to macrolides   116
  plasmid-mediated   138
  to sulphonamides   121, 226
  to tetracyclines   112, 227–8
  to infection, patient   400
  non-antibiotic antimicrobial   295–7
    bacteria   *296*, 297–303
    bacterial spores   302
    disinfection policies and   303–4
    fungi   *296*
    permeability and   297–300
    plasmids and   303
  to sterilization   404–7, 419
Respirator parts, sterilization   *443*
Respiratory tract
  entry for disease   85
  exit route for pathogens   95
  infection treatment   141–3
  natural microbial flora   360–1
  non-invasive pathogens in   88
Restriction endonucleases   *473–4*, 485
Reticuloendothelial system (RES)   308
Retroviruses   *73*
Reverse osmosis (RO), for water sterilization   358
Reverse transcriptase   278, 474, 485
Reye's syndrome   57
Rhabdoviruses   *72*
Rhesus incompatibility   327
Rheumatic fever   34
Rhinovirus   70, *73*
  effect of interferon   78
Rhizopus nigricans, progesterone conversion   499,
      500
Ribavirin   129, *130*
Riboflavin   *490*
Ribosomes
  effects of disinfectants on   291, *292*
  targets for antibiotics   *190*, 197–200
Rickettsia spp.   39
  drug sensitivity   *136*
Rideal–Walker (RW) coefficient, disinfectant
      powders   281
Rideal–Walker (RW) test   252, 267–8
  and preservative evaluation   282
Rifampicin   *112*–13, 123, 138, 141
  mode of action   *190*, 202
Rifamycins   *112–13*
Ringworm   51, 84
  treatment   119
RNA
  antimessenger   76
  mRNA   65
  tRNA   *198*
  viral   61
RNA polymerase   202
RNA viruses   *72–3*, 76
  oncogenic activity   79–80
  replication inhibition   492
Rods   4
  Gram-negative   36–40
  Gram-positive   34–6
Rotavirus   72
Roxithromycin   116
Royce's sachet   *463*
Rubber gloves, sterilization   *443*

Rubella
  vaccine 340, *342*
  virus *72*
Rugulosin 57
Rusts 52
Rye contamination 55

Sabbaromyces splendes 50
Sabouraud medium 29, 275
Saccharomyces cerevisiae 50
  as a cloning host 474, 479, *480*
  eukaryotic gene expression 58–9
Salicylic acid/talc mixture 281
Salmonella/microsome assay 496
Salmonella paratyphi, invasive mechanism 91
Salmonella spp. 37
  drug sensitivity *136*
  endotoxins 94
  partially invasive mechanism 89
  typing 311
  virulence 312
Salmonella typhi 86
  chloramphenicol resistance 221
  gut infection and treatment 147
  identification 28, 70
  invasive mechanism 91
  phagocytosis avoidance 87
  survival after phagocytosis 88
  treatment 141
  vaccine from *340*
Salmonella typhimurium
  invasive mechanism 91
  multiple drug resistance 212, *296*
  permeability 297, 298
  strains 298
Salmonellosis 89, 91
  from contaminated medicines 397, 399
  treatment 147
Sanitizer 259
Sarcina spp. 4, 360
Scarlet fever 34, 65, 93
Secretions, seromucous 307, 318
Seed lot 335
Sephadex 489
Sepsis, abdominal 138
Serial dilution test, measuring bacteriostasis *272, 273*
Serotonin 309
Serratia marcescens 11, 38, 116
  chlorhexidine resistance *296*, 301
  for sterilization monitoring 464
Serratia spp.
  contaminating pharmaceuticals 399
  drug sensitivity *136*
Serum albumin, recombinant *483*
Serum sickness 327
Sex strands, bacterial 10, 15
Shigella shiga 65
Shigella spp. 38
  drug sensitivity *136*
  gut infection 147
  partially invasive mechanism 89
  rise of a resistant strain 211–12
Shingles *71*
Shotgunning 473, 485
Sideromycins 227

Siderophores 491–2
Signal sequence 477, 485
Silver salts
  antimicrobial activity 249
  sites of antibacterial action *292*
Sinusitis, treatment 141
Sisomycin 115
Skin
  entry for microorganisms 84
  infections 84, 147–9
    see also Wounds
  innate immune system 307
  natural flora 360–1
  semi-solid disinfectants, evaluation 280
  tests for antimicrobial activity 271–2, 280
  treated with contaminated preparations 400
Slime, bacterial 11, 86
Slit sampler 282, *283*
Small-volume aqueous injections 433–4
Small-volume oily injections 434
Smallpox *71*
  cultivation of viruses 74
  prophylaxis 129
  vaccine 333, 337
    history 306
Smoke tests 355
Smuts 52
Sneezing 91
Soaps 266
Soft tissue infections 147–9
Solid dilution method, measuring bacteriostasis 274
Somatic cell hybridization 315
Somatostatin 57, 472, *478–9*
Somatotrophin, recombinant 481, 482
SOPHEA 185
Sorbic acid, antimicrobial activity 239
Sorbitol, microbial attack 373
Sordaria sp., toxins from 54
Spectinomycin 174
  inactivation *224*
Spermine 307
Spiramycin 115, 116
Spirochaetes 4, 40–1
Splicing 485
Spongiform encephalopathy 80–1
Spores
  bacterial *see* Bacteria
  fungal 47–8, 303
Sporesdesmin 54
Sporicidal activity, testing 271
Sporicide, definition 258
Sporofusarin 55
Sporotrichosis 52
Sputum
  bacteria in 34, 40
  culture 142
Stachybotrys alternans 54
Stains
  Gram staining 5–6
  spore staining 13
Staphylococcal phage *62*
Staphylococcus aureus 4, 33–4, 116, 118
  antimicrobial resistance *296*, 301
  brain abscess 150
  in cystic fibrosis 143

drug resistance 112, 136, 225
erythromycin sensitivity 115
food poisoning 93
fusidic acid resistance 138
identification by phages 70
lung abscess formation 143
neomycin resistance 217
oedema suppression 87
opsonization avoidance 87
pharmaceutical contamination from humans 360–1
pigment 11
skin infections 148
Staphylococcus pneumoniae, drug sensitivity *136*
Staphylococcus spp. 4, 33–4, 82
abscesses 90
antibiotic resistance 209
antibiotics against 113, 115, 117, 119
antimicrobial multiresistance 302
burn infection 149
food poisoning 93
identification 28
killing phagocytes 88
survival after phagocytosis 88
toxins 90
Staphylokinase 90
Steam sterilization (autoclaving) 410–15, *426*
Sterile products manufacturing unit 249–52
Sterilization 367, 403–4, *426*
filtration 423–5, *426*
monitoring 459–60, 464
gaseous 416–19
ethylene oxide 417–18, *426*
formaldehyde 419, *420, 426*
monitoring 459, *462*
heat 13–14, 408–16
dry heat 415–16, *426*
moist heat 410–15, *426*
monitoring 459, *462*
monitors 458–64
radiation 419, 421–3, *426*
monitoring 459, *463*
raw products for pharmaceuticals 362–3
resistance of microbes to 404–7
sterility assurance 406–7
testing *see* Sterility testing
Sterility assurance 406–7
Sterility testing 35, 464–9
microbial enzymes in 496–7
of vaccines 346–7
Sterilizers
dry heat 415–16
ethylene oxide 417–18
steam 412–15
Steroids
biological transformations of 499–500
contaminated 400
degradation by microbes 374
Stickland reaction 26
Sticky ends 473, 485
Stock-pots, contaminated 387
Strate-line *461*
Streptococcus faecalis, drug resistance 225, *296*
Streptococcus haemolyticus, benzylpenicillin
    sensitivity 209
Streptococcus mutans 296

Streptococcus pneumoniae 136
treatment 121, 142–3
vaccine from 334, 337, *339*
Streptococcus pyogenes 4, 34, 86
burn infection 149
drug sensitivity 135, *136*
eradication 141
scarlet fever from 93
skin infection 148–9
toxins 90
Streptococcus spp. 4, 34
abscesses 90
erythromycin resistance 225
haemolytic, antibiotics against 115, 121
killing phagocytes 88
phagocytosis avoidance 87
survival after phagocytosis 88
toxins 90
Streptodornase 90, 493
Streptokinase 90, 309, 492–3
Streptomyces aureofaciens 163
Streptomyces clavuligerus 162
clavam from 105
Streptomyces olivaceus, clavams from 109
Streptomyces pilosus, siderophore from 491
Streptomyces spp.
antibiotics from 100
tetracyclines from 111–12
Streptomycin 113, *114*, 123, 174
inactivation *224*–5
manufacture 155
mode of action 197
ribosomal resistance to 211
sources 100
Styes 34
Succinylsulphathiazone 122
sugars (syrups)
attack by microbes 375, 378
to protect pharmaceuticals 378
Sulphadiazine *121*, 122
Sulphadimidine *121*, 122
Sulphamethoxazole *121*, 122
    /trimethoprim combination 122, 132, *136*, 204
Sulphanilamide *121*
Sulphites
antimicrobial activity 239
mode of action 294
sites of antibacterial action *292*
Sulphomyxin sodium 116
Sulphonamides 120–2, 135, *136*
causing haemolysis 139
mode of action *190*, 203–4
neutralizing agent for *270, 466*
resistance to 226
Sulphur dioxide
antimicrobial activity 239
modes of action 291, 294
sites of antibacterial action *292*
Superinfection 140
Suppressor T cells (TS cells) 322–3
Surface-active agents
as antimicrobials 253–5
attacked by microbes 372–3
Surface antigens 312
Surgery, peri-operative antibiotics 140–1

Survivor curves 404–5
Susceptibility testing 135
Suspension tests, for disinfectants 267–71
Sutures, sterile 442–4
Swedish National Board of Health, pharmaceutical
    quality tests 397
Sweetening agents, attack by microbes 373
Synergism 131–2, 138, 284
Synthetic peptides, vaccines 335
Syphilis 41
Syringes, sterilization 421, *443*
Systemic lupus erythematosus (SLE) 326

T-cell leukaemia virus 79–80
T cells 313, 321–3
T-even phages *66*, 67
Tablets, spoilage 379
Talampicillin *101, 102*
Tar acids 252–3
Taurolidine 245
Tc cells 323
Teichoic acids 6, 7
Teichoplanin (teicoplanin) 117, 196
   mode of action *190*
Teicoplanin *see* Teichoplanin
Temocillin *101, 102*, 104
Temperature
   affecting disinfection 262–3
   for bacterial growth 17, 30
   and benzylpenicillin production 158
   storing pharmaceuticals 379
Temperature coefficient 263, 381
Temptubes *461–2*
Terbinafine *127*, 128
Tetanus 35, 92
   immunoglobulin *349*
   immunoserum 347–8
   vaccine *339–40*, 343, 345, 346
Tetrachlorosalicylanilide 289
Tetracycline *111*
   candidiasis development 140
   for cholera 147
   for impetigo treatment 149
Tetracyclines 111–12, 135, *136*, 139
   mode of action *190*, 199
   resistance to 227–8
   sources 100
Tetrahydrofolate (THF) synthesis 202–*3*
Tetramethylphenols *251*
Tetroxoprim *121*, 122
Tн cells 322
Thermalog S *462–3*
Thermophiles 17
Thiacetazone 123, *124*
Thiacycline *111*, 112
Thiatetracyclines *111*, 112
Thienamycin *109*, 220–1
Thioglycollic acid 290
Thiol groups, disinfectant action on 290, 291, *292*
Thiomersal *250*
   antimicrobial uses *241*
Throat, sore 34, 141
Thrush 119
Thucydidyes (quoted) 305–6
Thyroiditis 326

Thyrotoxicosis 326
Ticarcillin *101, 102*
   /clavulanic acid combination 105, 132
Timecards *461*
Tin, antimicrobial activity 249
Tissue culture 73–*4*
Tissue typing 329
Tobacco mosaic disease, virus *62*, 63
Tobramycin 114, *136*
   assay for 174–5
Togaviruses *73*
Tolerance, immunological 324–5
Tonsillitis 34
Torulopsis, antibiotics against 128
Total parenteral nutrition (TPN) 432–3
Toxicity
   drug *see* Drugs
   tissue, from antiseptics and preservatives 237–8
Toxicity tests
   for disinfectants 272
   for vaccines 347
Toxicosis 53
Toxins
   bacterial 14, 90, 91–4, 310–11
   estimation in pharmaceuticals 388–9
   fungal 53–7
     pharmaceutical uses 57
Toxoid vaccines 334
TPN *see* Total parenteral nutrition
Trachoma 39
Transamination 25
Transcription
   bacterial chromosome 201
   of cloned genes 475–6
   inhibition by interferon 78, 131
Transduction 16, 69, 216–17
Transformation 15, 217
Translation, of cloned genes 475–6
Transplantation, tissue 328–9
Transposons 215
Travellers' diarrhoea 147
Trench fever 39
Treponema pallidum
   penicillin sensitivity 209
   transmission between hosts 95
Treponema spp. 41
Triacetyloleandomycin 115, 116
Tricarbanilide 289
Trichlorocarbanilide 289
Trichlosan 253
Trichodermin 57
Trichomonas vaginalis, antibiotics against 128
Trichophyton mentagrophytes 84
   antimicrobial success 235
   sensitivity to chlorhexidine 296
Trichophyton spp. 360
   infection by 51
   treatment 119
Trimethoprim *121*, 122, 145, 146
   mode of action *190*, 204
   for prostate infection 146
   resistance to 225–6
   /sulphamethoxazole combination 122, 132, *136*,
    204
Trimethylphenols *251*

Triphenylmethane dyes, antimicrobial activity   255
Tryptophanase   24
Ts cells   322–3
Tuberculin test   327
Tuberculosis   40, 60
    disinfection of equipment   234
    treatment   113, 123, 138
    see also Mycobacterium tuberculosis
Tularaemia   37
Tumour necrosis factor, recombinant   *483*
Tumour viruses   79–80
Tumours, immune response to   329
Turkey X disease   54
Typhoid   37, 91, 141, 267, 311
    treatment   147
    vaccines   312, 333, 336, *340*
Typhus infections   39
Tyrothricin   35

Ulcers, dextran application   489
Ultraviolet (UV) light
    for air disinfection   356
    effect on viruses   64
    sterilization by   419, 423
    for water disinfection   360
Undulant fever   37
United States Pharmacopoeia   29, 284
Urea decomposition   38
Urease assay   174
Ureidopenicillins   104
Urethritis   39
Urinary tract infections   37, 38
    from contaminated bladder washouts   370
    pathogenesis   144
    treatment   124, 125, 144–6
Urine   85
Urinogenital tract, entry for disease   85
Uterine contractions, induced   55
UV light *see* Ultraviolet light

Vaccination
    history   306
    programmes   96
Vaccines
    bacterial   335–7, *339–40*
        safety tests   345–6
    classification   333–5
    manufacture   335–42
        quality control   342–7
        seed lot system   335
    mixed   340
    potency assay   344–5
    viral   337–8
        safety tests   346–7
Vaccinia   *71*, 76
    immunoglobulin   *349*
    inhibition   129
Vacuum exhaust systems   366
Vacuum packing   379
Vagina, infections in   85
Vaginitis   35
Valves   365
Vancomycin   113, 116–17, 140
    mode of action   *190*, 194, 196
Varicella-zoster   *71*

Variola   *71*
    inhibition   129
Variolation   306
Vectors
    of disease   95
    gene *see* Plasmids
Venereal disease   85
    gonorrhoea   34
    syphilis   41
Vi antigen   312
Viable count   31, 259–60, 269–71
    for pharmaceutical evaluation   388
Vibrio cholerae   36, 86
    effect on the gut   89, 93
    neuraminidase from   *493*
Vibrio parahaemolyticus   36
Vibrio spp.   4
Vidarabine   129
Vincent's angina   41
Viomycin   116
Viral subunit vaccines   334
Virucide (viricide), definition   258
Virulence   86
    definitions   306
Viruses   60
    antiviral disinfectants   64, 235–6
    antiviral drugs   129–31
    bacteriophage   16, 65–70
    chemotherapy   77–9
    effect of physical and chemical agents on   64–5
    general properties   60–1
    host–cell interactions   64–5
    human   70–5
        cultivation   70–5
        multiplication   75–7
    as insect pathogens   500–1
    interference   78
    oncogenic   79, 329
    structure   61–4
    temperate   16
    tests for virucidal preparations   276–8
    tumour   79–80
Vitamins
    bioassays   494, *495*
    from microorganisms   489–91
Voges–Proskauer (VP) reaction   30
Volutin granules, bacterial   10

Walls, microbes commonly on   363–4
Water
    essential for bacteria   16
    for injections   448
    microorganisms found in   40, 356, 379
    non-injectible and sterile (topical)   434, 448
    for pharmaceutical manufacture   356–60, 393–4, 448
    storage   393–4
Water activity (A_W), of pharmaceuticals   377–9, 384, 385
Weil's disease   41
White fluids (phenolic)   *251*, 252
    properties   *238*
Whooping cough   37, 88
    vaccines   334, 336, *340*
        testing   345

see also Bordetella pertussia
Wounds 90
  infected via pharmaceuticals 393, 399
  organisms infecting 34, 35, 36, 38, 148–9, 360–1
  treatment 124, 148–9

Xylenols *251*, 252

Yaws 41
Yeasts
  asexual reproduction 48
  for fermentation 49–50

present on the skin 360
  structure 44–5
Yellow fever
  vaccine 337–8, *342*
  virus *73*
Yersinia pseudotuberculosis sp. *pestis* 36

Z-value 14, 405–6
Zidovudine 129–30
Zinc, antimicrobial activity 249
Zinc-based pharmaceuticals, contamination 395
Zinc undecenoate 281
Zoospores 44, 47